MXenes

This book introduces MXenes and provides a summary of current discoveries in their synthesis, properties, characterization techniques, and emerging applications in several fields. It explores MXenes' distinctive electrical, mechanical, and biological features, as well as their applications. It discusses the various emerging applications of MXenes in a variety of fields, including regenerative medicine and tissue engineering, separation membranes, photocatalytic hydrogen production, environmental applications, and so forth.

Features:

- Provides comprehensive review of synthesis and application of MXenes in 2D materials.
- Explores greener approaches that reduce the environmental impact and cost of the process.
- Includes frameworks for assessing the balance between the synthesis and applications.
- Discusses an overview of MXenes with a focus on energy, health, and medical applications that can be improved with new encapsulation.
- Reviews the modern spectroscopic instruments that are utilized in sophisticated characterization methods.

This book is aimed at researchers and graduate students in materials engineering, nanomaterials, and 2D materials.

Emerging Materials and Technologies

Series Editor: Boris I. Kharissov

The *Emerging Materials and Technologies* series is devoted to highlighting publications centered on emerging advanced materials and novel technologies. Attention is paid to those newly discovered or applied materials with potential to solve pressing societal problems and improve quality of life, corresponding to environmental protection, medicine, communications, energy, transportation, advanced manufacturing, and related areas.

The series takes into account that, under present strong demands for energy, material, and cost savings, as well as heavy contamination problems and worldwide pandemic conditions, the area of emerging materials and related scalable technologies is a highly interdisciplinary field, with the need for researchers, professionals, and academics across the spectrum of engineering and technological disciplines. The main objective of this book series is to attract more attention to these materials and technologies and invite conversation among the international R&D community.

Smart Micro- and Nanomaterials for Pharmaceutical Applications
*Edited by Ajit Behera, Arpan Kumar Nayak, Ranjan K. Mohapatra,
and Ali Ahmed Rabaan*

Friction Stir-Spot Welding
Metallurgical, Mechanical and Tribological Properties
Edited by Jeyaprakash Natarajan and K. Anton Savio Lewise

Phase Change Materials for Thermal Energy Management and Storage
Fundamentals and Applications
Edited by Hafiz Muhammad Ali

Nanofluids
Fundamentals, Applications, and Challenges
Shriram S. Sonawane and Parag P. Thakur

MXenes
From Research to Emerging Applications
Edited by Subhendu Chakroborty

Biodegradable Polymers, Blends and Biocomposites
Trends and Applications
Edited by A. Arun, Kunyu Zhang, Sudhakar Muniyasamy and Rathinam Raja

For more information about this series, please visit: www.routledge.com/Emerging-Materials-and-Technologies/book-series/CRCEMT

MXenes

From Research to Emerging Applications

Edited by Subhendu Chakroborty

CRC Press
Taylor & Francis Group
Boca Raton London New York

CRC Press is an imprint of the
Taylor & Francis Group, an **informa** business

Designed cover image: © Shutterstock Images

First edition published 2025
by CRC Press
2385 NW Executive Center Drive, Suite 320, Boca Raton FL 33431

and by CRC Press
4 Park Square, Milton Park, Abingdon, Oxon, OX14 4RN

CRC Press is an imprint of Taylor & Francis Group, LLC

Library of Congress Cataloging-in-Publication Data
Names: Chakroborty, Subhendu, editor.
Title: MXenes : from research to emerging applications / edited by Subhendu Chakroborty.
Description: First edition. | Boca Raton : CRC Press, 2024. | Series: Emerging materials and
    technologies | Includes bibliographical references and index.
Subjects: LCSH: MXenes. | MXenes—Synthesis. | MXenes—Properties.
Classification: LCC QD172.T6 M84 2024 (print) | LCC QD172.T6 (ebook) |
    DDC 546/.34—dc23/eng/20240409
LC record available at https://lccn.loc.gov/2024004808
LC ebook record available at https://lccn.loc.gov/2024004809

ISBN: 978-1-032-41558-1 (hbk)
ISBN: 978-1-032-43219-9 (pbk)
ISBN: 978-1-003-36622-5 (ebk)

DOI: 10.1201/9781003366225

Contents

Chapter 4   MXene Materials as Efficient Separation Membranes 49

Shikha Indoria, Swati Awasthi, and Vickramjeet Singh

Chapter 5   Environmental Applications of MXene-Based Materials 64

Shubham Mishra, Dinesh Kumar Pati, R. Padhee,
Subhendu Chakroborty, and Nibedita Nath

Chapter 6   MXene-Based Biosensors: Next-Generation Emerging
                Materials for Detection 84

Akhilesh Babu Ganganboina, Indra Memdi Khoris,
and Kenshin Takemura

About the Editor

Subhendu Chakroborty received his Doctorate (Ph.D.) in Chemistry from Ravenshaw University (India). He is an associate professor and Deputy Registrar at Chandigarh University, India. He has received the Best Keynote Speaker award at the International Summit and Conference on Material Science Nanotechnology & Bio Manufacturing (ISCMNB), Malaysia, and was a winner of the International Picture Contest Award illustrating "Everything is Chemistry" conducted by the European Chemical Society. His research interests include the synthesis of perovskite and 2D nanomaterials for applications in biomedicine, sensors, biomaterials, functional materials, and drug delivery. With over 12 years of experience in teaching and research, he has produced more than 70 scientific publications, including original research papers, review articles, textbooks, and invited book chapters in material science and medicinal chemistry by internationally reputed major presses. He is a member (MRSC) of the Royal Society of Chemistry (RSC), UK, and the American Chemical Society (ACS). He served as a repeated scientific reviewer for several international journals.

Contributors

Laxmidhar Besra
Materials Chemistry Department
CSIR-Institute of Minerals and
  Materials Technology
Bhubaneswar-751013, Odisha, India

Subhendu Chakroborty
School of Applied Sciences
Chandigarh University
Lucknow, India

Prakash Chandra
School of Technology
Pandit Deendayal Energy University
Gandhinagar, Gujarat 382007, India

Priyanka Chandra
Department of Pharmaceutical Sciences
  and Technology
Birla Institute of Technology
Mesra, India

Jyh-Ping Chen
Department of Chemical and Materials
  Engineering
Chang Gung University
Kwei-San Taoyuan 33302, Taiwan

Banendu Sunder Dash
Department of Chemical and Materials
  Engineering
Chang Gung University
Kwei-San Taoyuan 33302, Taiwan

Dipak Dutta
Department of Electrical Engineering
National Central University
No. 300, Zhongda Road Taoyuan
  City-320317, Taiwan

Akhilesh Babu Ganganboina
International Center for Young
  Scientists
ICYS-NAMIKI National Institute for
  Materials Science
1–1 Namiki Tsukuba, Ibaraki 305-
  0044, Japan

Pratik Kumar Jagtap
Department of Chemistry
Faculty of Science The ICFAI
  University
Raipur, 490042, Chhattisgarh, India

N. Usha Kiran
Materials Chemistry Department
CSIR-Institute of Minerals and
  Materials Technology
Bhubaneswar-751013, Odisha, India

Mainak Majumder
Nanoscale Science and Engineering
  Laboratory (NSEL) Department
  of Mechanical and Aerospace
  Engineering
Monash University
Clayton, VIC 3800, Australia

Nirmal Mazumder
Department of Biophysics Manipal
  School of Life Sciences
Manipal Academy of Higher
  Education
Manipal, Udupi 576104

Swarna P. Mantry
Analytical Chemistry Division
Bhabha Atomic Research Centre
Mumbai, 400085, India

Nilima Priyadarsini Mishra
Department of Chemistry
Ravenshaw University
Cuttack-753003, Odisha, India

Anchit Modi
Department of Basic Sciences
IITM, IES University
Bhopal-462044, India

Srikanta Moharana
Department of Chemistry, School of
    Applied Sciences
Centurion University of Technology and
    Management
Odisha, India

Nibedita Nath
Department of Chemistry
D.S. Degree College
Laida, Sambalpur, 768214, Odisha, India

Manas Ranjan Panda
Nanoscale Science and Engineering
    Laboratory (NSEL) Department of
    Mechanical and Aerospace Engineering
Monash University
Clayton, VIC 3800, Australia

Jagabandhu Patra
Hierarchical Green-Energy Materials
    (Hi-GEM) Research Center
National Cheng Kung University 1
    University Road
Tainan 70101, Taiwan

Chita Ranjan Sahoo
Department of Health Research,
    Ministry of Health & Family
    Welfare, Govt. of India
ICMR-Regional Medical Research
    Centre
Bhubaneswar 751023, India

Dojalisa Sahu
Department of Chemistry, School of
    Applied Sciences
Centurion University of Technology and
    Management
Odisha, India

Sigamani Saravanan
Department of Physics (S&H)
    Swarnandhra College of
    Engineering & Technology
Narsapur-534 280, West Godavari (AP),
    India

Bhisham Narayan Singh
Department of Biotechnology
Manipal School of Life Sciences
    Manipal Academy of Higher
    Education Manipal
Udupi 576104, India

Vickramjeet Singh
Department of Chemistry
Dr. B. R. Ambedkar National Institute
    of Technology
Jalandhar-144027, Punjab, India

Preface

In the ever-evolving landscape of materials science, a remarkable family of 2D transition metal carbides and nitrides, known as MXenes, has emerged as a subject of intense research and innovation. This book, *MXenes: From Research to Emerging Applications*, delves into the exciting journey of MXenes from their inception in the laboratory to their promising applications in various fields of science and technology.

MXenes, initially discovered in 2011, have swiftly captured the attention of scientists, engineers, and researchers worldwide due to their exceptional properties and versatility. These two-dimensional nanomaterials have opened up a realm of possibilities, ranging from traditional synthesis methods to environmentally friendly routes, leading to a profound understanding of their structure and characteristics.

The chapters in this book provide a comprehensive overview of MXenes, authored by leading experts in the field. Each chapter explores a unique facet of MXenes, from their synthesis and characterization to their groundbreaking applications in fields such as regenerative medicine, separation membranes, environmental remediation, biosensors, cancer therapy, photocatalysis, antibacterial materials, electrochemical sensors, biomedical applications, water desalination, electrocatalysis, supercapacitors, photothermal therapy, sodium-ion batteries, flexible piezoresistive sensors, fuel cells, gas sensors, and even textiles.

The diverse range of topics covered in this book reflects the multifaceted nature of MXenes and their immense potential to revolutionize various industries. Whether you are an aspiring researcher, a seasoned scientist, or an industry professional, *MXenes: From Research to Emerging Applications* offers invaluable insights into the world of MXenes and their transformative role in shaping the future of materials science and technology.

We extend our sincere gratitude to all the contributors who have shared their knowledge, expertise, and passion for MXenes in the following chapters. Their dedication and commitment to advancing the field have made this book possible.

As we embark on this journey through the fascinating world of MXenes, we hope that this book will serve as a valuable resource for researchers, academics, and industry leaders who seek to harness the full potential of MXenes and propel them from the realm of research to the forefront of emerging applications.

Dr. Subhendu Chakroborty

1 MXenes

From Research to Emerging Applications—A Note

Subhendu Chakroborty

1.1 INTRODUCTION

Since their discovery in 2011, there has been a meteoric rise in the manufacture of 2D transition metal carbides and nitrides, more often referred to as MXenes. Consumers now have access to a selection of about 40 distinct MXene compounds to select from. In fact, there are a great deal more, and it's possible that they'll turn out to be the most extensive family of 2D materials ever found. MXenes have a broad variety of uses because of their remarkable properties, such as an electrical conductivity that is comparable to that of metal and may reach up to 20,000 S cm^{-1}. This makes MXenes particularly useful in a number of different contexts. To name a few examples, there is the field of optoelectronics, as well as energy storage, biology, communications, and environmental protection. There has been a rise in the number of patents and papers that cover MXenes over the course of the previous several years. In the realm of materials synthesis, one of the most important recent developments was the invention of a selective etching procedure that makes use of hydrofluoric acid (HF). At this moment, the concept of MXenes as a separate category of matter began to take shape in people's minds. This cutting-edge technique was used to strip layers of transition metal carbide and carbonitride from MAX phases, which resulted in the production of the first generation of MXene materials. After this momentous breakthrough, researchers set out to discover new ways to synthesize the compound, which ultimately led to the creation of numerous other MXene analogues. This endeavour has produced a multiplicity of unique techniques, each of which has its own set of benefits and the potential of adapting MXenes to particular applications. The results of this effort have been quite fruitful. One of these innovative methods involves the selective removal of the MAX phase layers by the use of a variety of etchants, such as hydrofluoric acid. In addition, researchers have investigated the possibility of producing MXenes with a variety of characteristics by employing a variety of precursors and synthesis conditions. This variety of synthesis techniques has resulted in an expansion of the spectrum of MXene materials that are currently accessible, making it possible to customize the composition, thickness, and surface chemistry of these materials. As a result of these improvements, MXenes have discovered uses in a variety of disciplines, including the storage of energy, the purification of water, and the production of electrical devices. The use of MXenes as anodes in lithium-ion

DOI: 10.1201/9781003366225-1

1

batteries, which, thanks to their high conductivity and huge surface area, allows for quicker charging and a longer battery life, is one of the most famous applications of MXenes in the field of energy storage. MXenes have also demonstrated promising results in supercapacitors due to their capacity to store enormous quantities of energy and release it in a short amount of time. MXenes have been exploited in the field of water purification due to the outstanding adsorption capabilities that they possess. As a result, heavy metals and other contaminants have been successfully removed from water sources. In addition, MXenes have demonstrated promise in the field of electronics as transparent conductive films, which makes them an excellent choice for use in touch screens and flexible displays. MXenes possess a tremendous deal of potential for both the progression of technology and the solution of environmental problems, provided that research and development on them is maintained. Selective etching in fluoride salts and other acid solutions, etching with non-aqueous etchants, etching with halogens, and etching with molten salts are all examples of these methods. These diverse methods of synthesis have not only expanded the MXene family but also revealed many new and interesting properties and applications. These approaches have made it possible to synthesize new MXenes with enhanced control over the surface chemistries of those MXenes [1].

In 2011, Naguib et al. demonstrated that Ti_3AlC_2 could be selectively etched to yield Ti_3C_2 for the first time by simply submerging it in hydrofluoric acid (HF) at room temperature [2]. A year later [3], using the aluminium A-layer, one of the more common A-elements in MAX phase, we demonstrated that the technique of selectively etching were suitable for numerous other MAX phases. At this point, we gave this new family the name MXenes to highlight their connection to the MAX phases and their dimensionality [3,4]. The number of MAX phases has increased from about 70 in 2011 to 150 today [5], with new ones being discovered on a regular basis, demonstrating the abundance of MXene precursors. The numerous MXene structures currently known are represented by atomistic models in Figure 1.1. Since then, more than a thousand patent applications have been made, and thousands of research papers, a book, and more have been published. The very wide range of applications that can be used, including electronics, medicine, sensing, communication, optoelectronics, and tribology among many others, is more significant than the quantity of patents that have been filed and issued.

Electrical conductivity, the size of the resultant MXene ensemble, defects, and surface termination groups all affect how conductivity behaves in MXenes, which are assumed to be conductive [6]. According to this point of view, the intercalation process might lead to a modification of the electronic properties as a consequence of the inter-flake effects (such as an increase or reduction in electrical resistance and inter-flake spacing). It was discovered that the influence of intercalating H_2O and organic ions into $Mo_2TiC_2T_x$ increased the spacing between the flakes as well as the electrical resistance [7]. In contrast to MXenes intercalated with a nitrile group, which display activity similar to that of a semiconductor, the presence of H_2O in the $Ti_3C_2T_x$ MXene indicates the intrinsic metallic properties of the complex [8]. Crystal structure, stacking phases, twisting angle, and compositional distribution are all material features that have a direct impact on the electrical and magnetic indicators of layered materials [9,10]. The structure of MXene is seen in Figure 1.2, and

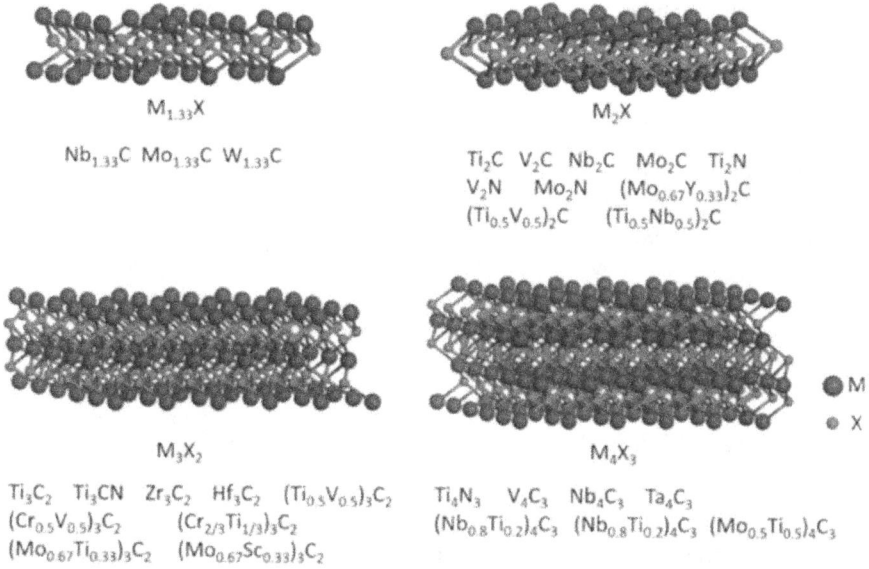

FIGURE 1.1   Common MXene compositions and structures. $M_{n+1}X_nT_x$, in which M is an early transition metal, X is either C or N, and T_x denotes surface terminations, is the formula for MXenes.

Adapted with permission from Ref. [1]. This work is licensed under the Creative Commons Attribution 4.0 International License, De Gruyter, Berlin/Boston.

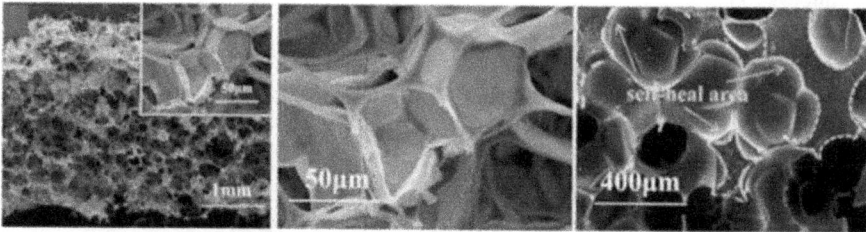

FIGURE 1.2   SEM image depicting the microstructure of polyurethane, melamine sponge, and MXene, which has tiny cages for EMI shielding.

Adapted with permission from Ref. [11]. Copyright 2021, Elsevier.

it resembles a porous sponge in terms of its shape. MXenes also possesses inherent shielding capabilities and an ability to mend themselves.

A new method for creating nanoscale controlled materials is made possible by using a two-dimensional MXene as the precursor for derived materials. From MXenes, a variety of materials, such as metal oxides, carbon material, metal sulphides, and metal–organic frameworks (MOFs), are prepared. These materials have excellent performance and tremendous potential and are used in batteries, adsorption, nitrogen fixation, photocatalysts, capacitors, and other applications [12–17]. These types of MXene-derived materials and their uses are currently very fascinating.

1.2 EMERGING APPLICATIONS OF MXENES

The three main areas of use for MXene derivatives are electrochemistry, photochemistry, and adsorption. Batteries made of sodium, potassium, or lithium, as well as supercapacitors are examples of electrochemistry. Pollutant degradation by photocatalysis, nitrogen fixation by photocatalysis, and hydrogen synthesis by photocatalysis are all examples of photochemistry. Pollutant adsorption, capacitive desalination, microwave absorption, and other processes are all examples of adsorption using MXenes in a variety of industries [18].

1.3 MXENES FOR ENERGY STORAGE DEVICES

Since their discovery, MXenes have received extensive attention for use in energy storage applications because of their exceptional electrical and electronic characteristics. MXenes have a low electron transport energy barrier, a high specific surface area, and a short ion-diffusion channel thanks to their crystalline nature, atomic thickness, and layered structure. Additionally, it has been demonstrated both theoretically and empirically that changing the surface terminations of MXenes can change their electrical characteristics. In order to achieve effective ESDs in the next 50 years, novel methods of employing MXenes and MXene-based resources have been explored in recent energy storage studies. Numerous ESD components and their interface areas now use MXenes and MXene-based technology [19–21].

Recent years have seen a flurry of research into ways to increase Coulombic efficiency with the cyclic ability of lithium-ion batteries by reducing the formation of Li dendrites using MXenes. Additionally, lithium–sulphur, aluminium, and zinc-ion batteries that use MXene-based materials are rapidly evolving [21,22]. MXenes can offer outstanding rate performance and cycling stability due to their high lithium capacity, broader interlayer spacing, low diffusion barrier for Li ions, high electrical conductivity, and low operating voltage (-0.2 to 0.6 V vs. Li/Li^+) [23]. As prospective anodes for LIBs, a variety of MXene materials, including $Ti_3C_2T_x$, $Mo_2TiC_2T_x$, Nb_2CT_x, V_2CT_x, and Mo_2CT_x, have been investigated.

MXene-based SCs have been the subject of extensive investigation for energy storage. $Ti_3C_2T_x$ electrode performances in SCs in both acidic and basic electrolytes were examined by Lukatskaya et al. They discovered that a $Ti_3C_2T_x$ clay electrode displayed a higher level of capacitance (900 F/cm^3) at the same scan rate in an H_2SO_4 electrolyte than a $Ti_3C_2T_x$ paper electrode did at 442 F/cm^3 and 2 mV/s in a KOH electrolyte [24,25]. These findings inspire innovative ideas. Interlayer spacers like metal oxides, carbon nanotubes, and reduced graphene oxides are employed in MXene electrodes to widen the distance between the MXene nanosheets.

Wang et al. discovered a polyaniline and V_2C MXene composite for the first time in addition to titanium carbide MXene. They were able to create SCs with a high-density and high-sensitivity ammonia sensor using this material combination [26].

1.4 MXENES FOR BIOMEDICAL APPLICATIONS

MXenes, with their fascinating physicochemical features, offer several intriguing uses in many disciplines. MXene-based materials are useful in biosensing, medical diagnostics, bioimaging, implants, and antibacterial medications. MXenes may open new doors and spur medical and biotech developments as research in this subject advances. High-performance receptors have sensitivity, high selectivity, response time, LOD, and linear range. For commercial scale-up manufacturing, they must also be cheap to make. MXene biosensors employ photonic, electrochemical, and biocompatible field-effect transistors [27–29]. MXene-based materials can be employed with MRI, CT, photoacoustic, and fluorescence imaging. Imaging with MXene-based chemicals avoids some of the drawbacks of traditional reagents. Two-dimensional MXene-based chemicals have brought exciting quantum size effects to imaging. These materials are suitable for improved cell imaging due to their photoluminescence (PL) capabilities. Quantum size effects in MXene-based reagents improve photoluminescence efficiency and sensitivity, allowing more accurate and detailed imaging of cells and cellular processes. MXene reagents also improve photothermal characteristics for photoacoustic (PA) imaging and elemental contrast for X-ray CT imaging. MXene-based reagents can improve biomedical imaging, diagnostics, and research in a range of fields due to their multifunctionality [30]. MXenes and their composites have been used in biosensors, photodynamic therapy (PDT), drug delivery systems, immuno-therapy, diagnostics, conventional photothermal therapy (PTT), and synergistic com-binations of multiple treatment technologies. MXene-based compounds fight cancer cells by improving payload absorption and drug release [31–34]. MXene-based immu-notherapy materials are potential cancer treatments. Their high specific surface area and biocompatibility make loading and delivering therapeutic medicines to tumour locations efficient. MXenes' specific accumulation in tumours boosts immunotherapy by activating the immune system to target cancer cells. MXene-based immunother-apy materials may revolutionize cancer treatment and patient outcomes as research continues. MXene-based materials have shown immunological benefits [35–38]. Ma et al. reported the first flexible and sensitive piezoresistive sensor based on $Ti_3C_2T_x$ for sensing delicate human actions and other light pressures [39]. Lei et al. created a sweat biosensor using a novel MXene/Prussian blue ($Ti_3C_2T_x$/PB) composite for non-invasive biomarker monitoring [40]. MXene-based materials have a large specific surface area, are easy to chemically modify and functionalize, and may carry a range of antimicrobial functional groups [41].

1.5 MXENES FOR ENVIRONMENTAL APPLICATIONS

MXenes are good dye-removing adsorbents due to their negatively charged surface, layered structure, and high hydrophilicity. This makes them the perfect materials for this purpose. Mashtalir's group was the first to reveal that a multi-layered MXene, also known as ML-$Ti_3C_2T_x$, efficiently absorbed cationic dyes like methylene blue (MB) [42,43]. MXene-COOH@(PEI/PAA)n adsorbs neutral red (NR, 42.50 mg g^{-1}), safranine T (ST, 33.76 mg g^{-1}), and MB (36.69 mg g^{-1}). MXenes and their derivatives are thought to have the potential to act as adsorbents for heavy metal cations due to the high number of surface functions that they possess in combination with their considerable surface areas. Several articles claim that MXenes show high adherence to a wide range of heavy metal ions [44,45].

It has been demonstrated that MXenes are rather helpful for the adsorption of radioactive heavy metal ions [46,47]. This might be because of the different capabilities they have for adsorption in addition to the redox ability of their surfaces. MXenes have been shown to be effective materials for use in membrane technology due to the fact that the surface chemistry of these materials can be handled, and the interlayer spacing between these materials can also be controlled. Researchers Ren et al. were the first to publish their results on the use of an MXene-based membrane in the purification of water [48]. MXenes have the potential to be used in the manufacture of complex co-catalysts as one-of-a-kind substrates, which is a possibility. These co-catalysts have the capacity to efficiently minimize charge–carrier recombination while simultaneously increasing the adsorption and dispersibility of photocatalysts in the process. As photocatalysts, MXenes are utilized in the process of degrading a variety of dyes [49,50]. In addition to this, an MXene-based sensor is utilized in order to detect toxic metal ions [51,52].

1.6   FUTURE RESEARCH

By changing the dimensions of MXenes, it is possible to more precisely regulate MXene-derived materials. As a result, in future studies, it may be possible to build derived materials with a 3D structure on-site from a purposefully produced 3D MXene material. In addition to the environment and energy storage, it is anticipated that more and more MXene-derived materials having cutting-edge characteristics and distinctive architectures will be investigated in the future [18]. The MXene scientific community is advancing because numerous issues are being addressed and numerous problems have been resolved over the last two years [53]. MXenes with homogeneous surface terminations were realized, as was previously mentioned. The MXene family currently includes high-entropy MXenes. It has been proven that oxycarbide MXenes exist. Several solid-solution MXenes have demonstrated effects of compositions on their characteristics. Research on MXene shelf life and oxidation stability has grown significantly, and numerous techniques are being examined to stop MXene oxidation [54,55] and create MXenes with greater oxidation stability [56]. As a result, single-layer MXenes in solution as well as the atmosphere became more environmentally stable, with Ti_3C_2's lifetime increasing from a few weeks to several months and V_2C's lifetime increasing.

REFERENCES

1. Wang, Yifan, Yanheng Xu, Menglei Hu, Han Ling, and Xi Zhu. "MXenes: Focus on optical and electronic properties and corresponding applications." *Nanophotonics* 9, no. 7 (2020): 1601–1620.
2. Naguib, Michael, Murat Kurtoglu, Volker Presser, Jun Lu, Junjie Niu, Min Heon, Lars Hultman, Yury Gogotsi, and Michel W. Barsoum. "Two-dimensional nanocrystals produced by exfoliation of Ti_3AlC_2." *Advanced Materials* 23, no. 37 (2011): 4248–4253.

3. Naguib, Michael, Olha Mashtalir, Joshua Carle, Volker Presser, Jun Lu, Lars Hultman, Yury Gogotsi, and Michel W. Barsoum. "Two-dimensional transition metal carbides." *ACS Nano* 6, no. 2 (2012): 1322–1331.

4. Naguib, Michael, V. N. Mochalin, M. W. Barsoum, and Y. Gogotsi. "MXenes: A new family of two-dimensional materials." *Advanced Materials* 26, no. 7 (2014): 992.

5. Sokol, Maxim, Varun Natu, Sankalp Kota, and Michel W. Barsoum. "On the chemical diversity of the MAX phases." *Trends in Chemistry* 1, no. 2 (2019): 210–223.

6. Khazaei, Mohammad, Masao Arai, Taizo Sasaki, Chan-Yeup Chung, Natarajan S. Venkataramanan, Mehdi Estili, Yoshio Sakka, and Yoshiyuki Kawazoe. "Novel electronic and magnetic properties of two-dimensional transition metal carbides and nitrides." *Advanced Functional Materials* 23, no. 17 (2013): 2185–2192.

7. Enyashin, A. N., and A. L. Ivanovskii. "Two-dimensional titanium carbonitrides and their hydroxylated derivatives: Structural, electronic properties and stability of MXenes Ti3C2-xNx (OH) 2 from DFTB calculations." *Journal of Solid State Chemistry* 207 (2013): 42–48.

8. Tahir, Muhammad, Azmat Ali Khan, Sehar Tasleem, Rehan Mansoor, and Wei Keen Fan. "Titanium carbide (Ti3C2) MXene as a promising co-catalyst for photocatalytic CO_2 conversion to energy-efficient fuels: A review." *Energy & Fuels* 35, no. 13 (2021): 10374–10404.

9. Sarycheva, Asia, and Yury Gogotsi. "Raman spectroscopy analysis of the structure and surface chemistry of Ti3C2T x MXene." *Chemistry of Materials* 32, no. 8 (2020): 3480–3488.

10. Cheng, Yongfa, Yanan Ma, Luying Li, Meng Zhu, Yang Yue, Weijie Liu, Longfei Wang et al. "Bioinspired microspines for a high-performance spray Ti3C2T x MXene-based piezoresistive sensor." *Acs Nano* 14, no. 2 (2020): 2145–2155.

11. Ma, Wenjie, Wenrui Cai, Wenhua Chen, Pengju Liu, Jianfeng Wang, and Zhuoxin Liu. "A novel structural design of shielding capsule to prepare high-performance and self-healing MXene-based sponge for ultra-efficient electromagnetic interference shielding." *Chemical Engineering Journal* 426 (2021): 130729.

12. Soren, S., S. Chakroborty, and K. Pal. "Enhanced in tunning of photochemical and electrochemical responses of inorganic metal oxide nanoparticles via rGO frameworks (MO/rGO): A comprehensive review." *Materials Science and Engineering: B* 278 (2022): 115632.

13. Tang, Jiayong, Xia Huang, Tongen Lin, Tengfei Qiu, Hengming Huang, Xiaobo Zhu, Qinfen Gu, Bin Luo, and Lianzhou Wang. "MXene derived TiS2 nanosheets for high-rate and long-life sodium-ion capacitors." *Energy Storage Materials* 26 (2020): 550–559.

14. Nath, N., A. Kumar, S. Chakroborty, S. Soren, A. Barik, K. Pal, and F. G. de Souza Jr. "Carbon nanostructure embedded novel sensor implementation for detection of aromatic volatile organic compounds: An organized review." *ACS Omega* 8 (2023): 4436–4452.

15. Liu, Qiuxia, Lunhong Ai, and Jing Jiang. "MXene-derived TiO 2@ C/gC 3 N 4 heterojunctions for highly efficient nitrogen photofixation." *Journal of Materials Chemistry A* 6, no. 9 (2018): 4102–4110.

16. Zhang, Peng, Lin Wang, Li-Yong Yuan, Jian-Hui Lan, Zhi-Fang Chai, and Wei-Qun Shi. "Sorption of Eu (III) on MXene-derived titanate structures: The effect of nano-confined space." *Chemical Engineering Journal* 370 (2019): 1200–1209.

17. Chang, Hui, Zhichao Shang, Qingqing Kong, Pingle Liu, Jikai Liu, and He'an Luo. "α-Fe2O3 nanorods embedded with two-dimensional {0 0 1} facets exposed TiO2 flakes derived from Ti3C2TX MXene for enhanced photoelectrochemical water oxidation." *Chemical Engineering Journal* 370 (2019): 314–321.

18. Yu, Lanlan, Baojun Liu, Yayi Wang, Fei Yu, and Jie Ma. "Recent progress on MXene-derived material and its' application in energy and environment." *Journal of Power Sources* 490 (2021): 229250.

19. Tripathi, Alok Kumar, and Rajendra Kumar Singh. "Application of ionic liquids as a green material in electrochemical devices." *Industrial Applications of Green Solvents* (2019): 106–147.

20. Wang, Zixing, Zhong Xu, Haichao Huang, Xiang Chu, Yanting Xie, Da Xiong, Cheng Yan, Haibo Zhao, Haitao Zhang, and Weiqing Yang. "Unraveling and regulating self-discharge behavior of Ti3C2T x MXene-based supercapacitors." *ACS Nano* 14, no. 4 (2020): 4916–4924.

21. VahidMohammadi, Armin, Ali Hadjikhani, Sina Shahbazmohamadi, and Majid Beidaghi. "Two-dimensional vanadium carbide (MXene) as a high-capacity cathode material for rechargeable aluminum batteries." *ACS Nano* 11, no. 11 (2017): 11135–11144.

22. Chen, Qian, and Yongji Gong. "Applications and challenges of 2D materials in lithium metal batteries." *Materials Lab* 1, no. 3 (2022): 220034.

23. Zhang, Haitao, Xiaojun Xin, Huan Liu, Haichao Huang, Ningjun Chen, Yanting Xie, Weili Deng, Chunsheng Guo, and Weiqing Yang. "Enhancing lithium adsorption and diffusion toward extraordinary lithium storage capability of freestanding Ti3C2T x MXene." *The Journal of Physical Chemistry C* 123, no. 5 (2019): 2792–2800.

24. Lukatskaya, Maria R., Olha Mashtalir, Chang E. Ren, Yohan Dall'Agnese, Patrick Rozier, Pierre Louis Taberna, Michael Naguib, Patrice Simon, Michel W. Barsoum, and Yury Gogotsi. "Cation intercalation and high volumetric capacitance of two-dimensional titanium carbide." *Science* 341, no. 6153 (2013): 1502–1505.

25. Lukatskaya, Maria R., Sankalp Kota, Zifeng Lin, Meng-Qiang Zhao, Netanel Shpigel, Mikhael D. Levi, Joseph Halim et al. "Ultra-high-rate pseudocapacitive energy storage in two-dimensional transition metal carbides." *Nature Energy* 2, no. 8 (2017): 1–6.

26. Wang, Xingwei, Dongzhi Zhang, Haobing Zhang, Likun Gong, Yan Yang, Wenhao Zhao, Sujing Yu, Yingda Yin, and Daofeng Sun. "In situ polymerized polyaniline/MXene (V2C) as building blocks of supercapacitor and ammonia sensor self-powered by electromagnetic-triboelectric hybrid generator." *Nano Energy* 88 (2021): 106242.

27. Koyappayil, Aneesh, Sachin Ganpat Chavan, Mohsen Mohammadniaei, Anna Go, Sei Young Hwang, and Min-Ho Lee. "β-Hydroxybutyrate dehydrogenase decorated MXene nanosheets for the amperometric determination of β-hydroxybutyrate." *Microchimica Acta* 187 (2020): 1–7.

28. Zhang, Qiuxia, Feng Wang, Huixin Zhang, Youyu Zhang, Meiling Liu, and Yang Liu. "Universal Ti3C2 MXenes based self-standard ratiometric fluorescence resonance energy transfer platform for highly sensitive detection of exosomes." *Analytical Chemistry* 90, no. 21 (2018): 12737–12744.

29. Xu, Bingzhe, Minshen Zhu, Wencong Zhang, Xu Zhen, Zengxia Pei, Qi Xue, Chunyi Zhi, and Peng Shi. "Ultrathin MXene-micropattern-based field-effect transistor for probing neural activity." *Advanced Materials* 28, no. 17 (2016): 3333–3339.

30. Li, Hui, Rangrang Fan, Bingwen Zou, Jiazhen Yan, Qiwu Shi, and Gang Guo. "Roles of MXenes in biomedical applications: Recent developments and prospects." *Journal of Nanobiotechnology* 21, no. 1 (2023): 1–39.

31. Yuan, Xiaoli, Ying Zhu, Shasha Li, Yiqun Wu, Zhongshi Wang, Rui Gao, Shiyao Luo, Juan Shen, Jun Wu, and Liang Ge. "Titanium nanosheet as robust and biosafe drug carrier for combined photochemo cancer therapy." *Journal of Nanobiotechnology* 20, no. 1 (2022): 154.

32. Dong, Yangjin, Shanshan Li, Xiaoyun Li, and Xiaoying Wang. "Smart MXene/agarose hydrogel with photothermal property for controlled drug release." *International Journal of Biological Macromolecules* 190 (2021): 693–699.

33. Liu, Zhen, Lan Xie, Jia Yan, Pengfei Liu, Huixiang Wen, and Huijun Liu. "Folic acid-targeted MXene nanoparticles for doxorubicin loaded drug delivery." *Australian Journal of Chemistry* 74, no. 12 (2021): 847–855.

34. Jin, Lin, Yanfei Ma, Ruiya Wang, Shuo Zhao, Zhishuai Ren, Shengnan Ma, Yupeng Shi, Bin Hu, and Yuqi Guo. "Nanofibers and hydrogel hybrid system with synergistic effect of anti-inflammatory and vascularization for wound healing." *Materials Today Advances* 14 (2022): 100224.

35. Rafieerad, Alireza, Weiang Yan, Glen Lester Sequiera, Niketa Sareen, Ejlal Abu-El-Rub, Meenal Moudgil, and Sanjiv Dhingra. "Application of Ti3C2 MXene quantum dots for immunomodulation and regenerative medicine." *Advanced Healthcare Materials* 8, no. 16 (2019): 1900569.

36. He, Chao, Luodan Yu, Heliang Yao, Yu Chen, and Yongqiang Hao. "Combinatorial photothermal 3D-printing scaffold and checkpoint blockade inhibits growth/metastasis of breast cancer to bone and accelerates osteogenesis." *Advanced Functional Materials* 31, no. 10 (2021): 2006214.

37. Bai, Lei, Wenhui Yi, Taiyang Sun, Yilong Tian, Ping Zhang, Jinhai Si, Xun Hou, and Jin Hou. "Surface modification engineering of two-dimensional titanium carbide for efficient synergistic multitherapy of breast cancer." *Journal of Materials Chemistry B* 8, no. 30 (2020): 6402–6417.

38. Liu, Kaiyuan, Yuxin Liao, Zifei Zhou, Li Zhang, Yingying Jiang, Hengli Lu, Tianyang Xu et al. "Photothermal-triggered immunogenic nanotherapeutics for optimizing osteosarcoma therapy by synergizing innate and adaptive immunity." *Biomaterials* 282 (2022): 121383.

39. Ma, Yanan, Nishuang Liu, Luying Li, Xiaokang Hu, Zhengguang Zou, Jianbo Wang, Shijun Luo, and Yihua Gao. "A highly flexible and sensitive piezoresistive sensor based on MXene with greatly changed interlayer distances." *Nature Communications* 8, no. 1 (2017): 1207.

40. Lei, Yongjiu, Wenli Zhao, Yizhou Zhang, Qiu Jiang, Jr-Hau He, Antje J. Baeumner, Otto S. Wolfbeis, Zhong Lin Wang, Khaled N. Salama, and Husam N. Alshareef. "A MXene-based wearable biosensor system for high-performance in vitro perspiration analysis." *Small* 15, no. 19 (2019): 1901190.

41. Rasool, Kashif, Mohamed Helal, Adnan Ali, Chang E. Ren, Yury Gogotsi, and Khaled A. Mahmoud. "Antibacterial activity of Ti3C2T x MXene." *ACS Nano* 10, no. 3 (2016): 3674–3684.

42. Mashtalir, Olha, Kevin M. Cook, Vadym N. Mochalin, Martin Crowe, Michel W. Barsoum, and Yury Gogotsi. "Dye adsorption and decomposition on two-dimensional titanium carbide in aqueous media." *Journal of Materials Chemistry A* 2, no. 35 (2014): 14334–14338.

43. Wei, Zheng, Zhang Peigen, Tian Wubian, Qin Xia, Zhang Yamei, and Sun ZhengMing. "Alkali treated Ti3C2Tx MXenes and their dye adsorption performance." *Materials Chemistry and Physics* 206 (2018): 270–276.

44. Peng, Qiuming, Jianxin Guo, Qingrui Zhang, Jianyong Xiang, Baozhong Liu, Aiguo Zhou, Riping Liu, and Yongjun Tian. "Unique lead adsorption behavior of activated hydroxyl group in two-dimensional titanium carbide." *Journal of the American Chemical Society* 136, no. 11 (2014): 4113–4116.

45. Ying, Yulong, Yu Liu, Xinyu Wang, Yiyin Mao, Wei Cao, Pan Hu, and Xinsheng Peng. "Two-dimensional titanium carbide for efficiently reductive removal of highly toxic chromium (VI) from water." *ACS Applied Materials & Interfaces* 7, no. 3 (2015): 1795–1803.

46. Hwang, Seung Kyu, Sung-Min Kang, Muruganantham Rethinasabapathy, Changhyun Roh, and Yun Suk Huh. "MXene: An emerging two-dimensional layered material for removal of radioactive pollutants." *Chemical Engineering Journal* 397 (2020): 125428.

47. Wang, Lin, Liyong Yuan, Ke Chen, Yujuan Zhang, Qihuang Deng, Shiyu Du, Qing Huang et al. "Loading actinides in multilayered structures for nuclear waste treatment: The first case study of uranium capture with vanadium carbide MXene." *ACS Applied Materials & Interfaces* 8, no. 25 (2016): 16396–16403.

48. Ren, Chang E., Kelsey B. Hatzell, Mohamed Alhabeb, Zheng Ling, Khaled A. Mahmoud, and Yury Gogotsi. "Charge-and size-selective ion sieving through Ti3C2T x MXene membranes." *The Journal of Physical Chemistry Letters* 6, no. 20 (2015): 4026–4031.

49. Gao, Yupeng, Libo Wang, Aiguo Zhou, Zhengyang Li, Jingkuo Chen, Hari Bala, Qianku Hu, and Xinxin Cao. "Hydrothermal synthesis of TiO2/Ti3C2 nanocomposites with enhanced photocatalytic activity." *Materials Letters* 150 (2015): 62–64.

50. Peng, C., X. Yang, Y. Li, H. Yu, H. Wang, and F. Peng. "Hybrids of two-dimensional Ti3C2 and TiO2 exposing {001} facets toward enhanced photocatalytic activity." *ACS Applied Materials & Interfaces* 8 (2016): 6051–6060.

51. Rasheed, P. Abdul, Ravi P. Pandey, Kashif Rasool, and Khaled A. Mahmoud. "Ultra-sensitive electrocatalytic detection of bromate in drinking water based on Nafion/Ti3C2Tx (MXene) modified glassy carbon electrode." *Sensors and Actuators B: Chemical* 265 (2018): 652–659.

52. Cheng, Haoliang, and Jurui Yang. "Preparation of Ti3C2-PANI composite as sensor for electrochemical determination of mercury ions in water." *International Journal of Electrochemical Science* 15 (2020): 2295–2306.

53. Gogotsi, Yury, and Qing Huang. "MXenes: Two-dimensional building blocks for future materials and devices." *ACS Nano* 15, no. 4 (2021): 5775–5780.

54. Matthews, Kyle, Teng Zhang, Christopher E. Shuck, Armin VahidMohammadi, and Yury Gogotsi. "Guidelines for synthesis and processing of chemically stable two-dimensional V2CT x MXene." *Chemistry of Materials* 34, no. 2 (2021): 499–509.

55. Cao, Fangcheng, Ye Zhang, Hongqing Wang, Karim Khan, Ayesha Khan Tareen, Wenjing Qian, Han Zhang, and Hans Ågren. "Recent advances in oxidation stable chemistry of 2D MXenes." *Advanced Materials* 34, no. 13 (2022): 2107554.

56. Mathis, Tyler S., Kathleen Maleski, Adam Goad, Asia Sarycheva, Mark Anayee, Alexandre C. Foucher, Kanit Hantanasirisakul, Christopher E. Shuck, Eric A. Stach, and Yury Gogotsi. "Modified MAX phase synthesis for environmentally stable and highly conductive Ti3C2 MXene." *ACS Nano* 15, no. 4 (2021): 6420–6429.

2 Synthesis and Characterization of MXenes through an Array of Conventional Methods and Green Synthesis Routes

N. Usha Kiran, Siba Soren, Sriparna Chatterjee, and Laxmidhar Besra

2.1 INTRODUCTION

MXenes are a class of two-dimensional (2D) nanomaterials that have gained significant attention in recent years due to their unique properties and potential applications. They are composed of transition metal carbides, nitrides, or carbonitrides. The name 'MXene' comes from their chemical formula of $M_{n+1}X_nT_x$, where M represents early transition metal, X is carbon and/or nitrogen, and T_x represents surface functional groups such as hydroxyl (-OH), fluoride (-F), etc. MXenes were first discovered in 2011 by researchers at Drexel University, USA, who synthesized $Ti_3C_2T_x$ by selectively etching the aluminum layer from its precursor Ti_3AlC_2 MAX phase material with hydrofluoric acid.[1,2] MAX phases are a class of layered materials that exhibit a unique combination of metallic and ceramic properties. These materials have a compact layered structure consisting of a transition metal (M) layer sandwiched between two layers of elements from groups IIIA or IVA (A), with carbon or nitrogen (X) atoms on the top and bottom layers. The transition metal layer (M) in MAX phases provides the metallic properties, while the 'A' element and 'X' layer provide the ceramic properties.[3,4] To date, more than 70 different MXenes have been synthesized and characterized. This number continues to grow rapidly as researchers develop new synthesis methods and investigate new compositions and structures.

MXenes have several unique properties that make them a promising class of materials for a wide range of applications. MXenes have a layered structure with large interlayer spacing, which results in a high surface area. The surface area can be as high as $300 \, m^2/g$, making MXenes attractive for use in energy storage applications

DOI: 10.1201/9781003366225-2

such as supercapacitors and batteries.[5] The large surface area also makes them useful for catalysis,[6] gas adsorption,[7] and water purification.[8] One of their most notable properties is their high electrical conductivity, which makes them useful for electronics applications. They have been used as electrodes in electrochemical devices such as batteries, supercapacitors, and sensors. MXenes also exhibit metallic conductivity at room temperature, which is useful for electromagnetic interference (EMI) shielding and other applications.[9] MXenes also possess good mechanical properties, including high strength, stiffness, and toughness. They have been used as reinforcements in polymer composites to improve their mechanical properties. MXenes also have excellent wear resistance and can be used as coatings or lubricants. They are chemically stable and resistant to oxidation and corrosion.[10] They can be synthesized with different compositions and structures, and their properties can be tuned by changing their composition, structure, and surface functionalization. This opens a range of possibilities for tailoring their properties for specific applications.[11]

MXene synthesis protocols reported in the literature vary significantly, and it is important to note that one method may be suitable for a specific application but not for others. The selective etching process to produce MXenes typically involves the use of strong acids such as HF or HCl, or a combination of acids and heat. The MAX phase precursor is initially synthesized through high-temperature solid-state reactions, followed by etching with acid to selectively remove the 'A' layer in the typical synthesis process. This leaves behind a 2D material with a layered structure, where transition metal atoms were sandwiched between two layers of carbon and/or nitrogen atoms. The resulting MXene can then be modified with various functional groups or delaminated into individual layers to alter its properties.[12] However, the use of HF as an etchant has some drawbacks, including high toxicity, corrosiveness, and safety concerns, which require proper safety measures and equipment to be used during the synthesis process. Therefore, researchers are exploring alternative etchants that are less toxic and more environmentally friendly to address these concerns. Two different methods are commonly used for the synthesis of 2D MXenes. One approach is the bottom-up approach, which includes a chemical vapor deposition (CVD) technique that can produce high-quality films on various substrates. However, this method is not suitable for producing single-layer MXene nanosheets. [13] The second approach for MXene synthesis is the top-down approach, which involves mechanical and chemical exfoliation of layered solids. However, it is challenging to mechanically exfoliate MXene layers from their precursor MAX phase because the 'M' atoms are covalently bonded to 'A' atoms, making it difficult to separate the layers. Therefore, chemical exfoliation is typically used to separate the MXene layers.[14]

2.2 TECHNIQUES OF MXENE SYNTHESIS

The synthesis of MXenes typically involves three main steps: (1) MAX phase synthesis, as shown in Figure 2.1, (2) selective etching of the 'A' layer to form multilayer MXene, and (3) delamination and exfoliation of the multilayer MXene to form a few/single-layer MXene.

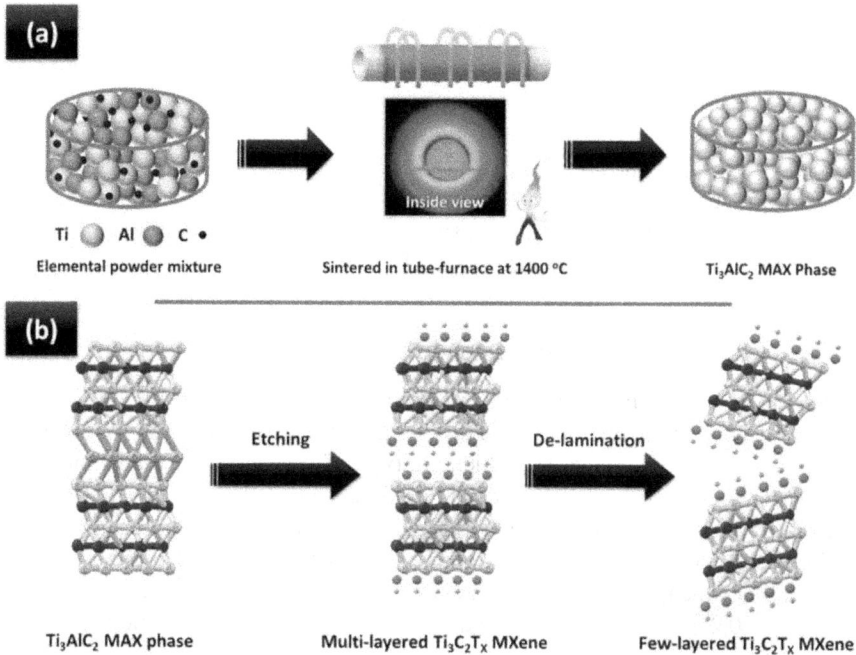

FIGURE 2.1 (a) Schematic representation of the formation of the Ti_3AlC_2 MAX phase (grey powder) using an elemental mixture of Ti (yellow atoms), Al (pink atoms), and C (black atoms) at a particular molar ratio, followed by heating in a tube furnace at 1400 °C; (b) Atomic structure representation shows that MXenes are formed after selective etching of 'A' group elements, i.e., Al (pink atoms), in this case using HF/HCl acid, formation of multilayered MXenes with surface terminations (red atoms), followed by delamination using intercalants like TBAOH, LiCl, etc., to form single/few-layered MXenes.

[Copyright taken from 19]

2.2.1 THE FIRST STEP (MAX PHASE SYNTHESIS)

MAX phases are a family of layered ternary carbides and nitrides with the general formula of $M_{n+1}AX_n$, where 'M' is a transition metal, 'A' is an A-group element, 'X' is either carbon or nitrogen, and n = 1, 2, or 3. These materials have unique properties that make them attractive for a wide range of applications. The synthesis of MAX phases can be achieved by various methods, including solid-state reactions, powder metallurgy, and chemical vapor deposition (CVD).[15] In solid-state synthesis, MAX phases are typically synthesized by heating metal powders with carbon or nitrogen sources at high temperatures (typically above 1400 °C) under a vacuum or inert atmosphere. This method is a common approach for synthesizing bulk MAX phase materials with high purity and homogeneity. [16] However, the high-temperature requirements can lead to issues with non-uniformity and impurities. In powder metallurgy, MAX phases are produced by

consolidating MAX phase powders through various techniques such as hot pressing, spark plasma sintering, self-propagating high-temperature synthesis, etc.
[17,18,19] This approach allows for the production of dense and complex-shaped
MAX phase components, which are useful for applications such as coatings, composites, and high-temperature structural materials. CVD is another method used
to synthesize MAX phases that involves the deposition of thin films of MAX
phases onto substrates using a precursor gas. This method allows for the precise
control of film thickness and composition, which is important for applications
such as electronics and catalysis.

2.2.2   THE SECOND STEP (MULTILAYER MXENE SYNTHESIS)

The second step involves the selective etching of the 'A' layer of the MAX phase
using a strong acid or fluoride salt. This etching process removes the 'A' layer and
leaves behind the MXene layers, which are typically composed of transition metal
carbides or nitrides. The specific etchant used depends on the type of MAX phase
being used and the desired MXene composition. The synthesis of MXenes is a relatively recent development, with the first MXene material ($Ti_3C_2T_x$) being reported
in 2011. Since then, researchers have developed several methods for the synthesis of
MXenes, [20,21] each with their own advantages and limitations. Some of them are
described in the following sections.

2.2.2.1   HF Etching

To synthesize a $Ti_3C_2T_x$ MXene, the process starts with the chemical exfoliation
of the Ti_3AlC_2 MAX phase using hydrofluoric acid (HF) at room temperature. The
resulting material is a multilayered MXene powder, with 2D layers held together by
hydrogen bonds and/or van der Waals forces. The powder is then washed multiple
times through centrifugation until the pH reaches a neutral value (~7) to remove any
remaining acid or reaction byproducts. The multilayered MXene flakes are then collected using vacuum filtration and dried for further processing. The first synthesis of
MXene was carried out by Naguib et al. [22] through the immersion of Ti_3AlC_2 fine
powders in a 50% concentration of hydrofluoric acid at ambient conditions for a total
of 2 hours. The reaction can be represented as:

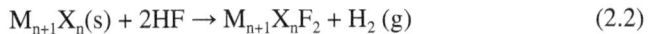

$$M_{n+1}AX_n + 3HF \rightarrow M_{n+1}X_n \text{ (s)} + AF_3 + 1.5H_2 \text{ (g)} \quad (2.1)$$

$$M_{n+1}X_n(s) + 2HF \rightarrow M_{n+1}X_nF_2 + H_2 \text{ (g)} \quad (2.2)$$

In Equation 2.1, the HF reacts with the 'A' layer of the MAX phase, leaving behind
the MX layer in the solution with evolution of H_2 gas. In the next reaction (Equation
2.2), the 'M' atoms, being highly reactive, form M–OH and M–F bonds upon reaction with water and HF. The synthesis of various MXenes depends on the bond
energy of the M–A bond in different MAX phases; consequently, there are significant differences in etching conditions implemented for their synthesis. For example,
a $Ti_3C_2T_x$ MXene was synthesized by Kiran et al.[23] from a Ti_3SiC_2 MAX phase
using oxidant-assisted selective etching of Si atoms. Recent studies have also revealed

that $Zr_3Al_3C_5$ and Mo_2Ga_2C can act as precursor materials for 2D MXene synthesis, despite not belonging to the MAX family.[24,25] The synthesis of Zr-based MXenes is challenging due to the difficulty in producing single-phase dense Zr–Al–C compounds. Among the layered ternary transition metal carbides, $Zr_3Al_3C_5$ is a typical member with an alternated crystal structure consisting of hexagonal MC intergrowth layers and Al_4C_3-like Al_3C_2 layers that share carbon monolayers at coupling boundaries. However, the etching of $Zr_3Al_3C_5$ can be achieved by mixing it with 50% HF, leading to the following reactions:

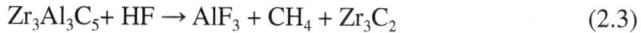

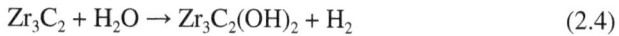

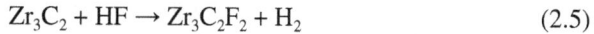

$$Zr_3Al_3C_5 + HF \rightarrow AlF_3 + CH_4 + Zr_3C_2 \qquad (2.3)$$

$$Zr_3C_2 + H_2O \rightarrow Zr_3C_2(OH)_2 + H_2 \qquad (2.4)$$

$$Zr_3C_2 + HF \rightarrow Zr_3C_2F_2 + H_2 \qquad (2.5)$$

The quality of MXene flakes depends heavily on different etching parameters such as particle size, temperature, concentration of the etchant, and etching time. Although the use of a strong etchant and an increase in temperature can significantly shorten the etching time, these parameters must be properly controlled to avoid the destruction of the layered structure and properties. However, HF procedures have limitations in scaling-up processes, including hazardous chemistry solutions, and they do not produce monolayer MXene materials.

2.2.2.2   In Situ HF Etching

Alternative etching methods that can form HF etchant in situ have been explored to address the corrosive issue of HF etching. In these in situ HF-forming systems, the F ion reacts with the Al atoms of MAX precursors, which have a high reactivity with F ions due to their Al-containing composition. This reaction produces fluoride, H_2, and the desired MXenes. The advantages of this procedure are: (1) mild corrosive environment of HF in comparison to the highly corrosive media on direct addition, (2) requires no sonication or low sonication time for the delamination of MXene flakes, (3) high exfoliation yield with single/few-layered MXenes, (4) less defective MXene sheets, and (5) highly flexible clay-like MXenes.[26] In order to enhance MXene exfoliation, the environmentally benign etchant LiF/HCl is used to facilitate the intercalation of Li^+ ions during etching. This weakens the interlayer interaction and increases the gap between the MXene nanosheets. The change in molar ratio of the reactants results in a mild route which produces delaminated, high-quality MXene nanosheets with high yield. This method is called the minimally intensive layer delamination (MILD) method. Currently, this method is most extensively being used to synthesize high-quality MXenes, particularly in the field of energy storage and opto-electronics.

To avoid using unnecessarily high HF concentration, Ghidiu et al.[27] explored the reaction of a Ti_3AlC_2 MAX phase with a mixture of LiF and HCl solution to generate HF in situ, which ultimately produces $Ti_3C_2T_x$ MXene flakes. The reaction mechanism is as follows:

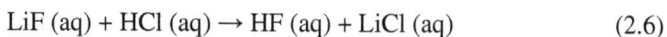

$$LiF \text{ (aq)} + HCl \text{ (aq)} \rightarrow HF \text{ (aq)} + LiCl \text{ (aq)} \qquad (2.6)$$

Using this method, it was possible to produce a highly conductive clay of $Ti_3C_2T_x$, which exhibited robust plasticity and could be easily fashioned into a film through roller pressing. The rolled MXene clay, which was obtained as a free-standing structure, could be directly utilized as the working electrode for supercapacitors without requiring any additional modifications. This electrode exhibited a high specific capacity of 900 F cm^{-3}, and after 10,000 cycles, it demonstrated minimal capacity loss, indicating exceptional stability over repeated use. Currently, the mature method for etching MAX phases involves mixing fluoride salts with acid. By varying the type of fluoride salt used, the resulting MXenes' interlayer spacing can be tailored to meet specific application requirements. Besides LiF, other types of fluoride salts have been included utilized as etchants. Commonly used fluoride sources include hydrofluoric acid (HF) and potassium fluoride (KF), while the buffering agent can be any ammonium salt, such as ammonium fluoride (NH_4F) or ammonium bifluoride (NH_4HF_2). The ammonium fluoride/bifluoride acts as a buffer, controlling the pH of the solution, while the fluoride source helps to dissolve the 'A' layer.[28] The etching conditions, such as temperature, time, and concentration of the etchant solution, also play a crucial role in the synthesis of MXene. Generally, a lower temperature and longer etching time result in a higher degree of 'A' layer removal and thinner MXene flakes. However, these conditions need to be optimized to prevent over-etching or under-etching of the precursor material. Liu et al.[29] developed a method that involves integrating HCl with different fluoride salts such as LiF, NaF, KF, and NH_4F to create a mixed solution for etching Ti_3AlC_2. The authors reported that the use of KF resulted in a more uniform and complete etching of the Ti_3AlC_2 MAX phase, leading to the synthesis of a $Ti_3C_2T_x$ MXene with a high degree of crystallinity and uniform morphology. HF salts like ammonium bifluoride (NH_4HF_2) were used by Halim et al.[30] in 2014 to etch a Ti_3AlC_2 MAX phase at room temperature or elevated temperature for a few hours to easily prepare a $Ti_3C_2T_x$ MXene. The larger d-spacing after drying is attributed to the intercalation of NH_4^+ ions in between the MXene layers. Unlike the HF-etched MXene, MXenes obtained by this method require a longer time and higher temperature of vacuum drying to completely remove the intercalated water molecules.

Of the various in situ HF etchants available, the HCl/LiF combination is the most used configuration. This combination enables easy etching and delamination of the MAX phase, unlike other salt combinations, which lead to the formation of accordion-like MXenes that have not been directly delaminated yet. Typically, in the preparation of $Ti_3C_2T_x$ using the HCl/LiF route, the molar ratio of LiF to MAX phase is set at 7.5:1, while the concentration of HCl ranges between 6 M and 12 M. Regarding the etching time, common Ti-based MXenes can be fully etched within 24 hours. However, for some newer types, the etching process may take longer due to the higher stripping energy of their MAX phase. Halim et al.[31] utilized HCl/LiF at 35 °C to etch Mo_2Ga_2C and successfully obtained Mo_2CT_x. However, a relatively long etching time of 16 days was required to achieve a reasonable yield.

Compared to the HF etching approach, the acid/fluoride salt etching method is less harsh and safer. Furthermore, the accordion-like MXenes that are obtained through the HCl/LiF method can be directly delaminated into single-layer MXene nanosheets by means of ultrasound treatment or hand shaking, simplifying the production of 2D MXene nanosheets. However, this approach can sometimes lead to un-etched MAX phase residues, in turn necessitating a systematic approach to enhance the yield of the final product.

2.2.2.3 Electrochemical Etching

The electrochemical etching technique involves the anodic etching of MAX phases, which selectively removes the 'A' layers and leaves behind the M-C layers to form 2D MXene flakes. In an electrochemical cell, the MAX phase acts as a working electrode immersed in an etchant solution, typically containing hydrofluoric acid and a supporting electrolyte, along with a counter electrode and a reference electrode. Upon the application of external voltage by cyclic voltammograms between 0 and 2.5 V, the M–A bond breaks, resulting in dissolution of the 'A' layer and the formation of MXene flakes. As the voltage gradually increases, the 'M' layer is further removed, ultimately resulting in the formation of amorphous carbon materials. By controlling the etching voltage window (i.e., the etching potential) within the range of the reaction potential between the 'A' and 'M' layers and selecting an appropriate etching time, it is possible to selectively remove 'A' atoms, thereby enabling precise control of the synthesized MXenes. However, it should be noted that the working electrode is typically composed of the MAX phase and, therefore, the etching process first occurs on the surface of the MAX electrode. This can result in the formation of surface CDCs (carbide-derived carbons), which can potentially hinder the subsequent etching process. Sun et al.[32] utilized a three-electrode system to carry out electrochemical etching of the Ti_2AlC MAX phase at a voltage of 0.6 V (vs Ag/AgCl). They took Pt as a counter electrode and HCl as an electrolyte in the electrolytic system. The electrochemical etching process demonstrated the conversion of Ti_2AlC into Ti_2CT_x and CDC layers. However, the presence of CDC layers on the surface of Ti_2AlC hindered further etching, resulting in the formation of MXene-covered MAX. The morphology of Ti_2AlC after etching revealed that the process carried out with 1 M HCl as an electrolyte led to incomplete etching, while the optimized etchant concentration of 2-M HCl with etchant time of 120 h developed an accordion-like structure. Similarly, Yang et al.[33] utilized a two-electrode system where Ti_3AlC_2 MAX pieces were used as both the working and counter electrode. The electrochemical etching was carried out using different electrolytes, including H_2SO_4, HNO_3, NaOH, NH_4Cl, and $FeCl_3$. The research emphasized how the etching process can be affected by various electrolytes. The etching yield in this case was approximately 40%, calculated based on the weight ratio of the etched product to the precursor. To enhance the internal accessibility of the MAX phase, intercalators can be used to expand the interlayer spacing, enabling the smooth diffusion of electrolyte ions. One instance is the utilization of a mixed electrolyte composed of 1 M NH_4Cl and 0.2 M TMAOH to produce $Ti_3C_2T_x$ through electrochemical etching at 5 V (vs SCE) for a duration of 5 hours. Using intercalators to reduce the interference of CDC layers during the etching process can be effective. However, one concern is the toxicity of intercalators, which can be a safety issue during experimentation.

Electrochemical etching offers a low-energy-consumption and environmentally friendly method for synthesizing MXenes. However, the presence of the CDC layer remains a challenge to be addressed in addition to the issue of insufficient yield. Although the MAX phase electrode can be recycled multiple times, the yield of MXenes obtained through a typical etching process is low.

2.2.2.4 Molten-Salt Etching

Due to unsuccessful production of nitride-based MXenes using HF or HCl-LiF as etchants, the Ti_4N_3-MXene was first synthesized in 2016 by the reaction of a Ti_4AlN_3 MAX phase with molten fluoride salt at 550 °C in an argon atmosphere.[34] The molten-salt method relies on the selective dissolution of the 'A' layer by the molten salt. A suitable molten salt is chosen based on its melting point, chemical stability, and ability to dissolve the 'A' layer. Commonly used molten salts include eutectic mixtures of LiCl–KCl or NaCl–KCl. The salt mixture is heated to a temperature above its melting point in a crucible or furnace. The precursor material is immersed in the molten-salt bath and heated for a specified time to facilitate the selective etching of the 'A' layer. The salt bath serves as a source of reactive ions, which react with the 'A' layer to form soluble fluoride or chloride compounds, leaving behind the 'M' and 'X' layers. Recently, a salt-templated method was implemented to produce Mo_2NT_x and V_2NT_x by the amination of MXene carbides Mo_2CT_x and V_2CT_x. [35] This is due to the less cohesive and high formation energies of $M_{n+1}N_n$ class of compounds. The delamination process using TBAOH was carried out successfully to obtain single/few-layered MXene nanosheets. The MXene obtained using fluoride-containing etchant had a higher fraction of fluoride (F)-terminated moieties, due to which the electrochemical performance of the material was reduced. Since HF is a corrosive chemical, other researchers have tried other methods to obtain high-quality, fluoride-free MXenes.

MXene materials can also be synthesized using fluoride-free etching methods, which involve the use of strong acids or other reagents to selectively remove the 'A' layer from the MAX phase, leaving the 'M' and 'X' layers intact.[34] Commonly used reagents for this method include hydrochloric acid, hydroiodic acid, nitric acid, etc. The choice of acid or reagent for the etching process can greatly affect the quality and properties of the resulting MXene material. For example, hydrochloric acid is commonly used due to its high selectivity and efficiency, but it may also lead to the formation of unwanted chlorides.[36] The etching conditions, such as temperature, time, and concentration of acid or reagent, must be carefully controlled to achieve the desired degree of 'A' layer removal and to prevent damage to the 'M' and 'X' layers.[37] For example, higher temperatures and concentrations may result in faster etching but may also lead to greater damage to the MXene nanosheets. Some recent reports described the synthesis of Cl-terminated Ti_2C and Ti_3C_2 MXene flakes from their corresponding Al-based MAX phases in a Lewis acidic melt, i.e., $ZnCl_2$. [38] In this molten-salt system, the etching of the MAX phase was achieved using $ZnCl_2$ as the etchant. The molten-salt bath was formed using NaCl and KCl in a 1:1 molar ratio to reduce the melting point of the eutectic system. The etching process is described by the following reaction mechanism:

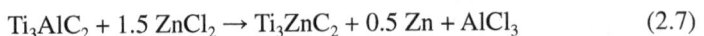

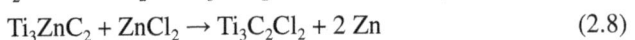

$$Ti_3AlC_2 + 1.5\ ZnCl_2 \rightarrow Ti_3ZnC_2 + 0.5\ Zn + AlCl_3 \qquad (2.7)$$

$$Ti_3ZnC_2 + ZnCl_2 \rightarrow Ti_3C_2Cl_2 + 2\ Zn \qquad (2.8)$$

Another study by Kamysbayev[39] and colleagues reported a two-step approach to synthesize Lewis acid-etched MXenes that allows for adjustable surface terminations, where the MAX phase was etched using molten chlorine or bromine salt to obtain Cl/Br-terminated MXenes. The application of molten inorganic salts in the etching process enables the substitution or elimination of surface terminations on MXenes, thus providing a novel approach for designing surface-selective MXenes suitable for advanced applications. However, this method is still in its early stages and requires further investigation into the physical and chemical characteristics of the produced MXenes, including their electrical conductivity, hydrophilicity, and mechanical properties.

2.2.2.5 Alkali Etching

Many of the etching methods discussed earlier involve the use of acids to etch the 'A' atom layers. However, it is possible to achieve selective etching of the MAX phase using alkalis. By utilizing a hydrothermal method and a highly concentrated KOH solution (~93 wt%), MXene nanoribbons were successfully obtained through the etching of Ti_3AlC_2 at 180 °C for 24 hours.[40] In another study, Ti_3AlC_2 was subjected to a two-step etching process by Xie et al.,[41] wherein it was first soaked in 1 M NaOH solution for 100 h and then in 1 M H_2SO_4 solution for 2 h at 80 °C. This method resulted in the surface etching of the MAX phase, producing a $Ti_3C_2T_x$ MXene. The described procedure enabled effective etching of the MAX phase with a low concentration of alkali as the etchant. However, only the superficial layer of the MAX phase could be etched, resulting in an extremely low yield of MXenes. Moreover, the formation of oxide/hydroxide layers over the MAX phase was another obstacle for alkali etching. For example, when Ti_3SiC_2 was etched at 200 °C using a hydrothermal reaction with 2 M KOH, a core–shell MAX@$K_2Ti_8O_{17}$ composite was formed, while the use of NaOH resulted in the formation of $Na_2Ti_7O_{15}$ on the MAX phase surface. These impurities hindered the process of producing pure MXenes.[42] The alkali concentration and temperature also have an adverse effect on the qualitative change in reaction between MAX phase and the alkali. Li et al.[43] explored the hydrothermal alkaline etching method using NaOH as an etchant, where it formed the Al-containing oxides and hydroxides, resulting in an up to 92% yield of fluoride-free high-purity $Ti_3C_2T_x$ MXene. The reaction mechanism during the etching process is follows:

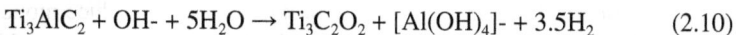

$$Ti_3AlC_2 + OH- + 5H_2O \rightarrow Ti_3C_2(OH)_2 + [Al(OH)_4]- + 2.5H_2 \qquad (2.9)$$
$$Ti_3AlC_2 + OH- + 5H_2O \rightarrow Ti_3C_2O_2 + [Al(OH)_4]- + 3.5H_2 \qquad (2.10)$$

Although the use of concentrated alkali for etching the MAX phase is highly efficient and can produce highly hydrophilic MXenes with F-free terminations, its applicability for large-scale MXene preparation is limited due to the dangers of using highly concentrated alkali and high temperature.

2.2.3 THE THIRD STEP (SINGLE/FEW-LAYER MXENE SYNTHESIS)

Delamination, the process of separating individual MXene layers from the multilayer MXene, is a critical step in the synthesis and processing of MXenes. This process can increase the distance between MXene layers, allowing for enhanced accessibility to the layers by chemical species such as water or other polar solvents, which enhances the surface area and exposes more active sites that can further increase the reactivity and potential applications of MXene nanosheets. The delamination of MXenes can be achieved through various techniques, including chemical intercalation, mechanical shearing, etc.[44,45] Recent advancements in MXene delamination techniques have allowed for the synthesis of MXene-based materials with enhanced electrochemical performance, catalytic activity, and mechanical properties, making them attractive for a wide range of applications in energy storage, catalysis, sensing, and electronics. Therefore, the delamination process is crucial for tailoring the properties and functionality of MXenes for various applications.

2.2.3.1 Chemical Intercalation

MXenes exhibit similar properties to clay minerals such as kaolinite, including rheological behavior, hydrophilicity, and plasticity. This similarity suggests that organic molecules that are commonly used as intercalators for clay may also be effective for MXenes. In 2013, Mashtalir et al. discovered that multilayer $Ti_3C_2T_x$ can be converted into single-layer MXene nanosheets by delamination using dimethyl sulfoxide (DMSO) as an intercalator.[46] The interlayer spacing enlargement could effectively decrease the van der Waals forces among the layers of the MXene, thus simplifying the ultrasonication process for the subsequent exfoliation of the DMSO-intercalated multilayer $Ti_3C_2T_x$ MXene to the single/few-layer $Ti_3C_2T_x$ MXene. While DMSO has been found to be an effective intercalation agent for $Ti_3C_2T_x$, it does not demonstrate similar effectiveness in intercalating other types of MXenes such as V_2CT_x and Mo_2CT_x. As alternatives to DMSO, many other organic intercalants such as hydrazine monohydrate (HM), N,N-dimethylformamide (DMF), TMAOH, TBAOH, urea, etc. have been investigated as intercalators for exfoliating multilayer $Ti_3C_2T_x$ MXenes.[47,48] Han et al.[49] proposed that the Ti–Ti and Ti–Al bonds present in the layers of $Ti_3C_2T_x$ MXene were fundamental obstacles to exfoliation, in addition to the van der Waals forces. Hence, a larger, bulky group such as TMAOH was used as an intercalant that diffused and intercalated the $Ti_3C_2T_x$ MXene nano-channels with subsequent delamination. Furthermore, TMAOH has been shown to intercalate and delaminate MXenes under microwave treatment, but the yield of monolayer MXene nanosheets was found to be relatively low, thereby limiting its use in preparative applications.[50] In another study, Xuan et al. suggested that TMAOH could function as both an etchant and an intercalant to produce monolayer $Ti_3C_2T_x$ MXenes from Ti_3AlC_2 MAX, owing to the high reactivity between TMAOH and Al atoms. This method enabled the synthesis of delaminated MXenes having surface-covered $[Al(OH)_4]^-$ groups without requiring sonication.[51] Tetrabutyl ammonium hydroxide (TBAOH) has also been used for the intercalation and delamination of multilayer MXenes. Liu et al.[52] reported the successful exfoliation of $Ti_3C_2T_x$ MXenes derived by the molten-salt (MS) etching method via the intercalation of TBAOH solvent. The mechanism behind the exfoliation of MXenes using TBAOH involves ion exchange between TBA+ ions and cations (protons/K+ ions) to access the interlayer space, which results in further exfoliation from multilayer to few-layer MXenes. The stability of the MS-$Ti_3C_2T_x$ nanosheet suspension treated with TBAOH was observed for a period of two weeks, with no observable precipitation. Another study by Kiran et al.[25] reported the synthesis of $Ti_3C_2T_x$ MXene from Ti_3SiC_2 MAX phase where they have used TBAOH as an intercalant to delaminate the multilayer MXene, producing high-quality few-layer MXenes to study the electron-emission behavior. The $Ti_3C_2T_x$ MXene exhibited excellent field emission property with a turn-on field of 4.7 V/μm. Currently, the delamination process of multilayer MXenes is typically applied to products obtained through HF etching or other aqueous etching methods. Therefore, the potential use of organic intercalators on non-aqueous systems requires further investigation.

The advancements made in the delamination procedure, leading to the production of substantial and superior-quality MXene nanosheets, have made $Ti_3C_2T_x$ the most extensively researched MXene for diverse applications. The use of inorganic Li^+ for delamination, either as a separate step or in situ, results in a significantly higher electrical conductivity with larger interlayer spacing in comparison to other MXenes that require bulkier organic intercalants such as TMAOH, TBAOH, etc. This enlarging of the interlayer spacing results in the weakening of interlayer van der Waals forces, allowing for the subsequent delamination of multilayer MXenes into monolayer MXene nanosheets through ultrasonic treatment. Ghidiu et al.[53] used a mixture of HF/LiCl solution to etch Ti_3AlC_2 and obtained an accordion-like $Ti_3C_2T_x$ structure with the intercalation of Li^+ ions. The interlayer Li^+ was then subsequently replaced with large $[(CH_3)_3NR]^+$ cations, where 'R' represents an alkyl chain. By selecting different alkyl chains, the interlayer spacing of MXenes can be adjusted within the range of ~5–28 Å, resulting in the fine tuning of MXenes' conductivity. In another study, Shekhirev et al.[54] reported the delamination of a $Ti_3C_2T_x$ MXene using NaCl and KCl instead of the commonly used LiCl. The MXene obtained through this process was highly comparable to the one obtained via conventional LiCl-assisted delamination, except that it doesn't have the negative effects associated with residual Li^+ ions. K^+ ions are larger and more hydrophobic than Na^+ and especially Li^+ ions, which makes them less prone to adsorbing water molecules from the environment. This results in MXene films prepared from KCl-delaminated MXenes having better conductivity retention when used at an elevated temperature in a high-humidity environment.

Currently, the LiF/HCl etching method remains the sole approach for the direct delamination of multilayer MXenes into monolayers using an ultrasonic or manual shaking process.[55] Monolayer MXenes obtained through delamination are typically stable in certain solvents such as water, DMF, and PC, forming a uniform colloidal dispersion[56] that can be further utilized in wet chemical synthesis or transformed into flexible fibers,[57] free-standing films,[58] vertical arrays[59] and other unique structures via directional induction assembly. Therefore, due to its simplicity and ease of use, the HCl/LiF etching method has become popular.

2.2.3.2 Mechanical Delamination

Mechanical delamination is a method to produce high-quality MXene flakes from bulk materials without using etchants or intercalators. This technique involves the mechanical exfoliation of layered materials, such as graphite or MoS_2, using a simple process of shearing or peeling the layers apart using adhesive tape or Scotch tape.[60,61] However, mechanical exfoliation is not efficient for large-scale production, and the resulting flakes are often limited in size and thickness. In past studies, Ti_2CT_x nanosheets were delaminated onto a 'Si' wafer by Xu et al.[61]and Lai et al.[62] through a process of delamination using adhesive tape. While some degree of successful delamination was achieved, it was limited to obtaining only few-layer MXene nanosheets. A novel approach for MXene delamination involves cyclic freezing–thawing, which has been recently reported. This technique exploits the expansion of water molecules upon freezing to increase the interlayer spacing of multilayer MXenes, thus reducing the strong van der Waals forces and facilitating easy delamination without the need for intercalators.[63]

2.3 CHARACTERIZATION OF MXENES

Characterizing a material involves describing its composition, size, thickness, crystallinity, defects, oxidation states, and electric properties. Until this point, there has been a significant amount of research dedicated to the characterization of 2D MXene nanomaterials. Various techniques such as XRD, SEM, TEM, XPS, Raman spectroscopy, and optical characterization have been employed to differentiate and understand these materials.

Here, we discuss some characterization techniques for investigations of MXenes. MXenes hold immense properties due to their 2D nature, containing various transition elements. For design and practical applications, it is necessary to examine its actual crystallographic structure and elemental composition through different characterization tools such as XRD, SEM, TEM, EDS, etc. The XRD patterns give information about the formation of MXenes. In some cases, when studying the etching rate and yield of MXene, it is necessary to quantify MXene XRD patterns.[44,29] In characterization by XRD, observed patterns reveal the degree of etching and exfoliation of MXene nanosheets.[64] The important changes after etching in the MAX phase are correlated with increase in c-lattice parameter (c-LP) and the intercalated layers of the MXene. The XRD plane (002) of the former parameter remains constant at low 2θ angles where the XRD plane of the latter parameter decreases or disappears, as shown in Figure 2.2.[65]

Generally, the morphology of bulk MAX phases and eventually the 2D MXene structure can be easily visualized by SEM imaging.[65,66,67] Because the MAX and MXene powders are predominantly analyzed after processing, the information is static. SEM imaging is a valuable tool for comprehending and validating the effectiveness of alternative processing methods. It provides valuable insights into the morphological transformations that occur when transitioning from MAX bulk materials to MXene sheets or flakes. Various techniques, such as wet chemical methods or heating, can be employed to study or improve the properties of MXenes, making them highly desirable for numerous applications.[68,69,70] The TEM investigations of MAX phases were performed long before MXenes were discovered. Extensive TEM observations of MAX phases reveal characteristic features, including the layered zig-zag appearance, that are remarkably different from competing phases such as the tetragonal Al_3Ti of Zr_3AlC_2.[18] Surface groups of MXenes can be characterized by Raman spectroscopy. This can give information on the stacking, surface composition, and quality of $Ti_3C_2T_x$ MXenes. Like other nanomaterials, MXenes

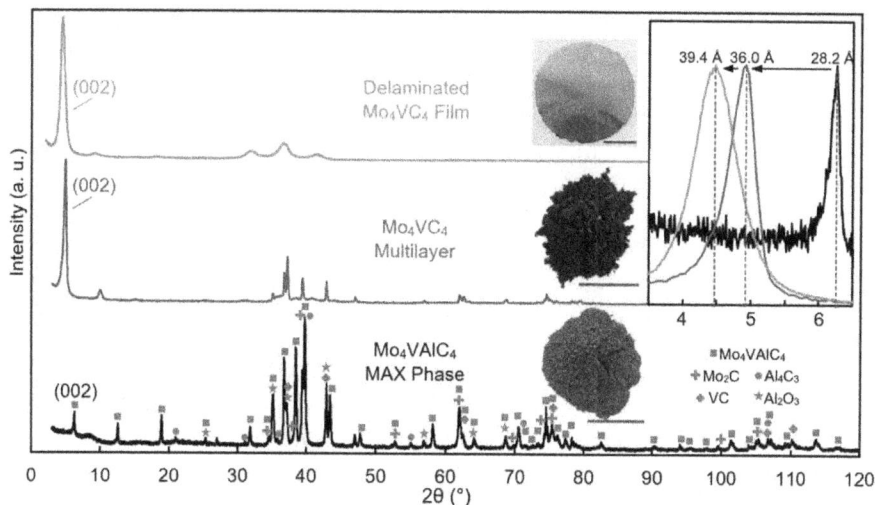

FIGURE 2.2 XRD patterns of Mo$_4$VAlC$_4$ MAX phase powder, Mo$_4$VC$_4$ multilayer MXene powder, and free-standing film of delaminated Mo$_4$VC$_4$ MXene.

[Copyright taken from 65].

produce two types of vibration peaks, consisting of in-plane (E$_g$) and out-of-plane (A$_{1g}$) vibrations. Moreover, when a 785 nm solid-state 440 laser is used, a Raman resonant peak for a Ti$_3$C$_2$T$_x$ MXene is seen at around 120 cm^{-1}.[71] In the wavenumber range of 100–800 cm^{-1}, a Ti$_3$C$_2$T$_x$ MXene shows four regions: (1) the resonant peak (120 cm^{-1}); (2) out-of-plane vibrations of Ti, C, and O; (3) the surface group vibration region (230–470 cm^{-1}); and (4) carbon vibration regions (580 and 730 cm^{-1}).[72] These Raman features change depending on the synthesis method and the environment surrounding the MXene flakes.

2.4 CONCLUSION AND OUTLOOK

In conclusion, the synthesis and characterization of MXenes have been extensively investigated using both conventional methods and green synthesis routes. Through these approaches, significant progress has been made in understanding the properties and potential applications of MXenes. Conventional methods such as the exfoliation of large crystals have allowed for the production of single-layered MXenes, albeit with concerns related to oxidation and degradation. The emergence of green synthesis routes has brought new possibilities for the synthesis of MXenes under milder conditions, reducing the reliance on high temperature and pressure. These environmentally friendly approaches offer promising alternatives that minimize the use of hazardous chemicals and energy-intensive processes, making MXene synthesis more sustainable.

Looking ahead, there are several key areas of focus for further research and development. Firstly, efforts should be directed towards optimizing the green synthesis routes to enhance the efficiency and yield of MXene production. This includes exploring novel precursors, solvents, and reaction conditions that can facilitate the synthesis process while maintaining the desired MXene properties. Additionally, comprehensive characterization techniques should continue to be employed to gain a deeper understanding of MXenes' structural, morphological, and electronic properties. This will enable researchers to fine tune the synthesis parameters and tailor MXene materials for specific applications. Furthermore, it is crucial to address the challenges related to the oxidation and degradation of MXenes. Developing protective strategies and surface modifications to prevent or mitigate the adverse effects of oxygen-mediated degradation will be essential for the practical utilization of MXenes in various fields.

In summary, the synthesis and characterization of MXenes through a range of conventional methods and green synthesis routes have provided valuable insights into their properties and potential applications. Continued research efforts and advancements in these areas will contribute to the further development and utilization of MXenes in diverse technological fields, opening up new opportunities for this fascinating class of materials.

REFERENCES

1 Anasori, B., Lukatskaya, M.R. and Gogotsi, Y., 2017. 2D metal carbides and nitrides (MXenes) for energy storage. Nature Reviews Materials, 2(2), pp. 1–17.
2 Gogotsi, Y. and Anasori, B., 2019. The rise of MXenes. ACS Nano, 13(8), pp. 8491–8494.
3 Sokol, M., Natu, V., Kota, S. and Barsoum, M. W., 2019. On the chemical diversity of the MAX phases. Trends in Chemistry, 1(2), pp. 210–223.
4 Radovic, M. and Barsoum, M.W., 2013. MAX phases: Bridging the gap between metals and ceramics. American Ceramics Society Bulletin, 92(3), pp. 20–27.
5 Xiong, D., Li, X., Bai, Z. and Lu, S., 2018. Recent advances in layered $Ti_3C_2T_X$ MXene for electrochemical energy storage. Small, 14(17), p. 1703419.
6 Liu, H., Zhu, Y., Ma, J., Zhang, Z. and Hu, W., 2020. Recent advances in atomic-level engineering of nanostructured catalysts for electrochemical CO_2 reduction. Advanced Functional Materials, 30(17), p. 1910534.
7 Zhou, X., Hao, Y., Li, Y., Peng, J., Wang, G., Ong, W.J. and Li, N., 2022. MXenes: An emergent material for packaging platforms and looking beyond. Nano Select, 3(7), pp. 1123–1147.
8 Al-Hamadani, Y.A., Jun, B.M., Yoon, M., Taheri-Qazvini, N., Snyder, S.A., Jang, M., Heo, J. and Yoon, Y., 2020. Applications of MXene-based membranes in water purification: A review. Chemosphere, 254, p. 126821.
9 Dong, J., Luo, S., Ning, S., Yang, G., Pan, D., Ji, Y., Feng, Y., Su, F. and Liu, C., 2021. MXene-coated wrinkled fabrics for stretchable and multifunctional electromagnetic interference shielding and electro/photo-thermal conversion applications. ACS Applied Materials & Interfaces, 13(50), pp. 60478–60488.
10 AhadiParsa, M., Dehghani, A., Ramezanzadeh, M. and Ramezanzadeh, B., 2022. Rising of MXenes: Novel 2D-functionalized nanomaterials as a new milestone in corrosion science-a critical review. Advances in Colloid and Interface Science, p. 102730.
11 Verger, L., Natu, V., Carey, M. and Barsoum, M.W., 2019. MXenes: An introduction of their synthesis, select properties, and applications. Trends in Chemistry, 1(7), pp. 656–669.

12 Lim, K.R.G., Shekhirev, M., Wyatt, B.C., Anasori, B., Gogotsi, Y. and Seh, Z.W., 2022. Fundamentals of MXene synthesis. Nature Synthesis, 1(8), pp. 601–614.

13 C. Xu, et al., 2015. Large-area high-quality 2D ultrathin Mo_2C superconducting crystals. Nature Materials, 14, pp. 1135–1141.

14 Khazaei, M., Ranjbar, A., Esfarjani, K., Bogdanovski, D., Dronskowski, R. and Yunoki, S., 2018. Insights into exfoliation possibility of MAX phases to MXenes. Physical Chemistry Chemical Physics, 20(13), pp. 8579–8592.

15 Haemers, J., Gusmão, R. and Sofer, Z., 2020. Synthesis protocols of the most common layered carbide and nitride MAX phases. Small Methods, 4(3), p. 1900780.

16 Kiran, N.U., Das, P., Chatterjee, S. and Besra, L., 2022. Effect of 'Ti' particle size in the synthesis of highly pure Ti_3SiC_2 MAX phase. Nano-Structures & Nano-Objects, 30, p. 100849.

17 Ghosh, N.C. and Harimkar, S.P., 2012. Consolidation and synthesis of MAX phases by Spark Plasma Sintering (SPS): A review. Advances in Science and Technology of $M_{n+1}AX_n$ Phases, pp. 47–80.

18 Lapauw, T., Halim, J., Lu, J., Cabioc'h, T., Hultman, L., Barsoum, M.W., Lambrinou, K. and Vleugels, J., 2016. Synthesis of the novel Zr_3AlC_2 MAX phase. Journal of the European Ceramic Society, 36(3), pp. 943–947.

19 Kiran, N.U., Choudhary, B., Trivedi, R., Chakraborty, B., Chatterjee, S. and Besra, L., 2022. Electric-field assisted ultrafast synthesis of Ti_3SiC_2 MAX phase. Journal of the American Ceramic Society, 105(12), pp. 7053–7063.

20 Ronchi, R.M., Arantes, J.T. and Santos, S.F., 2019. Synthesis, structure, properties, and applications of MXenes: Current status and perspectives. Ceramics International, 45(15), pp. 18167–18188.

21 Wei, Y., Zhang, P., Soomro, R.A., Zhu, Q. and Xu, B., 2021. Advances in the synthesis of 2D MXenes. Advanced Materials, 33(39), p. 2103148.

22 Naguib, M., Kurtoglu, M., Presser, V., Lu, J., Niu, J., Heon, M., Hultman, L., Gogotsi, Y. and Barsoum, M.W., 2011. Two-dimensional nanocrystals produced by exfoliation of Ti_3AlC_2. Advanced Materials, 23(37), pp. 4248–4253.

23 Kiran, N.U., Deore, A.B., More, M.A., Late, D.J., Rout, C.S., Mane, P., Chakraborty, B., Besra, L. and Chatterjee, S., 2022. Comparative study of cold electron emission from 2D $Ti_3C_2T_X$ MXene nanosheets with respect to its precursor Ti_3SiC_2 MAX phase. ACS Applied Electronic Materials, 4(6), pp. 2656–2666.

24 Zhou, J., Zha, X., Chen, F.Y., Ye, Q., Eklund, P., Du, S. and Huang, Q., 2016. A two-dimensional zirconium carbide by selective etching of Al3C3 from nanolaminated $Zr_3Al_3C_5$. AngewandteChemie International Edition, 55(16), pp. 5008–5013.

25 Meshkian, R., Näslund, L.Å., Halim, J., Lu, J., Barsoum, M.W. and Rosen, J., 2015. Synthesis of two-dimensional molybdenum carbide, Mo_2C, from the gallium based atomic laminate Mo_2Ga_2C. Scripta Materialia, 108, pp. 147–150.

26 Gentile, A., Marchionna, S., Balordi, M., Pagot, G., Ferrara, C., Di Noto, V. and Ruffo, R., 2022. Critical analysis of MXene production with in situ HF forming agents for sustainable manufacturing. ChemElectroChem, 9(23), p. e202200891.

27 Ghidiu, M., Lukatskaya, M.R., Zhao, M.Q., Gogotsi, Y. and Barsoum, M.W., 2014. Conductive two-dimensional titanium carbide 'clay' with high volumetric capacitance. Nature, 516(7529), pp. 78–81.

28 Zhao, X., Radovic, M. and Green, M.J., 2020. Synthesizing MXene nanosheets by water-free etching. Chem, 6(3), pp. 544–546.

29 Liu, F., Zhou, A., Chen, J., Jia, J., Zhou, W., Wang, L. and Hu, Q., 2017. Preparation of Ti_3C_2 and Ti_2C MXenes by fluoride salts etching and methane adsorptive properties. Applied Surface Science, 416, pp. 781–789.

30 Halim, J., Lukatskaya, M.R., Cook, K.M., Lu, J., Smith, C.R., Näslund, L.Å., May, S.J., Hultman, L., Gogotsi, Y., Eklund, P. and Barsoum, M.W., 2014. Transparent conductive two-dimensional titanium carbide epitaxial thin films. Chemistry of Materials, 26(7), pp. 2374–2381.

31 Halim, J., Kota, S., Lukatskaya, M.R., Naguib, M., Zhao, M.Q., Moon, E.J., Pitock, J., Nanda, J., May, S.J., Gogotsi, Y. and Barsoum, M.W., 2016. Synthesis and characterization of 2D molybdenum carbide (MXene). Advanced Functional Materials, 26(18), pp. 3118–3127.

32 Sun, W., Shah, S.A., Chen, Y., Tan, Z., Gao, H., Habib, T., Radovic, M. and Green, M.J., 2017. Electrochemical etching of Ti_2AlC to Ti_2CT_X (MXene) in low-concentration hydrochloric acid solution. Journal of Materials Chemistry A, 5(41), pp. 21663–21668.

33 Yang, S., Zhang, P., Wang, F., Ricciardulli, A.G., Lohe, M.R., Blom, P.W. and Feng, X., 2018. Fluoride-free synthesis of two-dimensional titanium carbide (MXene) using a binary aqueous system. Angewandte Chemie, 130(47), pp. 15717–15721.

34 Urbankowski, P., Anasori, B., Makaryan, T., Er, D., Kota, S., Walsh, P.L., Zhao, M., Shenoy, V.B., Barsoum, M.W. and Gogotsi, Y., 2016. Synthesis of two-dimensional titanium nitride Ti_4N_3 (MXene). Nanoscale, 8(22), pp. 11385–11391.

35 Urbankowski, P., Anasori, B., Hantanasirisakul, K., Yang, L., Zhang, L., Haines, B., May, S.J., Billinge, S.J. and Gogotsi, Y., 2017. 2D molybdenum and vanadium nitrides synthesized by ammoniation of 2D transition metal carbides (MXenes). Nanoscale, 9(45), pp. 17722–17730.

36 Xiu, L.Y., Wang, Z.Y. and Qiu, J.S., 2020. General synthesis of MXene by green etching chemistry of fluoride-free Lewis acidic melts. Rare Metals, 39(11), pp. 1237–1238.

37 Wong, A.J.Y., Lim, K.R.G. and Seh, Z.W., 2022. Fluoride-free synthesis and long-term stabilization of MXenes. Journal of Materials Research, pp. 1–10.

38 Li, M., Lu, J., Luo, K., Li, Y., Chang, K., Chen, K., Zhou, J., Rosen, J., Hultman, L., Eklund, P. and Persson, P.O., 2019. Element replacement approach by reaction with Lewis acidic molten salts to synthesize nanolaminated MAX phases and MXenes. Journal of the American Chemical Society, 141(11), pp. 4730–4737.

39 Kamysbayev, V., Filatov, A.S., Hu, H., Rui, X., Lagunas, F., Wang, D., Klie, R.F. and Talapin, D.V., 2020. Covalent surface modifications and superconductivity of two-dimensional metal carbide MXenes. Science, 369(6506), pp. 979–983.

40 Zhang, B., Zhu, J., Shi, P., Wu, W. and Wang, F., 2019. Fluoride-free synthesis and microstructure evolution of novel two-dimensional $Ti_3C_2(OH)_2$ nanoribbons as high-performance anode materials for lithium-ion batteries. Ceramics International, 45(7), pp. 8395–8405.

41 Xie, X., Xue, Y., Li, L., Chen, S., Nie, Y., Ding, W. and Wei, Z., 2014. Surface Al leached Ti_3AlC_2 as a substitute for carbon for use as a catalyst support in a harsh corrosive electrochemical system. Nanoscale, 6(19), pp. 11035–11040.

42 Zou, G., Zhang, Q., Fernandez, C., Huang, G., Huang, J. and Peng, Q., 2017. Heterogeneous Ti_3SiC_2@ C-containing $Na_2Ti_7O_{15}$ architecture for high-performance sodium storage at elevated temperatures. ACS Nano, 11(12), pp. 12219–12229.

43 Li, T., Yao, L., Liu, Q., Gu, J., Luo, R., Li, J., Yan, X., Wang, W., Liu, P., Chen, B. and Zhang, W., 2018. Fluorine-free synthesis of high-purity $Ti_3C_2T_X$ (T= OH, O) via alkali treatment. Angewandte Chemie International Edition, 57(21), pp. 6115–6119.

44 Alhabeb, M., Maleski, K., Anasori, B., Lelyukh, P., Clark, L., Sin, S. and Gogotsi, Y., 2017. Guidelines for synthesis and processing of two-dimensional titanium carbide ($Ti_3C_2T_X$ MXene). Chemistry of Materials, 29(18), pp. 7633–7644.

45 Inman, A., Šedajová, V., Matthews, K., Gravlin, J., Busa, J., Shuck, C.E., VahidMohammadi, A., Bakandritsos, A., Shekhirev, M., Otyepka, M. and Gogotsi, Y., 2022. Shear delamination of multilayer MXenes. Journal of Materials Research, pp. 1–11.

46 Mashtalir, O., Naguib, M., Mochalin, V.N., Dall'Agnese, Y., Heon, M., Barsoum, M.W. and Gogotsi, Y., 2013. Intercalation and delamination of layered carbides and carbonitrides. Nature Communications, 4(1), p. 1716.

47 Naguib, M., Unocic, R.R., Armstrong, B.L. and Nanda, J., 2015. Large-scale delamination of multi-layers transition metal carbides and carbonitrides 'MXenes'. Dalton Transactions, 44(20), pp. 9353–9358.

48 Levitt, A.S., Alhabeb, M., Hatter, C.B., Sarycheva, A., Dion, G. and Gogotsi, Y., 2019. Electrospun MXene/carbon nanofibers as supercapacitor electrodes. Journal of Materials Chemistry A, 7(1), pp. 269–277.

49 Han, F., Luo, S., Xie, L., Zhu, J., Wei, W., Chen, X., Liu, F., Chen, W., Zhao, J., Dong, L. and Yu, K., 2019. Boosting the yield of MXene 2D sheets via a facile hydrothermal-assisted intercalation. ACS Applied Materials & Interfaces, 11(8), pp. 8443–8452.

50 Wu, W., Xu, J., Tang, X., Xie, P., Liu, X., Xu, J., Zhou, H., Zhang, D. and Fan, T., 2018. Two-dimensional nanosheets by rapid and efficient microwave exfoliation of layered materials. Chemistry of Materials, 30(17), pp. 5932–5940.

51 Xuan, J., Wang, Z., Chen, Y., Liang, D., Cheng, L., Yang, X., Liu, Z., Ma, R., Sasaki, T. and Geng, F., 2016. Organic-base-driven intercalation and delamination for the production of functionalized titanium carbide nanosheets with superior photothermal therapeutic performance. Angewandte Chemie, 128(47), pp. 14789–14794.

52 Liu, L., Orbay, M., Luo, S., Duluard, S., Shao, H., Harmel, J., Rozier, P., Taberna, P.L. and Simon, P., 2021. Exfoliation and delamination of $Ti_3C_2T_X$ MXene prepared via molten salt etching route. ACS Nano, 16(1), pp. 111–118.

53 Ghidiu, M., Kota, S., Halim, J., Sherwood, A.W., Nedfors, N., Rosen, J., Mochalin, V.N. and Barsoum, M.W., 2017. Alkylammonium cation intercalation into Ti_3C_2 (MXene): Effects on properties and ion-exchange capacity estimation. Chemistry of Materials, 29(3), pp. 1099–1106.

54 Shekhirev, M., Ogawa, Y., Shuck, C.E., Anayee, M., Torita, T. and Gogotsi, Y., 2022. Delamination of $Ti_3C_2T_X$ nanosheets with NaCl and KCl for improved environmental stability of MXene films. ACS Applied Nano Materials, 5(11), pp. 16027–16032.

55 Zhang, T., Pan, L., Tang, H., Du, F., Guo, Y., Qiu, T. and Yang, J., 2017. Synthesis of two-dimensional $Ti_3C_2T_X$ MXene using HCl+ LiF etchant: Enhanced exfoliation and delamination. Journal of Alloys and Compounds, 695, pp. 818–826.

56 Maleski, K., Mochalin, V.N. and Gogotsi, Y., 2017. Dispersions of two-dimensional titanium carbide MXene in organic solvents. Chemistry of Materials, 29(4), pp. 1632–1640.

57 Shin, H., Eom, W., Lee, K.H., Jeong, W., Kang, D.J. and Han, T.H., 2021. Highly electroconductive and mechanically strong $Ti_3C_2T_X$ MXene fibers using a deformable MXene gel. ACS Nano, 15(2), pp. 3320–3329.

58 Zhao, M.Q., Xie, X., Ren, C.E., Makaryan, T., Anasori, B., Wang, G. and Gogotsi, Y., 2017. Hollow MXene spheres and 3D macroporous MXene frameworks for Na-ion storage. Advanced Materials, 29(37), p. 1702410.

59 Cao, Z., Zhu, Q., Wang, S., Zhang, D., Chen, H., Du, Z., Li, B. and Yang, S., 2020. Perpendicular MXene arrays with periodic interspaces toward dendrite-free lithium metal anodes with high-rate capabilities. Advanced Functional Materials, 30(5), p. 1908075.

60 DiCamillo, K., Krylyuk, S., Shi, W., Davydov, A. and Paranjape, M., 2018. Automated mechanical exfoliation of MoS_2 and $MoTe_2$ layers for two-dimensional materials applications. IEEE Transactions on Nanotechnology, 18, pp. 144–148.

61 Xu, J., Shim, J., Park, J.H. and Lee, S., 2016. MXene electrode for the integration of WSe2 and MoS2 field effect transistors. Advanced Functional Materials, 26(29), pp. 5328–5334.

62 Lai, S., Jeon, J., Jang, S.K., Xu, J., Choi, Y.J., Park, J.H., Hwang, E. and Lee, S., 2015. Surface group modification and carrier transport properties of layered transition metal carbides (Ti$_2$CT$_X$, T:—OH,—F and—O). Nanoscale, 7(46), pp. 19390–19396.

63 Huang, X. and Wu, P., 2020. A facile, high yield, and freeze and thaw assisted approach to fabricate MXene with plentiful wrinkles and its application in on chip micro-super-capacitors. Advanced Functional Materials, 30(12), p. 1910048.

64 Naguib, M., Mashtalir, O., Carle, J., Presser, V., Lu, J., Hultman, L., Gogotsi, Y. and Barsoum, M.W., 2012. Two-dimensional transition metal carbides. ACS Nano, 6(2), pp. 1322–1331.

65 Deysher, G., Shuck, C.E., Hantanasirisakul, K., Frey, N.C., Foucher, A.C., Maleski, K., Sarycheva, A., Shenoy, V.B., Stach, E.A., Anasori, B. and Gogotsi, Y., 2019. Synthesis of Mo$_4$VAlC$_4$ MAX phase and two-dimensional Mo4VC4 MXene with five atomic layers of transition metals. ACS Nano, 14(1), pp. 204–217.

66 Anasori, B., Xie, Y., Beidaghi, M., Lu, J., Hosler, B.C., Hultman, L., Kent, P.R., Gogotsi, Y. and Barsoum, M.W., 2015. Two-dimensional, ordered, double transition metals carbides (MXenes). ACS Nano, 9(10), pp. 9507–9516.

67 Luo, J., Zhang, W., Yuan, H., Jin, C., Zhang, L., Huang, H., Liang, C., Xia, Y., Zhang, J., Gan, Y. and Tao, X., 2017. Pillared structure design of MXene with ultralarge interlayer spacing for high-performance lithium-ion capacitors. ACS Nano, 11(3), pp. 2459–2469.

68 Xiu, L., Wang, Z., Yu, M., Wu, X. and Qiu, J., 2018. Aggregation-resistant 3D MXene-based architecture as efficient bifunctional electrocatalyst for overall water splitting. ACS Nano, 12(8), pp. 8017–8028.

69 Zhao, D., Clites, M., Ying, G., Kota, S., Wang, J., Natu, V., Wang, X., Pomerantseva, E., Cao, M. and Barsoum, M.W., 2018. Alkali-induced crumpling of Ti$_3$C$_2$T$_X$ (MXene) to form 3D porous networks for sodium ion storage. Chemical Communications, 54(36), pp. 4533–4536.

70 Wang, X., Fu, Q., Wen, J., Ma, X., Zhu, C., Zhang, X. and Qi, D., 2018. 3D Ti$_3$C$_2$T$_X$ aerogels with enhanced surface area for high performance supercapacitors. Nanoscale, 10(44), pp. 20828–20835.

71 Lioi, D.B., Neher, G., Heckler, J.E., Back, T., Mehmood, F., Nepal, D., Pachter, R., Vaia, R. and Kennedy, W.J., 2019. Electron-withdrawing effect of native terminal groups on the lattice structure of Ti$_3$C$_2$T$_X$ MXenes studied by resonance raman scattering: Implications for embedding MXenes in electronic composites. ACS Applied Nano Materials, 2(10), pp. 6087–6091.

72 Hu, T., Hu, M., Li, Z., Zhang, H., Zhang, C., Wang, J. and Wang, X., 2015. Covalency-dependent vibrational dynamics in two-dimensional titanium carbides. The Journal of Physical Chemistry A, 119(52), pp. 12977–12984.

3 Recent Developments in MXene-Based Materials for Biomedical Applications in Regenerative Medicine

*Swathi Sadashiva, Ishita Chakraborty,
Sanchi S. Havanagi, Sonali, Nirmal
Mazumder, and Bhisham Narayan Singh*

3.1 INTRODUCTION

MXenes (pronounced as "maxenes") are 2D early transition materials made of metal nitrides and carbides. They is a said to be a fastest growing family of 2D materials due to their unique features like good conductivity, mechanical flexibility, and many more. MXenes have emerged as a unique class of layered, structured materials where layers of early transition metals (M layer) such as titanium (Ti), vanadium (V), chromium (Cr), yttrium (Y), zirconium (Zr), niobium (Nb), molybdenum (Mb), hafnium (Hf), thallium (Ta), and tungsten (W) are interwoven with layers of nitrogen and carbon (X) with the general formula $M_{n+1}X_nT_x$. The T_x in the MXene formula represents terminal entities such as F, Cl, O, or OH that are bonded to the outer surface of the M layer [1]. The organization of MXenes provides an opportunity to effectively control the materials and form single layers, multiple layers, or nanoparticles. With structure designs, MXenes exhibit larger surface area, mechanical flexibility, and stretchability in wearable sensors, energy storage, electromagnetic shielding, and bone regeneration [2].

The applications of MXenes includes ferroelectricity, piezoelectricity, thermoelectricity, superconductivity, catalysis, photocatalysis, gas sensors, hydrogen storage, [3] etc., spanning various field of research. Because of their exceptional physicochemical, optical, biological, optical-electronic, and structural capabilities, MXenes have recently shown great potential in tissue engineering. Several studies show that MXene scaffolds have stimulatory effects on tissue regeneration, controlled discharge behavior, and the capacity to regenerate. This chapter focuses on the synthesis methods of various kinds of MXenes and their functionalization and applications, majorly in biomedical areas.

DOI: 10.1201/9781003366225-3

3.2 PROPERTIES

3.2.1 STRUCTURAL PROPERTIES

MXenes with great structural and chemical stability can work as functional materials for many applications, including energy conversion and storage, due to their superior conductivity and rich surface functions [4]. On the other hand, MXenes have a significant tendency to restack, and the absence of a restricted porosity structure is a disadvantage of ultrathin 2D materials. MXenes have been continuously designed and modified to achieve optimum porosity [5]. To date, several porous MXenes with suitable topologies have been created utilizing a variety of synthetic processes and used for a wide range of applications, all of which have shown improved performances.

3.2.2 OPTICAL PROPERTIES

Ti_3C_2T MXenes have exceptional transparency and a photothermal effect as optical characteristics. Strong surface plasmonic effects are a crucial characteristic of $Ti_3C_2T_x$ films. Regardless of the fact that transparency decreases as the spin rate of the coating increases, Ying et al. were able to achieve transparency of more than 75% in the visible and near-infrared bands of V_2CT_x, which is very transparent at 550 nm, and this transparency diminishes with increasing film thickness. An 11 nm thick layer with a maximum transmittance of 89% was achieved. At 550 nm, the film's absorbance coefficient was 1.22 0.05 105/cm [6].

Since MXenes have great reflectivity in the UV region, transparency in the visible region, and metallic conductivity, they can be considered for the development of flexible transparent electrodes and anti-UV coating materials [7]. However, aspects such as emission colors and plasmonic and non-lineal optical properties, including luminescence efficiency, need to be reported in order to expand the application of MXenes in the optics field.

3.2.3 THERMAL AND ELECTRICAL PROPERTIES

Ti_3AlC_2 has a strong thermal conductivity and thermal oxidation. In comparison to the $Ti_3C_2O_x$ monolayer with 8–10W/m, the $Ti_3C_2F_x$ monolayer has a greater thermal conductivity of around 11 W/m K. However, in the literature, it was discovered to be roughly 2.84 W/(mK) for $Ti_3C_2T_x$ films [8]. $Ti_3C_2T_x$ sheets have a 2400 S cm^{-1} metallic conductivity and are very flexible. A kind of MXene called V_2C has been examined, and it was discovered to have a conductivity of 3300 S cm^{-1}. Because of its adaptability, it is ideal for uses like wearable electronics. The layer thickness and calcination temperature have a significant impact on the conductivity of the sheets. According to Ying et al., there was rise in conductivity after the process of annealing. And it was also reported that thicker films had less sheet resistance than thinner films [6].

Only $Ti_3C_2T_x$'s thermal conductivity was assessed experimentally; as a result, more substances need to be examined. The relationship between particle size and thermal conductivity further emphasizes the significance of optimizing and controlling the morphology during the MXene production process [9].

Single MXene $Ti_3C_2T_x$ flakes were electrically characterized by Miranda et al., who also showed that the conductivities were metallic in nature. A minor but distinct field effect was seen after applying a gate voltage, allowing the carrier densities and field-effect mobilities to be calculated. It was discovered that the mobility was just $1 \text{ cm}^2/\text{V s}$, and the associated carrier densities were $5 \times 10^{14} \text{ cm}^2$. Additionally, when a magnetic field was applied at low temperatures, a slight but clearly discernible quadratic rise in conductance was noticed [10].

3.2.4 MAGNETIC PROPERTIES

Many virgin compounds, including Ti_4C_3, Ti_3CN, Fe_2C, Cr_2C, Ti_3N_2, Ti_2N, Zr_2C, and Zr_3C_2, are projected to have magnetic moments. Upon termination, however, each MXene and functionalization group must be examined separately. One study showed that OH and F groups result in a ferromagnetic nature when attached to Cr_2CT_x and Cr_2NT at ambient temperature, whereas Mn_2NT_x is ferromagnetic independently, and Ti_3CNT_x and $Ti_4C_3T_x$ show a non-magnetic nature when linked with functional groups [11]. The reported magnetic moments, however, are still just computer forecasts and have not yet been seen in an experimental setting. This is attributed to the restricted production of MXene compounds (particularly clean ones) and the absence of surface chemistry control.

3.3 FUNCTIONALIZATION OF MXENES

In order to maximize the potential of MXenes in many sectors, numerous surface functionalization techniques, including single heteroatom doping and surface-initiated polymerization, have been successfully created recently. By manipulating surface-active initiators, surface terminations, polymers, and small molecules, abundant -O- and -OH-containing groups on the MXene surfaces can be the probable sites for covalent binding. Because MXenes are hydrophilic, their performance can be easily improved by appropriate functional groups.

3.3.1 CONTROLLING SURFACE TERMINATIONS

The surfaces of MXenes are changed by modifying the T groups and the physical properties, including the types of electronic structures and electrochemical active sites. Using common etching techniques like HF etching or using fluoride-containing salts, the surface terminations -O and -F can be easily produced. The concentrations and kinds of surface terminations can be changed, which is more important, by adjusting the etching time and etchant concentration, as well as by using different post-treatments (hydrazine and annealing) in different atmospheres. In addition, MXenes' performance traits, including as their electrical properties, mechanical stability, and solution stability, can be improved even more by surface modifying them with a few small molecules that are convenient to use and reasonably priced.

MXenes' surfaces are covered in a variety of active functional groups that enable *ex situ* blending or *in situ* polymerization of polymers. *Ex situ* blending is typically used to create MXene/polymer composites, which have certain benefits including adjustable compositions and clearly defined polymer architectures. Hydrogen bonding and electrostatic interactions can also effectively enhance the interactions between polymers and MXenes. By using monomers like pyrrole and ε-caprolactone, in situ polymerization can create strong covalent connections between MXenes and polymers. Due to the abundance of catechol and amino groups in PDA backbone, it results in covalent and non-covalent interactions with both organic and inorganic surfaces. Polydopamine (PDA) is used in the surface modification of MXenes. Also, strong bonds between organic and inorganic entities can be created using silane reagents with various functional groups, including phenyl, chloride, vinyl, epoxy, methacrylate, amine, and diamine. In a recent study, a unique MXene (Ti_3C_2@IMIZ) in which a polyimidazole chain overlies the surface of the MXene, forming a hybrid composite of imidazole and MXene, was synthesized. By employing chitosan as a renewable reactant in a simple multicomponent reaction, the *in situ* growth of imidazole can be achieved. Based on the results of the characterization, it was shown that a thin layer of imidazole with an ordered chain structure was implanted on the Ti_3C_2 MXene surface, leading to the development of a unique imidazole–MXene hybrid composite [12]. With effective multifunctionality and strong biocompatibility, several MXene- and graphene-based (nano)composites have been used for biomedical applications. It is possible to augment the MXene-based materials by adding the right bioactive and biocompatible agents to the surface to increase their biocompatibility, pharmacokinetics, stability, and targeting properties (leading to high specificity and fewer off-target effects). The decline in their stability owing to unintended occurrences like aggregation or accumulation, which might reduce their performance/functionality and surface area, is another significant factor that should be highlighted for their potential clinical and biological applications [13]. To increase the biocompatibility, stability, and multifunctionality of MXene composites, hybridization and surface functionalization should be further investigated.

3.4 SYNTHESIS

3.4.1 ETCHING APPROACH

In the etching process, the MAX phase can be changed from a two-dimensional (2D) structure to a MXene. During the etching of the MAX phase, the A element is etched, and the chemical bond between M and X is broken. The different synthetic methods of etching are HP etching, fluoride-based acid etching, and alkaline solution etching. HF etching is the most common and effective; moreover, it produces a high yield [14].

HF etching was the first method for MXenes. One of the MXenes produced by HF etching is Ti_3AlC_2. As seen in Figure 3.1 [15], the Ti_3AlC_2 powders were deposited in HP solutions and bubbles were released, which were assumed to be hydrogen (H_2). When the reaction product was viewed under diffractogram after ultrasonication, the diffraction signal in the out-of-plane region was reduced due to exfoliation. After

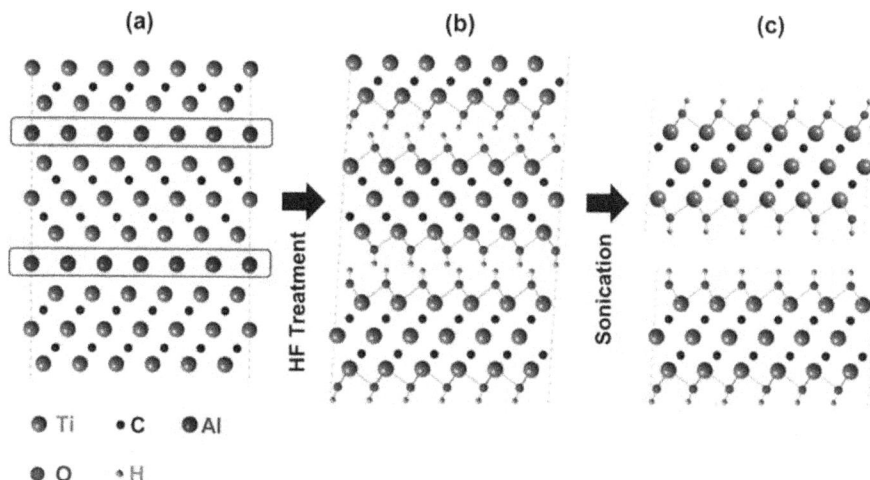

FIGURE 3.1 Schematic for exfoliation for Ti_3AlC_2.

Reused with permission from [15].

the HP treatment, the presence of the OH group was confirmed by Fourier transform infrared (FITR) spectroscopy.

An alternative method for the synthesis of Ti_3C_2TZ is the use of organic solvents in the absence of water. In March 2020, Varun Natu and his coworkers used propylene carbonate (PC) as a solvent, synthesizing double the quantity of MXenes compared to water [16]. Luo et al. prepared a cobalt–MXene hybrid, $Ti_3C_2T_x$:Co, by the molten-salt etching method as shown in Figure 3.2 [17]. There is a lack of an -OH group in molten-salt etching, which causes MXenes to be hydrophobic [18]. Hence, the MXenes have electrochemical properties, allowing them to be electrocatalysts in alkaline conditions. The MXenes showed excellent catalytic activity as a result of Co doping and free Co incorporation in the MXene matrix [17].

As shown by Wan et al., hydrothermal etching is a treatment which is conducted at 150°C for a duration of 4–6 hours [19]. As observed in an experiment, using $NaBF_4$ and HCl in hydrothermal etching results in two MXenes, Ti_3C_2 and Nb_2C, being produced. Peng et al. observed that the 2D MXenes produced by the method offered more interlayer distance and a higher degree of removal of Al layers [20].

$Ti_3C_2T_x$ MXenes have become an important family of MXenes, and an effective strategy for preparing them is by electrochemical etching. Electrochemical etching is a safe etching method which is free of fluorine and can be stripped directly on the MXenes [14]. As depicted in Figure 3.3, the addition of a hydroxide molecule during the removal of Al atoms leads to the formation of MXene sheets that show good qualities of being a supercapacitor. They can be used to store energy and do not use any harsh etching methods, making the method superior [21].

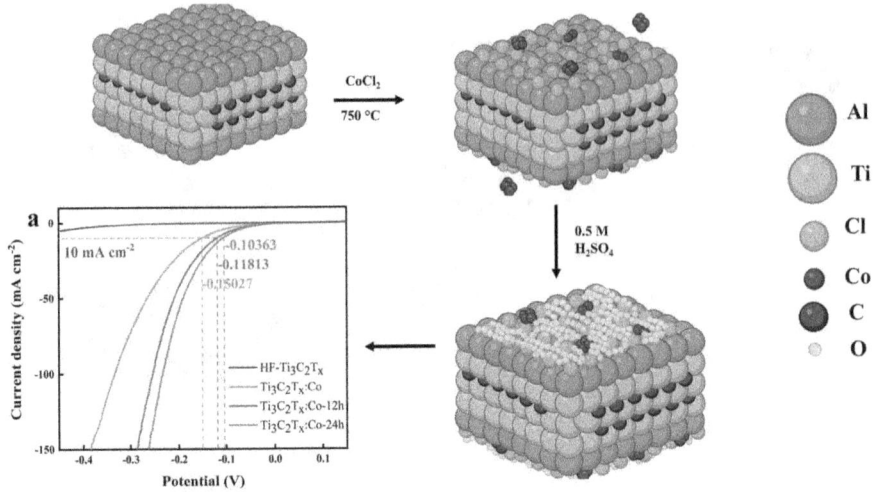

FIGURE 3.2 Preparation of a Co-modified MXene hybrid ($Ti_3C_2T_x$:Co) by the reaction of Ti_3AlC_2 with Lewis acid $CoCl_2$ at 750°C.

Reused with permission from [17].

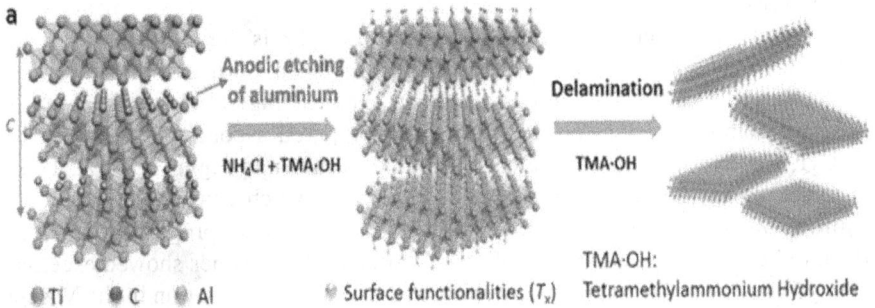

FIGURE 3.3 Schematic of the etching and delamination process.

Reused with permission from [21].

3.4.2 TOP-DOWN APPROACH

MXene preparation can be divided based on two main methods, called top-down and bottom-up.

For the top-down approach, MXenes are derived from the precursor MAX, having the formula $M_{n+1}AX_n$ (n = 1, 2, 3).

By etching with hydrofluoric acid (HF), MXenes can be obtained by selective etching of the "A" layers of the MAX system with strong acids. Layer A can be chemically selected without damaging the M–X bond by taking advantage of the differences in properties and strength between the M–A and M–X bonds.

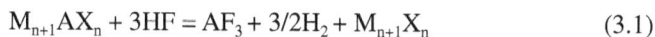

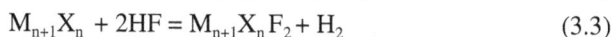

$$M_{n+1}AX_n + 3HF = AF_3 + 3/2H_2 + M_{n+1}X_n \tag{3.1}$$

$$M_{n+1}X_n + 2H_2O = M_{n+1}X_n(OH)_2 + H_2 \tag{3.2}$$

$$M_{n+1}X_n + 2HF = M_{n+1}X_nF_2 + H_2 \tag{3.3}$$

HF acid removes "A" element from the MAX phase. MXenes are more commonly represented as $M_{n+1}X_nT_x$, where T represents the abundant terminations. From their MAX counterparts, nonstoichiometric MXenes containing double transition metals and carbonitrides can be produced, along with ternary stoichiometric MXenes [22].

Taking Ti_3AlC_2 as an example, the etching reactions are as follows:

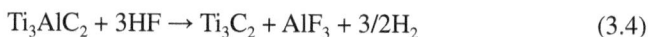

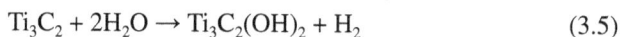

$$Ti_3AlC_2 + 3HF \rightarrow Ti_3C_2 + AlF_3 + 3/2H_2 \tag{3.4}$$

$$Ti_3C_2 + 2H_2O \rightarrow Ti_3C_2(OH)_2 + H_2 \tag{3.5}$$

$$Ti_3C_2 + 2HF \rightarrow Ti_3C_2F_2 + H_2 \tag{3.6}$$

The MXenes phase is primarily produced by reaction (3.4), and depending on whether reaction (3.5) or reaction (3.6) comes next, distinct surface terminations (-OH or -F) are produced [23].

Hydrofluoric acid (HF) is used to change Al atoms from the surface (-O), or fluorine-terminated (-F), from layers Ti_3AlC_2 $M_{n+1}X_n$, that is, MXenes. The materials used are immersed in the etching solution, i.e., HF or acid fluorides of the appropriate concentration, and then the solution is injected or filtered until the water is solid. The pH is maintained between 4 and 6. To achieve monolayer assembly, the solution is prepared with shear force or sonication. The collection of layer "A" obtained from a multilayer MXene ends with a different function (i.e., T_x). In the subsequent delamination process, various cations (e.g., Li$^+$) are introduced between the multilayered A-depleted MXenes to increase the interlayer space [24]. As shown in Figure 3.4, etching method of separating monolayer or few-layer from multilayer MXenes is employed (where a few layers are chosen). Weakening the connections between adjacent layers is essential for successful MXene delamination. Dimethyl sulfoxide (DMSO), hydrazine, urea, and other compounds that make multilayered MXenes expand have all been used to destroy MXenes [25].

An airtight apparatus is necessary to keep water and oxygen out of the delaminated sheets during the sonication process. Microscopic intercalated multilayered flakes develop after even brief, vigorous sonication that totally exfoliates multilayered MXenes several times [26].

The combination of fluoride salts (NH_4HF_2, LiF, NaF, KF, and FeF$_3$, etc.) and hydrochloric acid can replace the use of hydrofluoric acid directly, and the process is known as *in situ* hydrofluoric acid etching. When using ultrasound, one can more readily laminate MXenes produced by the *in situ* HF production procedure because the intercalation of cations and water molecules causes the layer gap to increase and the interlayer strength to decrease. This technique is frequently used to create one or more layers of MXenes since it is gentler than using hydrofluoric acid directly. It should be mentioned that whether you receive one or a low level of ultrasound, the timing and power of the

M_2AX

M_3AX_2

M_4AX_3

Etching "A" layer from $M_{n+1}AX_n$ + Sonication → MXene

M_2X

M_3X_2

M_4X_3

MAX phases are layered ternary carbides, nitrides, and carbonitrides consisting of "M", "A", and "X" layers

HF treatment

MAX phase

Sonication

Selective HF etching only of the "A" layers from the MAX phase

Physically separated 2-D MXene sheets after sonication

MXene sheets

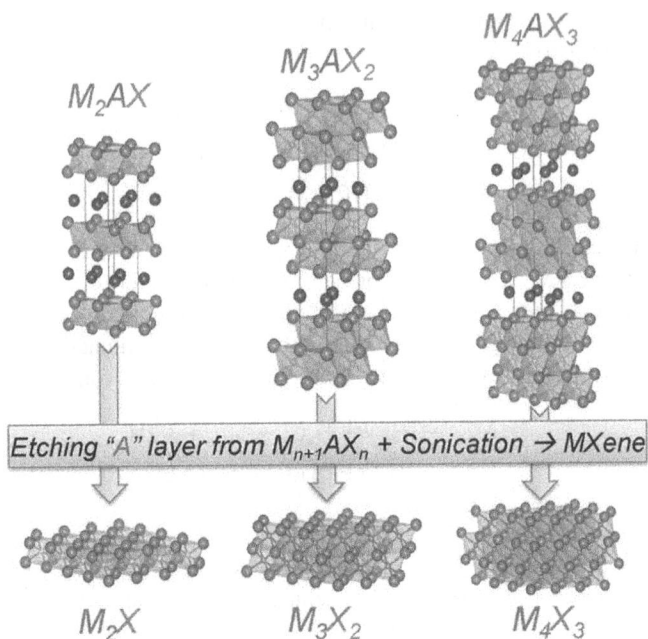

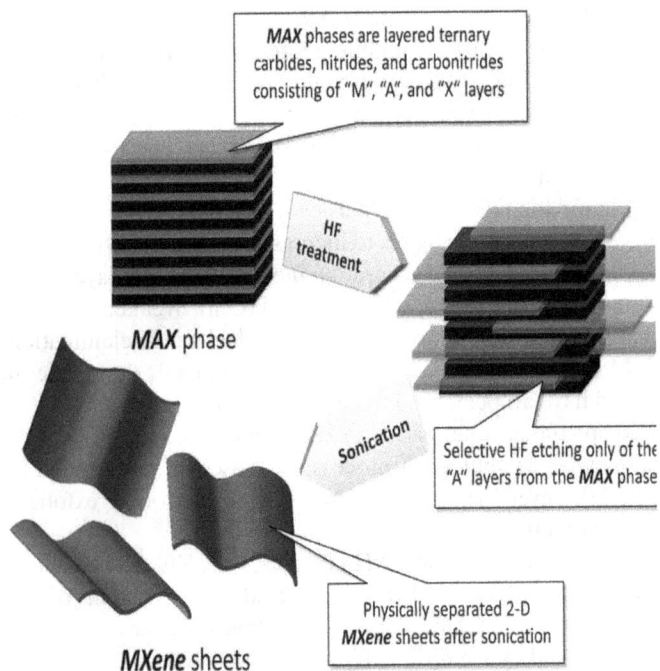

FIGURE 3.4 (a) Etching of MXene layers to individual layers; (b) schematic showing the exfoliation of MAX phases and the formation of MXenes.

Reused with permission from [27,28].

ultrasound must be adjusted. Ultrasonic waves of high power and duration will produce an excessive number of MXene laminates and needless flaws [29].

The anodic corrosion method is a fluorine-free corrosion method based on the anodic corrosion of titanium–aluminum carbides in a binary aqueous solution. Titanium carbide (Ti_3C_2) is becoming an important member of the MXene family. The diffusion of aluminum followed by the intercalation conditions of ammonium hydroxide results in carbide flakes ($Ti_3C_2T_x$, T=O, OH) with a size of around 18 μm and yield more than 90% of mono- and bi-layers [21].

Another method is chemical vapor deposition (CVD). Xu et al. synthesized -Mo_2C utilizing methane as the carbon source and a Cu/Mo sheet as the substrate, and this technique is frequently used to produce MXenes. Here, Mo_2C's dimensions and thickness can be changed, and by altering the experimental setup, various shapes can be produced. MoO_2 was used as the Mo source and diandiamide as the C/N source in the successful production of N-doped Mo_2C flakes. Furthermore, utilizing conventional 2D materials as substrates, the CVD method can be used to produce MXene-based heterojunctions [30].

With methane (CH_4) as the carbon source and a Mo–Cu alloy substrate, CVD is frequently utilized to create high-quality 2D ultrafine Mo_2C crystals at temperatures above 1085°C. For the development of Mo_2C, several CH_4 rates have been discovered separately. As shown in Figure 3.5, CH_4 produced an odd mixture of triangular, rectangular, pentagonal, and other shaped crystals at lower flow rates on a Mo substrate.

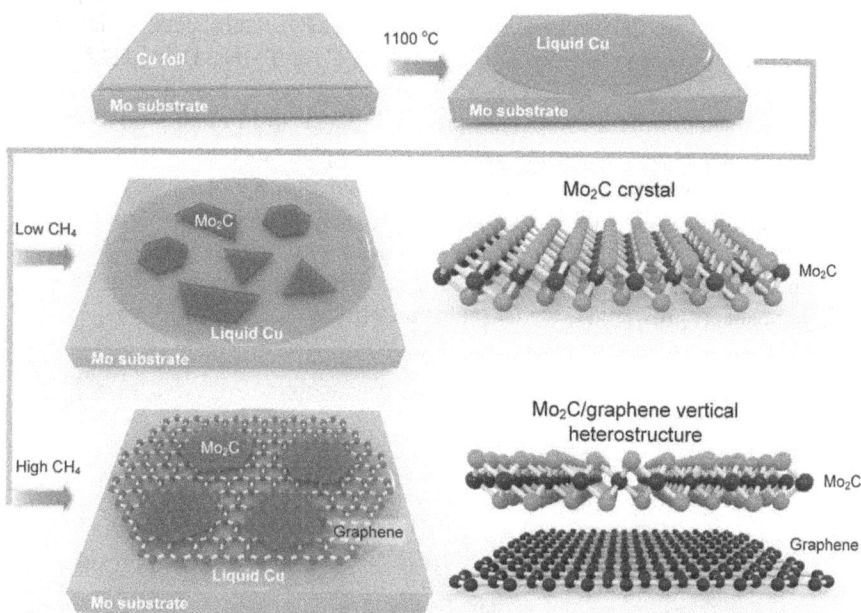

FIGURE 3.5 Schematic of the growth of Mo_2C crystals using low and high flow rates of CH_4. Reused from permission from [31].

Due to the influence of graphene, the Mo_2C crystal became uniform in shape, thickness, and size as the amount of CH_4 grew [23].

3.4.3 Bottom-Up Approach

Growing hexagonal TiC single adlayers on a Ti_3C_2 substrate allows for the successful preparation of Ti_4C_3 and Ti_5C_4, allowing for the controlled manufacturing of unique MXenes with a large surface area and good quality.

a) CVD Method—Xu et al. fabricated a two-way method to grow vertical 2D Mo_2C heterostructures/graphene in which 2D Mo_2C is found under the graphene layer. As shown in Figure 3.6, an intact graphene film is first grown on Cu at a temperature below the melting point of Cu. After that, Mo_2C crystals under the graphene layer are grown by increasing the temperature above copper's melting point in liquid Cu. Graphene and Mo_2C crystals are well suited for lattice rearrangement and heterostructures, suggesting epitaxial growth of Mo_2C and graphene [33].

b) Atomic Layer Deposition (ALD)—Halim et al. reported direct magnetron sputtering (DCMC) results in the successful deposition of Ti_3AlC_2 films from three elements (Ti, Al, C). By using NH_4HF_2 as etchants, a thin continuous epitaxial film of Ti_3C_2 can be obtained by selectively removing Al layers, and NH_3 and NH_4^+ are well combined with $Ti_3C_2T_x$ interlayers [34].

c) Template Methods—Template methods with good yield have also been developed to assemble 2D TMCs and TMNs. The templates for all template techniques are 2D transition metal oxide (TMO) nanosheets, which are then carbonized or nitrided to create 2D TMCs or TMNs. Using 2D MoO_3 nanosheets as templates, 2D h-MoN nanosheets were created by covering the crystals of the molybdenum precursor NaCl. Joshi et al. produced 2D hexagonal MoN nanosheets utilizing the template approach, in which hot Mo filaments interacting with oxygen served as a template for the vertical growth of 2D MoO_3 nanosheets on a substrate depicted in Figure 3.7. The process of heating 2D MoO_3 nanosheet with NH_3 at 800 °C resulted in 2D MoN, which is given in Figure 3.8 and completely resembles MoO_3 nanosheets grown on a vertical substrate [36].

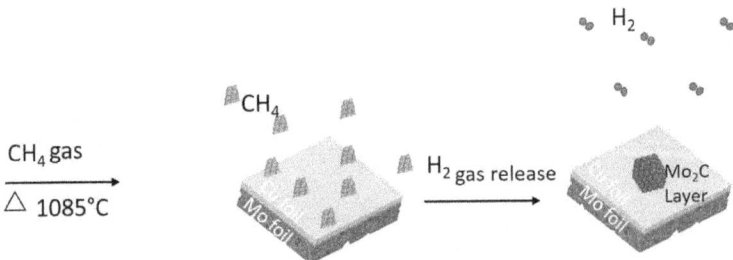

FIGURE 3.6 Schematic of one-step CVD growth of 2D α-Mo_2C vertical heterostructures.
Reused with permission from [32].

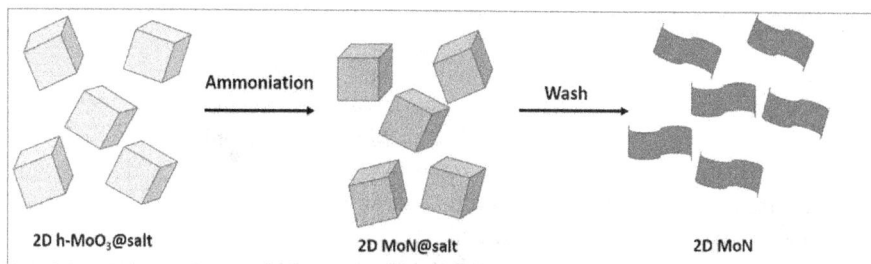

FIGURE 3.7 Schematic of the salt-assisted template synthesis process of ultrathin MoN nanosheets.

Reused with permission from [35].

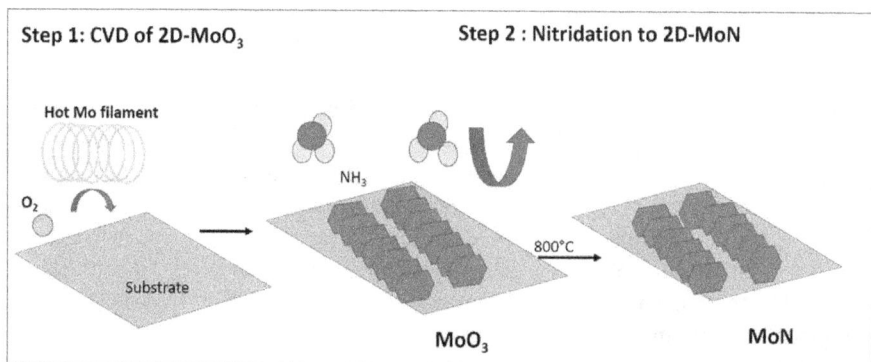

FIGURE 3.8 Schematic representation of template synthesis process of 2D layers of δMoN.

Reused with permission from [36].

MoO_2 nanosheets can also be utilized as templates to create 2D TMCs, in addition to MoO_3 nanosheets. Using dicyandiamide as a carbon and nitrogen source and MoO_2 nanosheets as a matrix, Verger et al. created ultrathin N-doped Mo_2C nanosheets over the course of two hours at 450°C and two hours at 700°C. Heating MoO_3 particles at 900°C in an Ar/N_2 combination produced hexagonal MoO_2 nanosheets (9:1). [37].

d) Plasma Enhanced Pulsed Laser Deposition—With this technique, the benefits of pulsed laser deposition and improved chemical deposition are combined. A PEPLD method was created by Zhang et al. to create extremely thin FCC Mo_2C films on sapphire. This synthesis uses CH_4 plasma as a source of C to react with the Mo vapor generated by a pulsed laser to deposit a high-quality Mo_2C film on sapphire at a temperature of 700°C. The production of Mo_2C is favored by plasma CH_4. By altering the laser pulse rate, the diameter can be regulated within the range of 2 to 25 nm [23].

3.5   APPLICATIONS OF MXENES

MXenes have strong electrical conductivity, film-forming ability, and hydrophilicity. They are frequently utilized in lithium-ion batteries, supercapacitors, and electromagnetic radiation shielding. Even though MXenes have only been used in the biomedical field for two years, research on theranostics, biosensors, dialysis, and brain electrodes has made them one of the trendiest study subjects. Electromagnetic applications such as printed antennae and electromagnetic interference shielding are further examples where MXene research is replacing other nanomaterials [38].

MXenes have recently been used in many ecological applications such as to remediate contaminated industrial water, municipal wastewaters, surface water, and ground water and for desalination. When used for electrochemical applications, MXene composites can deionize and adsorb a variety of organic and inorganic pollutants [39].

Further, Tunesi et al. were the first to describe the effective cationic dye adsorption by a multilayered MXene, or ML-$Ti_3C_2T_x$, for dyes like methylene blue (MB). The report discussed the adsorptive characteristics of the ML-$Ti_3C_2T_x$ MXene against MB and acid blue 80 (AB80), a cationic and anionic dye, respectively. According to the findings, MB may form an irreversible connection to $Ti_3C_2T_x$ (with an adsorption capacity of about 39 mg g^{-1}) [40].

Electromagnetic applications, such as printed antennae and electromagnetic shielding interference, are another area where MXene find wide applications. Theoretical and experimental studies of research in various disciplines, such as electronic and structural applications, are progressing. Numerous expected characteristics of MXenes, such as ferromagnetism or topological insulators, have not yet been examined or experimentally verified [41].

Additionally, due to their exceptional mechanical strength, amazing hydrophilicity, and unique surface chemistry, which adds a new degree of adaptability, MXenes are of particular relevance in hydrogel-based applications. In addition to enabling the creation of MXene-based soft materials with size-dependent features, the inclusion of MXenes in hydrogels considerably increases the stability of MXenes, which is frequently a limiting factor in many of their applications. Additionally, a variety of MXene hydrogel derivatives, including aerogels, may be produced utilizing straightforward techniques, broadening their range of applications [42].

3.5.1   SKIN TISSUE ENGINEERING

The skin is the largest organ and the outer covering of the body that interacts with the environment and plays a major role in immunity [43]. It consists of three layers, the epidermis, dermis, hypodermis, with multiple cell types that coordinate with each other in repair response to injury or any other inflammation. Hemostasis, inflammation, proliferation, and remodeling events occur in the overlapping phases [44]. Constant temperature, cell migration and proliferation, and antibacterial properties of the dressing material play a major role in wound dressing. Due to their moisture-holding capacity, hydrogel-based dressings are majorly applied in wound-healing applications. They also have strong attachment to the native tissue, greater oxygen permeability, and the ability to mimic the biological environment to facilitate tissue regeneration. The major disadvantage of the hydrogel-based scaffolds is that they show uncontrolled behavior during the wound-healing process [45]. To overcome such disadvantages, hydrogels were combined with synthetic materials; one such attempt was to combine hydrogel regenerated from bacterial cellulose and a MXene [46]. It was reported that the MXene in the fabricated dressing assisted in the wound healing on electrical stimulation.

3.5.2 Wound Healing

MXenes have been synthesized into various forms such as nanosheets, nanobelts, and MXene-integrated microneedle patches for applications in wound healing. As stated in a review written by Maleki et al., hydrogel-based dressings have shown desirable action for the steps of wound healing, as they have high oxygen permeability and are adhesive to the nearby tissues [47]. As shown in Figure 3.9, a hydrogel using regenerated bacterial cellulose (rBC) and an MXene ($Ti_3C_2T_x$) was developed by Mao et al. that exhibited desirable properties for wound healing and was biodegradable.

Two-dimensional (2D) $Ti_3C_2T_x$ MXenes, which were the first MXenes to be produced, also show wound-healing properties. $Ti_3C_2T_x$ MXenes were integrated with HPEM scaffolds for methicillin-resistant *Staphylococcus aureus* (MRSA)-infected wound healing. These scaffolds presented all the properties required for wound healing as well as antibacterial activity. The antibacterial activity was evaluated for *E. coli* (Gram-negative bacteria), *S. aureus* (Gram-positive bacteria), and MRSA, and it was observed that the HPEM scaffolds inhibited more than 98% of growth in all three. The MXene nanosheets in the HPEM scaffolds increased the thermal stability and conductivity of the HPEM. Moreover, the excellent antibacterial property of HPEM was due to the presence of MXene nanosheets and PGE polymer [49].

Nanobelts showing high mass loading and high surface area were also prepared, and the drug release could be controlled using near infrared (NIR) spectroscopy. The MXene nanosheet/nanobelt has a unique structure with a large width-to-thickness

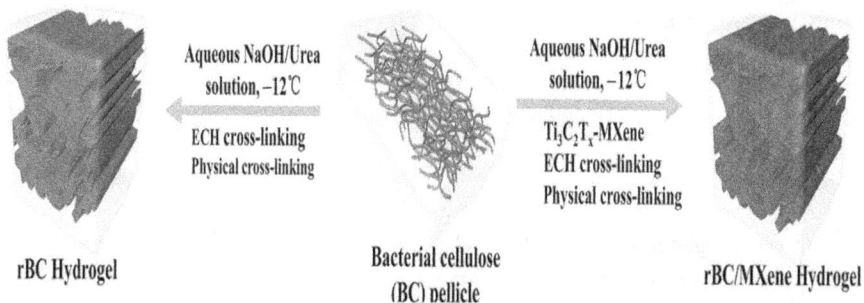

Aqueous NaOH/Urea solution, -12°C

ECH cross-linking
Physical cross-linking

Aqueous NaOH/Urea solution, -12°C

$Ti_3C_2T_x$-MXene
ECH cross-linking
Physical cross-linking

rBC Hydrogel

Bacterial cellulose (BC) pellicle

rBC/MXene Hydrogel

FIGURE 3.9   Schematic sketch showing the fabrication of rBC-based hydrogels.
Reused with permission from [48].

ratio, making it suitable for easy contact with the surface of the skin in order to control drug release using temperature. In the following experiment, vitamin E was stably released using MXene nanosheets over long periods of time using temperature control due to the presence of a PAAV layer copolymer [50].

An interesting utilization of conductive MXene hydrogels was done in flexible epidermic sensors for wound healing and wearable human–machine interaction. The MXene hydrogels possessed properties such as antibacterial activity, degradability, reliable injectability, and biocompatibility required for wound healing. When the properties were paired with the epidermic sensors, the machine was able to detect human movements and electrophysical signals, which can be used for the rehabilitation and diagnosis of muscle-related diseases [51].

Chitin/MXene composite sponges were prepared by incorporating different MXene-based nanomaterials into CH. The sponges showed mechanical properties required for wound healing, hemostatic performance, and water/blood absorption due to the morphology of the MXenes, hence promoting wound healing [52].

3.5.3 CARDIAC TISSUE ENGINEERING

Infarcted regions in cardiac tissue after MI were treated with patches of tissue that were engineered in labs and look promising for the treatment or management of MI. Studies on various conductive materials are being reported for fabricating cardiac patches [52,53,54,55]. A recent report from Ye and colleagues suggested that the cardiac patches can be developed using titanium carbide MXene (Ti_2CT_x) gel with rat cardiomyocytes [55]. Basara et al. showed that the patterning of the MXene nano-flakes with PEG (polyethylene glycol), forming a hydrogel, would help in the orientation of the cells when integrated in the MI region along with the healthy tissue. This results in the successive transfer of electrochemical signals from the healthy heart to the integrated patch because of the conductive properties of the MXene composite hydrogel [56].

3.5.4 BONE TISSUE ENGINEERING

Pan et al. studied the effects of 3D printed composite of MXene with bioglass. The printed scaffold was successfully incorporated in the tumor resection region in the hind limb of nude mice. The scaffold was able to show wound healing and regeneration in the region [57]. In another study by Zhang et al., $Ti_3C_2T_x$ MXene films has good biocompatibility and showed cell proliferation, spreading, and osteogenic differentiation of preosteoblasts *in vitro*. These MXene films showed minimum inflammatory responses upon integration in a rat calvarial defect model, declaring themselves as a novel 2D material in bone tissue engineering applications and guided bone regeneration [58]. The 3D printing of a MXene Ti_3C_2 composite with hydroxyapatite and sodium alginate was shown in one of the studies by Mi et al. The printed structure was able to show biomimetic roughness and appropriate mechanical strength. On the other hand, *in vitro* results suggested that it has good biocompatibility and osteoconductivity with bone marrow stem cells. *In vivo* experiments showed that the regeneration of calvarial bone was enhanced when treated with

MXene composite scaffolds [59]. A Nb_2C MXene was found to be highly angiogenic in nature, increasing the regeneration potential of the bone in both *in vitro* and *in vivo* experiments [60]. Sun et al. showed that a Nb_2C MXene inhibits osteolysis caused during arthroplasty using ultrahigh-molecular-weight polythene. Being antibacterial in nature with high biocompatibility, MXenes have high potential for being applied in bone tissue regeneration [61]. Although MXenes are cytotoxic at higher concentrations, they show no toxicity and promote osteogenic differentiation at lower concentrations [62], indicating their excellent antibacterial properties, biocompatibility, and ability to induce osteogenesis.

3.5.5 OTHER APPLICATIONS

MXenes were also explored in various other biological applications in regenerative medicine. MXenes in the nervous system, muscle, etc. are some of them. Guoa et al. showed the regeneration capacity and neural differentiation of primary mouse neural stem cells when cultured on a laminin-coated $Ti_3C_2T_x$ MXene, with stable adhesion and the spreading of terminal extensions [63]. MXenes were shown to maximize the efficiency of synaptic devices [64]. Since the surgical reconstruction of muscle tissue damage is limited by the lack of donors, tissue engineering provides an efficient way of treating such damage. In that context, Boularaoui et al. mixed $Ti_3C_2T_x$ MXene nanosheets with gelatin methacrylate for skeletal muscle bioprinting. This showed an improvement in the printability of the bioink as well as the attachment of skeletal muscle in C2C12 cells.

3.6 FUTURE PERSPECTIVE AND CONCLUSION

MXenes are exciting 2D materials gaining importance in biomedical applications for their excellent properties. MXenes were noted to have potential applications in various fields such as regenerative medicine, cancer and infection therapy, and biosensors. Furthermore, MXenes can be blended with other materials, which can significantly augment their performance better than its individual materials. For example, as mentioned earlier, MXenes were reinforced with bioglass and subjected to 3D printing, which resulted in building patterned structures, introducing greater potential in guided tissue engineering. The fabrication of MXene hydrogels led to breakthrough discoveries in the field of bionics, biosensors (wearable), and tissue engineering. The current development of MXene composites is highly encouraging due to the promising outcomes. However, there are greater challenges in bringing the idea to clinical applications, as there is a lack of standard operating procedures to investigate the safety of the materials. There have been various studies on different cell lines and animal models that are considered appropriate for individual studies. Further, toxicity assays could be performed in comparison with healthy and diseased conditions. Different composites having different chemical entities should be evaluated, and correlation should be established. There seems to be less attention given to the biodegradation of the MXenes and their composites, which may lead to the development of non-biodegradable composites posing issues during the integration of scaffolds *in vivo*. MXenes having better electrical and thermal properties represent an area of research that needs greater attention in terms of scaffold development for regenerative medicine.

REFERENCES

1. Anasori, B., Lukatskaya, M. R., and Gogotsi, Y., 2017. 2D metal carbides and nitrides (MXenes) for energy storage. *Nature Reviews Materials*, 2, p. 16098. https://doi.org/10.1038/natrevmats.2016.98

2. Anasori, B. and Gogotsi, Û.G., 2019. *2D Metal Carbides and Nitrides (MXenes)* (Vol. 2549). Berlin: Springer. https://doi.org/10.1007/978-3-030-19026-2

3. Frey, N.C., Wang, J., Vega Bellido, G.I., Anasori, B., Gogotsi, Y. and Shenoy, V.B., 2019. Prediction of synthesis of 2D metal carbides and nitrides (MXenes) and their precursors with positive and unlabeled machine learning. *ACS Nano*, 13(3), pp. 3031–3041. https://doi.org/10.1021/acsnano.8b08014

4. Khazaei, M., Mishra, A., Venkataramanan, N.S., Singh, A.K. and Yunoki, S., 2019. Recent advances in MXenes: From fundamentals to applications. *Current Opinion in Solid State and Materials Science*, 23(3), pp. 164–178. https://doi.org/10.1016/j.cossms.2019.01.002

5. Liu, A., Liang, X., Ren, X., Guan, W., Gao, M., Yang, Y., Yang, Q., Gao, L., Li, Y. and Ma, T., 2020. Recent progress in MXene-based materials: Potential high-performance electrocatalysts. *Advanced Functional Materials*, 30(38), p. 2003437. https://doi.org/10.1002/adfm.202003437

6. Lim, K.R.G., Handoko, A.D., Nemani, S.K., Wyatt, B., Jiang, H.Y., Tang, J., Anasori, B. and Seh, Z.W., 2020. Rational design of two-dimensional transition metal carbide/nitride (MXene) hybrids and nanocomposites for catalytic energy storage and conversion. *ACS Nano*, 14(9), pp. 10834–10864. https://doi.org/10.1021/acsnano.0c05482

7. Ying, G., Kota, S., Dillon, A.D., Fafarman, A.T. and Barsoum, M.W., 2018. Conductive transparent V2CTx (MXene) films. *FlatChem*, 8, pp. 25–30. https://doi.org/10.1016/j.flatc.2018.03.001

8. Ronchi, R.M., Arantes, J.T. and Santos, S.F., 2019. Synthesis, structure, properties and applications of MXenes: Current status and perspectives. *Ceramics International*, 45(15), pp. 18167–18188. https://doi.org/10.1016/j.ceramint.2019.06.114

9. Xu, X.W., Li, Y.X., Zhu, J.Q. and Mei, B.C., 2006. High-temperature oxidation behavior of Ti3AlC2 in air. *Transactions of Nonferrous Metals Society of China*, 16, pp. s869–s873. https://doi.org/10.1016/S1003-6326(06)60318-X

10. Bao, Z., Bing, N., Zhu, X., Xie, H. and Yu, W., 2021. Ti3C2Tx MXene contained nanofluids with high thermal conductivity, super colloidal stability and low viscosity. *Chemical Engineering Journal*, 406, p. 126390. https://doi.org/10.1016/j.cej.2020.126390

11. Miranda, A., Halim, J., Barsoum, M.W. and Lorke, A., 2016. Electronic properties of freestanding Ti3C2Tx MXene monolayers. *Applied Physics Letters*, 108(3), p. 033102. https://doi.org/10.1063/1.4939971

12. Tan, H., Wang, C., Duan, H., Tian, J., Ji, Q., Lu, Y., Hu, F., Hu, W., Li, G., Li, N. and Wang, Y., 2021. Intrinsic room-temperature ferromagnetism in V2C MXene nanosheets. *ACS Applied Materials & Interfaces*, 13(28), pp. 33363–33370. https://doi.org/10.1021/acsami.1c07906

13. Mostafavi, E. and Iravani, S., 2022. Mxene-graphene composites: A perspective on biomedical potentials. *Nano-Micro Letters*, 14(1), p. 130. https://doi.org/10.1007/s40820-022-00880-y

14. Chen, J., Ding, Y., Yan, D., Huang, J. and Peng, S., 2022. Synthesis of MXene and its application for zinc-ion storage. *SusMat*, 2(3), pp. 293–318. https://doi.org/10.1002/sus2.57

15. Naguib, M., Kurtoglu, M., Presser, V., Lu, J., Niu, J., Heon, M., Hultman, L., Gogotsi, Y. and Barsoum, M.W., 2011. Two-dimensional nanocrystals produced by exfoliation of Ti3AlC2. *Advanced Materials*, *23*(37), pp. 4248–4253. https://doi.org/10.1002/adma.201102306

16. Natu, V., Pai, R., Sokol, M., Carey, M., Kalra, V. and Barsoum, M.W., 2020. 2D Ti3C2Tz MXene synthesized by water-free etching of Ti3AlC2 in polar organic solvents. *Chem*, *6*(3), pp. 616–630. https://doi.org/10.1016/j.chempr.2020.01.019

17. Luo, R., Li, R., Jiang, C., Qi, R., Liu, M., Luo, C., Lin, H., Huang, R. and Peng, H., 2021. Facile synthesis of cobalt modified 2D titanium carbide with enhanced hydrogen evolution performance in alkaline media. *International Journal of Hydrogen Energy*, *46*(64), pp. 32536–32545. https://doi.org/10.1016/j.ijhydene.2021.07.110

18. Arole, K., Blivin, J.W., Saha, S., Holta, D.E., Zhao, X., Sarmah, A., Cao, H., Radovic, M., Lutkenhaus, J.L. and Green, M.J., 2021. Water-dispersible Ti3C2Tz MXene nanosheets by molten salt etching. *Iscience*, *24*(12), p. 103403. https://doi.org/10.1016/j.isci.2021.103403

19. Wan, J., Liu, R., Tong, Y., Chen, S., Hu, Y., Wang, B., Xu, Y. and Wang, H., 2016. Hydrothermal etching treatment to rutile TiO 2 nanorod arrays for improving the efficiency of CdS-sensitized TiO 2 solar cells. *Nanoscale Research Letters*, *11*, pp. 1–9. https://doi.org/10.1186/s11671-016-1236-9

20. Peng, C., Wei, P., Chen, X., Zhang, Y., Zhu, F., Cao, Y., Wang, H., Yu, H. and Peng, F., 2018. A hydrothermal etching route to synthesis of 2D MXene (Ti3C2, Nb2C): Enhanced exfoliation and improved adsorption performance. *Ceramics International*, *44*(15), pp. 18886–18893. https://doi.org/10.1016/j.ceramint.2018.07.124

21. Yang, S., Zhang, P., Wang, F., Ricciardulli, A.G., Lohe, M.R., Blom, P.W. and Feng, X., 2018. Fluoride-free synthesis of two-dimensional titanium carbide (MXene) using a binary aqueous system. *Angewandte Chemie*, *130*(47), pp. 15717–15721. https://doi.org/10.1002/ange.201809662

22. Gao, L., Zhao, Y., Chang, X., Zhang, J., Li, Y., Wageh, S., Al-Hartomy, O.A., Al-Sehemi, A.G., Zhang, H. and Ågren, H., 2022. Emerging applications of MXenes for photodetection: Recent advances and future challenges. *Materials Today*. https://doi.org/10.1016/j.mattod.2022.10.022

23. Zhang, T., Zhang, L. and Hou, Y., 2022. MXenes: Synthesis strategies and lithium-sulfur battery applications. *eScience*, *2*, pp. 164–182. https://doi.org/10.1016/j.esci.2022.02.010

24. Gautam, R., Marriwala, N. and Devi, R., 2022. A review: Study of Mxene and graphene together. *Measurement: Sensors*, p. 100592. https://doi.org/10.1016/j.measen.2022.100592

25. Abdolhosseinzadeh, S., Jiang, X., Zhang, H., Qiu, J. and Zhang, C.J., 2021. Perspectives on solution processing of two-dimensional MXenes. *Materials Today*, *48*, pp. 214–240. https://doi.org/10.1016/j.mattod.2021.02.010

26. Parajuli, D., Murali, N., KC, D., Karki, B., Samatha, K., Kim, A.A., Park, M. and Pant, B., 2022. Advancements in MXene-polymer nanocomposites in energy storage and biomedical applications. *Polymers*, *14*(16), p. 3433. https://doi.org/10.3390/polym14163433

27. Naguib, M., Mashtalir, O., Carle, J., Presser, V., Lu, J., Hultman, L., Gogotsi, Y. and Barsoum, M.W., 2012. Two-dimensional transition metal carbides. *ACS Nano*, *6*(2), pp. 1322–1331. https://doi.org/10.1021/nn204153h

28. Naguib, M., Mochalin, V.N., Barsoum, M.W. and Gogotsi, Y., 2014. 25th anniversary article: MXenes: A new family of two-dimensional materials. *Advanced Materials*, *26*(7), pp. 992–1005. https://doi.org/10.1021/nn204153h

29. Wang, Q., Han, N., Shen, Z., Li, X., Chen, Z., Cao, Y., Si, W., Wang, F., Ni, B.J. and Thakur, V.K., 2022. MXene-based electrochemical (bio) sensors for sustainable applications: Roadmap for future advanced materials. *Nano Materials Science*. https://doi.org/10.1016/j.nanoms.2022.07.003

30. Gao, L., Zhao, Y., Chang, X., Zhang, J., Li, Y., Wageh, S., Al-Hartomy, O.A., Al-Sehemi, A.G., Zhang, H. and Ågren, H., 2022. Emerging applications of MXenes for photodetection: Recent advances and future challenges. *Materials Today*. https://doi.org/10.1016/j. mattod.2022.10.022

31. Geng, D., Zhao, X., Chen, Z., Sun, W., Fu, W., Chen, J., Liu, W., Zhou, W. and Loh, K.P., 2017. Direct synthesis of large area 2D Mo2C on in situ grown graphene. *Advanced Materials*, 29(35), p. 1700072. https://doi.org/10.1002/adma.201700072

32. Zamhuri, A., Lim, G.P., Ma, N.L., Tee, K.S. and Soon, C.F., 2021. MXene in the lens of biomedical engineering: Synthesis, applications and future outlook. *Biomedical Engineering Online*, 20(1), pp. 1–24. https://doi.org/10.1186/s12938-021-00873-9

33. Xin, M., Li, J., Ma, Z., Pan, L. and Shi, Y., 2020. MXenes and their applications in wearable sensors. *Frontiers in Chemistry*, 8, p. 297. https://doi.org/10.3389/fchem.2020.00297

34. Li, X., Ran, F., Yang, F., Long, J. and Shao, L., 2021. Advances in MXene films: Synthesis, assembly, and applications. *Transactions of Tianjin University*, 27, pp. 217–247. https://doi.org/10.1007/s12209-021-00282-y

35. Xiao, X., Yu, H., Jin, H., Wu, M., Fang, Y., Sun, J., Hu, Z., Li, T., Wu, J., Huang, L. and Gogotsi, Y., 2017. Salt-templated synthesis of 2D metallic MoN and other nitrides. *ACS Nano*, 11(2), pp. 2180–2186. https://doi.org/10.1021/acsnano.6b08534

36. Joshi, S., Wang, Q., Puntambekar, A. and Chakrapani, V., 2017. Facile synthesis of large area two-dimensional layers of transition-metal nitride and their use as insertion electrodes. *ACS Energy Letters*, 2(6), pp. 1257–1262. https://doi.org/10.1021/acsenergylett.7b00240

37. Verger, L., Xu, C., Natu, V., Cheng, H.M., Ren, W. and Barsoum, M.W., 2019. Overview of the synthesis of MXenes and other ultrathin 2D transition metal carbides and nitrides. *Current Opinion in Solid State and Materials Science*, 23(3), pp. 149–163. https://doi. org/10.1016/j.cossms.2019.02.001

38. Gogotsi, Y. and Anasori, B., 2019. The rise of MXenes. *ACS Nano*, 13(8), pp. 8491–8494. https://doi.org/10.1021/acsnano.9b06394

39. Dixit, F., Zimmermann, K., Alamoudi, M., Abkar, L., Barbeau, B., Mohseni, M., Kandasubramanian, B. and Smith, K., 2022. Application of MXenes for air purification, gas separation and storage: A review. *Renewable and Sustainable Energy Reviews*, 164, p. 112527. https://doi.org/10.1016/j.rser.2022.112527

40. Tunesi, M.M., Soomro, R.A., Han, X., Zhu, Q., Wei, Y. and Xu, B., 2021. Application of MXenes in environmental remediation technologies. *Nano Convergence*, 8(1), pp. 1–19. https://doi.org/10.1186/s40580-021-00255-w

41. Sun, S., Liao, C., Hafez, A.M., Zhu, H. and Wu, S., 2018. Two-dimensional MXenes for energy storage. *Chemical Engineering Journal*, 338, pp. 27–45. https://doi.org/10.1016/j. cej.2017.12.155

42. Damiri, F., Rahman, M.H., Zehravi, M., Awaji, A.A., Nasrullah, M.Z., Gad, H.A., Bani-Fwaz, M.Z., Varma, R.S., Germoush, M.O., Al-Malky, H.S. and Sayed, A.A., 2022. MXene (Ti3C2Tx)-embedded nanocomposite hydrogels for biomedical applications: A review. *Materials*, 15(5), p. 1666. https://doi.org/10.3390/ma15051666

43. Rodrigues, M., Kosaric, N., Bonham, C.A. and Gurtner, G.C., 2019. Wound healing: A cellular perspective. *Physiological Reviews*, 99(1), pp. 665–706. https://doi.org/10.1152/ physrev.00067.2017

44. Guo, S.A. and DiPietro, L.A., 2010. Factors affecting wound healing. *Journal of Dental Research*, 89(3), pp. 219–229. https://doi.org/10.1177/0022034509359125

45. Diniz, F.R., Maia, R.C.A., de Andrade, L.R.M., Andrade, L.N., Vinicius Chaud, M., da Silva, C.F., Corrêa, C.B., de Albuquerque Junior, R.L.C., Pereira da Costa, L., Shin, S.R. and Hassan, S., 2020. Silver nanoparticles-composing alginate/gelatine hydrogel improves wound healing in vivo. *Nanomaterials*, 10(2), p. 390. https://doi.org/10.3390/nano10020390

46. Li, S., Gu, B., Li, X., Tang, S., Zheng, L., Ruiz-Hitzky, E., Sun, Z., Xu, C. and Wang, X., 2022. MXene-enhanced chitin composite sponges with antibacterial and hemostatic activity for wound healing. *Advanced Healthcare Materials*, *11*(12), p. 2102367. https://doi.org/10.1002/adhm.202102367

47. Maleki, A., Ghomi, M., Nikfarjam, N., Akbari, M., Sharifi, E., Shahbazi, M.A., Kermanian, M., Seyedhamzeh, M., Nazarzadeh Zare, E., Mehrali, M. and Moradi, O., 2022. Biomedical applications of MXene-integrated composites: Regenerative medicine, infection therapy, cancer treatment, and biosensing. *Advanced Functional Materials*, *32*(34), p. 2203430. https://doi.org/10.1002/adfm.202203430

48. Mao, L., Hu, S., Gao, Y., Wang, L., Zhao, W., Fu, L., Cheng, H., Xia, L., Xie, S., Ye, W. and Shi, Z., 2020. Biodegradable and electroactive regenerated bacterial cellulose/MXene (Ti3C2Tx) composite hydrogel as wound dressing for accelerating skin wound healing under electrical stimulation. *Advanced Healthcare Materials*, *9*(19), p. 2000872. https://doi.org/10.1002/adhm.202000872

49. Zhou, L., Zheng, H., Liu, Z., Wang, S., Liu, Z., Chen, F., Zhang, H., Kong, J., Zhou, F. and Zhang, Q., 2021. Conductive antibacterial hemostatic multifunctional scaffolds based on Ti3C2T x MXene nanosheets for promoting multidrug-resistant bacteria-infected wound healing. *ACS Nano*, *15*(2), pp. 2468–2480. https://doi.org/10.1021/acsnano.0c06287

50. Jin, L., Guo, X., Gao, D., Wu, C., Hu, B., Tan, G., Du, N., Cai, X., Yang, Z. and Zhang, X., 2021. NIR-responsive MXene nanobelts for wound healing. *NPG Asia Materials*, *13*(1), p. 24. https://doi.org/10.1038/s41427-021-00289-w

51. Li, M., Zhang, Y., Lian, L., Liu, K., Lu, M., Chen, Y., Zhang, L., Zhang, X. and Wan, P., 2022. Flexible accelerated-wound-healing antibacterial MXene-based epidermic sensor for intelligent wearable human-machine interaction. *Advanced Functional Materials*, p. 2208141. https://doi.org/10.1002/adfm.202208141

52. Shin, S.R., Jung, S.M., Zalabany, M., Kim, K., Zorlutuna, P., Kim, S.B., Nikkhah, M., Khabiry, M., Azize, M., Kong, J. and Wan, K.T., 2013. Carbon-nanotube-embedded hydrogel sheets for engineering cardiac constructs and bioactuators. *ACS Nano*, *7*(3), pp. 2369–2380. https://doi.org/10.1021/nn305559j

53. Martins, A.M., Eng, G., Caridade, S.G., Mano, J.F., Reis, R.L. and Vunjak-Novakovic, G., 2014. Electrically conductive chitosan/carbon scaffolds for cardiac tissue engineering. *Biomacromolecules*, *15*(2), pp. 635–643. https://doi.org/10.1021/bm401679q

54. Zhao, L., 2019. A novel graphene oxide polymer gel platform for cardiac tissue engineering application. *3 Biotech*, *9*(11), p. 401. https://doi.org/10.1007/s13205-019-1912-4

55. Ye, G., Wen, Z., Wen, F., Song, X., Wang, L., Li, C., He, Y., Prakash, S. and Qiu, X., 2020. Mussel-inspired conductive Ti2C-cryogel promotes functional maturation of cardiomyocytes and enhances repair of myocardial infarction. *Theranostics*, *10*(5), p. 2047. https://doi.org/10.7150/thno.38876

56. Basara, G., Saeidi-Javash, M., Ren, X., Bahcecioglu, G., Wyatt, B.C., Anasori, B., Zhang, Y. and Zorlutuna, P., 2022. Electrically conductive 3D printed Ti3C2Tx MXene-PEG composite constructs for cardiac tissue engineering. *Acta Biomaterialia*, *139*, pp. 179–189. https://doi.org/10.1016/j.actbio.2020.12.033

57. Pan, S., Yin, J., Yu, L., Zhang, C., Zhu, Y., Gao, Y. and Chen, Y., 2020. 2D MXene-integrated 3D-printing scaffolds for augmented osteosarcoma phototherapy and accelerated tissue reconstruction. *Advanced Science*, *7*(2), p. 1901511. https://doi.org/10.1002/advs.201901511

58. Zhang, J., Fu, Y. and Mo, A., 2019. Multilayered titanium carbide MXene film for guided bone regeneration. *International Journal of Nanomedicine*, pp. 10091–10103. http://doi.org/10.2147/IJN.S227830

59. Mi, X., Su, Z., Fu, Y., Li, S. and Mo, A., 2022. 3D printing of Ti3C2-MXene-incorporated composite scaffolds for accelerated bone regeneration. *Biomedical Materials, 17*(3), p. 035002. https://doi.org/10.1088/1748-605X/ac5ffe

60. Yin, J., Pan, S., Guo, X., Gao, Y., Zhu, D., Yang, Q., Gao, J., Zhang, C. and Chen, Y., 2021. Nb 2 C MXene-functionalized scaffolds enables osteosarcoma phototherapy and angiogenesis/osteogenesis of bone defects. *Nano-Micro Letters, 13*, pp. 1–18. https://doi.org/10.1007/s40820-020-00547-6

61. Sun, K.Y., Wu, Y., Xu, J., Xiong, W., Xu, W., Li, J., Sun, Z., Lv, Z., Wu, X.S., Jiang, Q. and Cai, H.L., 2022. Niobium carbide (MXene) reduces UHMWPE particle-induced osteolysis. *Bioactive Materials, 8*, pp. 435–448. https://doi.org/10.1016/j.bioactmat.2021.06.016

62. Jang, J.H. and Lee, E.J., 2021. Influence of MXene particles with a stacked-lamellar structure on osteogenic differentiation of human mesenchymal stem cells. *Materials, 14*(16), p. 4453. https://doi.org/10.3390/ma14164453

63. Guo, R., Xiao, M., Zhao, W., Zhou, S., Hu, Y., Liao, M., Wang, S., Yang, X., Chai, R. and Tang, M., 2022. 2D Ti3C2TxMXene couples electrical stimulation to promote proliferation and neural differentiation of neural stem cells. *Acta Biomaterialia, 139*, pp. 105–117. https://doi.org/10.1016/j.actbio.2020.12.035

64. Wei, H., Yu, H., Gong, J., Ma, M., Han, H., Ni, Y., Zhang, S. and Xu, W., 2021. Redox MXene artificial synapse with bidirectional plasticity and hypersensitive responsibility. *Advanced Functional Materials, 31*(1), p. 2007232. https://doi.org/10.1002/adfm.202007232

4 MXene Materials as Efficient Separation Membranes

Shikha Indoria, Swati Awasthi,
and Vickramjeet Singh

4.1 INTRODUCTION

A new class of two-dimensional (2D) material called MXenes, which are transition metal nitrides, carbonitrides, and carbides, have gained attention due to promising optical, chemical, structural, and electrical properties [1–2]. The MXene nanoparticles have a general formula represented as $M_{n+1}X_nT_x$, where M is for transition metal, X is for carbon and/or nitrogen, T_x is for ligands, in which T is a termination group such as hydroxyl, oxygen, fluorine, chlorine, x is a surface functionality, and the value of n can be 1, 2, 3, or 4 [1–4]. MXenes can be prepared from the precursor MAX (i.e. $M_{n+1}AX_n$), where M is the early transition metal from group A (group 13/14) [5–6].

Ever since the first discovery of MXenes ($Ti_3C_2T_x$), more than 70 MXene compositions have been prepared [7–8]. The synthesis of MXenes has led to development not only in the field of MXenes but also in the design/development of MXene precursors (MAX phases). Not only the development but also the properties of MAX phases have been studied to understand the chemical and physical behaviour [7]. A recent development in the field of MXenes is to perform fluoride-free MXene synthesis to reduce the environmental impact due to fluoride-based chemicals. MXene preparation requires fluoride-based chemicals in molten form or in aqueous solutions. Recently, fluoride-free synthesis was performed by an electrochemical method utilizing dilute hydrochloric acid for the synthesis of a Ti_2CT_x MXene [7–8]. Due to their exceptional properties, several applications of MXenes have been explored. The properties and applications of MXene are shown in Figure 4.1.

The surface functional groups present in MXenes influence various properties of MXenes, including the absorbent behaviour [1–2]. The surface terminal functional groups on MXenes are obtained during the synthesis and rely on the synthesis process [1–2]. MXenes obtained by etching in presence of HF acid have -OH, -O-, and -F surface functional groups. Similarly, -Cl/-Br functional groups appear when MXenes are prepared by etching in the presence of molten $CdBr_2/CdCl_2$ [9]. The functionalized 2D MXenes have been explored as membranes for pervaporation, waste-water treatment, desalination, and gas purification [1, 10]. For separation, MXenes bearing

DOI: 10.1201/9781003366225-4

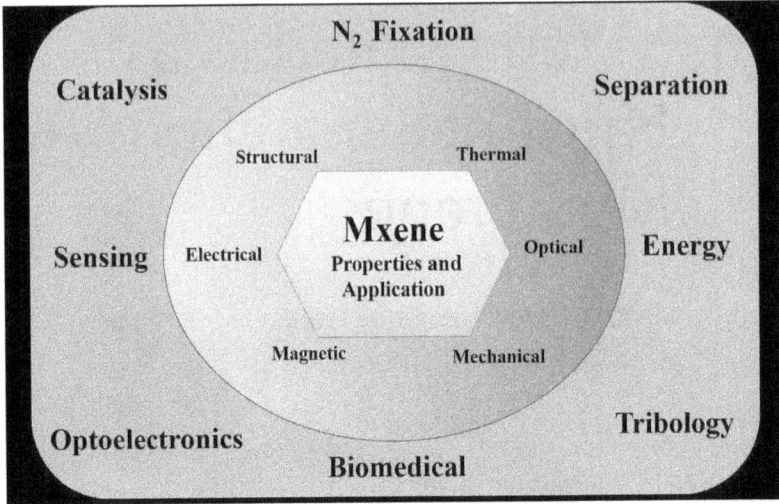

FIGURE 4.1   The properties and prospect applications of MXenes.

surface functional groups (-F, O, -OH) with high negative surface charges have been
employed. These surface groups contribute towards the stable distribution of MXenes
in organic as well as aqueous solvents [1, 10]. The MXene-based membranes have
been synthesized by wet-chemical, hot-press, dip-coating, layer-by-layer, electrospinning, spray-coating, electrophoretic deposition, and casting methods [1, 10–11].

The membrane-based technology has been extensively used for purification of
waste-water. The membrane separation process is based on size sieving or the exclusion effect. The regular even or uneven width distribution with organized and regular interlayer channels provides the opportunity for membrane-based permeation or
separation [1, 10–11]. Thus, membrane-based technology shows high efficiency and
low power consumption. The membranes are used for oil–water separation so as to
recover oil by the selective permeation of oil through the oleophobic membrane. For
oil/water (O/W) separation, nanomaterial-decorated membranes prove to be effective
separating meshes, demonstrating high O/W separation efficiency. Nanomaterials
such as graphene, reduced graphene, graphene oxide, metal–organic frameworks,
metal oxides, and MXenes have been used for modulating the surface wettability
of supported meshes. Although MXene-decorated membranes were shown to have
separated oil–water mixtures, they are still in in the development stage. Owing to
tuneable wetting behaviour, unique physicochemical properties (based on surface
functional groups) and tuneable channels, MXenes seem to be promising materials
for oil–waste-water treatment [8, 10–11]. In addition to oil–water separation, membranes and technology based on MXenes have been used for gas separation, molecular separation, waste-water treatment, and biomedical applications. In this chapter,
the synthesis and applications of MXene-based materials for applications in the field
of oil–water, gas, and molecular separation haven been discussed. The needs and
future prospective of such materials have also been highlighted.

4.1.1 SYNTHESIS AND PROPERTIES OF MXENE-BASED MEMBRANES

The general chemical formula of MXenes is $M_{n+1}X_n$ or $M_{n+1}TX_n$ {M = early transition metals, X = carbon/nitrogen, T = surface moieties mixed through etching (e.g. -O, -F, -OH etc)}. MXenes can be synthesized by 1) Chemical vapour deposition (CVD) through a bottom-up approach or 2) etching of MAX-phase precursor molecules followed by exfoliation through a top-down approach [12, 13]. In the literature, only a few MXenes are prepared through the CVD method to synthesize fine 2D MXenes [14], but these are not suitable to produce large-area membranes, except few examples of carbide MXenes [15]. Therefore, the second method, i.e. the top-down approach, has been of great interest to synthesize MXenes using MAX phases $(M_{n+1}AX_n)$ as parent molecules, where M is in a closed-packed array containing octahedral sites available for C/N atoms to form $M_{n+1}X_n$ layers (n = 1 to 4) stacked with layers of an A-group element (group 13/14 elements) in an alternate multilayer fashion [14]. These MAX-based MXenes have a greater advantage over graphite/graphene. In comparison to graphite, MAX phases are intercalated through strong M–A metallic bonds. However, this makes MAX exfoliation very difficult with only mechanical shearing [16]. The exfoliation is attained by delaminating the layers by increasing the M–A bond reactivity compared to that of M–X bonds by selective expulsion of A, forming A-atom-depleted $M_{n+1}X_n$ layers. Strong etching agents can do the job for strong M–A bonds without influencing the 2D structure of $M_{n+1}X_n$ stacked layers. At this point, only MXenes with low concentrations have been formed. To obtain a high concentration of multi-layered MXenes, the last step involves the ultrasonic exfoliation of $M_{n+1}X_n$ layers [17]. In addition, high-temperature treatment facilitates the MXene 3D framework [15]. Nowadays, the most commonly used MAX phases are Ti_3AlC_2 (M = Ti, A = Al, X = C) to synthesize $Ti_3C_2T_x$ MXene nanosheets (M= Ti, X = C, T_x = surface moieties mixed through the etching of -O, -F, -OH) [18]. The etchant used in these is ammonium hydrogen difluoride (NH_4HF_2), or ammonium fluoride (NH_4F) and lithium fluoride (LiF) are replaced by conventional hydrofluoric acids [15,19–22]. The main advantage to using these etchants are that the Li^+ cations can intercalate $M_{n+1}X_n$ layers, which leads to enhancing the exfoliation process through the enlargement of interlayer spacing to weaken the interactions [15, 23].

Because of several desirable physicochemical properties such as 2D morphology [24–26], hydrophilicity [27–32], nanoscale channels [14, 33–35], tuneable interspace [28, 36, 37], flexibility [26, 36, 37], chemical stability [38, 39–41], and electrical conductivity [15, 23–24], MXene nanosheets are beneficial for the fabrication of MXene membranes for high-performance separation processes. Presently, there are three types of membranes reported which are utilized in the separation processes: 1) nanolaminate membranes, 2) mixed-matrix membranes, and 3) thin-film nanocomposites. Being new entrants to the research field, most of the MXene membranes are pristine nanosheets. To prepare nanolaminate membranes, vacuum- or pressure-assisted filtration methods are used. Nanolaminate membranes are made up of 2D materials such as graphene oxides and MXenes. These membranes have shown great potential for molecular sieving via size-restricted diffusion 2D capillaries. To fabricate nanolaminate membranes, MXenes are dissolved in solvents, most likely water, at a suitable pH regime to form stable, uniform, and homogeneous colloidal dispersions.

Then, these dispersions are spread over the support membranes. The choice of support membranes and lateral dimension of the flakes are two important parameters which need to be considered to fabricate efficient membranes. Other parameters such as concentration/volume ratio, temperature, and pressure applied can also impact the compactness, thickness, and integrity of the resulting nanolaminate membranes. In 2015, Ren and co-workers synthesized the first MXene-based separation membranes [24]. These are fabricated by dissolving $Ti_3C_2T_x$ nanosheets to form a uniform homogenous dispersion upon filtration on polyvinylidene fluoride (PVDF) support membrane with a 450 nm pore size. A MXene membrane of controlled thickness (100 nm to several micrometres) with enhanced mechanical strength and flexibility was then isolated from the PVDF support membrane after complete drying. One of the important parameters for the fabrication of NL membranes is the solution processibility of MXenes, which in turn depends on the rheological properties of colloidal dispersions. $Ti_3C_2T_x$ dispersions show variable viscoelastic behaviour at different concentrations. Luo et al. [29] reported nanolaminated MXene membranes supported on flat polyacrylonitrile (PAN) sheets prepared via spin coating. Similarly, MXene membranes supported on ceramic α-Al_2O_3 hollow fibres were also fabricated [29]. To enhance the functionality of nanolaminate membranes, various chemical modulations have been performed. For example, various support membranes have been taken into consideration such as titania, ferric hydroxide borate, and silver ion nanoparticles in addition to graphene oxide nanosheets [26, 36, 37, 42, 43].

In the second category of MXenes, mixed-matrix MXene membranes were also reported to be fabricated through various methods such as spin coating and dip coating [44–46]. Herein, MXene nanosheets were found to act as a filler material in a doped polymeric solution to lessen the mass transfer confrontation and interlayer spacing and to improve the transport performance. In comparison with conventional matrix materials, MXene nanosheets have edging properties such as fillers to improve the feasibility of scale-up and the cost effectiveness. There are some other important factors to be considered for MMM fabrication such as compatibility, uniformity, and stability. The homogenous mixing of filler material into the polymeric solution before casting the membrane and solvent vapourization/non-solvent-induced coagulation are crucial factors to enhance the dispersion stability, uniformity, state of aggregation, and chemical compatibility of the filler/polymer interface [47]. MXene-based materials are promising candidates to fulfil all the requirements needed because of their polarity, hydrophilicity, and unique surface chemistry, which make them versatile for further functionalization and strong interactions with a variety of solvents. It was reported that $Ti_3C_2T_x$ nanosheets showed long-term chemical stability and dispersion in organic solvents such as N-methyl-2-pyrrolidone and dimethylformamide [48]. The hydroxyl (-OH) moiety of the MXene nanosheets also enhanced the functionalization and chemical stability at the filler/polymer interfaces by adopting silanol chemistry for moderating non-ideal interfacial morphologies. These $Ti_3C_2T_x$-based MMMs with various polymer matrixes demonstrated efficient water/organic solvent purification [49, 50–52]. Other important functions of MXene fillers in MMMs were also reported to enlarge 2D morphology, constrict interlayer spacing for nano pore size, enhance transportation performance for selective solute transport, and functionalize surface chemistry [47]. These factors further helped to improve the transportation performance and packing of polymer chains to enhance mechanical properties and separation processes. In addition, to make them more unique in their performance, another factor that is high electrical conductivity is also currently under investigation [51]. The preliminary results showed that $Ti_3C_2T_x$/PVA mixed-matrix membranes exhibited enhanced mechanical strength for repetitive performances with moderate electrical conductivity to facilitate water fluxes and electrostatically controlled rejection rates at different electric potentials [51]. More recently, MXene/chitosan mixed-matrix membranes were fabricated for solvent dehydration applications via spin coating [29]. Likewise, these membranes were also fabricated using a blade casting method to study their antifouling activity [45]. In addition, lamellar $Ti_3C_2T_x$–GO MMMs were also reported with 90 nm pore size on porous supports using VAF [53].

The third category of MXenes includes thin-film nanocomposite (TFN) membranes made up of selective ultrathin polyamide layers supported on a permeable material to achieve uniform structure. TFN membranes are hybrid membranes of thin-film composites (TFCs) and MMMs. Since TFCs to date are the most popular membranes for waste-water treatment and desalination processes, TFNs are considered smart, performance-driven membranes and therefore attract many scientific communities to explore them in an elaborated way. The promising method used for preparing TFN membranes is an interfacial polymerization (IP) technique. Presently, the polyamide-based TFC membranes have shown a wide range of applications in separation processes at the industrial level. It is highly recommendable to fabricate IP-based TFNs with enhanced hydrophilicity, surface homogeneity, antifouling activity, and increased chlorine resistance. Indeed, TFNs have therefore established a great revolution in this field for a wide range of applications in osmotically driven methods, e.g. reverse and forward osmosis, vacuum-assisted osmosis, and nanofiltration. Recently, different research groups have reported on the synthesis of TFNs fabricated by the intercalation of MXenes with TMC in n-hexane using IP techniques [47, 54–57]. However, there is still need for great accuracy in the design process in view of several parameters like choice of monomers and their concentrations, crosslinking reaction time, filler loading ratio, etc.

4.1.2 GAS SEPARATION

Although research on MXene membranes is in quite early stages, these membranes have established an important place as a promising candidate in membrane-based gas separation processes [58–60]. Ultrathin MXene nanosheets with one or a few layers can be easily obtained through the size exclusion mechanism and used as building blocks to fabricate membranes with inter-spacings in nanometres to entail a massive prospective for the specific separation of gas molecules. It has been observed that due to a lack of uniformity and scalability in intrinsic and artificial pores of MXene nanosheets membranes, it is more feasible to utilize laminated MXene membranes to separate gas. To acquire high permeability and selectivity, laminated membranes essentially require uniformly distributed sub-nanometre channels without any defects. It has been reported from gas permeation analysis that freestanding MXene nanolaminated membranes showed high mechanical robustness [34].

These membranes have shown high permeance to small gas molecules such as helium (2164 barrer) and hydrogen (2402 barrer) but less permeance to large gaseous molecules such as CO_2, O_2, N_2, and CH_4 at 25 °C and 1 bar and demonstrated more selectivity for several small/large gas pairs [61]. Huang et al. fabricated these MXene laminated membranes [36]. For fabrication, the process was started by constructing the 2D laminar sieving membrane via a top-down method using exfoliated MXene nanosheets of 1.5 nm in thickness and a lateral size of 1–2 μm (obtained after etching with LiF and HCl) as building blocks to obtain ultra-permeable membranes for H_2 and enhanced H_2/CO_2 selectivity in the gas separation process. These MXene membranes showed a potential mechanical strength of approx. 50 MPa and a Young's modulus of 3.8 GP with high reproducibility. From a comparative study, it has been noted that permeability of the membranes drastically decreases for large molecules. Permeability reported for CO_2 was only 10 barrer in comparison to H_2 (2402 barrer) and He (2164 barrer). This happened because of the strong interactions of CO_2 dye with large quadrupole moments. Because of the unparallel selectivity of H_2/CO_2, these membranes showed enhanced activity for hydrogen purification processes. These outcomes were also found to be consistent with molecular dynamic (MD) simulations and offered significant perspective to precisely understand the gas transportation performances of MXene-based membranes, specifically 2D laminar membranes. Ren et al. [36] also reported similar results and concluded that a H_2/CO_2 gas pair showed high ideal and mixed gas selectivity (234.4 and 166.6 respectively) compared with that of ideal selectivity for the CO_2/N_2 permeability of CO_2 (10 barrer) and N_2 (19 barrer). The trapping effect of MXenes on the permeation of CO_2 molecules was considered to occur due to strong interactions in between them because of oxygen-containing functionalities in the membranes. This behaviour of MXene membranes makes them available as promising candidates required for the suppression of CO_2 permeability. On the other hand, there have been numerous reports to improve the permeation of CO_2 molecules by reducing transport resistance. Shen et al. [62] successfully fabricated thinner NLMs with well-defined interlayer spacings by altering the crosslinking of MXene nanosheets with borate and polyethylene. The smooth and continuous MXene membrane was synthesized by filtering it on a dopamine-modified anodic aluminium oxide (AAO) substrate, which led to increased adhesion between them. The inclusion of borate and borate/polyethylene ions in these MXene membranes helped to distinguish between CO_2 and CH_4 (N_2) molecules and also enhanced the mechanical strength by bonding with oxygen functionalities. Therefore, these modified MXene nanofilms facilitate not only reduced interlayer spacing but also the release of trapped CO_2 by enhancing the CO_2 permeation with pristine MXene (CO_2: 10 barrer), which results in overcoming the upper boundary for MOF membranes. Recently, Petukhov et al. [63] reported $Ti_3C_2T_x$-based nanolaminate membranes as promising candidates for the fabrication of an ammonia-selective membrane to attain NH_3/H_2 selectivity over 50 with NH_3 permeance of 3.9 m^3 (STP)·m^{-2}·bar^{-1}·h^{-1} in moisture-sensitive medium and a H_2O/N_2 selectivity of 1000 with water permeance of 30 m^3 (STP)·m^{-2}·bar^{-1}·h^{-1} for sorption-type activity. Two-dimensional laminated MXene membranes were also reported to perform at high temperatures [64]. The thermal stability led a significant impact on various applications and operations. These membranes demonstrated high-temperature tolerance and firm gas separation performance at 320 °C with moderate H_2 permittivity and a selectivity of 41 for H_2/N_2 gaseous mixtures. In addition, these MXene membranes have shown excellent sustainability of over 200 hours without any significant changes or deformations in membranes. However, a continuous increase in temperature results in decreased selectivity, necessitating much attention to explore the high-temperature tolerance of MXene membranes.

Mixed-matrix membranes (MMMs) were also found to be efficient for gas separation. The enhanced absorption ability of MXene membranes with additional nanochannels provided these as promising candidates for this process. Liu et al. [44] fabricated MMMs using MXenes as nanofillers and poly(ether-*block*-amide) (PEBA) as the matrix. When the loading of $Ti_3C_2T_x$ increased to 0.15% wt on PAN (polyacronitrile) supports by spin coating, the permeance of CO_2 was raised by 81% and the selectivity of CO_2/N_2 was raised by 73%, with the achievement of high CO_2 permeance of 21.6 GPU. In another report, Shamsabadi et al. [65] managed to fabricate MXene nanonosheets in Pabex-1657 with less loading to synthesize thin-film MMMs with better performance of CO_2 permeance, high CO_2/N_2 selectivity, and an economically cheaper process. More recently, Guan et al. [66] also fabricated MMMs with greatly improved permeance of CO_2 and selectivity of the CO_2/N_2 gas pair.

4.1.3 MOLECULAR SEPARATION

The development of two-dimensional unique MXenes has a great potential for applications in molecular separation. With advantages of effective interlayer spacing and active surface charge, MXene-based membranes have led towards an indispensable technology for the fine separation of molecules [67]. Gogotsi and co-workers synthesized and investigated the interaction of $Ti_3C_2T_x$ membrane with aqueous dye molecules, e.g. methylene blue (cationic) and acid blue (anionic). The cationic methylene blue exhibited more adsorption on the membrane as compared to that of acid blue. This can be explained by the presence of a negatively charged $Ti_3C_2T_x$ membrane surface which undergoes electrostatic interactions with the cationic dye. However, with the comparatively lesser adsorption of acid blue, they showed relatively faster photodegradation under UV light [68]. More two-dimensional $Ti_3C_2T_x$ MXene membranes were prepared, and their separation performance was tested by applying external voltage. By taking the benefit of conductive MXene membranes and negative surface charge, the ionic separation of inorganic metal ions (Na^+ and Mg^{2+}) and organic dyes (cationic, methylene blue) can be controlled by a voltage-gated mechanism. Basically, the purpose of varying applied potential is attributed to modifying the interlayer spacing of two-dimensional nanosheets of MXene membranes and to control the intercalated inorganic ions or organic dye molecules for the molecular sieving process. However, with continuous applied potential, there is an increase in the flow of aqueous organic dye solution, due to which the nanochannels of the MXene membrane may swell, resulting in an increased interlayer distance. This causes a decrease in the separation performance of the MXene membrane. Therefore, this limitation of $Ti_3C_2T_x$-based MXenes can be overcome by incorporating some other material which exhibits less permeance to water molecules [69].

Kang et al. [70] synthesized MXene–graphene oxide ($Ti_3C_2T_x$–GO) composite membranes for carrying out the pressure-driven separation of organic dyes. Out of negatively charged (brilliant blue, 7.98 Å and rose Bengal, 5.88 Å), positively charged (methylene blue, 5.04 Å) and neutrally charged (methyl red, 4.87Å) compounds, the rejection rate by $Ti_3C_2T_x$–GO membrane was highest (100%) for brilliant blue and lowest (68%) for methyl red. The lower rejection for methyl red may be attributed to its neutral charge and relatively small hydrated radius. The surface-active sites of $Ti_3C_2T_x$ having polar oxygen groups such as Ti–OH are responsible for the electrostatic interaction with the charged dye, which causes the rejection of brilliant blue, rose Bengal, and methylene blue. Therefore, the composite membranes with nanochannels are effective in separating the molecules with sizes larger than 5 Å in radius, and their active surfaces are accountable for the separation of ionic dye molecules [70]. Han et al. have incorporated 70% MXene into graphene oxide nanosheets to prepare separation membranes which showed an excellent rejection rate 98.56% for methyl red [71].

For industrial-scale applications, a large-area MXene film was fabricated using a slot-die coating method. The aqueous solution of $Ti_3C_2T_x$ was uniformly placed on the target substrate using coater's slot head with a coating speed of 6 mm·s^{-1}. By this method, the membrane's thickness can be efficiently altered from the micrometre to nanometre scale. The alignment of $Ti_3C_2T_x$ nanosheets can also be controlled by moving the slot-die head, which induces the shear force on the interlayers of the nanosheets. Various organic dyes, including methyl red, Evans blue, rose Bengal, methylene blue, and brilliant blue G, were tested for slot-die-coated $Ti_3C_2T_x$ membrane separation performance under oxidizing conditions for one month, and further results were compared with a $Ti_3C_2T_x$ membrane fabricated by a state-of-the-art vacuum filtration approach. The slot-die-coated $Ti_3C_2T_x$ membrane exhibited an excellent rejection rate for all dyes, as compared to that of vacuum-filtered MXene membrane. For example, the rejection rates in the case of methyl red and Evans blue were 93.2% and 87.8% for the slot-die-coated membrane and 77.9% and 86.3% for the vacuum-filtered membrane, respectively. Hence, such large-area MXene films have potential to be used as molecular filters for the nanofiltration process [72].

Thin-film $Ti_3C_2T_x$ MXene nanosheets were also modified via interfacial polymerization using polyamide as the intercalate reagent and polyethyleneimine as the crosslinker. Due to the presence of polyamide in MXene sheets, a negative surface charge density will increase more with -COOH groups on surface. For positively charged polyethyleneimine, the interaction with MXene sheets altered the surface charge density, which led towards the lesser agglomeration of MXene nanosheets. The final fabricated thin-film nanocomposite membrane possessed an excellent dye-rejection rate [73].

Another MXene-based nanofiltration membrane was fabricated using negatively charged MXene nanosheets and positively charged polyethyleneimine dopamine (strong adhesion) to improve the membrane separation performance. All the tested organic dye molecules, e.g. reactive blue, methylene blue, Congo red, and Coomassie brilliant blue, demonstrated excellent rejection rates by the MXene nanosheet-decorated separating membrane [74]. New heterostructured CNT-MXene membranes were developed, where the interlayer spacing of the lamellar MXene membrane was modified using surface-functionalized multiwalled carbon nanotubes. Such CNTs will fuse with the MXene membrane within a smaller interlayer spacing of 1 nm to form unique fusiform channels for molecular sieving. These heterostructured CNT-MXene membranes with a 45% content of CNTs having a layer thickness of 820 nm proved to be outstanding for the rejection of small dye molecules as compares to other lamellar MXene membranes with lesser layer thickness [75]. The $Ti_3C_2T_x$ MXene lamellar nanofiltration membranes were fabricated and used for separation performance by employing an external electric field. In the case of cationic dye molecules, if a negative electric field is applied to lamellar $Ti_3C_2T_x$, then the separation performance will improve, but on the other hand, a positive voltage will reduce the sieving ability. However, in case of anionic dye molecules, the lamellar $Ti_3C_2T_x$ membrane showed opposite patterns as compared to those for cationic dye molecules. Therefore, the separation performance of the lamellar $Ti_3C_2T_x$ membrane can be altered by modulating the external electric field which induces the electrostatic interactions among the dye molecules and $Ti_3C_2T_x$. The separation of molecules via lamellar membranes is possible due to the presence of nano-transport channels having horizontal interlayer and vertical inter-edge gaps and the electrostatic modulation of their surface charge [76].

4.1.4 OIL–WATER (O/W) SEPARATION

The simple process of operation and high efficiency are the key requirements for membrane-based oil–water separation. To separate oil from water, the separation membrane must show selective permeation of oil or water. Surface wettability plays an important role in this regard. The wettability is characterized by measuring the static or dynamic contact angle. For a pure water drop, surfaces showing a contact angle (balance of interfacial forces) above 150° are termed as superhydrophobic. Such superhydrophobic surfaces (membranes), when used for oil–water separation, tend to easily repel water and allow oil to permeate [77–79]. The surface showing oil CA above 150° are termed as superoleophobic. For membranes to retain their super repellent properties, various techniques have been developed to coat suitable surface chemicals. The combination of a surface chemical with surface structure provides the opportunity to tune the surface wetting to become extremely repellent. For instance, superhydrophobic surfaces can be prepared by using low-surface-energy chemicals (silanes, fluorine, hydrophobic alkyl chains, etc.) with surface textures (nanoparticles). The superhydrophobic membranes which allow for the permeation of oil are oleophilic and can be used for recovering oil. Such membrane-based systems have to be mechanically and chemically stable for real-life applications.

The lack of mechanical, chemical, and thermal stability hinders the long-term use of membrane-based systems. Further, oil fouling due to continuous oil permeation can also hinders the operation of the mesh. To overcome this, suitable stable chemical modifications are required. Such modifications depend upon the kind of membrane (polymer, metallic meshes, etc.) used for separation. Among the modifications, the growth of nanostructures has shown promising applications towards separating the oil–water mixtures. The nanostructures not only provide the surface textures but also the desired chemistry. For instance, MXene-based coatings are highly hydrophilic and demonstrate the underwater oleophobic wetting regime. Further, the fascinating properties of MXenes such as their abundance of surface functional groups, flexibility, and hydrophilicity enable these 2D materials to be used as appropriate surface coatings.

Underwater oleophobic MXene membranes selectively block oil drops. The MXene hydrophilicity can trap a water layer on membrane surface, which tends to repel oil drops. The water can easily permeate. For stable emulsions, i.e. oil–salt–water mixtures, separation efficiency is enhanced owing to the ion intercalation behaviour of MXene membranes, giving higher oil rejection [8–10]. The MXene-based separation membranes can be fabricated by a phase-inversion method, electrospinning, vacuum-assisted filtration, and by extrusion and moulding [8]. The vacuum-assisted filtration of an oil–water mixture and emulsion was recently reported [80]. A polyacrylonitrile (PAN) membrane was modified using a 2D $Ti_3C_2T_x$ MXene. The MXene was synthesized by etching using 40% wt HF solution from a Ti_3AlC_2 precursor phase [80]. The membrane was fabricated by electrospinning a DMF solution of PAN and drying to obtain a nanofibrous PAN membrane. The ethanolic solution of the $Ti_3C_2T_x$ MXene was vacuum filtered onto the nanoporous PAN membrane and washed thoroughly with ethanol to remove any unetched MXene. The membrane was dried at a high temperature (80 °C) to form a $Ti_3C_2T_x$ MXene–PAN membrane utilized for oil–water separation. The water flux (J) and separation efficiency (% R) were evaluated by separating the oil-water mixture through the MXene-decorated PAN membrane. The working of the MXene–PAN membrane was also checked by measuring the intrusion pressure, and the membrane was found to withstand an intrusion pressure of 5533 Pa for toluene [80]. The intrusion pressure P (Pa) of the oil phase through the separating mesh can be estimated using the equation:

$$P = \rho\, g\, h \qquad\qquad (4.1)$$

Here, ρ, g, and h respectively represent the density of the oil, gravitational force, and maximum height the oil phase can achieve inside the column attached to membrane, i.e. maximum height withstood by the membrane without oil permeation [80]. The filtration of an oil–water mixture can be achieved under the influence of gravity by applying external pressure. In addition to intrusion pressure, the flux has also been evaluated to understand the separation ability of the fabricated meshes. For instance, a separation flux of 354.6 L/m²·h·bar was achieved for separating an emulsion by a superhydrophobic hollow hemispherical MXene. The hollow MXene was synthesized by thermal annealing of cationic polystyrene spheres decorated with MXene. The electrostatic interaction between the cationic polystyrene sphere and MXene was responsible for MXene-decorated spheres. The hollow hemispherical MXene (HSMX) was coated onto a polyvinylidene fluoride membrane to prepare the separating mesh [81]. The optical images of HSMXs and their superhydrophobic behaviour is depicted in Figure 4.2 [81].

Similarly, Moghaddasi et al. [82] fabricated copolyimide electrospun multi- and single-layered MXenes ($Ti_3C_3T_x$) for separating various oil–water emulsions. The nanofibres bearing an electrospun membrane of copolyimide 6,10 were fabricated firstly using an electrospinning device. The membrane was further decorated with

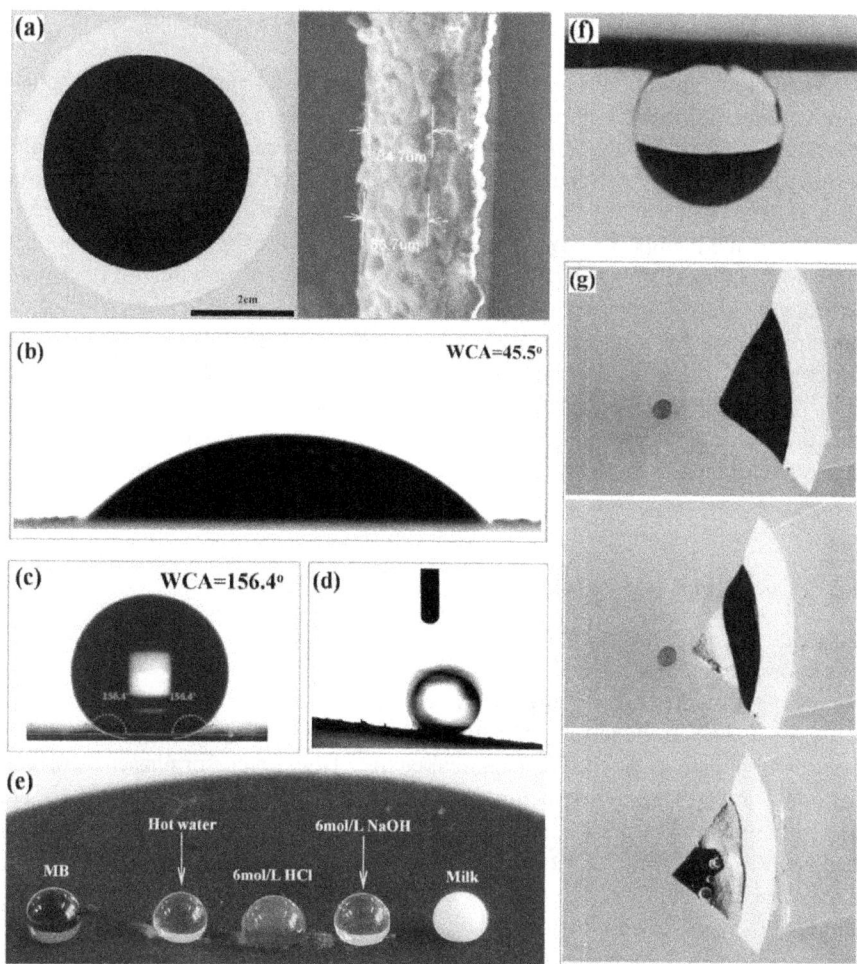

FIGURE 4.2 (a) Optical image of an HSMX and its SEM image showing the cross-section and contact angle for water drop for (b) MXene, (c) HSMX surface; (d) sliding angle (roll off) for water drop on the tilted HSMX surface; (e) various liquid drops on the HSMX membrane; (f) water drop adhered to the HSMX surface shown as an upside-down image; (g) oil–water separation images with HSMXs showing the silver mirror phenomena [81].

multi- and single-layered $Ti_3C_2T_x$. The presence of -NH and -CO functionalities on copolyimide were responsible for the hydrophilic behaviour, as evident from the water CA of ~42°; a maximum CA of 130° for a pure water drop was shown by the single-layered MXene-decorated polyamide membrane (coPA/SL-$Ti_3C_2T_x$). Underwater oil captive bubble angles of 124° and 141° were reported for ML-$Ti_3C_2T_x$ and SL-$Ti_3C_2T_x$ membranes, respectively. The underwater oleophobic membranes were able to separate the oil–water mixtures with a separation efficiency of 99.5%

and flux of over 11,000 L/m²·h for the single-layered MXene-decorated copolyimide membrane [82].

For the separation of oil–water mixtures, MXene-based membrane systems have been designed for real-life applications. The membranes are made either superhydrophobic or underwater oleophobic for ease of oil–water separation [8, 82]. A detailed review on the use of MXene-decorated membranes for oil–water separation has been reported recently [8]. The separation ability of the membranes can be tuned by modulating the membrane properties related to surface functionalities (hydrophilic/oleophilic), charge, and surface texture (roughness) [8]. The membranes have demonstrated exceptional separation ability with high oil/water permeation with exception flux; still, the challenges related to environmental impact needs to be resolved. Among these, the use of HF for etching to form MXene, biocompatibility, biodegradability, reusability, large-scale MXene synthesis, stability against oil fouling, harsh chemicals (acids, surfactants), and reusability need to be addressed. Nonetheless, the decoration of membranes with 2D nanoparticles seems to be promising approach for creating oil–water separation membranes.

4.2 CONCLUSION AND FUTURE PERSPECTIVES

This chapter highlights the importance of new class of 2D material called MXenes in gas, molecule, and oil–water separation. The MXenes combine with supporting membranes for the separation process. The development of a fabrication process for MXenes has provided the opportunity to tune the surface functional groups for targeted applications [1–8]. The chemical etching from precursor phases can forms MXenes with sponge-like structures with pores of varying sizes [12–14, 83]. The MXene membrane-based separation devices face challenges associated with the fabrication of MXene with adjustable nanochannels and the robustness of the prepared membrane [11–13, 83]. The chemical stability of MXenes provides another challenge, as they are susceptible to oxidation in the presence of water [12–14, 83]. The use of HF further complicates the synthesis process due to environmental risks associated with fluorinated chemicals. To overcome these issues, firstly, the replacement of HF with green etchants or less toxic etchants is being studied. The scale-up of MXenes needs to be addressed as well. Environmentally friendly, scalable, efficient, and cheap fabrication processes can be designed to produce MXenes with exceptional stability in both ambient and humid conditions.

REFERENCES

1. N. H. Solangi, R. R. Karri, N. M. Mubarak, S. A. Mazari, A. S. Jatoi, J. R. Koduru, Desalination **2023**, 549116314
2. W. Sun, S. Shah, Y. Chen, Z. Tan, H. Gao, T. Habib, M. Radovic, J. Mater. Chem. A **2017**, 5, 21663
3. C. E. Shuck, K. Ventura-Martinez, A. Goad, S. Uzun, M. Shekhirev, Y. Gogotsi, ACS Chem. Health & Safety **2021**, 28, 326
4. A. Szuplewska, D. Kulpińska, A. Dybko, M. Chudy, A. M. Jastrzębska, A. Olszyna, Z. Trends Biotechnol. **2020**, 38, 264

5. N. H. Solangi, S. A. Mazari, N. M. Mubarak, R. R. Karri, N. Rajamohan, D.-V. N. Vo, Environ. Res. **2023**, 222,115337

6. M. Naguib, M. Kurtoglu, V. Presser, J. Lu, J. Niu, M. Heon, L. Hultman, Y. Gogotsi, M. W. Barsoum, Adv. Mater. **2011**, 23, 4248

7. Y. Gogotsi, B. Anasori, ACS Nano **2019**, 13, 8491

8. I. Ihsanullah, M. Bilal, App. Mat. Today **2022**, 29, 101674.

9. V. Kamysbayev, A. S. Filatov, H. Hu, X. Rui, F. Lagunas, D. Wang, R. F. Klie, D. V. Talapin, Sci. **2020**, 369, 979

10. Z. Ahmed, F. Rehman, U. Ali, A. Ali, M. Iqbal, K. H. Thebo, ChemBioEng. Rev. **2021**, 8, 110

11. G. P. Lim, C. F. Soon, A. Al-Gheethi, M. Morsin, K. S. Tee, Ceram. Int. **2022**, 48, 16477

12. M. M. Pendergast, E. M. V. Hoek, Energy Environ. Sci. **2011**, 4, 1946

13. X. Qu, J. Brame, Q. Li, P. J. J. Alvarez, Acc. Chem. Res. **2013**, 46, 834

14. C. Xu, L. Wang, Z. Liu, L. Chen, J. Guo, N. Kang, X.-L. Ma, H.-M. Cheng, W. Ren, Nat. Mater. **2015**, 14, 1135

15. B. Anasori, M. R. Lukatskaya, Y. Gogotsi, Nat. Rev. Mater. **2017**, 2, 16098

16. A. M. Naguib, V. N. Mochalin, M. W. Barsoum, Y. Gogotsi, Adv. Mater. **2014**, 26, 992

17. M. Malaki, A. Maleki, R. S. Varma, J. Mater. Chem. A **2019**, 7, 10843

18. M. Sokol, V. Natu, S. Kota, M. W. Barsoum, Trends Chem. **2019**, 1, 210

19. J. Halim, M. R. Lukatskaya, K. M. Cook, J. Lu, C. R. Smith, L.-Å. Näslund, S. J. May, L. Hultman, Y. Gogotsi, P. Eklund, M. W. Barsoum, Chem. Mater. **2014**, 26, 237

20. L. Wang, H. Zhang, B. Wang, C. Shen, C. Zhang, Q. Hu, A. Zhou, B. Liu, Electron. Mater. Lett. **2016**, 12, 702

21. L. H. Karlsson, J. Birch, J. Halim, M. W. Barsoum, P. O. Å. Persson, Nano Lett. **2015**, 15, 4955

22. P. Lakhe, E. M. Prehn, T. Habib, J. L. Lutkenhaus, M. Radovic, M. S. Mannan, M. J. Green, Ind. Eng. Chem. Res. **2019**, 58, 1570

23. C. J. Zhang, S. Pinilla, N. McEvoy, C. P. Cullen, B. Anasori, E. Long, S.-H. Park, A. Seral-Ascaso, A. Shmeliov, D. Krishnan, C. Morant, X. Liu, G. S. Duesberg, Y. Gogotsi, V. Nicolosi, Chem. Mater. **2017**, 29, 4848

24. C. E. Ren, K. B. Hatzell, M. Alhabeb, Z. Ling, K. A. Mahmoud, Y. Gogotsi, J. Phys. Chem. Lett. **2015**, 6, 4026

25. L. Ding, Y. Wei, Y. Wang, H. Chen, J. Caro, H. Wang, Angew. Chem. Int. Ed. **2017**, 56, 1825

26. J. Shen, G. Liu, Y. Ji, Q. Liu, L. Cheng, K. Guan, M. Zhang, G. Liu, J. Xiong, J. Yang, W. Jin, Adv. Funct. Mater. **2018**, 28, 1801511

27. M. A. Hope, A. C. Forse, K. J. Griffith, M. R. Lukatskaya, M. Ghidiu, Y. Gogotsi, C. P. Grey, Phys. Chem. Chem. Phys. **2016**, 18, 5099

28. X. Wu, L. Hao, J. Zhang, X. Zhang, J. Wang, J. Liu, J. Membr. Sci. **2016**, 515, 175

29. J. Luo, W. Zhang, H. Yuan, C. Jin, L. Zhang, H. Huang, C. Liang, Y. Xia, J. Zhang, Y. Gan, X. Tao, ACS Nano **2017**, 11, 2459

30. Y. Xia, J. Zhang, Y. Gan, X. Tao, ACS Nano **2017**, 11, 2459

31. L. Hao, H. Zhang, X. Wu, J. Zhang, J. Wang, Y. Li, Compos. Part A **2017**, 100, 139

32. Z. Ling, C. E. Ren, M.-Q. Zhao, J. Yang, J. M. Giammarco, J. Qiu, M. W. Barsoum, Y. Gogotsi, Proc. Natl. Acad. Sci. USA **2014**, 111, 16676

33. L. Ding, Y. Wei, L. Li, T. Zhang, H. Wang, J. Xue, L.-X. Ding, S. Wang, J. Caro, Y. Gogotsi, Nat. Commun. **2018**, 9, 155

34. V. Natu, M. Sokol, L. Verger, M. W. Barsoum, J. Phys. Chem. C **2018**, 122, 27745

35. R. P. Pandey, K. Rasool, V. E. Madhavan, B. Aïssa, Y. Gogotsi, K. A. Mahmoud, J. Mater. Chem. A **2018**, 6, 3522

36. C. E. Ren, K. B. Hatzell, M. Alhabeb, Z. Ling, K. A. Mahmoud, Y. Gogotsi, J. Phys. Chem. Lett. **2015**, 6, 4026

37. L. Ding, Y. Wei, Y. Wang, H. Chen, J. Caro, H. Wang, Angew. Chem. Int. Ed. **2017**, 56, 1825

38. C. J. Zhang, M. P. Kremer, A. Seral-Ascaso, S.-H. Park, N. McEvoy, B. Anasori, Y. Gogotsi, V. Nicolosi, Adv. Funct. Mater. **2018**, 28, 1705506

39. K. Wang, Y. Zhou, W. Xu, D. Huang, Z. Wang, M. Hong, Ceram. Int. **2016**, 42, 8419

40. Y. Lee, S. J. Kim, Y.-J. Kim, Y. Lim, Y. Chae, B.-J. Lee, Y.-T. Kim, H. Han, Y. Gogotsi, A. C. Won, J. Mater. Chem. A **2019**, 8, 573.

41. T. Habib, X. Zhao, S. A. Shah, Y. Chen, W. Sun, H. An, J. L. Lutkenhaus, M. Radovic, M. J. Green, npj 2D Mater. Appl. **2019**, 3, 8

42. K. M. Kang, D. W. Kim, C. E. Ren, K. M. Cho, S. J. Kim, J. H. Choi, Y. T. Nam, Y. Gogotsi, H.-T. Jung, ACS Appl. Mater. Interfaces **2017**, 9, 44687

43. B. Akuzum, K. Maleski, B. Anasori, P. Lelyukh, N. J. Alvarez, E. C. Kumbur, Y. Gogotsi, ACS Nano **2018**, 12, 2685

44. G. Liu, L. Cheng, G. Chen, F. Liang, G. Liu, W. Jin, Chem. Asian J. **2020**, 15, 2364

45. R. P. Pandey, P. A. Rasheed, T. Gomez, R. S. Azam, K. A. Mahmoud, J. Membr. Sci. **2020**, 607, 11813

46. S. S. Li, J. Dai, X. Geng, J. D. Li, P. Li, J. D. Lei, L. Y. Wang, J. He, Sep. Purif. Technol. **2020**, 235, 1

47. C. Y. Chuah, K. Goh, Y. Yang, H. Gong, W. Li, H. E. Karahan, M. D. Guiver, R. Wang, T.-H. Bae, Chem. Rev. **2018**, 118, 8655

48. K. Maleski, V. N. Mochalin, Y. Gogotsi, Chem. Mater. **2017**, 29, 1632

49. Z. Xu, Y. Sun, Y. Zhuang, W. Jing, H. Ye, Z. Cui, J. Membr. Sci. **2018**, 564, 35

50. L. Hao, H. Zhang, X. Wu, J. Zhang, J. Wang, Y. Li, Compos. Part A **2017**, 100, 139

51. C. E. Ren, M. Alhabeb, B. W. Byles, M.-Q. Zhao, B. Anasori, E. Pomerantseva, K. A. Mahmoud, Y. Gogotsi, ACS Appl. NanoMater. **2018**, 1, 3644

52. Q. Luan, Y. Xie, D. Teng, R. Han, S. Zhang, Desalin. Water Treat. **2018**, 108, 90

53. K. M. Kang, D. W. Kim, C. E. Ren, K. M. Cho, S. J. Kim, J. H. Choi, Y. T. Nam, Y. Gogotsi, H.-T. Jung, ACS Appl. Mater. Interfaces **2017**, 9, 44687

54. R. Mahajan, W. J. Koros, M. Thundyil, Membr. Technol. **1999**, 1999, 6.

55. D. Qadir, H. Mukhtar, L. K. Keong, Sep. Purif. Rev. **2017**, 46, 62

56. M. A. Aroon, A. F. Ismail, T. Matsuura, M. M. Montazer-Rahmati, Sep. Purif. Technol. **2010**, 75, 229

57. T.-S. Chung, L. Y. Jiang, Y. Li, S. Kulprathipanja, Prog. Polym. Sci. **2007**, 32, 483

58. F. Dixit, K. Zimmermann, M. Alamoudi, L. Abkar, B. Barbeau, M. Mohseni, B. Kandasubramanian, K. Smith, Renew. Sust. Energ. Rev. **2022**, 164, 112527

59. T. Amrillah, A. R. Supandi, V. Puspasari, A. Hermawan, Z. W. Seh, Trans. Tianjin Univ. **2022**, 28, 307

60. H. E. Karahan, K. Goh, C. Zhang, E. Yang, C. Yıldırım, C. Y. Chuah, M. Göktuğ Ahunbay, J. Lee, Ş. B. T. Ersolmaz, Y. Chen, T. H. Bae, Adv. Mater. **2020**, 1906697

61. L. M. Robeson, J. Membr. Sci. **2008**, 320, 390

62. J. Shen, G. Liu, Y. Ji, Q. Liu, L. Cheng, K. Guan, M. Zhang, G. Liu, J. Xiong, J. Yang, W. Jin, Adv. Funct. Mater. **2018**, 28, 1801511

63. D. I. Petukhov, A. S. Kan, A. P. Chumakov, O. V. Konovalov, R. G. Valeev, A. A. Eliseeva, J. Membr. Sci. **2021**, 621, 118994

64. Y. Fan, L. Wei, X. Meng, W. Zhang, N. Yang, Y. Jin, X. Wang, M. Zhao, S. Liu, J. Membr. Sci. **2019**, 569, 117

65. A. A. Shamsabadi, A. P. Isfahani, S. K. Salestan, A. Rahimpour, B. Ghalei, E. Sivaniah, M. Soroush, ACS Appl. Mater. Interfaces **2020**, 12, 3984

66. W. Guan, X. Yang, C. Dong, X. Yan, W. Zheng, Y. Xi, X. Ruan, Y. Dai, G. He, J. Appl. Polym. Sci. **2020**, 138, e49895

67. Y. Sun, S. Li, Y. Zhuang, G. Liu, W. Xing, W. Jing, J. Memb. Sci. **2019**, 591, 117350

68. O. Mashtalir, K. M. Cook, V. N. Mochalin, M. Crowe, M. W. Barsoum, Y. Gogotsi, J. Mat. Chem. A **2014**, 2, 14334

69. C. E. Ren, M. Alhabeb, B. W. Byles, M. Q. Zhao, B. Anasori, E. Pomerantseva, K. A. Mahmoud, Y. Gogotsi, ACS Appl. Nano Mater. **2018**, 1(7), 3644

70. K. M. Kang, D. W. Kim, C. E. Ren, K. M. Cho, S. J. Kim, J. H. Choi, Y. T. Nam, Y Gogotsi, H. T. Jung, ACS App. Mat. Interf. **2017**, 9(51), 44687

71. S. Wei, Y. Xie, Y. Xing, L. Wang, H. Ye, X. Xiong, S Wang, K. Han, J. Memb. Sci. **2019**, 582, 414

72. J. H. Kim, G. S. Park, Y. J. Kim, E. Choi, J. Kang, O. Kwon, S. J. Kim, J. H. Cho, D. W. Kim, ACS Nano. **2021**, 15, 58860

73. J. Li, L. Li, Y. Xu, J. Zhu, F. Liu, J. Shen, J. Lin, Chem. Eng. J. **2022**, 427, 132070

74. S. Gu, Y. Ma, T. Zhang, Y. Yang, Y. Xu, J. Li, ACS ES&T Water **2022**, 3, 1756

75. M. Ding, H. Xu, W. Chen, Q. Kong, T. Lin, H. Tao, K. Zhang, Q. Liu, K. Zhang, Q. Liu, K. Zhang, Z. Xie, J. Mater. Chem. A **2020**, 8(43), 22666

76. J. Li, C. Xu, J. Long, Z. Ding, R. Yuan, Z. Li, ACS Appl. Nano Mater. **2022**, 5, 7373

77. M. Bala, V. Singh, Chem. Pap. (2023). https://doi.org/10.1007/s11696-023-02710-w

78. S.-W. Hu, V. Singh, Y.-J. Sheng, H.-K. Tsao, J. Taiwan Inst. Chem. Eng. **2020**, 107, 182

79. M. Bala, V. Singh, J. Mol. Liq. **2023**, 375, 121361

80. R. Imsong, D. D. Purkayastha, Sep. Purif. Techn. **2023**, 306, 122636

81. H. Chen, R. Wang, W. Meng, F. Chen, T. Li, D. Wang, C. Wei, H. Lu, W. Yang, Nanomaterials **2021**, 112866. https://doi.org/10.3390/nano11112866

82. A. Moghaddasi, P. Sobolčiak, A. Popelka, I. Krupa, Materials **2020**, 13, 3171

83. R. A. Soomro, P. Zhang, B. Fan, Y. Wei, B. Xu, Nano-Micro Lett., **2023**, 15, 108.

5 Environmental Applications of MXene-Based Materials

Shubham Mishra, Dinesh Kumar Pati, R. Padhee,
Subhendu Chakroborty, and Nibedita Nath

5.1 INTRODUCTION

The ecological system has been severely harmed by the economy and industry's continued growth, and there is severe water contamination. The protection of water sources and a reduction in water pollution are the two biggest challenges facing modern society [1, 2]. Inorganic and organic molecules could be used to categorise environmental water pollutants. As a result of industrial output and domestic trash, hazardous heavy metals as well as organic dyestuffs have emerged as the primary sources of wastewater among them [3, 4]. In addition, radioactive materials, leftover antibiotics, and waste gas constitute threats to the health of humans [5, 6].

For the removal of these environmental pollutants, numerous techniques, including adsorption, membrane separation, photocatalysis, redox, and chemical precipitation, have been extensively explored recently [7–9]. The creation of nanomaterials has lately demonstrated a huge potential application in environmental clean-up because of intriguing properties like high specific surface area, environmental versatility, and outstanding biocompatibility [10–12].

After the initial finding of single-layered graphene in 2004, various 2D nanomaterials became the focus of extensive research because of their outstanding optical, electrical, and mechanical capabilities [13–16]. At Drexel University in Pennsylvania, United States, Yury Gogotski and his team created a unique group of 2D material called MXenes in 2011 [17]. Due to their distinctive properties, MXenes have since attracted research interest [18, 19]. $M_{n+1}X_nT_x$ (n = 1–3) is the typical formula for MXenes, where M is the transition metal, X is carbon or nitrogen, T is -OH, -O, -F, or -C, and x is the number of surface functional groups [20, 21]. The width typically fluctuates between 1 nm and 2 nm depending on the value of n in MXenes [22].

5.2 2D MXENE MATERIALS

A brand-new class of 2D-TM carbide or carbonitride material called MXenes has a graphene-like 2D structure. The extraordinary 2D layered structure, sizeable particular surface area, high electrical conductivity, and superior mechanical strength and stability of

DOI: 10.1201/9781003366225-5

these new 2D materials all contribute to their exceptional performance. These characteristics have helped MXenes become prominent novel substrate materials for the study of a variety of applications, such as energy preservation and transformation, photothermal treatment, drug delivery, environmental adsorption, or catalytic degradation [23].

The structural design of MXenes can be explained as an $(MX)_nM$ pattern with n+1 layers of element X (carbon/nitrogen) and n layers of transition elements of metal M. The MXene structural layout is depicted in Figure 5.1. MXenes, which include M_2X, M_3X_2, and M_4X_3, have been given several formulae [24]. The layered MXene pattern is confirmed by scanning photos. It is incredibly feasible that MXenes containing a double layer of transition metals were created; these compounds can be identified by the way they appear in either ordered or solid solution form, in which the transition elements stack up in a particular order rather than randomly filling the M sites. For instance, although $(Cr_2V) C_2$ creates an ordered solution, $(TiV)_3C_2$ is available as a solid solution [25]. Physical and chemical properties of MXene-based materials are heavily influenced by surface functional groups, which have an impact on their environmental applications. Numerous functional groups like -O, -F, and -OH commonly render their creation on MXene surfaces. As a result, MXenes can be expressed as $M_{n+1}X_nT_x$, where T stands for the functional groups that were exfoliated from the surface [26]. A Ti_3C_2 MXene, for instance, can have at least three alternative notations, including $Ti_3C_2O_2$, $Ti_3C_3(OH)$, and $Ti_3C_2F_2$. According to Tran et al. (2018), MXenes often contain a mixture of functional groups such that the associated quantities are comparatively impacted by different synthesis processes [27].

Because of their exceptional biocompatibility, hydrophilic properties, large active sites with different functional groups, particular 2D layered framework, higher surface area, large interlayer spacing, and environmentally friendly nature, MXenes have generated a great deal of interest in expanding their environmental remediation applications [25, 28, 29].

The "A" layers of the MAX phase $(M_{n+1}AX_n)$ precursor are typically chemically etched to produce MXenes, as shown in Figure 5.1a and 5.1b, where M stands for

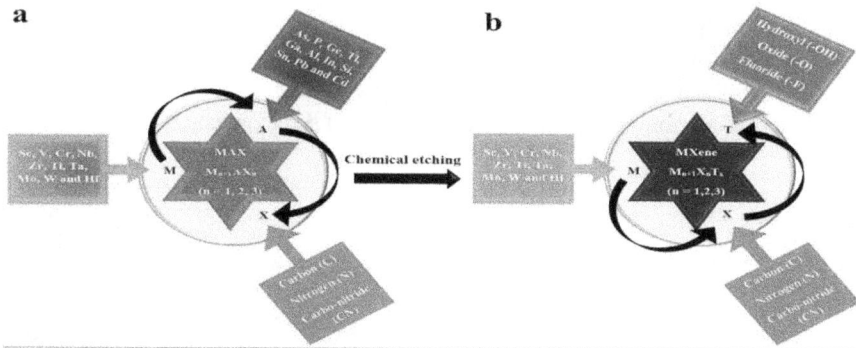

FIGURE 5.1 MAX phases in MXenes. (a) MAX phases having the formula $M_{n+1}AX_n$, and (b) MXenes having the formula $M_{n+1}X_nT_x$.

Copyright 2021, Elsevier. Reprinted with permission from Ref. [22].

the transition metal, A is any element from groups 12 to 16, X is C or/and N, and n = 1–3 [22, 30]. In the MAX phase, the "A" layer is sandwiched between the $M_{n+1}X_n$ with a weak M–A bond and a comparatively strong M–X bond [31, 32]. The first MXene (titanium carbide, Ti_3C_2) was created by etching the Al atoms out of Ti_3AlC_2 in hydrofluoric acid (HF) at normal temperature [17]. Since then, some 30 MXene materials—including $Ti_3N_3T_x$ [33], $Ti_3C_2T_x$ [34], Nb_2CT_x [35], V_2CT_x [30], Zr_3C_2 [36], and Lu_2CT_x [37]—have been reported, based on the current MAX precursors, and several more were theoretically predicted [36, 38, 39]. Because of their abundant elemental distribution and non-hazardous breakdown products, MXenes are widely used in environmental cleaning [40, 41]. This chapter covers the different environmental uses for materials based on MXenes.

5.3 MXENE-BASED MATERIALS FOR ENVIRONMENTAL APPLICATIONS

MXenes are particularly well suited to a range of environmental applications due to their hydrophilicity, high electric conductivity, and adsorptive, reductive, and antibacterial properties. According to Table 5.1, MXenes have demonstrated enormous potential in applications such as water treatment, desalination, sensing, photocatalysis, antimicrobial protection, and biofouling. For the successful detection of diverse contaminants, MXene-based adsorbents [32], photocatalysts [42], membranes [43], and sensors [44] have been used.

5.3.1 MXENES AS PHOTOCATALYSTS

Large functional groups; a negative surface charge; a variable bandgap energy (0.92–1.75 eV); a high surface/volume ratio; chemical, plasmonic, and thermoelectric capabilities; and other characteristics make these materials unique, as well as their high stability and visible light activity [53, 54]. MXenes represent a new and fascinating family of 2D semiconductor materials that has garnered a lot of research attention. MXenes are the most widely used photocatalytic materials due to their non-toxicity, superior electronic structure, great optical properties, and considerably efficient anisotropic mobility of electrons (e^-) and holes (h^+). Furthermore, despite having enormous potential for photocatalysis, MXenes are not currently being employed as photocatalytic materials because they lack the characteristics of semiconductors [55, 56].

TABLE 5.1
Various Environmental Applications of MXenes

MXene	Use	Reference
Ti_3C_2/TiO_2	Photocatalyst	[45]
$Ti_3C_2T_x$	Adsorbent	[46, 47]
$Ti_3C_2T_x$	Membrane	[48, 49]
$Ti_3C_2T_x/Ti_3C_2$	Sensor	[47, 50]
$Ti_3C_2T_x$	Solar desalination	[51, 52]

A 2D $Ti_3C_2T_x$ membrane was created to be used in a synergistic photocatalysis method to recover Ag^+ ions and break down rhodamine B (RhB). The photothermal concentration impact of the solution enhanced the rate of the photocatalytic reaction. In the meantime, the $Ti_3C_2T_x$ MXene (TCM) membrane was uniformly coated with Ag nanoparticles that had been photo-reduced from Ag^+ ions. The surface plasma impact caused by Ag nanoparticles considerably increased the efficiency of converting solar energy into heat. It simultaneously boosted photocatalysis by improving the absorption of light and the migration of photoinduced charges. Because of the synergistic photo-catalysis/thermal contributions, both the recovery of Ag^+ ions and the degradation of RhB were greatly enhanced. Figure 5.2d proposes a synergistic photocatalytic/thermal impact for simultaneous Ag^+ recovery and RhB elimination on the membrane. The initial XRD patterns shown in Figure 5.2a showed that the Ti_3AlC_2 precursor crystallised and that a TCM membrane without a PTFE substrate was formed. The TEM image in Figure 5.2b shows the graphene-like $Ti_3C_2T_x$ nanosheets with creases at the edges. The SEM photos Figure 5.2c may be used to further support the $Ti_3C_2T_x$ and TCM sample's planar characteristics. According to a hypothetical reaction mechanism proposed in Figure 5.2d, Ag^+ recovery and RhB degradation using MXene membranes may be done simultaneously by photocatalysis/thermal contributions [57].

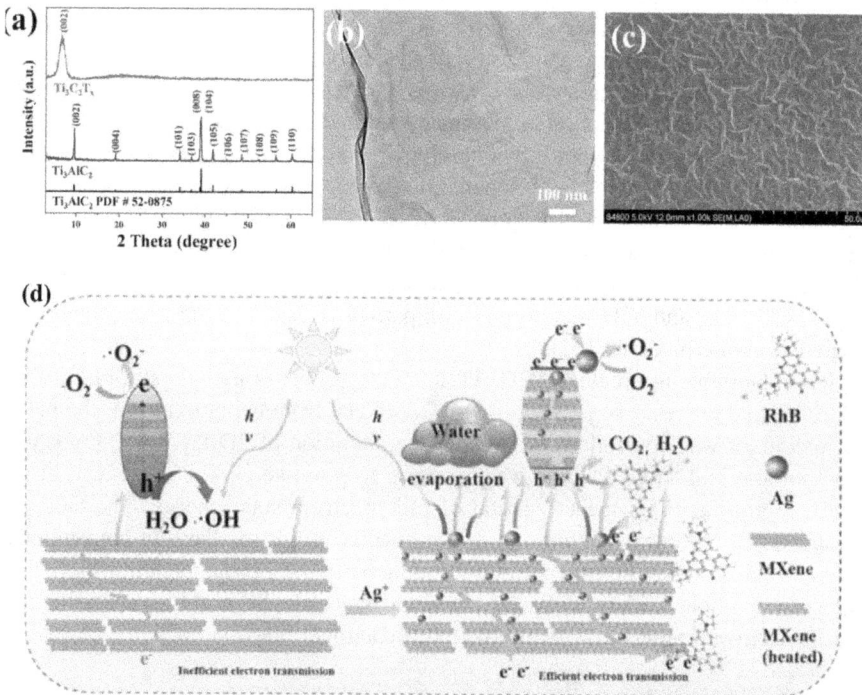

FIGURE 5.2 (a) XRD image of a Ti_3AlC_2-TCM membrane with no PTFE. (b) TEM images of $Ti_3C_2T_x$ nanosheets. (c) SEM images of TCM membranes. (d) Synergistic impact on the MXene membrane for the removal of Ag^+ ions and RhB.

A variety of ternary photocatalysts adorned with Au plasmonic nanoparticles and made of graphitic carbon nitride (gCN) and Ti_3C_2 MXenes have been developed to break down the colourless pharmaceutical contaminant cefixime when exposed to visible light. The quantity of Ti_3C_2 MXene used to create these photocatalysts was altered. When used as co-catalysts, plasmonic Au NPs and Ti_3C_2 MXenes can significantly improve gCN's photocatalytic efficiency in the elimination of the colourless pharmaceutical contaminant [58].

A simple hydrothermal post-annealing process was used to transform $Ti_3C_2T_x$ into a $TiO_2/Ti_3C_2O_x$ composite with a characteristic urchin-like shape. The urchin-like $TiO_2/Ti_3C_2O_x$ hybrid with a large specific surface area could be produced by partially removing the surface groups of $Ti_3C_2T_x$ and uniting them into -O adsorption. TiO_2 nanowires could be synthesised in situ using $Ti_3C_2T_x$ as a titanium source. Both the photocatalytic hydrogen synthesis and levofloxacin (LEV) degradation processes exhibit exceptional performance from the generated TiO2/Ti3C2Ox. The disordered peak (D-band) and graphite structure peak (G-band) in the $Ti_3C_2T_x$ Raman spectra were both connected to carbon and were found at 1362 and 1602 cm^{-1}, respectively [59]. The abundance of defects in $TiO_2/Ti_3C_2O_x$ were indicated by the D-band's high intensity, which made active H_2O adsorption sites possible and promoted the photocatalytic process [60]. The e- and $O_2^{\cdot-}$ were significant ROS in the levofloxacin (LEV) degradation reaction due to the likelihood that the e$^-$ concentrated on $Ti_3C_2O_x$ may react with dissolved oxygen to create superoxide radical anions ($O_2^{\cdot-}$) [61]. To fully utilise solar energy and hasten the decomposition of tetracycline, Gao and colleagues designed and created an effective Z-scheme heterojunction photocatalyst based on the cooperative action of MXene quantum dots (MQDs) and oxygen vacancies (OVs). A straightforward ethylene glycol solvothermal technique was used to introduce the MXene into the Z-scheme heterojunction of gC_3N_4 and BiOBr. The formation of a structural coupling process between MQDs and OV-BiOBr was anticipated to create an asynergistic effect, inhibit electron reflux, improve charge separation efficiency, and speed up carrier transport. The degradation process of TC-HCl by H$^+$, $O^{\cdot2-}$, $^1O_2$, and $\cdot OH$ was proposed in particular based on the determination of degradation intermediates [62].

It was possible to create 2D/2D $BiOIO_3/Ti_3C_2$ MXenes using a straightforward hydrothermal process. To prevent the Ti_3C_2 MXene from being oxidised, the pH of the precursor was controlled. In situ surface alteration of 2D layered $BiOIO_3$ in a 2D photothermal Ti_3C_2 MXene was expected to promote carrier movement at the interface and facilitate the utilisation of full-spectrum solar energy. The tactic of using a Ti_3C_2 MXene as a co-catalyst in particular avoids the flaws of conventional heterostructures with the expense of the conductivity of electron reduction as well as valence band hole oxidation of semiconductor materials. Several organic pollutants can be destroyed photocatalytically with this chemical [63].

5.3.2 MXENES AS ADSORBENTS

MXenes have already been employed as new adsorbents for removing cationic dyes from aqueous solutions since they have a lot of specific surface area, layers, and negatively charged surfaces [64, 35]. Table 5.2 provides a quick analysis of the

TABLE 5.2

Adsorption Performance of Various 2D Adsorbents

Adsorbent	Co (mg L^{-1})	adsorption capacity, qm (mg g^{-1})	Reference
Metal–organic framework	277	27.7	[66]
Activated charcoal	0–200	181	[67]
Powdered activated carbon	5–20	231	[68]
Sepiolite clay	180	106	[69]
Graphene	50	200	[70]
MXene	0–22.2	241	[70]

efficiency of MXene-based adsorbents in contrast to conventional 2D materials. The superior stability of MXene-based membranes and adsorbents is what accounts for their superior performance when compared with other 2D materials. On the surface of MXenes, there are a number of active functional groups, like -O, -OH, and -F groups. Compared with other 2D materials, these groups have a great chance to trap the most targeted pollutants [65].

MXenes have recently been used to remove dyes. For example, Kadhom et al. used a $Ti_3C_2T_x$ MXene as an adsorbent to remove malachite green (MG) from a simulated dye solution. Figure 5.3 displays SEM pictures of the MXene at various sizes, with (a) being an organoid material at a scale of 10 µm and (b) and (c) being images at 5 and 3 µm, respectively. The MXene particles in the image could be anywhere between 3 and 10 µm. According to image (c), the layer may have a thickness of a few nanometres, providing a large contact area for adsorption. In Figure 5.3d, where quantities of the adsorbent varied from 0 to 0.1 g, the dosage relationship between the MXene and the dye removal ratio was analysed. The graphic makes it clear that the adsorption ratio rose as the amount of MXene increased. This might be explained by the rise in adsorption sites as the MXene mass rose. But as the adsorbent mass increased, the removal ratio enhancement was reduced. It first increased quickly in the beginning and then slowed down [71].

MXene@NiFe-LDH (MNFL), a unique three-dimensional, flower-shaped composite, was created using a straightforward hydrothermal process and used to remove common harmful contaminants like Cr^{6+} and 1-naphthol from water. The physicochemical parameters of MNFL were changed by the addition of a particular amount of LDH, which enhanced the F-MXene's specific surface area, efficiency of removal, and eco-friendly attributes. Particularly, MNFL-60 demonstrated superior solid–liquid separation capabilities, quick kinetics, and effective regeneration. The MNFL-60 showed excellent efficacy and removing capability for 1-naphthol via the interaction with -N/O-containing groups and for Cr^{6+} by electrostatic attraction, complexation, as well and partial reduction of Cr^{6+} to Cr^{3+}. The interfering particular surface area (64.04 m^2/g) with the profusion of -C-, -N-, or -O-containing groups allowed for this [72].

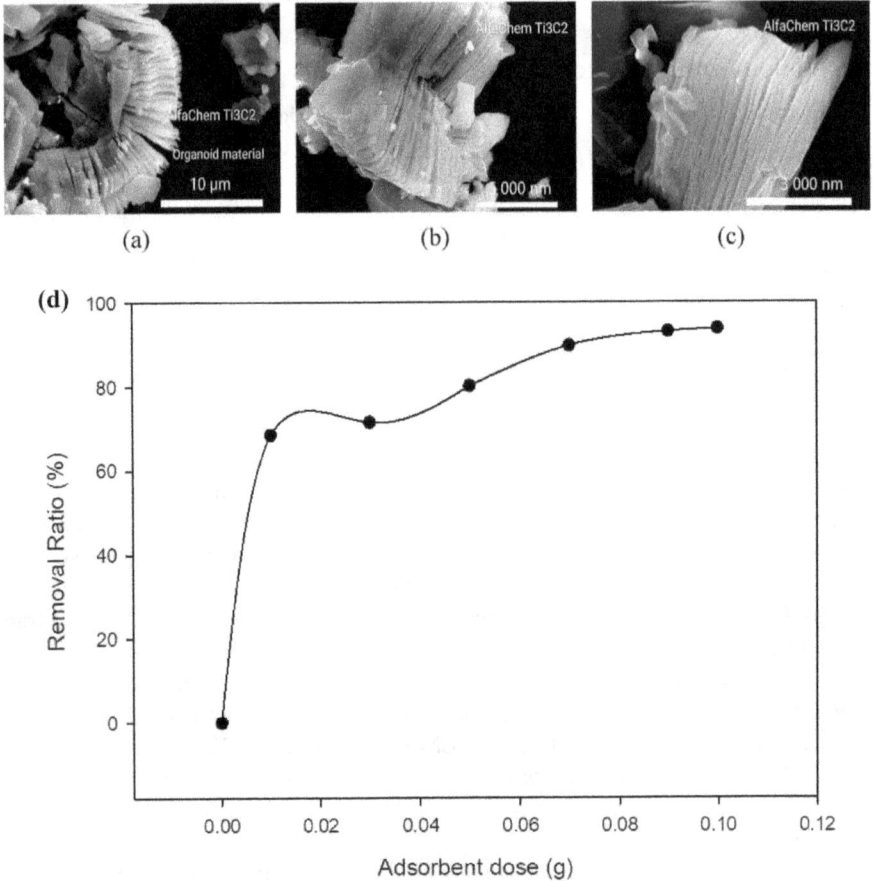

FIGURE 5.3   (a–c) Various $Ti_3C_2T_x$ MXene SEM images. (d) The link between the removal ratio and adsorbent dosage. The tests were carried out with a pH of 7, a dye concentration of 5 ppm, and a duration of 60 minutes.

By crosslinking EDTA with the MXene, a new adsorbent was created. The C–N, C–F, OH, and C–O groups in the MXene polymer composite act as sites of sorption for lead, holmium, and hydroquinone. Lead, holmium, and hydroquinone removal from the EDTA-cross-linked MXene polymer composite exhibited remarkable efficiency. Lead (II), holmium (III), and hydroquinone adhered to the Langmuirand PSO kinetic models when solubilised on composite, according to the sorption isotherm and kinetics [73].

Ijaz et al. developed the composite of $Ni_3(HITP)_2$/MXene/CS as an adsorbent for lead and methylene blue (MB). The MXene and a metal–organic framework were added to increase active sites even further. Due to -OH, -O, -F, and N groups present, the $Ni_3(HITP)_2$/MXene/CS composite efficiently absorbed MB and Pb^{2+}

by H-bonding, electrostatic attraction, and complexation interaction. The greatest adsorption capacity of lead (II) was measured at 298 K, 400 K, and 318 K, respectively. It was 448.93 mg g^{-1} at 400 K. On the other hand, methyl blue had a maximum adsorption capacity of 424.99 mg g^{-1} at 298 K, 354.03 mg g^{-1} at 308 K, and 251.78 mg g^{-1} at 318 K [74].

It is claimed that MXene/PNIPAM composite hydrogels with cyclodextrin encapsulation exhibit superior mechanical qualities, a high adsorption efficiency, and are oxidation resistant. In order to create a physical barrier that inhibited the oxidative degradation of a MXene, the silane coupling agent 2-cyanoethyltriethoxysilane (CTES) was used to encase the MXene. A thermally sensitive composite hydrogel adsorbent having a high rate of adsorption and no secondary contamination was generated using this technique. This improved the amount of active groups and the oxidation resistance of the MXenes. The modified MXene was in situ polymerised with the N-isopropylacrylamide (NIPAM) monomer to produce a thermosensitive MXene–CTES–CD/PNIPAM nanocomposite hydrogel adsorbent. This hydrogel can be employed as a thermosensitive adsorbent and has good phenol adsorption efficiency [75].

5.3.3 MXENES AS MEMBRANES

Recent research has demonstrated the significant potential of MXene-based membrane applications in organic solvent filtering and water purification that results from the efficient manufacturing of a variety of MXenes, as shown Table 5.3 [76].

For the first time, an MXene, BiOBr, and Bi_2MoO_6 were coupled to create a ternary heterojunction of BiOBr/Bi_2MoO_6@MXene in situ using a one-pot hydrothermal technique, as shown in Figure 5.4a. Following interface self-assembly, the photocatalytic membrane was built. The ternary heterojunction's band gap was lower (2.13 eV) than the binary heterojunction as a result of the interaction between the MXene, BiOBr, and Bi_2MoO_6. Because of this, the composite membrane was very good at separating and transferring charge carriers as well as capturing visible light. Therefore, compared to pure MXene membrane and the Bi_2MoO_6@MXene composite membrane, the BiOBr/Bi_2MoO_6@MXene composite membrane showed greater photocatalytic performance in the degradation of organic contaminants. TEM and HRTEM were used to examine the heterojunction microstructure. The binary heterojunction's TEM picture in Figure 5.4b shows that several Bi_2MoO_6 molecules were

TABLE 5.3
Recent Uses of MXenes as Membranes

MXene application	Year	Ref.
Organic solvent recovery, antibiotic separation, wastewater treatment	2020	[77]
Water purification and pollutant removal	2021	[78]
Oil/water separation, toxic metal removal from wastewater		[79]

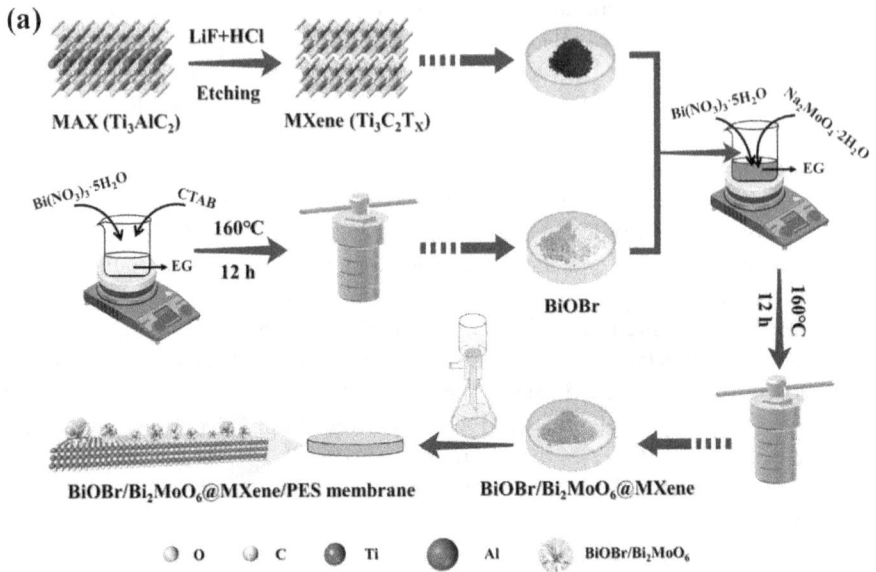

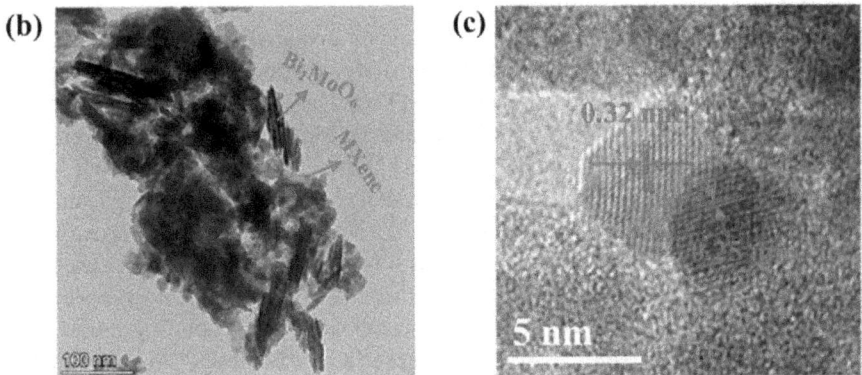

FIGURE 5.4 (a) BiOBr/Bi$_2$MoO$_6$@MXene/PES composite membrane synthesis. Bi$_2$MoO$_6$@MXene TEM image (b) and (c) HRTEM image.

embedded in or covered by the MXene nanosheets. Bi$_2$MoO$_6$ and the MXene's close interfacial contact was further illustrated by the HRTEM image showing the binary heterojunction's lattice distributions Figure 5.4c. The permeability of the membrane was increased by adding a mixture of Bi$_2$MoO$_6$ and BiOBr and reached 1296.91 Lm^{-2} h^{-1} bar^{-1}. Otherwise, the complete removal ratio for the antibiotic TC and CIP was below 90%, and the complete removal rate for the dye CR was above 98% for the BiOBr/Bi$_2$MoO$_6$@MXene composite membrane. The membrane showed great self-cleaning skills through the self-cleaning performance test and maintained outstanding permeability as well as selectivity during five cycles [80].

The MXene–PANI/PES composite membranes were made using a MXene and PANI. Li et al. firstly developed the MXene and PANI using their separate forms before joining them by electrostatic assembly. As a result, PANI was able to adhere to the MXene and lay down an effective conducting network onto the membrane. The membrane was created by mixing MXene–PANI/PES with PES and using non-solvent-induced phase separation (NIPS). The MXene–PANI/PES membrane surpassed the PES membrane in terms of the penetrating flux of pure water by 200.9%, while those for Congo Red and methyl blue dyes remained at 99.1% and 98.4%, respectively. Bovine serum albumin (BSA) was maintained above 99% [81].

Because of the effects of this group of "forever chemicals" on both people and water resources, per- and polyfluoroalkyl substances (PFASs) have gained international attention. The use of membrane technologies to separate PFASs from water supplies is extremely difficult. For PFAS, high permeance as well as a rejection membrane are necessary. In one study, it was proposed to facilitate the avoidance of short-chain PFASs by interfacial polymerisation of negative-charged polyamide (PA) membranes. By enhancing surface hydrophilicity and surface roughness, layered MXene has been demonstrated to be essential in enhancing membrane performance. As a result, water permeability rose from 8.65 L m^{-2} h^{-1} bar^{-1} to 12.16 L m^{-2} h^{-1} bar^{-1}. High rejection rates of 96.85% and 93.35% for PFHxS and PFHxA, respectively, were observed for the MXene-modified PA membranes [82].

The goal of the Zeng research group is to introduce CN to the MXene nanosheet structure in order to increase the membrane's hydrophilic properties and anti-fouling capability by harnessing the synergistic effects of 2D materials. Due to its exceptional metal conductivity, MXenes can be used as adjunct catalysts to boost CN's photocatalytic efficiency and enhance its potential to be recycled for the removal of colours and antibiotics from wastewater. The development of the g-C_3N_4@MXene/PES (CN-MX) 2D/2D composite membrane using vacuum filtration raises ideas for further multi-dimensional water treatment membranes. The CN-MX membrane had a permeability of up to 1790 Lm^{-2} h^{-1} bar^{-1}, which was higher than that of the pure MXene membrane. Because of the metal conductivity of the MXene, this membrane demonstrated exceptional removal effects on various contaminants in water [83].

In order to adsorb and dynamically intercept Sb ions from wastewater, Wan et al. recommended employing a self-cleaning, free-standing lamellar MXene@CNF@FeOOH (MCF) membrane. The $Ti_3C_2T_x$ MXenes and -FeOOH nanorods were in situ anchored onto an interlayer of MXenes. CNFs were used as functional scaffolds to link with MXenes. In addition to providing a wealth of active sites for the efficient collection of target pollutants. The interlaced CNFs and MXenes function as multilayered substrates with exceptional mechanical strength and flexibility. The greatest adsorption capabilities of the as-assembled MCF membranes, 19.9 and 18.1 mg/g for Sb^{3+} and Sb^{5+}, respectively, show the stable static removal effects of the membranes. The rapid photocatalytic breakdown of the MB dye and the continuance of water flow after decontamination confirm the exceptional capacity of the membranes to self-clean by using reactive free radicals created in the nanoconfined gaps [84].

5.3.4 MXenes as Electrochemical Sensors

Because of the excellent electrical and optical properties of MXenes, MXene-based nanomaterials have generated a lot of attention in the field of sensing for numerous applications, including sensing of pesticides, heavy metals, etc. [85, 86]. MXenes' electrical conductivity, ion transport capabilities, hydrophilicity, and distinctive surface chemistry have all been established in studies [87]. Numerous emerging toxins, including heavy metals, pesticides, organic contaminants, and other ions, have been detected using biosensing technology. Table 5.4 includes MXene-based electrochemical sensors for the detection of environmental pollutants.

Electrospun $MnMoO_4$ nanofibers and single-layered, delaminated nanosheets of an MXene were combined to form hybridised 1D–2D $MnMoO_4$–MXene nanocomposites. This nanocomposite enhanced a decrease in the overpotential of a glassy carbon electrode (GCE) by increasing the oxidation peak current density, allowing for effective electrochemical determination of HQ and CC. This nanocomposite demonstrates exceptional catalytic strength, dependability, repeatability, and robust adsorption between the electrode material and hazardous pollutants (HQ and CC) under aqueous circumstances. Figure 5.5 depicts the hybridised 1D–2D $MnMoO_4$–MXene–GCE construction approach for the electrochemical analysis of HQ and CC. Following the standard synthesis procedure, surface functionalisation and hydrothermal treatment were used to tag the multilayer MXene with $MnMoO_4$ NFs. Additionally, the fibre network was tagged as a result of surface functionality caused by the MPA. Combining 1D and 2D networks prohibited the construction of multilayer structures from aggregating. The prepared hybrid nanocomposites were successfully loaded onto a GCE as well. To investigate the morphological benefits of sensing interfaces, $MnMoO_4$ NFs and MXene nanocomposites were imaged using SEM and TEM [93].

By employing Cu_2O octahedrons as sacrificial templates and employing a galvanic replacement and disproportionation process, octahedral nanosheet-assembled PtCu alloy nanocages were created. Then, by utilising its diverse surface chemistries, Ti3C$_2$T$_x$ was selected as a conductive reinforcement addition to immobilise the PtCu through electrostatic attraction. For very sensitive detection of endocrine-disrupting chemicals, the glassy carbon electrode (GCE) was coated with the PtCu–Ti$_3$C$_2$T$_x$ nanocomposite. The produced Cu_2O polyhedrons exhibited a normal octahedral

TABLE 5.4
MXene-Based Sensors for the Detection of Various Pollutants

Sensor	Pollutants	Limit of detection (µM)	Ref.
AuNPs-PDDA-Ti$_3$C$_2$T$_x$	Nitrite	0.059	[88]
2D BP-Ti$_3$C$_2$T$_x$	Naphthalene acetic acid	0.0016	[89]
PtNPs/Ti$_3$C$_2$T$_x$	Bisphenol A	0.032	[90]
Nb$_2$CT$_x$/Zn-Co-NC	4-Nitrophenol	0.07	[91]
MWCNTs-Ti$_3$C$_2$	Hydroquinone catechol	0.0066 0.0039	[92]
Alk-Ti$_3$C$_2$	Cd^{2+}, Pb^{2+}, Cu^{2+}, Hg^{2+}	0.098, 0.041, 0.032, 0.130	[54]

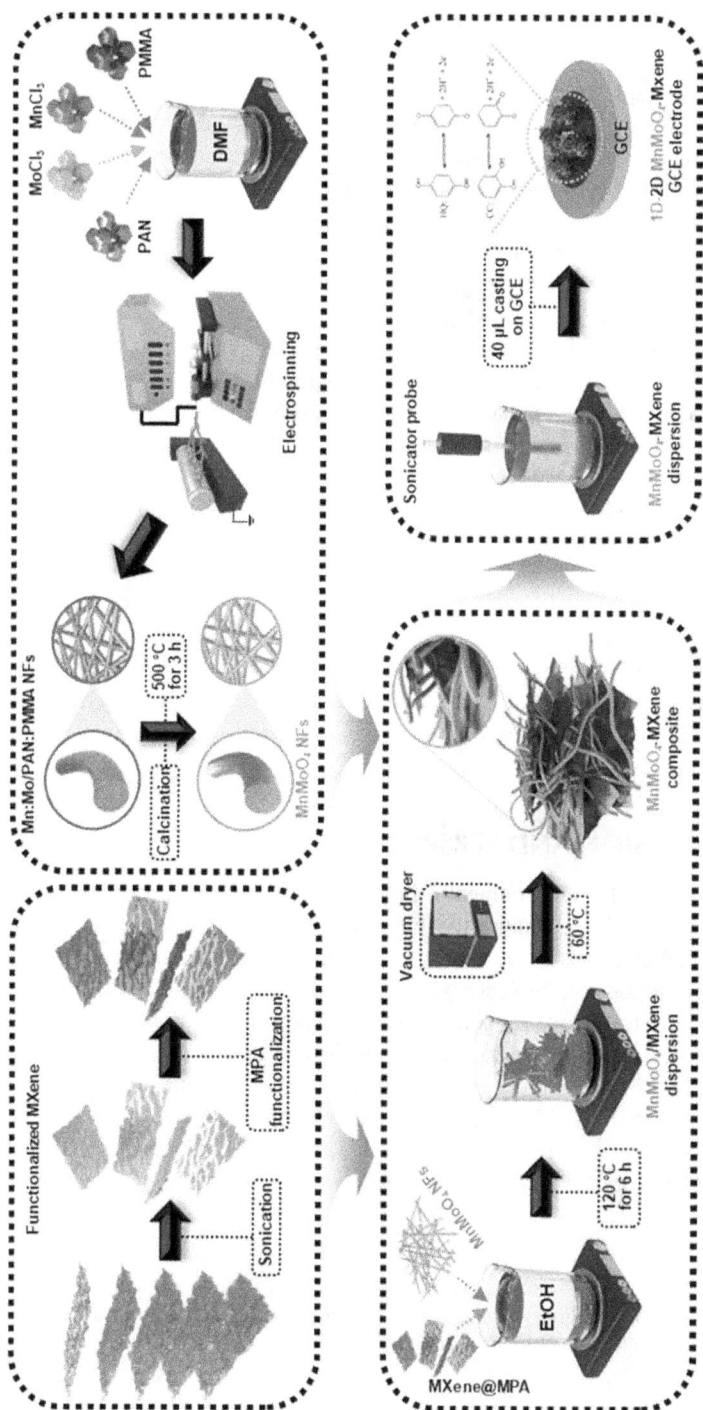

FIGURE 5.5 Steps involved in developing a 1D–2D MnMoO$_4$–MXene-based sensor. Copyrights 2022, Elsevier. Reprinted with permission from Ref. [93].

shape with smooth surfaces and solid internal structure, as seen by the TEM images. Each Cu_2O particle was uniformly enclosed in a dendritic-like PtCu constructed shell when the Cu_2O was mixed with an H_2PtCl_6 aqueous solution as a result of a spontaneous galvanic replacement reaction. With ultrawide linear ranges as well as sub-nanomole detection limits, the hybrid sensor's sensitivity for electrochemical sensing of endocrine-disrupting contaminants in water was significantly increased [94].

For the purpose of detecting CO_2 at room temperature, Thomas et al. constructed Mo_2CT_x MXene sensors mounted on various substrates (glass, crystalline, or porous silicon). The working temperature ranged from 30 to 250 °C and CO_2 concentrations from 50 to 150 ppm for the gas sensing studies. In comparison to crystalline silicon sensors, porous silicon and glass sensors showed superior room temperature sensing responses and rapid reaction and recovery times under 50 ppm CO_2 gas [95]. Fe-MOF was in situ deposited using a straightforward hydrothermal technique on a Ti_3-C_2T_x-based MXene matrix, and the resulting material exhibited a potent electrochemical reaction to As^{3+}. There hasn't been any published study that tries to synthesise Fe-MOF/MXene via a hydrothermal process, as far as we know. This sensor based on MXene/MOF was capable of identifying As^{3+} using square wave anodic stripping voltammetry (SWASV) with exceptional sensitivity due to the presence of active sites and functional groups. The Fe-MOF/MXene electrode exhibited a potent As–OH bond and Fe–O–As coordination, as shown by XPS. At the Fe-MOF/MXene/glassy carbon electrode (GCE), the synergistic impact of Fe-MOF and MXene dramatically boosted the sensing of As^{3+}. Using square wave anodic stripping voltammetry, the Fe-MOF/MXene-modified electrode displayed an exceedingly high current response to As^{3+} with an incredibly low limit of detection (0.58 ng L^{-1}) and an incredibly high sensitivity (8.94 A (ng L^{-1})$^{-1}$ cm^2) [96].

5.4 CONCLUSION AND PERSPECTIVE

Future applications for environmental remediation should favour materials based on MXenes. MXenes are among the most innovative 2D materials created for a wide range of applications because of their distinctive characteristics, and adjustable structure and composition. MXenes were extensively investigated for uses in adsorption, membrane filtration, sensing, and photocatalytic degradation in environmental remediation. MXenes face two major challenges: a high production cost and a limited yield. MXenes are currently only generated in limited quantities in laboratories. In order to develop this field of study and open new avenues for the use of MXenes on a commercial scale, it is essential to design a system that generates cost-effective, efficient, and environmentally friendly MXenes on a bigger scale.

REFERENCES

1. Ali, Imran. "New generation adsorbents for water treatment." Chemical Reviews 112, no. 10 (2012): 5073–5091.
2. Santhosh, Chella, Venugopal Velmurugan, George Jacob, Soon Kwan Jeong, Andrews Nirmala Grace, and Amit Bhatnagar. "Role of nanomaterials in water treatment applications: A review." Chemical Engineering Journal 306 (2016): 1116–1137.

3. Laws, E. A. "Aquatic pollution: An introductory text 4th edition." (2017). ISBN: 978-1-119-30450-0.

4. Jin, Xinliang, Cui Yu, Yanfeng Li, Yongxin Qi, Liuqing Yang, Guanghui Zhao, and Huaiyuan Hu. "Preparation of novel nano-adsorbent based on organic—Inorganic hybrid and their adsorption for heavy metals and organic pollutants presented in water environment." Journal of Hazardous Materials 186, no. 2–3 (2011): 1672–1680.

5. Gothwal, Ritu, and Thhatikkonda Shashidhar. "Antibiotic pollution in the environment: A review." Clean—Soil, Air, Water 43, no. 4 (2015): 479–489.

6. Burgess, Joanna E., Simon A. Parsons, and Richard M. Stuetz. "Developments in odour control and waste gas treatment biotechnology: A review." Biotechnology Advances 19, no. 1 (2001): 35–63.

7. Zhang, Lei, Ying Li, Han Guo, Huihui Zhang, Ning Zhang, Tasawar Hayat, and Yubing Sun. "Decontamination of U (VI) on graphene oxide/Al2O3 composites investigated by XRD, FT-IR and XPS techniques." Environmental Pollution 248 (2019): 332–338.

8. Zhang, Zexin, Haibo Liu, Lei Liu, Wencheng Song, and Yubing Sun. "Effect of staphylococcus epidermidis on U (VI) sequestration by Al-goethite." Journal of Hazardous Materials 368 (2019): 52–62.

9. Li, Jian, Xin Li, and Bart Van der Bruggen. "An MXene-based membrane for molecular separation." Environmental Science: Nano 7, no. 5 (2020): 1289–1304.

10. Wang, Min, Wen Cheng, Tian Wan, Baowei Hu, Yuling Zhu, Xiaofei Song, and Yubing Sun. "Mechanistic investigation of U (VI) sequestration by zero-valent iron/activated carbon composites." Chemical Engineering Journal 362 (2019): 99–106.

11. Mei, Peng, Huihui Wang, Han Guo, Ning Zhang, Sailun Ji, Yapeng Ma, Jiaqi Xu et al. "The enhanced photodegradation of bisphenol a by TiO_2/C_3N_4 composites." Environmental Research 182 (2020): 109090.

12. Sun, Yubing, Dingkun Peng, Ying Li, Han Guo, Ning Zhang, Huihui Wang, Peng Mei, Alhadi Ishag, Hamed Alsulami, and Mohammed Sh Alhodaly. "A robust prediction of U (VI) sorption on Fe3O4/activated carbon composites with surface complexation model." Environmental Research 185 (2020): 109467.

13. Zhan, Xiaoxue, Chen Si, Jian Zhou, and Zhimei Sun. "MXene and MXene-based composites: Synthesis, properties and environment-related applications." Nanoscale Horizons 5 (2020): 235–258.

14. Guo, Shiying, Yupeng Zhang, Yanqi Ge, Shengli Zhang, Haibo Zeng, and Han Zhang. "2D V-V binary materials: Status and challenges." Advanced Materials 31, no. 39 (2019): 1902352.

15. He, Junshan, Lili Tao, Han Zhang, Bo Zhou, and Jingbo Li. "Emerging 2D materials beyond graphene for ultrashort pulse generation in fiber lasers." Nanoscale 11, no. 6 (2019): 2577–2593.

16. Pei, Jiajie, Jiong Yang, Tanju Yildirim, Han Zhang, and Yuerui Lu. "Many-body complexes in 2D semiconductors." Advanced Materials 31, no. 2 (2019): 1706945.

17. Naguib, Michael, Murat Kurtoglu, Volker Presser, Jun Lu, Junjie Niu, Min Heon, Lars Hultman, Yury Gogotsi, and Michel W. Barsoum. "Two-dimensional nanocrystals produced by exfoliation of Ti3AlC2." Advanced Materials 23, no. 37 (2011): 4248–4253.

18. Ming, Fangwang, Hanfeng Liang, Gang Huang, Zahra Bayhan, and Husam N. Alshareef. "MXenes for rechargeable batteries beyond the lithium-ion." Advanced Materials 33, no. 1 (2021): 2004039.

19. Wu, Leiming, Xiantao Jiang, Jinlai Zhao, Weiyuan Liang, Zhongjun Li, Weichun Huang, Zhitao Lin et al. "MXene-based nonlinear optical information converter for all-optical modulator and switcher." Laser & Photonics Reviews 12, no. 12 (2018): 1800215.

20. Yang, Sheng, Panpan Zhang, Faxing Wang, Antonio Gaetano Ricciardulli, Martin R. Lohe, Paul W.M. Blom, and Xinliang Feng. "Fluoride-free synthesis of two-dimensional titanium carbide (MXene) using a binary aqueous system." Angewandte Chemie 130, no. 47 (2018): 15717–15721.

21. Yu, Xue-fang, Yan-chun Li, Jian-bo Cheng, Zhen-bo Liu, Qing-zhong Li, Wen-zuo Li, Xin Yang, and Bo Xiao. "Monolayer Ti2CO2: A promising candidate for NH3 sensor or capturer with high sensitivity and selectivity." ACS Applied Materials & Interfaces 7, no. 24 (2015): 13707–13713.

22. Jaffari, Zeeshan Haider, Salahaldin MA Abuabdou, Ding-Quan Ng, and Mohammed JK Bashir. "Insight into two-dimensional MXenes for environmental applications: Recent progress, challenges, and prospects." Flat Chem 28 (2021): 100256.

23. Chen, Junyu, Qiang Huang, Hongye Huang, Liucheng Mao, Meiying Liu, Xiaoyong Zhang, and Yen Wei. "Recent progress and advances in the environmental applications of MXene related materials." Nanoscale 12, no. 6 (2020): 3574–3592.

24. Liang, Xiao, Arnd Garsuch, and Linda F. Nazar. "Sulfur cathodes based on conductive MXene nanosheets for high-performance lithium—Sulfur batteries." Angewandte Chemie 127, no. 13 (2015): 3979–3983.

25. Zhang, Peng, Lin Wang, Li-Yong Yuan, Jian-Hui Lan, Zhi-Fang Chai, and Wei-Qun Shi. "Sorption of Eu (III) on MXene-derived titanate structures: The effect of nanoconfined space." Chemical Engineering Journal 370 (2019): 1200–1209.

26. Kajiyama, Satoshi, Lucie Szabova, Keitaro Sodeyama, Hiroki Iinuma, Ryohei Morita, Kazuma Gotoh, Yoshitaka Tateyama, Masashi Okubo, and Atsuo Yamada. "Sodium-ion intercalation mechanism in MXene nanosheets." ACS Nano 10, no. 3 (2016): 3334–3341.

27. Tran, Minh H., Timo Schäfer, Ali Shahraei, Michael Dürrschnabel, Leopoldo Molina-Luna, Ulrike I. Kramm, and Christina S. Birkel. "Adding a new member to the MXene family: Synthesis, structure, and electrocatalytic activity for the hydrogen evolution reaction of V4C3T x." ACS Applied Energy Materials 1, no. 8 (2018): 3908–3914.

28. Jun, Byung-Moon, Min Jang, Chang Min Park, Jonghun Han, and Yeomin Yoon. "Selective adsorption of Cs+ by MXene (Ti3C2Tx) from model low-level radioactive wastewater." Nuclear Engineering and Technology 52, no. 6 (2020): 1201–1207.

29. Cao, Yang, Yu Fang, Xianyu Lei, Bihui Tan, Xia Hu, Baojun Liu, and Qianlin Chen. "Fabrication of novel CuFe2O4/MXene hierarchical heterostructures for enhanced photocatalytic degradation of sulfonamides under visible light." Journal of Hazardous Materials 387 (2020): 122021.

30. Wang, Lin, Liyong Yuan, Ke Chen, Yujuan Zhang, Qihuang Deng, Shiyu Du, Qing Huang et al. "Loading actinides in multilayered structures for nuclear waste treatment: The first case study of uranium capture with vanadium carbide MXene." ACS Applied Materials & Interfaces 8, no. 25 (2016): 16396–16403.

31. Szuplewska, Aleksandra, Dominika Kulpińska, Artur Dybko, Michał Chudy, Agnieszka Maria Jastrzębska, Andrzej Olszyna, and Zbigniew Brzózka. "Future applications of MXenes in biotechnology, nanomedicine, and sensors." Trends in Biotechnology 38, no. 3 (2020): 264–279.

32. Rasool, Kashif, Ravi P. Pandey, P. Abdul Rasheed, Samantha Buczek, Yury Gogotsi, and Khaled A. Mahmoud. "Water treatment and environmental remediation applications of two-dimensional metal carbides (MXenes)." Materials Today 30 (2019): 80–102.

33. Pandey, Ravi P., Kashif Rasool, Vinod E. Madhavan, Brahim Aïssa, Yury Gogotsi, and Khaled A. Mahmoud. "Ultrahigh-flux and fouling-resistant membranes based on layered silver/MXene (Ti3C2Tx) nanosheets." Journal of Materials Chemistry A 6, no. 8 (2018): 3522–3533.

34. Peng, Chao, Ping Wei, Xin Chen, Yongli Zhang, Feng Zhu, Yonghai Cao, Hongjuan Wang, Hao Yu, and Feng Peng. "A hydrothermal etching route to synthesis of 2D MXene (Ti_3C_2, Nb_2C): Enhanced exfoliation and improved adsorption performance." Ceramics International 44, no. 15 (2018): 18886–18893.

35. Cui, Ce, Ronghui Guo, Erhui Ren, Hongyan Xiao, Xiaoxu Lai, Qin Qin, Shouxiang Jiang, Hong Shen, Mi Zhou, and Wenfeng Qin. "Facile hydrothermal synthesis of rod-like Nb2O5/Nb2CTx composites for visible-light driven photocatalytic degradation of organic pollutants." Environmental Research 193 (2021): 110587.

36. Zhou, Jie, Xianhu Zha, Fan Y. Chen, Qun Ye, Per Eklund, Shiyu Du, and Qing Huang. "A two-dimensional zirconium carbide by selective etching of Al3C3 from nano-laminated Zr3Al3C5." Angewandte Chemie International Edition 55, no. 16 (2016): 5008–5013.

37. Bai, Xiaojing, Xian-Hu Zha, Yingjie Qiao, Nianxiang Qiu, Yiming Zhang, Kan Luo, Jian He et al. "Two-dimensional semiconducting Lu 2 CT 2 (T= F, OH) MXene with low work function and high carrier mobility." Nanoscale 12, no. 6 (2020): 3795–3802.

38. Ronchi, Rodrigo Mantovani, Jeverson Teodoro Arantes, and Sydney Ferreira Santos. "Synthesis, structure, properties and applications of MXenes: Current status and perspectives." Ceramics International 45, no. 15 (2019): 18167–18188.

39. Persson, Per OÅ, and Johanna Rosen. "Current state of the art on tailoring the MXene composition, structure, and surface chemistry." Current Opinion in Solid State and Materials Science 23, no. 6 (2019): 100774.

40. Shahzad, Asif, Mohsin Nawaz, Mokrema Moztahida, Jiseon Jang, Khurram Tahir, Jiho Kim, Youngsu Lim, Vassilios S. Vassiliadis, Seung Han Woo, and Dae Sung Lee. "Ti3C2Tx MXene core-shell spheres for ultrahigh removal of mercuric ions." Chemical Engineering Journal 368 (2019): 400–408.

41. Xie, Xiuqiang, Chi Chen, Nan Zhang, Zi-Rong Tang, Jianjun Jiang, and Yi-Jun Xu. "Microstructure and surface control of MXene films for water purification." Nature Sustainability 2, no. 9 (2019): 856–862.

42. Xie, Xiuqiang, and Nan Zhang. "Positioning MXenes in the photocatalysis landscape: Competitiveness, challenges, and future perspectives." Advanced Functional Materials 30, no. 36 (2020): 2002528.

43. Ihsanullah, Ihsanullah. "Potential of MXenes in water desalination: Current status and perspectives." Nano-Micro Letters 12 (2020): 1–20.

44. Kim, Seon Joon, Hyeong-Jun Koh, Chang E. Ren, Ohmin Kwon, Kathleen Maleski, Soo-Yeon Cho, Babak Anasori et al. "Metallic Ti3C2T x MXene gas sensors with ultra-high signal-to-noise ratio." ACS Nano 12, no. 2 (2018): 986–993.

45. Peng, Chao, Xianfeng Yang, Yuhang Li, Hao Yu, Hongjuan Wang, and Feng Peng. "Hybrids of two-dimensional Ti3C2 and TiO2 exposing {001} facets toward enhanced photocatalytic activity." ACS Applied Materials & Interfaces 8, no. 9 (2016): 6051–6060.

46. Shahzad, Asif, Kashif Rasool, Waheed Miran, Mohsin Nawaz, Jiseon Jang, Khaled A. Mahmoud, and Dae Sung Lee. "Two-dimensional Ti3C2T x MXene nanosheets for efficient copper removal from water." ACS Sustainable Chemistry & Engineering 5, no. 12 (2017): 11481–11488.

47. Fard, Ahmad Kayvani, Gordon Mckay, Rita Chamoun, Tarik Rhadfi, Hugues Preud'Homme, and Muataz A. Atieh. "Barium removal from synthetic natural and produced water using MXene as two dimensional (2-D) nanosheet adsorbent." Chemical Engineering Journal 317 (2017): 331–342.

48. Lu, Zong, Yanying Wei, Junjie Deng, Li Ding, Zhong-Kun Li, and Haihui Wang. "Self-crosslinked MXene ($Ti_3C_2T_x$) membranes with good antiswelling property for monovalent metal ion exclusion." ACS Nano 13, no. 9 (2019): 10535–10544.

49. Ren, Chang E., Kelsey B. Hatzell, Mohamed Alhabeb, Zheng Ling, Khaled A. Mahmoud, and Yury Gogotsi. "Charge-and size-selective ion sieving through Ti3C2T x MXene membranes." The Journal of Physical Chemistry Letters 6, no. 20 (2015): 4026–4031.
50. Zhu, Xiaolei, Bingchuan Liu, Huijie Hou, Zhenying Huang, Kemal Mohammed Zeinu, Long Huang, Xiqing Yuan, Dabin Guo, Jingping Hu, and Jiakuan Yang. "Alkaline intercalation of Ti3C2 MXene for simultaneous electrochemical detection of Cd (II), Pb (II), Cu (II) and Hg (II)." Electrochimica Acta 248 (2017): 46–57.
51. Li, Renyuan, Lianbin Zhang, Le Shi, and Peng Wang. "MXene Ti3C2: An effective 2D light-to-heat conversion material." ACS Nano 11, no. 4 (2017): 3752–3759.
52. Zhao, Jianqiu, Yawei Yang, Chenhui Yang, Yapeng Tian, Yan Han, Jie Liu, Xingtian Yin, and Wenxiu Que. "A hydrophobic surface enabled salt-blocking 2D Ti 3 C 2 MXene membrane for efficient and stable solar desalination." Journal of Materials Chemistry A 6, no. 33 (2018): 16196–16204.
53. Peng, Jiahe, Xingzhu Chen, Wee-Jun Ong, Xiujian Zhao, and Neng Li. "Surface and heterointerface engineering of 2D MXenes and their nanocomposites: Insights into electro-and photocatalysis." Chem 5, no. 1 (2019): 18–50.
54. Sun, Yuliang, Xing Meng, Yohan Dall'Agnese, Chunxiang Dall'Agnese, Shengnan Duan, Yu Gao, Gang Chen, and Xiao-Feng Wang. "2D MXenes as co-catalysts in photocatalysis: Synthetic methods." Nano-Micro Letters 11 (2019): 1–22.
55. Kuang, Panyong, Jingxiang Low, Bei Cheng, Jiaguo Yu, and Jiajie Fan. "MXene-based photocatalysts." Journal of Materials Science & Technology 56 (2020): 18–44.
56. Zhang, Ke, Danqing Li, Hongyang Cao, Quihui Zhu, Christos Trapalis, Pengfei Zhu, Xinhua Gao, and Chuanyi Wang. "Insights into different dimensional MXenes for photocatalysis." Chemical Engineering Journal 424 (2021): 130340.
57. Qingxiao, Ziping Zhang, Danyang Zhao, Lei Wang, Hui Li, Fang Zhang, Yuning Huo, and Hexing Li. "Synergistic photocatalytic-photothermal contribution enhanced by recovered Ag+ ions on MXene membrane for organic pollutant removal." Applied Catalysis B: Environmental 320 (2023): 122009.
58. Kumar, Ajay, Palak Majithia, Priyanka Choudhary, Ian Mabbett, Moritz F. Kuehnel, Sudhagar Pitchaimuthu, and Venkata Krishnan. "MXene coupled graphitic carbon nitride nanosheets based plasmonic photocatalysts for removal of pharmaceutical pollutant." Chemosphere 308 (2022): 136297.
59. Ran, Jingrun, Guoping Gao, Fa-Tang Li, Tian-Yi Ma, Aijun Du, and Shi-Zhang Qiao. "Ti3C2 MXene co-catalyst on metal sulfide photo-absorbers for enhanced visible-light photocatalytic hydrogen production." Nature Communications 8, no. 1 (2017): 13907.
60. Li, Ning, Yue Jiang, Chuanhong Zhou, Yan Xiao, Bo Meng, Ziya Wang, Dazhou Huang, Chenyang Xing, and Zhengchun Peng. "High-performance humidity sensor based on urchin-like composite of Ti3C2 MXene-derived TiO2 nanowires." ACS Applied Materials & Interfaces 11, no. 41 (2019): 38116–38125.
61. Wang, Zirong, Yue Zhang, Yiming Chen, Ping Wei, Hongjuan Wang, Hao Yu, Jianbo Jia, Kun Zhang, and Chao Peng. "Surface-O terminated urchin-like TiO2/Ti3C2Ox (MXene) as high performance photocatalyst: Interfacial engineering and mechanism insight." Applied Surface Science (2023): 156343.
62. Gao, Kexuan, Li-an Hou, Xiaoqiang An, Doudou Huang, and Yu Yang. "BiOBr/MXene/gC3N4 Z-scheme heterostructure photocatalysts mediated by oxygen vacancies and MXene quantum dots for tetracycline degradation: Process, mechanism and toxicity analysis." Applied Catalysis B: Environmental 323 (2023): 122150.

63. Liu, Bing, Xia Zhang, Jialong Chu, Fei Li, Caixia Jin, and Jing Fan. "2D/2D $BiOIO_3$/Ti_3C_2 MXene nanocomposite with efficient charge separation for degradation of multiple pollutants." Applied Surface Science 618 (2023): 156565.

64. Li, Kaikai, Guodong Zou, Tifeng Jiao, Ruirui Xing, Lexin Zhang, Jingxin Zhou, Qingrui Zhang, and Qiuming Peng. "Self-assembled MXene-based nanocomposites via layer-by-layer strategy for elevated adsorption capacities." Colloids and Surfaces A: Physicochemical and Engineering Aspects 553 (2018): 105–113.

65. Solangi, Nadeem Hussain, Rama Rao Karri, Nabisab Mujawar Mubarak, Shaukat Ali Mazari, Abdul Sattar Jatoi, and Janardhan Reddy Koduru. "Emerging 2D MXene-based adsorbents for hazardous pollutants removal." Desalination 549 (2023): 116314.

66. Lin, Shuo, Yufeng Zhao, and Yeoung-Sang Yun. "Highly effective removal of nonsteroidal anti-inflammatory pharmaceuticals from water by Zr (IV)-based metal—Organic framework: Adsorption performance and mechanisms." ACS Applied Materials & Interfaces 10, no. 33 (2018): 28076–28085.

67. Zhao, Yufeng, Jong-Won Choi, John Kwame Bediako, Myung-Hee Song, Shuo Lin, Chul-Woong Cho, and Yeoung-Sang Yun. "Adsorptive interaction of cationic pharmaceuticals on activated charcoal: Experimental determination and QSAR modelling." Journal of Hazardous Materials 360 (2018): 529–535.

68. Real, Francisco J., F. Javier Benitez, Juan L. Acero, and Francisco Casas. "Adsorption of selected emerging contaminants onto PAC and GAC: Equilibrium isotherms, kinetics, and effect of the water matrix." Journal of Environmental Science and Health, Part A 52, no. 8 (2017): 727–734.

69. Ngulube, Tholiso, Jabulani Ray Gumbo, Vhahangwele Masindi, and Arjun Maity. "An update on synthetic dyes adsorption onto clay based minerals: A state-of-art review." Journal of Environmental Management 191 (2017): 35–57.

70. Abu-Nada, Abdulrahman, Ahmed Abdala, and Gordon McKay. "Removal of phenols and dyes from aqueous solutions using graphene and graphene composite adsorption: A review." Journal of Environmental Chemical Engineering 9, no. 5 (2021): 105858.

71. Kadhom, Mohammed, Khairi Kalash, and Mustafa Al-Furaiji. "Performance of 2D MXene as an adsorbent for malachite green removal." Chemosphere 290 (2022): 133256.

72. Wang, Junyi, Yucheng Li, Njud S. Alharbi, Changlun Chen, and Xuemei Ren. "Coupling few-layer MXene nanosheets with NiFe layered double hydroxide as 3D composites for the efficient removal of Cr (VI) and 1-naphthol." Journal of Molecular Liquids 371 (2023): 121082.

73. Bukhari, Aysha, Irfan Ijaz, Ezaz Gilani, Ammara Nazir, Hina Zain, Sajjad Hussain, and Ayesha Imtiaz. "Highly rapid and efficient removal of heavy metals, heavy rare earth elements, and phenolic compounds using EDTA-cross-linked MXene polymer composite: Adsorption characteristics and mechanisms." Chemical Engineering Research and Design 194 (2023): 497–513.

74. Ijaz, Irfan, Aysha Bukhari, Ezaz Gilani, Ammara Nazir, Hina Zain, Awais Bukhari, Attia Shaheen, Sajjad Hussain, and Ayesha Imtiaz. "Functionalization of chitosan biopolymer using two dimensional metal-organic frameworks and MXene for rapid, efficient, and selective removal of lead (II) and methyl blue from wastewater." Process Biochemistry 129 (2023): 257–267.

75. Wang, Qian, Yukun Xiong, Jing Xu, Fuping Dong, and Yuzhu Xiong. "Oxidation-resistant cyclodextrin-encapsulated-MXene/Poly (N-isopropylacrylamide) composite hydrogel as a thermosensitive adsorbent for phenols." Separation and Purification Technology 286 (2022): 120506.

76. Raheem, Ijlal, Nabisab Mujawar Mubarak, Rama Rao Karri, Nadeem Hussain Solangi, Abdul Sattar Jatoi, Shaukat Ali Mazari, Mohammad Khalid, Yie Hua Tan, Janardhan Reddy Koduru, and Guilherme Malafaia. "Rapid growth of MXene-based membranes for sustainable environmental pollution remediation." Chemosphere (2022): 137056.

77. Feng, Xiaofang, Zongxue Yu, Runxuan Long, Xiuhui Li, Liangyan Shao, Haojie Zeng, Guangyong Zeng, and Yiheng Zuo. "Self-assembling 2D/2D (MXene/LDH) materials achieve ultra-high adsorption of heavy metals Ni2+ through terminal group modification." Separation and Purification Technology 253 (2020): 117525.

78. Ul Hassan, Muhmood, Sujeong Lee, Muhammad Taqi Mehran, Faisal Shahzad, Syed M. Husnain, and Ho Jin Ryu. "Post-decontamination treatment of MXene after adsorbing Cs from contaminated water with the enhanced thermal stability to form a stable radioactive waste matrix." Journal of Nuclear Materials 543 (2021): 152566.

79. Assad, Humira, Ishrat Fatma, Ashish Kumar, Savas Kaya, Dai-Viet N. Vo, Adel Al-Gheethi, and Ajit Sharma. "An overview of MXene-based nanomaterials and their potential applications towards hazardous pollutant adsorption." Chemosphere (2022): 134221.

80. Yang, Zhaomei, Qingquan Lin, Guangyong Zeng, Simiao Zhao, Guilong Yan, Micah Belle Marie Yap Ang, Yu-Hsuan Chiao, and Shengyan Pu. "Ternary hetero-structured BiOBr/Bi2MoO6@ MXene composite membrane: Construction and enhanced removal of antibiotics and dyes from water." Journal of Membrane Science 669 (2023): 121329.

81. Li, Nan, Tian-Jiao Lou, Wenyi Wang, Min Li, Li-Chao Jing, Zhi-Xian Yang, Ru-Yu Chang, Jianxin Li, and Hong-Zhang Geng. "MXene-PANI/PES composite ultrafiltration membranes with conductive properties for anti-fouling and dye removal." Journal of Membrane Science 668 (2023): 121271.

82. Ma, Jun, Yuanyuan Wang, Hang Xu, Mingmei Ding, and Li Gao. "MXene (Ti3T2CX)-reinforced thin-film polyamide nanofiltration membrane for short-chain perfluorinated compounds removal." Process Safety and Environmental Protection 168 (2022): 275–284.

83. Zeng, Guangyong, Zhenzhen He, Tao Wan, Tairan Wang, Zhaomei Yang, Yongcong Liu, Qingquan Lin, Yiheng Wang, Arijit Sengupta, and Shengyan Pu. "A self-cleaning photocatalytic composite membrane based on g-C3N4@ MXene nanosheets for the removal of dyes and antibiotics from wastewater." Separation and Purification Technology 292 (2022): 121037.

84. Wan, Keming, Timing Fang, Wenliang Zhang, Guomei Ren, Xiao Tang, Zhezheng Ding, Yan Wang, Pengfei Qi, and Xiaomin Liu. "Enhanced antimony removal within lamellar nanoconfined interspaces through a self-cleaning MXene@ CNF@ FeOOH water purification membrane." Chemical Engineering Journal 465 (2023): 143018.

85. Lakhdari, Delloula, Abderrahim Guittoum, Nassima Benbrahim, Ouafia Belgherbi, Mohammed Berkani, Yasser Vasseghian, and Nadjem Lakhdari. "A novel non-enzymatic glucose sensor based on NiFe (NPs)—Polyaniline hybrid materials." Food and Chemical Toxicology 151 (2021): 112099.

86. Vasseghian, Yasser, Elena-Niculina Dragoi, Masoud Moradi, and Amin Mousavi Khaneghah. "A review on graphene-based electrochemical sensor for mycotoxins detection." Food and Chemical Toxicology 148 (2021): 111931.

87. Deshmukh, Kalim, Tomáš Kovářík, and SK Khadheer Pasha. "State of the art recent progress in two dimensional MXenes based gas sensors and biosensors: A comprehensive review." Coordination Chemistry Reviews 424 (2020): 213514.

88. Wang, Yuhuan, Zhixing Zeng, Jianyu Qiao, Shuqing Dong, Qing Liang, and Shijun Shao. "Ultrasensitive determination of nitrite based on electrochemical platform of AuNPs deposited on PDDA-modified MXene nanosheets." Talanta 221 (2021): 121605.

89. Zhu, Xiaoyu, Lei Lin, Ruimei Wu, Yifu Zhu, Yingying Sheng, Pengcheng Nie, Peng Liu, Lulu Xu, and Yangping Wen. "Portable wireless intelligent sensing of ultra-trace phytoregulator α-naphthalene acetic acid using self-assembled phosphorene/Ti3C2-MXene nanohybrid with high ambient stability on laser induced porous graphene as nanozyme flexible electrode." Biosensors and Bioelectronics 179 (2021): 113062.

90. Rasheed, P. Abdul, Ravi P. Pandey, Khadeeja A. Jabbar, and Khaled A. Mahmoud. "Platinum nanoparticles/Ti3C2Tx (MXene) composite for the effectual electrochemical sensing of Bisphenol A in aqueous media." Journal of Electroanalytical Chemistry 880 (2021): 114934.

91. Huang, Runmin, Dan Liao, Zhenhua Liu, Jingang Yu, and Xinyu Jiang. "Electrostatically assembling 2D hierarchical Nb2CTx and zifs-derivatives into Zn-Co-NC nanocage for the electrochemical detection of 4-nitrophenol." Sensors and Actuators B: Chemical 338 (2021): 129828.

92. Huang, Runmin, Sisi Chen, Jingang Yu, and Xinyu Jiang. "Self-assembled Ti_3C_2/MWCNTs nanocomposites modified glassy carbon electrode for electrochemical simultaneous detection of hydroquinone and catechol." Ecotoxicology and Environmental Safety 184 (2019): 109619.

93. Ranjith, Kugalur Shanmugam, AT Ezhil Vilian, Seyed Majid Ghoreishian, Reddicherla Umapathi, Seung-Kyu Hwang, Cheol Woo Oh, Yun Suk Huh, and Young-Kyu Han. "Hybridized 1D–2D $MnMoO_4$—MXene nanocomposites as high-performing electrochemical sensing platform for the sensitive detection of dihydroxybenzene isomers in wastewater samples." Journal of Hazardous Materials 421 (2022): 126775.

94. Liu, Xian, Like Chen, Yang Yang, Liping Xu, Junyong Sun, and Tian Gan. "MXene-reinforced octahedral PtCu nanocages with boosted electrocatalytic performance towards endocrine disrupting pollutants sensing." Journal of Hazardous Materials 442 (2023): 130000.

95. Thomas, Tijin, Jesus Alberto Ramos Ramon, V. Agarwal, A. Álvarez-Méndez, JA Aguilar Martinez, Y. Kumar, and K. C. Sanal. "Highly stable, fast responsive Mo_2CTx MXene sensors for room temperature carbon dioxide detection." Microporous and Mesoporous Materials 336 (2022): 111872.

96. Xiao, Ping, Guodong Zhu, Xiaohong Shang, Bin Hu, Boshuang Zhang, Zhaoyu Tang, Jianmao Yang, and Jianyun Liu. "An Fe-MOF/MXene-based ultra-sensitive electrochemical sensor for arsenic (III) measurement." Journal of Electroanalytical Chemistry 916 (2022): 116382.

6 MXene-Based Biosensors
Next-Generation Emerging Materials for Detection

Akhilesh Babu Ganganboina, Indra Memdi Khoris, and Kenshin Takemura

6.1 INTRODUCTION TO MXENES

A type of two-dimensional material, MXenes have received a lot of interest recently because of their remarkable features and prospective uses in a variety of sectors, including biosensing. MXenes are a novel family of two-dimensional (2D) materials distinguished by their unusual mix of metallic conductivity, large surface area, and variable surface chemistry [1–3]. They were discovered in 2011 by Drexel University researchers in the United States who produced a novel class of materials by selectively etching the A element from stacked ternary carbides and nitrides known as MAX phases [4, 5].

6.1.1 Definition and Properties of MXenes

MXenes include 2D transition metals with a general formula $M_{n+1}X_nT_x$ with carbides, nitrides, and carbonitrides, in which M represents a transition metal, followed by X as carbon or nitrogen; n can be 1, 2, or 3, and T represents the surface functional groups such as hydroxyl, oxygen, or fluorine [6]. MXenes are multilayered, having transition metal atoms sandwiched between two carbon or nitrogen layers. Upon the removal of the transition metal, an empty layer forms between the transition metal and the carbon or nitrogen layers, giving rise to the peculiar features of MXenes [7, 8].

MXenes contain a number of features that make them appealing for a variety of applications. They have a large surface area, for example, which makes them suitable for energy storage, catalysis, and sensing applications. They are also metallically conductive, making them interesting for electrical and optoelectronic devices. Moreover, MXenes have variable surface chemistry, which enables for the alteration of surface characteristics to improve functioning [9–11].

6.1.2 Synthesis Methods of MXenes

Various techniques for synthesizing MXenes have been established, including etching, exfoliation, and hydrothermal synthesis. The etching process is the most frequent, and it entails selectively etching the A element from MAX phases using a

DOI: 10.1201/9781003366225-6

strong acid or fluoride salt. The MXene may then be cleaned and delaminated to create a 2D material with a large surface area.

Two-dimensional MXene materials are generated from three-dimensional layered structures of transition metals, forming MAX phases of carbides and nitrides. Layered MAX phases, with the general formula M_n+1AX_n, are arranged in a structure of hexagon with P63/mmc space group symmetry and exhibit remarkable metallic and ceramic characteristics. More than 70 MAX phases have been documented up to this point [12, 13]. In addition, around 40 MXenes have been generated experimentally, starting from $Ti_3C_2T_x$ and Ti_2CT_x, followed by V_2CT_x, Nb_2CT_x, $Nb_4C_3T_x$, $Ta_4C_3T_x$, $(Mo_2/3Sc_{1/3})_2AlC$, $Hf_3C_2T_x$, and dozens of other types of MXenes [14]. MXenes have an empiric expression of $M_{n+1}X_nT_x$, where "M" indicates the transition metals, such as Ti, Nb, V, and so on, and "X" and "T_x" symbolize the carbide and/or nitride and the surface's functional residue, respectively. The value n = 1, 2, 3 denotes the number of layers, resulting in 21, 32, and 43 MXene structures. Thus far, MXenes have already been synthesized from MAX precursors using a top-down method and a wet chemical etching process [15]. One of the widely known methods is the minimum intensive layer delamination (MILD) method. It is done by in situ fluoride salts (HCl + LiF). It is known as water-absent etching in a polar solvent. Another major method for delaminating the MAX phase is by reacting with various hydrofluoric (HF) acid solutions. During the etching process, the 3D MAX phase is typically submerged in HF-containing solutions with continuous stirring. Several factors such as temperature, reaction time, pH, and etchant concentration affect the end products [16]. Several other alternatives such as alkali etching, halogen-based etching, and solid-state etching with molten salts have been used to create MXenes. For example, an Nb_2C MXene is made by etching Nb_2AlC MAX powders in mild HF solution with an elevated temperature and incubation stirring time, succeeded by a washing phase. The accordion-like architecture and microscopic flakes of a fluoride-based etching process are positives, but the downside is its high toxicity and dangerous nature. While avoiding fluoride-based etching, some oxides and hydroxides in alkali etching hinder the aluminum removal, lowering the quality of MXenes [17]. As a result, depending on the desired quality of MXenes and the eventual use, there are several approaches to their synthesis.

Figures 6.1a and 6.1b provide an illustration for the direct delamination of MAX phase either by the MILD technique (route 1) or by liquid salt etching (route 2). The resulting product is a few-layered MXene sheet with nanometer thickness but with a wide lateral dimension down to a few microns [18]. Through intercalation of various organic molecules, then, the thickness was reduced down to a few layers or a single layer of MXene. The bonding of M–X layers can be ionic, covalent, and metallic, whereas the bonds in M–A layers are merely metallic [19]. The relative difference between the bonding of the layer atoms and the reactivity of the metals to the fluoride, it is feasible to chemically remove the A layer from the interlayered 3D phase.

Foreign species (e.g., H_2O, Li^+ ions, etc.) can interleave between the structure of the MXene layers and modulate the inter-spacing due to the layered structure and wettability. It is consequently critical to comprehend the distinctions between c-LP, interlaying space, and d-spacing. When addressing the structure of MXenes, the basic concept of multiple lattice parameters might be confusing in some circumstances. An SEM picture of Ti_3AlC_2 MAX is shown in Figure 6.1ci, whereas an SEM

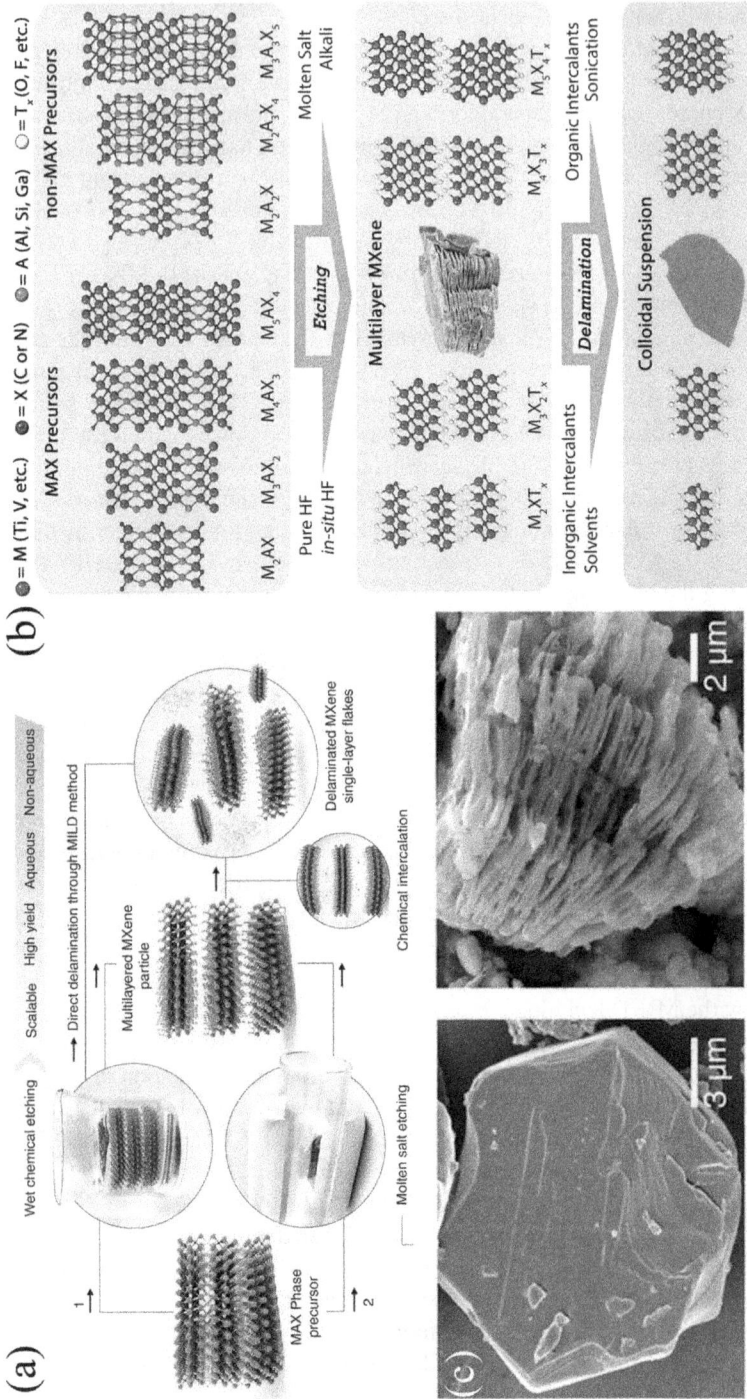

FIGURE 6.1 (a) Schematic illustrating several MXene etching techniques, as well as depictions of future intercalation and delamination phases [18]. (b) MAX phase selective etching followed by multilayer MXene synthesis [19]. (c) Scanning electron microscope (SEM) picture of a hexagonal-shaped Ti₃AlC₂ MAX phase crystal and SEM image of a Ti₃C₂Tₓ MXene particle generated from Ti₃SiC₂ by selective etching of Si layers in molten salt [18].

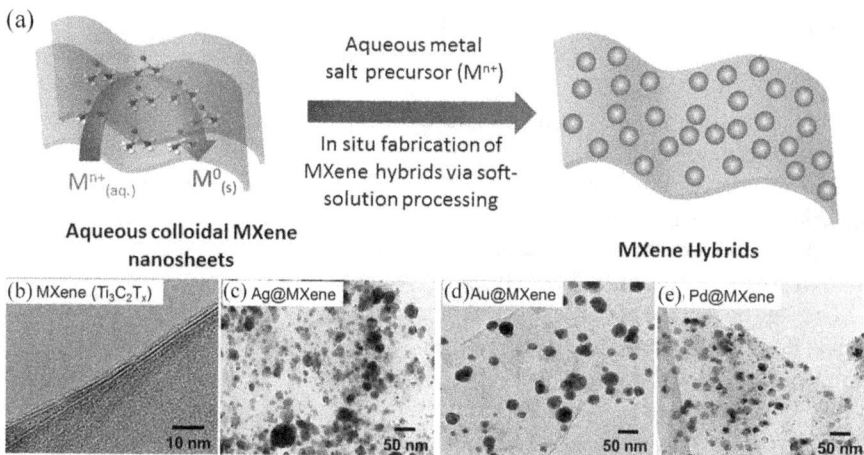

FIGURE 6.2 (a) Illustration of the hybrid formation of the MXene by soft-solution processing for the preparation of silver, gold, or palladium@MXene (Ag, Au, or Pd@$Ti_3C_2T_x$). (b) TEM analysis of the morphological observation of (c) Ag, (d) Au, and (e) Pd in MXenes forming hybrid nanosheets [20].

image of the matching $Ti_3C_2T_x$ MXene in Figure 6.1cii reveals the typical accordion-like shape of MXenes, which is another sign of MXene creation.

The process of exfoliating the nanosheet of the Ti_3C_2 titanium carbide (MXene) had been carried out using Ag, Au, and Pd sourcing from the soft solution reduced onto these nanosheets, which is depicted in Figure 6.2a. Figures 6.2b–6.2d illustrate the cross-section of the multilayered MXene framework and the scattered formed nanoparticles of Ag, Au, or Pd on the MXene 2D structure [20]. It is obvious from the TEM analysis with high-resolution image that the NPs were firmly bonded to the MXene. Several metals produced in this work have varied NP sizes and shapes. The diameters of the rounded Ag and Au NPs range from 10 to 70 and 40 to 50 nm, respectively. As expected from their reactiveness, silver ions (Ag^+) underwent a fast nanoparticle formation during ultrasonication, even when employing a modest precursor concentration (0.1 mM $AgNO_3$ solution), making it more difficult to regulate the size of Ag NPs in MXenes. However, under the same test conditions, a more uniform size range of Au NPs (average size: 40–50 nm) was observed, indicating a slower decrease in Au^{3+} ions [20]. Excitingly, it was clearly observed that the Pd@MXene hybrid structure showed a planar nanoparticle resembling a sheet-like structure on the MXene, which is different from Ag or Au. This indicates the distinct formation of soft solution depending on its metal ion precursor.

6.2 MXENE-BASED BIOSENSORS

6.2.1 MXene-Based Biosensors: Working Principle and Advantages

Due to their distinct qualities, such as high conductivity, large surface area, and good biocompatibility, MXenes, a relatively new family of two-dimensional (2D) materials, have attracted a lot of interest lately. Because of these characteristics, MXenes are excellent candidates for use in biosensors, which are tools that identify and measure the presence of certain biomolecules in a sample. MXene-based biosensors operate on a rather straightforward fundamental premise: a biomolecule of interest is immobilized on the surface of the MXene material, and when the target analyte attaches to the immobilized biomolecule, it causes a change in the MXene material's electrical characteristics [21]. Many methods, including electrochemical impedance spectroscopy, cyclic voltammetry, and field-effect transistor tests, can be used to assess and quantify these changes.

One key advantage of MXene-based biosensors is their high sensitivity and selectivity. Because of their unique properties, MXenes are capable of detecting very small quantities of target analytes with high accuracy. This makes them ideal for use in a wide range of applications, from environmental monitoring to clinical diagnostics. Another advantage of MXene-based biosensors is their fast response time [22]. MXenes can detect target analytes fast and precisely due to their large surface area and conductivity, making them ideal for real-time monitoring applications [23].

MXene-based biosensors are also relatively low in cost, which is important for the widespread adoption of these devices. Compared to other types of biosensors, MXene-based biosensors can be manufactured using simple, inexpensive processes, which makes them more accessible to researchers and clinicians. Lastly, MXene-based biosensors might be employed in a variety of applications, including environmental control, food safety, and clinical diagnostics [24–26]. Because they are compatible with a wide range of biomolecules, MXenes can be used to detect various targets with high accuracy. Moreover, MXene-based biosensors can monitor biomolecules in complex biological matrices like blood and saliva, which is crucial for clinical and diagnostic applications.

Altogether, MXene-based biosensors are a fascinating new technology with the potential to transform biosensing. MXene-based biosensors are set to have a big impact on a variety of disciplines due to their high sensitivity, selectivity, and quick reaction time, as well as their inexpensive cost and broad range of possible applications.

6.2.2 MXene-Based Biosensors: Types and Their Applications

MXene-based biosensors represent a rapidly growing field of research that has the potential to revolutionize various areas of applications, ranging from healthcare to environmental monitoring. These biosensors are based on MXenes, which are layered transition metal carbides, nitrides, and carbonitrides with a high surface area, high conductivity, and good biocompatibility. MXenes' unique features make them good candidates for building biosensors capable of identifying an extensive variety of biomolecules with excellent selectivity and sensitivity.

6.2.2.1 MXene-Based Electrochemical Biosensors

MXene-based electrochemical biosensors, which make use of MXenes' electrochemical characteristics, can also be used to detect biomolecules. In this type of biosensor, the target biomolecule is immobilized onto the surface of the MXene material. The target analyte changes the electrical characteristics of the MXene material after binding to an adsorbed biomolecule [27]. This change can be measured and analyzed using different electrochemical techniques, including electrochemical impedance spectroscopy, amperometry, and cyclic voltammetry. Electrochemical biosensors based on MXenes have been used to detect metabolites such as cholesterol, glucose, and lactate, as well as neurotransmitters in the brain.

Electrochemical biosensors have indeed been recognized as potentially useful diagnostic tools for cancer diagnosis. These biosensors use various electronic analyses, including cyclic voltammetry (CV), amperometry, potentiometry, DPV, EIS, and SWV, to give a variety of advantages such as dependability, affordability, and high sensitivity [28, 29]. Moreover, miniaturized biosensors, which are simple in size and portability, have therapeutic potential in cancer biomarker research. Surface nanoarchitectures utilized in this form of detection have appealing properties such as high LODs, resilience, small volumetric analyte, and a capability to operate in complicated biomatrix fluids containing optically high backgrounds of absorbance and fluorescence chemicals.

Bio-receptors were covalently bonded to amine-functionalized siliconized MXene (Ti_3C_2) nanosheets using APTES chemical to detect the active biomarker carcinoembryonic antigen (CEA), a routinely used tumor marker [30]. A single- to few-layer 2D nanosheet Ti_3C_2 MXene exhibits a high density of functional groups, which allows for a higher degree of antibody attachment and quicker analyte access (Figure 6.3a). The biosensor (BSA/anti-CEA/f-Ti_3C_2-MXene/GC) based on a label-free aminosilane and bio-functionalized f-Ti_3C_2 MXene showcased a low LOD of 18 fg mL^{-1} and a sensitivity of 37.9 A ng mL cm^2 per decade, with a linear detection range of 100 fg mL^{-1} to 2000 ng mL^{-1}, for the determination of the biomarker (Figure 6.3b) [30].

Carcinoembryonic antigen (CEA) is an extensively glycosylated cell surface protein whose levels rise in the event of the growing cancers such as breast cancer, pancreatic cancer, and more. Serum, as one of the most common example of a biomatrix fluid with the presence of CEA, is used in clinical studies to establish an early diagnostic method to detect the presence of cancer cells, track tumor recurrence, or determine the metastatic phase. Serum CEA levels in healthy nonsmokers are less than 2.5 ng mL^{-1}, whereas levels in smokers are less than 5.0 ng mL^{-1} [30]; however, these levels increase quickly when normal cells become malignant.

MXenes are employed to provide the conjugation site for the targeting aptamer and give superior electronic properties to the biosensor platform because of the electron transfer active properties of the MXenes, high ratio of specific surface area, and numerous possible attaching binding sites. Wang et al. created an MUC1 biosensor by coating the electrode with a MXene and attaching a ferrocene-labeled complementary DNA as a detecting probe for the electrochemical method to improve [31]. To prevent non-specific interactions, gold nanoparticles (AuNPs) bearing MUC1 aptamer were linked to the modified electrode through the affinity between the Au–S bond, and the surface of the AuNPs/MXene hybrids was passivated using general blocking protein albumin. The aptamer-conjugated MXene hybridized to the surface of the working substrate using duplex formation between the probing DNA

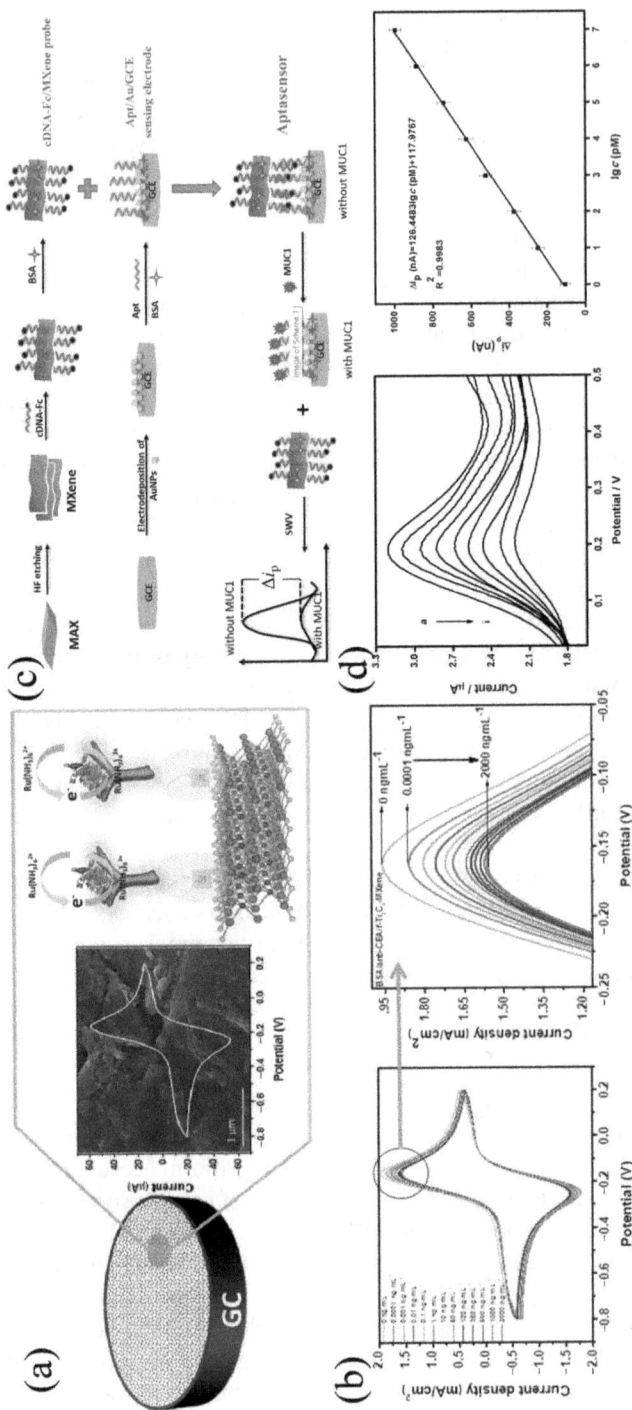

FIGURE 6.3 (a) Electrochemical CEA detection technique, depicted schematically [30]. (b) Electrochemical response tests of a BSA/anti-CEA/f-Ti$_3$C$_2$-MXene/GC electrode at various concentrations of CEA (0.0001–2000 ng mL^{-1}) and an expanded perspective of the response studies [30]. (c) Construction technique for the competitive electrochemical aptasensor, depicted schematically. (d) SWV of an aptasensor acquired after detecting various concentrations of MUC1, as well as the standard curve for MUC1 detection [31].

and an MUC1 aptamer. As MUC1 made contact with the area of interest (working electrode), the MUC1-specific aptamer became detached because of the high affinity and specific binding of the aptamer to MUC1 compared to that of the aptamer to the nanosheet. This resulted in a decrease in the electrochemical signal, which indicated the presence of MUC1. This method showcased a highly sensitive concept with a detection limit down to the molar level around 330 aM with linear range up to the millimolar level, which is considered a wide dynamic range. The aptasensor's peak current difference response had an RSD of 1.43%, suggesting strong repeatability, and the response did not vary significantly over 10 days, indicating satisfactory stability [31]. Up to recently, MUC genes could express up to 20 different kinds of mucins. Although glycoproteins are expressed in epithelial cells, glycosylated mucins are produced abnormally in cancer cells and could be targeted as oncogenic agents.

A label-free method based on altering a glassy carbon electrode (GCE) with MoS_2/Ti_3C_2, AuNPs, and ssRNA was used to detect miRNA-182 [32]. The modification was through the electrostatic interaction between the component and the non-covalent bonding from van der Waals forces. The surface of loose gold particles was saturated with blocking protein to avoid nonspecific adsorption. Using the DPV approach, the biosensor detected miRNA-182 with an LOD of 0.43 fM (with a linearity response from 1 fM to 0.1 nM) [32]. The biosensor detected miRNA-182 in actual samples with a recovery of 105%, 95.3%, and 93.0% at analyte concentrations of 10^{-10} M, 10^{-12} M, and 10^{-14} M, respectively [32].

A zero-dimensional (0D)/2D nanohybrid of $Ti_3C_2T_x$ nanosheets embellished with iron phthalocyanine quantum dots (FePcQDs), known as $Ti_3C_2T_x@FePcQDs$, was created by Duan et al. to detect the miRNA-155 (Figure 6.4a). The linear concentration range of the impedimetric biosensor for miRNA-155 detection was 0.01 fM to 10 pM, and the LOD was 4.3 aM (S/N = 3). (Figure 6.4b). The signal remained at 104% of the original after 15 days of storage, demonstrating adequate stability of the suggested aptasensor, and the RSD of five biosensors for detecting miRNA-155 was 2.98%, indicating good repeatability [33].

By using a non-industrial-scale screen-printed electrode, a pairing of a 2D MXene and gold nanoparticles of 5 nm in size was utilized as the composite modifier on the electrode interface. Then, duplex-specific nuclease (DSN) was employed as an amplification method to assess onco microRNAs in whole plasma [34]. First, a pair of active magnetic particles was conjugated to a single-stranded DNA (ssDNA) that was labeled with a different active electrochemical probe, methylene blue (MB), and ferrocene (Fc). These ssDNAs were partly complementary to the different side of the individual target miRNA. After the breakage of the hybridization by the nuclease enzyme, the released un-cleaved DNA sequences with redox labels were captured on the surface of the sensor platform to the target invasion and amplification cycle. The SPGE was functionalized with $Ti_3C_2T_x$, co-modified with AuNPs, and then conjugated with the plentiful ssDNAs (base) to generate a substantially larger electrochemical signal than conventional AuNP-modified electrodes (up to four times greater in magnitude). For miRNA-21 and miRNA-141 detection, the LODs of the biosensor were 204 aM and 138 aM, which demonstrates the simultaneous detection, an antifouling property, and capability of identifying a single mutation identification,

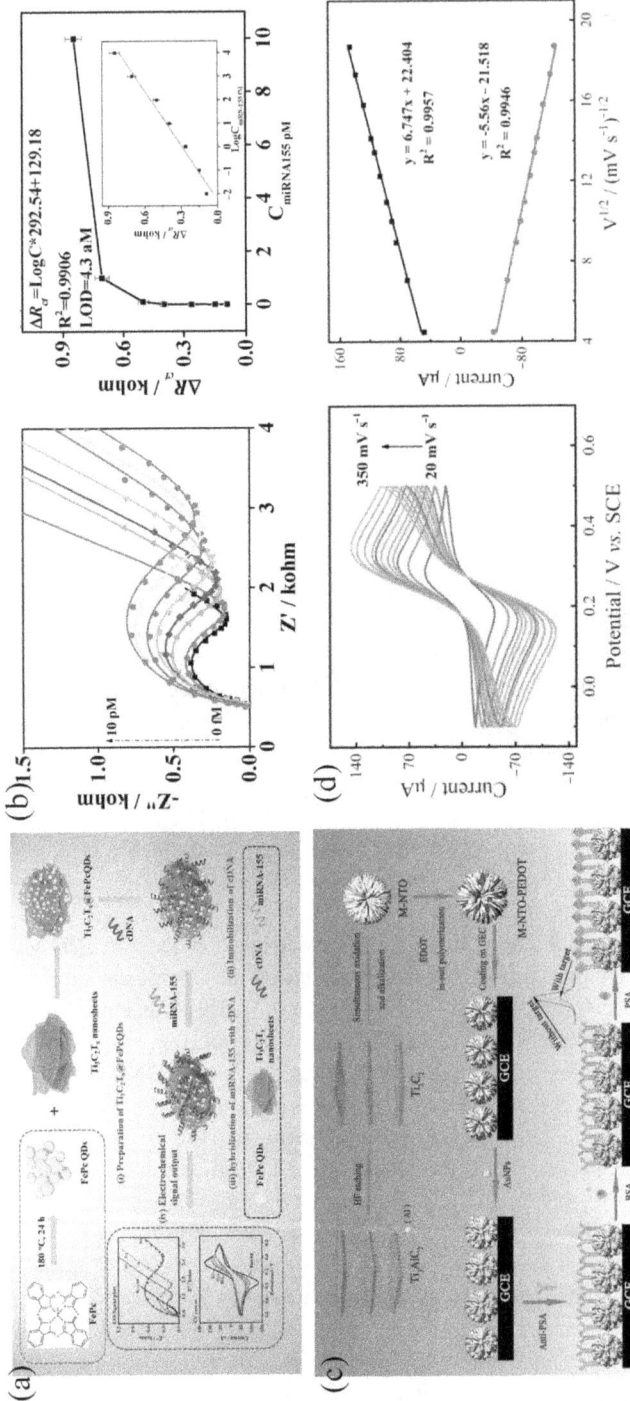

FIGURE 6.4 (a) Aptasensor for miRNA-155 detection using $Ti_3C_2T_x$@FePcQDs with a stepwise method starting from (i) $Ti_3C_2T_x$@FePcQDs nanohybrid preparation, (ii) aptamer conjugation to the nanohybrid, (iii) hybridization of the miRNA-155 to cDNA, and (iv) electrochemical signal measurement [33]. (b) EIS Nyquist plots for detecting various quantities of miRNA-155 using the $Ti_3C_2T_x$@FePcQDs-based aptasensor, as well as the dependency of Rct on miRNA-155 concentration [33]. (c) The immunosensor fabrication and detection stages for prostate specific antigen detection [35]. (d) CVs of the electrode surface at various scan rates, as well as a linear connection between redox peak currents and scan rates [35].

respectively. After four weeks of storage, the synergistic property of integrating the highly active MXene-modified electrochemical setup and the cyclic amplification from the specific-targeting DSN maintained a good assay reproducibility with relative activity around 95.2% and 97.1% and an average relative error of 4.7% with respect to the initial signal values assigned to MB and Fc [34].

Another strategy was introduced by Xu and colleagues [35] utilizing a 3D sodium titanate nanoribbon (M-NTO) composite that was prepared by alkali and peroxide treatment to a Ti_3C_2 MXene in a Teflon-lined stainless-steel autoclave. Those small molecules helped to etch the stacking MXene and prevented the recurrence of the flakes (Figure 6.4c). The composite demonstrated great biocompatibility, a high specific surface area, and rapid electron transfer capabilities. And, further, a combination of these nanoribbon and conductive polymer poly(3,4-ethylenedioxythiophene) (PEDOT) could improve its electron transfer rate. Then, similar to the other methods, introducing AuNPs by electrodepositing on the surface of M-NTO-PEDOT supported the immobilization of prostate specific antigen (PSA)-specific antibodies for PSA detection. After two weeks of storage at 4°C, the biosensor demonstrated great assay repeatability, with an RSD of 1.89% and good stability of 84.2% of its original response. Using DPV, the label-free immunosensor detected PSA with an LOD of 0.03 pg. L^{-1} (S/N = 3) using DPV (Figure 6.4d) [35].

6.2.2.2 MXene-Based Optical Biosensors

The optical biosensor is another form of MXene-based biosensor that employs the optical characteristics of MXenes to detect biomolecules. The biomolecule of interest is immobilized on the surface of the MXene material in optical biosensors, and when the target analyte attaches to the immobilized biomolecule, it causes a change in the optical characteristics of the MXene material. Several optical methods, such as surface plasmon resonance, fluorescence, and absorbance spectroscopy, can be used to detect this shift. These biosensors have been used to detect a variety of biomolecules such as proteins, DNA, and environmental contaminants.

MXenes have a number of benefits, including high conductivity, electron transfer rate, a high specific surface area, and considerable fluorescence efficiency [36]. MXenes have these features, which make them excellent for application in fluorescence biosensors, where they provide additional benefits like chemical stability, ease of surface functionalization, and high hydrophilicity [37]. As a result, there is a significant interest in developing MXene nanosheets for optical biosensors. Lately, fresh synthesis processes for MXene nanosheets and MQDs have been established, and researchers have investigated effective surface modification approaches to improve their hydrophilicity and fluorescence-measuring performance in water circumstances [38].

In recent investigations, Ti_3C_2 MXene nanosheets in conjunction with Cy3-labeled CD63 aptamer were used in the creation of biosensing devices based on fluorescence resonance energy transfer (FRET) for the detection of exosomes [41]. CD63 is an exosome transmembrane protein that contains essential macromolecules such as DNA and RNA, making it an appealing target molecule for non-destructive cellular monitoring [42]. The sensing probe was based on the competitiveness between the metal chelation of Cy3-labeled CD63 aptamer to the MXene nanosheet

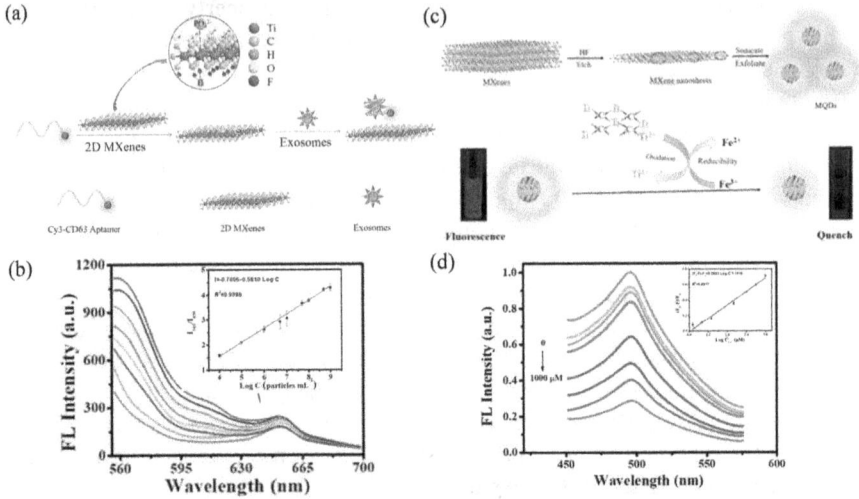

FIGURE 6.5 (a) Exosome detection using an optical biosensing system based on the competitive interaction between a MXene nanosheet coupled with fluorescence reporter Cy3-labeled CD63 aptamer and a target exosome is shown in the schematic image [39]; (b) its linear calibration curve [39]. (c) Illustration of MXene quantum dot (MQD) synthesis and utilization for the Fe^{3+} detection principle in schematic form [40], and (d) the fluorescence emission of MQDs with varying Fe^{3+} levels [40].

and the hydrogen bond between the aptamer to the target exosome. When the fluorescence probe of Cy3 is near the MXene nanosheets, the fluorescence was suppressed (Figure 6.5a). However, when exosomes were introduced, the aptamer, which has higher binding affinity to the exosome surface protein (CD63) than to the MXene, will detach from the nanosheet, resulting in the detachment of the Cy3-labeled CD63 aptamer from the nanosheets, restoring the quenched Cy3 signal and enabling the fluorescent detection of exosomes. With a detection limit of 1.4×10^3 exosomes/mL and a detection range of 10^4 to 10^9 exosomes/mL, our FRET-based biosensing device displayed very sensitive exosome sensing capabilities (Figure 6.5b). The detection limit was about 1000 times lower than that of traditional enzyme-linked immunosorbent assay (ELISA) techniques. The biosensing device was also stable under a wide range of pH settings and retained its sensing capability for more than nine days (pH 6.4–8.4).

Using Ti_3C_2 MXene nanosheets and fluorescein-labeled ssDNA, researchers developed a fluorescent DNA biosensor for diagnosing the human papillomavirus (HPV) [43]. Similarly to a prior work, researchers discovered that the interaction between the MXene and FAM-ssDNA resulted in the dampening of the fluorescence signal. When only a portion of the HPV gene sequence target DNA was used, hybridization with FAM-ssDNA led to the synthesis of dsDNA and release from the MXene, restoring the fluorescence signal. Exonuclease III (ExoIII) was developed to improve detection sensitivity by cleaving FAM from dsDNA, resulting in ssDNA breakdown without FAM or target DNA. The segregated target DNA could then link to unreacted FAM-ssDNA on the MXene surface, causing the fluorescent signal to be amplified. The linear response range without ExoIII was 1.0 nM to 50 nM, with a detection limit of 800 pM. ExoIII increased the detection range (0.5 nM to 50 nM) and the detection limit to 100 pM. Additionally, by utilizing the MXene for its capability to quench the fluorescence based on its interaction, these 2D nanomaterials were used to develop a high-performing optical biosensing device for essential protein identification.

Researchers have studied the possibility of using MQDs to produce such biosensors in addition to using MXene nanosheets as a quenching material in fluorescence biosensors. One work used Ti_3C_2 MQDs, which are recognized for their high fluorescence emission capabilities, to create a fluorescent biosensor for Fe^{3+} detection [41]. Because Fe^{3+} may cause a variety of health problems when present in excessive or insufficient levels in the body, precisely assessing its content is critical [44]. MQDs with a size of 1.75 nm were created by exfoliating MXene nanosheets etched with hydrofluoric acid using ultrasonic vibrations (Figure 6.5c). These MQDs were shown to be stable in a variety of solutions with physiological pH around 6–8 and considerable salt ion strength (0.1 M NaCl), with a QY of 7.7%.

Two different approaches were used to detect Fe^{3+} by using the quenching of MQDs. First, as Fe^{3+} ions could absorb light in the region similar to the excitation wavelength of the MQDs (300–400 nm), they lowered the fluorescence signal generated by the MQDs. The other approach is the interaction between the metal ions on the surface of the MQDs, which is attributed to the electrostatic interaction between the change in the surface charge of the MQDs to positively-charged and the Fe^{3+} ions. The interaction resulted in the quenching of the optical signal of the MQDs that was proportional to the concentration of Fe^{3+}. This developed detection of Fe^{3+} was evaluated in a seawater sample and a serum sample, obtaining a linear range of 5 μM to 1000 μM. (Figure 6.5d). This biosensor's detection limit was 310 nM, which was lower than the drinking water quality requirement (0.2 mg/L), demonstrating its practical application potential. Moreover, the biosensor was highly reproducible, with RSD values of 1.1% and 1.2% at 10 M and 250 M of Fe^{3+}, respectively.

Colorimetric biosensors based on MXene nanocomposites have received a lot of interest because of their ease of use, cheap cost, sensitive detection with naked eyes, and simplicity, making them good candidates for developing point-of-care devices [46, 47]. MXenes have peroxidase-like characteristics that can be used to create colorimetric optical biosensors. For example, a heterostructure was developed by integrating $Ti_3C_2T_x$ MXene nanosheets into heteroatom (Ni, Fe)-layered double hydroxide (NiFe-LDH), forming MXene/NiFe-LDH for- detection of small biomolecules such as glutathione (GSH) using peroxidase-like activity (Figure 6.6a) [45]. This activity was exhibited by both the MXene and NiFe-LDHs. Because of the surface configuration from the conjugation of those nanomaterials, it gives higher specific surface area, high accessibility to the diffusion of ion species, and a higher rate of the electron transfer. Owing to those enhanced properties, the MXene/NiFe-LDH nanocomposite exhibits higher catalytic characteristics than its precedent state, like MXene or NiFe-LDH as individual nanomaterial (Figure 6.6b). As mentioned earlier, the concentration of GSH was determined based on the catalytic reaction between the

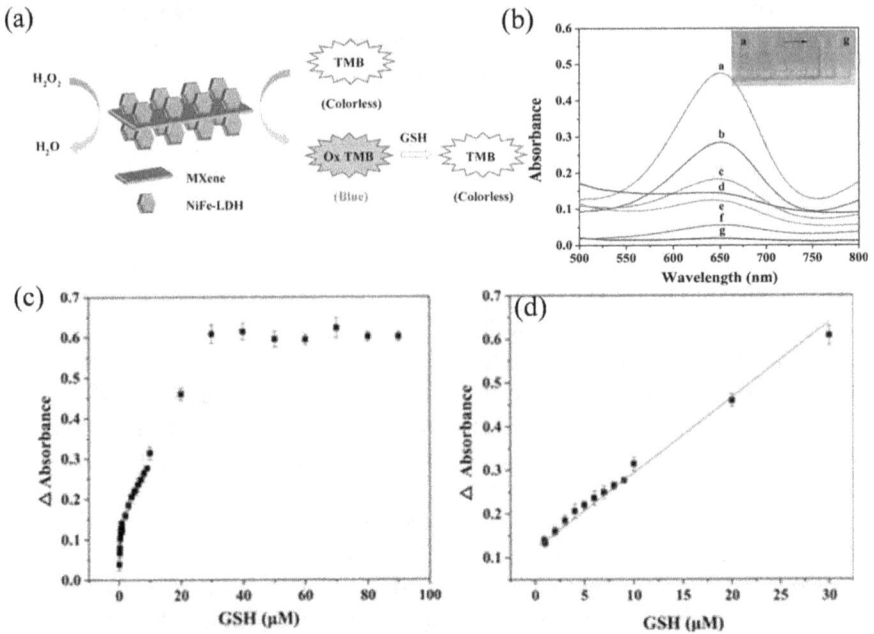

FIGURE 6.6 (a) The catalytic oxidation of TMB by MXene@NiFe-LDH nanocomposite and its application in GSH detection. (b) UV-Vis absorbance spectroscopy analysis of the catalytic oxidation of TMB with various peroxidase-exhibiting nanomaterials. (c) Concentration-dependent graph of the GSH concentration in response to the TMB oxidation as a value of absorbance. (d) Linear region of the wide dynamic range of the concentration-dependent graph of (c) in the GSH detection [45].

nanocomposite, MXene/NiFe-LDH, the chromogenic substrate, 3,3′,5,5′-tetramethylbenzidine (TMB), and hydrogen peroxide (H_2O_2). MXene/NiFe-LDH would catalytically oxidize TMB by peroxide substrate forming a complex quinone of TMB$^{•+}$, giving a bluish hue in the translucent aqueous solution of reduced TMB. But, when GSH was added, the oxidized TMB oxidized the GSH and affected the complex quinone formation, resulting in the loss of the blue color in the solution. GSH detection was conducted based on this colorimetric shift by assessing the change in the saturation of the blueish solution of the oxidized TMB (Figure 6.6c, d). With a detection limit of 84 nM, the technique demonstrated a linear response from 0.9 M to 30 M in concentration. Furthermore, when evaluated on GSH concentrations of 1 M, 10 M, and 20 M in human blood samples, the MXene/NiFe-LDH-based detection approach was extremely reliable, with a recovery rate ranging from 92% to 107% and an RSD value ranging from 0.93% to 2.36%.

MXenes, as noted before in this section, have particular surface chemical characteristics that make them adaptable for the development of fluorescent and optical biosensors. Despite significant limitations, such as the complex synthesis method and limited oxidative stability, several research efforts are now being done to precisely alter the characteristics of MXenes. These studies are expected to result in major breakthroughs in the creation of fluorescent and optical biosensors based on MXenes, perhaps transcending present constraints in the near future.

6.2.2.3 MXene-Based Field-Effect Transistor

MXene-based field-effect transistor (FET) biosensors are another type of biosensor that uses the FET properties of MXenes to detect biomolecules [27, 48]. This change can be detected using various FET techniques, such as current–voltage measurements and capacitance–voltage measurements. These biosensors have been used to detect a variety of biomolecules such as DNA, proteins, and environmental contaminants.

6.2.2.4 MXene-Based Piezoelectric Biosensors

MXene-based piezoelectric biosensors are another type of biosensor that uses the piezoelectric properties of MXenes to detect biomolecules. In piezoelectric biosensors, the biomolecule of interest is immobilized onto the surface of the MXene material, and when the target analyte binds to the immobilized biomolecule, it induces a mechanical deformation of the MXene material [26, 49]. This deformation can be detected using various piezoelectric techniques, such as quartz crystal microbalance and surface acoustic wave devices. These biosensors have been used to detect a wide range of biomolecules, including antigens, antibodies, and environmental contaminants.

MXene-based biosensors might be used in a variety of domains, including clinical diagnostics, environmental monitoring, food safety, and drug development. MXene-based biosensors can be utilized in clinical diagnostics to diagnose illnesses such as cancer, diabetes, and cardiovascular disease by delivering quick, accurate, and cost-effective detection of biomolecules in blood, urine, and other physiological fluids. MXene-based biosensors can be utilized in environmental monitoring for real-time monitoring of water and air quality, as well as the detection of contaminants in soil and sediment.

6.3 FUTURE PROSPECTS AND CONCLUDING REMARKS

6.3.1 ADVANCEMENTS AND POTENTIAL OF MXENE-BASED BIOSENSORS

MXene-based biosensors are an intriguing and exciting technology that has the potential to transform numerous disciplines, including medical diagnostics, environmental monitoring, and food safety. MXenes' strong electrical conductivity is one among the qualities that makes them ideal for biosensors. As biological molecules interact with the MXene surface, changes in the material's electrical conductivity may be detected and quantified. As a result, MXene-based biosensors are very sensitive and accurate, detecting even trace levels of biological substances.

MXene-based biosensors may also be easily combined with a variety of signal transduction modalities, including electrochemical, optical, and piezoelectric. Electrochemical biosensors, which employ changes in electrical conductivity to detect the presence of biological molecules, are one of the most prevalent forms of

MXene-based biosensors. Optical biosensors, on the other hand, detect the presence of biological substances by measuring changes in light absorption or emission. Piezoelectric biosensors detect the presence of biological substances by detecting variations in mechanical stress.

One of the most exciting potential applications of MXene-based biosensors is in the field of medical diagnostics. By using MXene-based biosensors, doctors and medical professionals could rapidly and accurately detect a wide range of diseases, from cancer to infectious diseases like COVID-19. MXene-based biosensors could also be used for point-of-care testing, which means that tests could be performed quickly and easily right in the doctor's office, eliminating the need for lengthy and expensive lab tests.

MXene-based biosensors also have a lot of potential in other fields, such as environmental monitoring and food safety. By using MXene-based biosensors, we could detect pollutants and other harmful substances in our environment more quickly and accurately, which could lead to better health outcomes for people and animals alike. Similarly, by using MXene-based biosensors to detect foodborne pathogens, we could prevent outbreaks of illness and ensure that the food we eat is safe.

Overall, MXene-based biosensors are an exciting and promising technology with many potential applications. As research in this field continues, we can expect to see even more innovative uses of MXene-based biosensors in the future.

6.3.2 Challenges and Future Directions

While MXene-based biosensors have a lot of potential, there are also some challenges that must be addressed in order for this technology to reach its full potential. Here are some of the key challenges that researchers in this field are currently working to overcome:

Fabrication: One of the biggest challenges in developing MXene-based biosensors is the fabrication process. MXene synthesis and processing require harsh chemicals and high temperatures, which can be challenging to work with. Researchers are working on developing new methods for synthesizing and processing MXenes that are more efficient, cost-effective, and environmentally friendly.

Stability: Another challenge is ensuring the stability of MXene-based biosensors over time. MXenes can be prone to oxidation and degradation, which can affect their performance as biosensors. Researchers are exploring various strategies for improving the stability of MXene-based biosensors, such as protective coatings and encapsulation.

Specificity: A key requirement for any biosensor is specificity—the ability to detect only the target molecule or analyte and no other molecules that may be present in the sample. MXene-based biosensors must be designed to be highly specific in order to avoid false positives and false negatives. Researchers are working on developing new methods for enhancing the specificity of MXene-based biosensors, such as using aptamers or antibodies to selectively bind to the target analyte.

Despite these challenges, the potential of MXene-based biosensors is enormous, and researchers are actively exploring new directions for this technology. Here are some of the future directions that are being pursued:

Multimodal biosensors: One direction for future MXene-based biosensors is the development of multimodal biosensors that can detect multiple analytes or molecules simultaneously. This could be particularly useful for medical diagnostics, where multiple biomarkers may be needed to accurately diagnose a disease.

Wearable biosensors: Another avenue of research is the creation of wearable biosensors that may be worn on the body and used to monitor various health factors such as heart rate, blood glucose levels, and breathing rate. MXene-based biosensors have the potential to be incorporated into wearable devices, allowing for non-invasive and easy real-time health monitoring.

Environmental monitoring: MXene-based biosensors could also be used for environmental monitoring, such as detecting pollutants in water or air. This could help to identify and mitigate environmental hazards and protect public health.

6.3.3 CONCLUSION

This chapter gives an overview of MXenes' possibilities as biosensor materials. MXenes are a novel family of two-dimensional materials with distinct features such as strong electrical conductivity, huge surface area, and exceptional biocompatibility. These properties make them attractive for biosensor applications, which require high sensitivity, specificity, and selectivity. The chapter addresses the benefits and limits of several types of MXene-based biosensors, such as electrochemical, optical, and piezoelectric biosensors. It also offers an overview of current developments in MXene-based biosensors, such as their integration with other materials and approaches to improve sensitivity and selectivity.

Finally, MXene-based biosensors show tremendous promise as next-generation developing materials for detecting a variety of biomolecules such as proteins, nucleic acids, and tiny compounds. Further study is required to improve their efficacy and solve their limits, but MXenes' unique features make them a fascinating and viable option for biosensor development. Notwithstanding the obstacles and limits, recent advances in MXene-based biosensors provide significant prospects for the creation of extremely sensitive, selective, and cost-effective biosensors for a variety of applications.

REFERENCES

[1] M. Naguib, Y. Gogotsi, Synthesis of two-dimensional materials by selective extraction, Accounts of Chemical Research, 48 (2015) 128–135.
[2] N.K. Chaudhari, H. Jin, B. Kim, D. San Baek, S.H. Joo, K. Lee, MXene: An emerging two-dimensional material for future energy conversion and storage applications, Journal of Materials Chemistry A, 5 (2017) 24564–24579.

[3] H. Shi, P. Zhang, Z. Liu, S. Park, M.R. Lohe, Y. Wu, A. Shaygan Nia, S. Yang, X. Feng, Ambient-stable two-dimensional titanium carbide (MXene) enabled by iodine etching, Angewandte Chemie International Edition, 60 (2021) 8689–8693.

[4] M. Naguib, O. Mashtalir, J. Carle, V. Presser, J. Lu, L. Hultman, Y. Gogotsi, M.W. Barsoum, Two-dimensional transition metal carbides, ACS Nano, 6 (2012) 1322–1331.

[5] G. Deysher, C.E. Shuck, K. Hantanasirisakul, N.C. Frey, A.C. Foucher, K. Maleski, A. Sarycheva, V.B. Shenoy, E.A. Stach, B. Anasori, Synthesis of Mo4VAlC4 MAX phase and two-dimensional Mo4VC4 MXene with five atomic layers of transition metals, ACS Nano, 14 (2019) 204–217.

[6] B. Anasori, Y. Xie, M. Beidaghi, J. Lu, B.C. Hosler, L. Hultman, P.R. Kent, Y. Gogotsi, M.W. Barsoum, Two-dimensional, ordered, double transition metals carbides (MXenes), ACS Nano, 9 (2015) 9507–9516.

[7] M. Naguib, M.W. Barsoum, Y. Gogotsi, Ten years of progress in the synthesis and development of MXenes, Advanced Materials, 33 (2021) 2103393.

[8] P.P. Michałowski, M. Anayee, T.S. Mathis, S. Kozdra, A. Wójcik, K. Hantanasirisakul, I. Jóźwik, A. Piątkowska, M. Możdżonek, A. Malinowska, Oxycarbide MXenes and MAX phases identification using monoatomic layer-by-layer analysis with ultralow-energy secondary-ion mass spectrometry, Nature Nanotechnology, 17 (2022) 1192–1197.

[9] X. Li, Z. Huang, C.E. Shuck, G. Liang, Y. Gogotsi, C. Zhi, MXene chemistry, electrochemistry and energy storage applications, Nature Reviews Chemistry, 6 (2022) 389–404.

[10] J.T. Lee, B.C. Wyatt, G.A. Davis Jr, A.N. Masterson, A.L. Pagan, A. Shah, B. Anasori, R. Sardar, Covalent surface modification of Ti3C2T x MXene with chemically active polymeric ligands producing highly conductive and ordered microstructure films, ACS Nano, 15 (2021) 19600–19612.

[11] M. Benchakar, L. Loupias, C. Garnero, T. Bilyk, C. Morais, C. Canaff, N. Guignard, S. Morisset, H. Pazniak, S. Hurand, One MAX phase, different MXenes: A guideline to understand the crucial role of etching conditions on Ti3C2Tx surface chemistry, Applied Surface Science, 530 (2020) 147209.

[12] X. Li, C. Wang, Y. Cao, G. Wang, Functional MXene materials: Progress of their applications, Chemistry—An Asian Journal, 13 (2018) 2742–2757.

[13] R. Meshkian, Q. Tao, M. Dahlqvist, J. Lu, L. Hultman, J. Rosen, Theoretical stability and materials synthesis of a chemically ordered MAX phase, Mo2ScAlC2, and its two-dimensional derivate Mo2ScC2 MXene, Acta Materialia, 125 (2017) 476–480.

[14] W. Huang, C. Ma, C. Li, Y. Zhang, L. Hu, T. Chen, Y. Tang, J. Ju, H. Zhang, Highly stable MXene (V2CTx)-based harmonic pulse generation, Nanophotonics, 9 (2020) 2577–2585.

[15] K. Maleski, M. Alhabeb, Top-down MXene synthesis (selective etching), 2D Metal Carbides and Nitrides (MXenes) Structure, Properties and Applications, (2019) 69–87.

[16] K. Gong, K. Zhou, X. Qian, C. Shi, B. Yu, MXene as emerging nanofillers for high-performance polymer composites: A review, Composites Part B: Engineering, 217 (2021) 108867.

[17] F. Kamarulazam, S. Bashir, S. Ramesh, K. Ramesh, Emerging trends towards MXene-based electrolytes for electrochemical applications, Materials Science and Engineering: B, 290 (2023) 116355.

[18] A. VahidMohammadi, J. Rosen, Y. Gogotsi, The world of two-dimensional carbides and nitrides (MXenes), Science, 372 (2021) eabf1581.

[19] M. Shekhirev, C.E. Shuck, A. Sarycheva, Y. Gogotsi, Characterization of MXenes at every step, from their precursors to single flakes and assembled films, Progress in Materials Science, 120 (2021) 100757.

[20] E. Satheeshkumar, T. Makaryan, A. Melikyan, H. Minassian, Y. Gogotsi, M. Yoshimura, One-step solution processing of Ag, Au and Pd@ MXene hybrids for SERS, Scientific Reports, 6 (2016) 32049.

[21] S. Ullah, F. Shahzad, B. Qiu, X. Fang, A. Ammar, Z. Luo, S.A. Zaidi, MXene-based aptasensors: Advances, challenges, and prospects, Progress in Materials Science, (2022) 100967.

[22] A. Parihar, N.K. Choudhary, P. Sharma, R. Khan, MXene-based aptasensor for the detection of aflatoxin in food and agricultural products, Environmental Pollution, (2022) 120695.

[23] M. Song, Y. Ma, L. Li, M.-C. Wong, P. Wang, J. Chen, H. Chen, F. Wang, J. Hao, Multiplexed detection of SARS-CoV-2 based on upconversion luminescence nano-probe/MXene biosensing platform for COVID-19 point-of-care diagnostics, Materials & Design, 223 (2022) 111249.

[24] S. Kang, K. Zhao, D.-G. Yu, X. Zheng, C. Huang, Advances in biosensing and environmental monitoring based on electrospun nanofibers, Advanced Fiber Materials, 4 (2022) 404–435.

[25] K. Nemčeková, J. Labuda, Advanced materials-integrated electrochemical sensors as promising medical diagnostics tools: A review, Materials Science and Engineering: C, 120 (2021) 111751.

[26] M. Mathew, C.S. Rout, Electrochemical biosensors based on $Ti_3C_2T_x$ MXene: Future perspectives for on-site analysis, Current Opinion in Electrochemistry, 30 (2021) 100782.

[27] J. Yoon, M. Shin, T. Lee, J.-W. Choi, Highly sensitive biosensors based on biomolecules and functional nanomaterials depending on the types of nanomaterials: A perspective review, Materials, 13 (2020) 299.

[28] L. Li, H. Zhao, Z. Chen, X. Mu, L. Guo, Aptamer biosensor for label-free square-wave voltammetry detection of angiogenin, Biosensors and Bioelectronics, 30 (2011) 261–266.

[29] B. Mekassa, M. Tessema, B.S. Chandravanshi, P.G. Baker, F.N. Muya, Sensitive electrochemical determination of epinephrine at poly (L-aspartic acid)/electro-chemically reduced graphene oxide modified electrode by square wave voltammetry in pharmaceutics, Journal of Electroanalytical Chemistry, 807 (2017) 145–153.

[30] S. Kumar, Y. Lei, N.H. Alshareef, M. Quevedo-Lopez, K.N. Salama, Biofunctionalized two-dimensional Ti_3C_2 MXenes for ultrasensitive detection of cancer biomarker, Biosensors and Bioelectronics, 121 (2018) 243–249.

[31] H. Wang, J. Sun, L. Lu, X. Yang, J. Xia, F. Zhang, Z. Wang, Competitive electrochemical aptasensor based on a cDNA-ferrocene/MXene probe for detection of breast cancer marker Mucin1, Analytica Chimica Acta, 1094 (2020) 18–25.

[32] L. Liu, Y. Wei, S. Jiao, S. Zhu, X. Liu, A novel label-free strategy for the ultrasensitive miRNA-182 detection based on MoS_2/Ti_3C_2 nanohybrids, Biosensors and Bioelectronics, 137 (2019) 45–51.

[33] F. Duan, C. Guo, M. Hu, Y. Song, M. Wang, L. He, Z. Zhang, R. Pettinari, L. Zhou, Construction of the 0D/2D heterojunction of $Ti_3C_2T_x$ MXene nanosheets and iron phthalocyanine quantum dots for the impedimetric aptasensing of microRNA-155, Sensors and Actuators B: Chemical, 310 (2020) 127844.

[34] M. Mohammadniaei, A. Koyappayil, Y. Sun, J. Min, M.-H. Lee, Gold nanoparticle/MXene for multiple and sensitive detection of oncomiRs based on synergetic signal amplification, Biosensors and Bioelectronics, 159 (2020) 112208.

[35] Q. Xu, J. Xu, H. Jia, Q. Tian, P. Liu, S. Chen, Y. Cai, X. Lu, X. Duan, L. Lu, Hierarchical Ti_3C_2 MXene-derived sodium titanate nanoribbons/PEDOT for signal amplified electrochemical immunoassay of prostate specific antigen, Journal of Electroanalytical Chemistry, 860 (2020) 113869.

[36] Q. Guan, J. Ma, W. Yang, R. Zhang, X. Zhang, X. Dong, Y. Fan, L. Cai, Y. Cao, Y. Zhang, Highly fluorescent Ti 3 C 2 MXene quantum dots for macrophage labeling and Cu 2+ ion sensing, Nanoscale, 11 (2019) 14123–14133.

[37] G. Xu, Y. Niu, X. Yang, Z. Jin, Y. Wang, Y. Xu, H. Niu, Preparation of Ti3C2Tx MXene-derived quantum dots with white/blue-emitting photoluminescence and electrochemiluminescence, Advanced Optical Materials, 6 (2018) 1800951.

[38] W. Kong, Y. Niu, M. Liu, K. Zhang, G. Xu, Y. Wang, X. Wang, Y. Xu, J. Li, One-step hydrothermal synthesis of fluorescent MXene-like titanium carbonitride quantum dots, Inorganic Chemistry Communications, 105 (2019) 151–157.

[39] Q. Zhang, F. Wang, H. Zhang, Y. Zhang, M. Liu, Y. Liu, Universal Ti3C2 MXenes based self-standard ratiometric fluorescence resonance energy transfer platform for highly sensitive detection of exosomes, Analytical Chemistry, 90 (2018) 12737–12744.

[40] Q. Zhang, Y. Sun, M. Liu, Y. Liu, Selective detection of Fe 3+ ions based on fluorescence MXene quantum dots via a mechanism integrating electron transfer and inner filter effect, Nanoscale, 12 (2020) 1826–1832.

[41] H. Wang, H. Li, Y. Huang, M. Xiong, F. Wang, C. Li, A label-free electrochemical biosensor for highly sensitive detection of gliotoxin based on DNA nanostructure/MXene nanocomplexes, Biosensors and Bioelectronics, 142 (2019) 111531.

[42] J.-H. Lee, J.-H. Choi, S.-T.D. Chueng, T. Pongkulapa, L. Yang, H.-Y. Cho, J.-W. Choi, K.-B. Lee, Nondestructive characterization of stem cell neurogenesis by a magneto-plasmonic nanomaterial-based exosomal mirna detection, ACS Nano, 13 (2019) 8793–8803.

[43] X. Peng, Y. Zhang, D. Lu, Y. Guo, S. Guo, Ultrathin Ti3C2 nanosheets based "off-on" fluorescent nanoprobe for rapid and sensitive detection of HPV infection, Sensors and Actuators B: Chemical, 286 (2019) 222–229.

[44] S. Li, Y. Li, J. Cao, J. Zhu, L. Fan, X. Li, Sulfur-doped graphene quantum dots as a novel fluorescent probe for highly selective and sensitive detection of Fe3+, Analytical Chemistry, 86 (2014) 10201–10207.

[45] H. Li, Y. Wen, X. Zhu, J. Wang, L. Zhang, B. Sun, Novel heterostructure of a MXene@ NiFe-LDH nanohybrid with superior peroxidase-like activity for sensitive colorimetric detection of glutathione, ACS Sustainable Chemistry & Engineering, 8 (2019) 520–526.

[46] M. Liu, Y. He, J. Zhou, Y. Ge, J. Zhou, G. Song, A "naked-eye" colorimetric and ratiometric fluorescence probe for uric acid based on Ti3C2 MXene quantum dots, Analytica Chimica Acta, 1103 (2020) 134–142.

[47] B. Shao, Z. Xiao, Recent achievements in exosomal biomarkers detection by nanomaterials-based optical biosensors-a review, Analytica Chimica Acta, 1114 (2020) 74–84.

[48] B. Xu, C. Zhi, P. Shi, Latest advances in MXene biosensors, Journal of Physics: Materials, 3 (2020) 031001.

[49] H. Riazi, G. Taghizadeh, M. Soroush, MXene-based nanocomposite sensors, ACS Omega, 6 (2021) 11103–11112.

7 Theranostics Approach of MXenes in Cancer Therapy

Nilima Priyadarsini Mishra, Bighnanshu Kumar Jena, Pramod K. Singh, Tarun Yadav, and N.P. Yadav

7.1 INTRODUCTION

Since the 19th century, cancer has led as the second most common cause of death among all life-threatening diseases [1]. Every sixth death, there is a death due to cancer, as reported by the World Health Organization [2]. High financial costs as well as the emotional breakdown of patients due to cancer severely affect the entire society [3]. Nowadays, the foremost standards of living such as abundant intake of alcohol, unhealthy diet, smoking, etc., are tremendous causes of cancer [4]. To target cancer, various types of remedial and diagnostic equipment such as surgery, chemotherapy, radiotherapy, targeted therapy, photoacoustic magnetic resonance imaging, and immunotherapy are rapidly used [5]. Due to the straightforward treatment of cancer cells and more rapid recovery, radiotherapy and chemotherapy are extensively used. But extreme exploitation of chemotherapeutic doses and high exposure to radiation of adjacent healthy tissues are also affected significantly [6].

In previous times, an outstanding number of nanostructure- and nanoscale-based remedial as well as diagnostic agents have been explored [7]. In comparison with traditional modalities, nanoparticles demonstrate a high response to the healing of cancer and its diagnosis [8]. Mainly the tiny particle size improves the permeation and straightforward penetration against various cells. Although several therapeutic pathways are available, their efficiencies and performance are still too poor to provide 100% results [9]. Among all types of nanomaterials, MXenes, a type of 2D material, possess significant biomedical applications such as tumor detection, cancer therapy, antimicrobial effect, and drug delivery [10–14]. In MXene, the M stands for an early transition metal, X represents carbon or nitrogen, and -ene is just like in graphene, and the first MXene 2D material is Ti_3C_2 [15]. The additional analysis of MXenes revealed them to have surface plasmon resonance effects and photothermal effects [16]. MXenes with a size of 180 nm can arrive at the cancerous microenvironment and improve the cell permeability and retention of EPR [17]. The extraordinary electronic structure and large surface areas of MXenes enhance the electronic and energy application and become the major exhilarating areas of

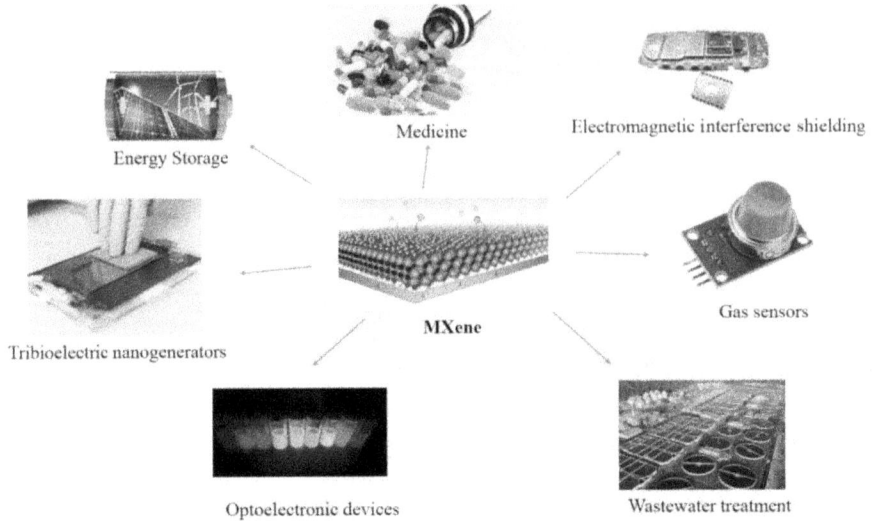

FIGURE 7.1 A diverse range of applications of MXenes.

Reprinted with permission from Ref. [21] (b). Copyright 2020, Elsevier.

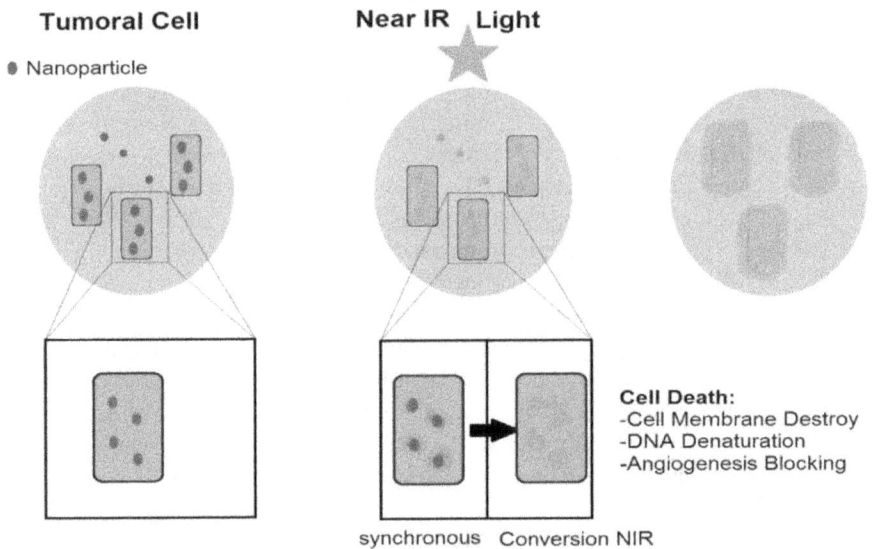

FIGURE 7.2 Photothermal therapy in cancer treatment.

Reprinted with permission from Ref. [21] (b). Copyright 2020, Elsevier.

curiosity in scientific field [18]. Along with this, MXenes show evidence for a hydrophilic nature, adsorption ability, and high surface reactivity due to their large surface area [19]. Diverse applications of MXenes were found in triboelectric nanogenerators, energy storage, wastewater treatment, optoelectronic devices, gas sensors, medicine, and electromagnetic interference shielding, which is well represented in Figure 7.1 [20–27].

Here, we discuss the theranostic applications of MXenes in cancer therapy along with therapeutic applications in diagnostic imaging (i.e., CT, PA, MR, and photoluminescence imaging), and biomedical application (i.e., PTT and drug delivery) in Figure 7.2. Also, we explain the detailed mechanistic approach towards cancer cells. By screening a deeper insight into the chemistry of 2D MXenes based on the existing and capable therapeutic modalities, we provide a perspective on future chances and potential to support the multiple functionalization and clinical translation of MXenes in cancer theranostics.

7.2 MECHANISTIC PATHWAYS OF MXENE APPLICATIONS AGAINST CANCER

Currently, nanomaterials have entered the research mainstream due to their various therapeutical applications such as cancer therapy, imaging, sensors, diagnosis, etc. [28]. In comparison with various bulk materials, the quantum electronic effects of nanomaterials along with a high surface area-to-volume ratio provide a very unique characteristic [29]. Among different types of nanomaterials, MXene nanomaterials appear as significant nanomaterials for their cancer theranostics applications [30]. There are different types of MXenes, which are made by the selective removal of an A-element from their consequent layered MAX phases and termed as $M_{n+1}AX_n$. Here, M stands for transition metal elements such as Ti, Zr, V, Nb, Ta, Mo, Sc, or W; A refers to an element from group 13 or 14, i.e., Al, In, Si, Ga, or Ge; and X stands for C/N [31]. Among these different types of MXenes, here we have discussed the theranostic approach of tantalum carbides, titanium carbides, niobium carbides, vanadium carbides, and molybdenum carbides towards cancer therapy. Also, 2D MXene materials have superior capabilities and can be used for bio-imaging, targeted drug delivery, antibacterial studies, photothermal therapy, biosensing, etc. [32]. *In vitro* and *in vivo* analyses using MXenes have guaranteed tumor-specific accumulation, easy vascular transport, destruction of malignant cells, and drug release behavior in cancer cells [33]. Also, there are pioneering reports on photothermal tumor suppression via MXenes based on NIR-I and NIR-II bio-windows [34]. Photothermal tumor ablation was exhibited by MXenes through phototherapy using NIR laser radiation where a general heating effect was exploited to treat tumor cells with an inadequately vascularized microenvironment [35]. In this direction, the photothermal agents can absorb the light that is distributed to cancer tissues, and later on, this absorbed laser light is transformed into thermal energy to destroy the tumor cells.

When the adverse side effects are minimized, the incorporation of NIR light as a source of external and isolated stimulus has been carried out for the benefit of elevated temporal and spatial control of general heating. Along with this, the NIR irradiation requires tissue scattering and absorption at a lower point to formulate efficient light penetration and photothermal conversion of photo-absorbers.

7.2.1 TWO-DIMENSIONAL TANTALUM CARBIDE (MXENE) THERANOSTIC APPROACH TOWARDS CANCER

Among different types of MXenes, the Dai research group in 2017 [36] discussed the tantalum carbide (Ta_4C_3) MXene, which was considered for numerous studies on imaging-guided photothermal hyperthermia of cancer ablation and contrast-enhanced CT imaging. Here, by following the surface engineering strategy, they anticipated *in situ* enlargement of manganese oxide nanoparticles on the surface of tantalum carbide to synthesize a strongly oxidative MnO_x/Ta_4C_3 composite. This composite became an attractive contrast agent for photoacoustic (PA) imaging.

Photoacoustic (PA) imaging can rupture the penetration limit of the optical image that is detected by induced pressure waves of laser-irradiated tissues. When the surface is further modified by soybean phospholipid (SP), then the biocompatible composite is systematically evaluated for biosafety and theranostic performance against cancer. MnO_x/Ta_4C_3-SP composite nanosheets were established as having high biocompatibility and biosafety by systematic *in vivo* evaluation [36].

Another research group first time worked on the dual-mode action of photoacoustic/computed tomography imaging and extremely efficient *in vivo* photothermal ablation of tumor cells by using a Ta_4C_3 MXene. Also, these types of MXenes are considered ultrathin nanosheets, and within the atomic thickness of this MXene, tumor cell ablation occurred. Hence, they focused on the advanced photothermal conversion with efficiency η of 44.7% and *in vitro/in vivo* photothermal ablation of tumor cells by using biocompatible soybean phospholipid-modified Ta_4C_3 nanosheets. The composition and nanostructures of Ta_4C_3 MXenes are considered to achieve the requirements of biomedicine [37].

A superparamagnetic MXene-based theranostic nanoplatform was developed for breast cancer therapy by utilizing tantalum carbide (Ta_4C_3), and its surface was functionalized as a Ta_4C_3-IONP-SPs composite. These composite nanosheets were established for three specific functionalization applications in breast cancer theranostics. In this case, tantalum was employed for integrated superparamagnetic IONPs and CT imaging used for T2-weighted MR imaging. Also, these composites accomplished inclusive tumor suppression without reoccurrence and were extremely proficient in breast tumor hyperthermia performance [38].

7.2.2 TWO-DIMENSIONAL TITANIUM CARBIDE (MXENE) THERANOSTIC APPROACH TOWARDS CANCER

Like tantalum-based MXenes, titanium-based MXenes considerably widen the application of 2D materials. Wu's group in 2017 first reported the Ti_3C_2-based composite MnO_x/Ti_3C_2 for its spectacular applications in the treatment of cancer cells. This type of MXene complex is used for various types of theranostic applications in photoacoustic dual-modality imaging-guided photothermal therapy and proficient magnetic resonance against cancer cell lines. MnO_x/Ti_3C_2-SP composite nanosheets also demonstrated *in vivo* biocompatibility [39].

Again in the same year, the fluorine-free method was adopted for the development of titanium carbide MXene quantum dots by the Yu group. Here, they modified the

MXene quantum dots' surface by the fluorine-free method, which gifted the quantum dots with a very strong and wide range of absorption in the NIR region. These titanium MXene QDs also illustrated immense biocompatibility devoid of any conspicuous toxicity *in vitro* and *in vivo*, which specified their elevated perspective for biomedical applications [40].

Another research group, Peng et al., in 2018 reported the emergence of diverse applications of two-dimensional nanosystems of 2D MXenes in the field of nanomedicine. They explained that superparamagnetic two-dimensional TI_3C_2 MXenes were highly proficient for cancer treatment therapy and also the surface chemistry of TI_3C_2 MXenes designed for *in situ* development of superparamagnetic Fe_3O_4 nanocrystals. It is worth mentioning that these superparamagnetic MXenes have been thoroughly demonstrated both *in vivo* and *in vitro*, which revealed high photothermal destruction of cancer cells and suppression of tumor tissues. Also, TI_3C_2 MXene composites assure their additional prospective clinical translation due to the high biocompatibility of these MXenes. These new functionalization applications of MXene-based 2D nanosheets also broaden their application in theranostic nanomedicine [41].

Inspired by the theranostic applications of different types of titanium-based MXenes, in 2018, the Chen group first reported the derivatization of 2D Ti_3C_2 MXene nanosheets by assimilation of GdW10-based polyoxometalates (POMs). Also, these multifunctional GdW10@Ti_3C_2 composites provided hyperthermal management with magnetic resonance and computed tomography for tumor cells. Due to the unique composition of the GdW10@Ti_3C_2 composite, it served as a contrast agent for CT and MR imaging and showed potent diagnostic imaging for monitoring tumor hyperthermia. In the field of medicinal chemistry, GdW10@Ti_3C_2 composite nanosheets also provided high *in vivo* biocompatibility in the coming era [42].

Very recently, Liu's group worked on two-dimensional MXenes and their applications in the field of theranostic biomedicine. In this work, they first reported the development of Ti3C2MXene nanosheets for near-infrared and magnetic resonance imaging-guided photothermal treatment. Here, NaErF$_4$@ Ti_3C_2 composites emitted in the NIR-IIB region at 1530 nm, which conferred NIR-IIb imaging guidance and monitoring competence throughout tumor hyperthermia. The NaErF$_4$@Ti_3C_2 composite was highly efficient for T_2-weighted MR imaging, where the inherent magnetic properties of Er^{3+} played a vital role. Also, this composite could be used against tumor ablation with an inhibition ratio of 92.9% both *in vitro* and *in vivo* owing to photothermal stability and high photothermal conversion capability. These composites were not found to be toxic at the injected dose, so they could be used for versatile applications in cancer therapy [43].

The Gai group in 2022 developed Ti-based MXene nanocomposites ($Ti_3C_2T_xPt$-PEG) that were decorated with Pt artificial nano enzymes on the Ti_3C_2 nanosheets. These types of 2D ultrathin nanosheets have high absorbance in the near-infrared-II region, so that they could be used as a substrate to attach purposeful components, i.e., nanodrugs and nanoenzymes. Here, Pt nanoparticles exhibited peroxide-like (POD-like) activity inside the tumor microenvironment, where they catalyzed hydrogen peroxide to produce hydroxyl radicals to trigger cell apoptosis and necrosis in situ.

Along with this, these nanocomposites also showed desirable photothermal effects and improved POD-like activity significantly [44].

7.2.3 TWO-DIMENSIONAL NIOBIUM CARBIDE (MXENE) THERANOSTIC APPROACH TOWARDS CANCER

In 2018, Li's group worked on a two-dimensional Nb_2C MXene that could be considered a "therapeutic mesopore" and could be used for chemotherapy due to its enhanced photothermal character. These MXenes were innovatively designed, as they worked in situ by self-assembled mesopore-making agent CTAC surrounded by the mesopores for chemotherapy in comparison with other traditional chemo drug loading. The "theranostic mesopore"-coated 2D Nb_2C provided systematic *in vivo* and *in vitro* evaluations with improved therapeutic effectiveness against U87 brain cancer cells [45].

7.2.4 TWO-DIMENSIONAL MOLYBDENUM CARBIDE (MXENE) THERANOSTIC APPROACH TOWARDS CANCER

In 2019, Yan's group worked on a molybdenum carbide MXene and studied its application in both cancer treatment and real-time imaging. They investigated Mo_2C nanospheres as "one-for-all" theranostic agents. These nanosheets have potential photothermal and photodynamic behavior, so they could be used for synergistic phototherapy in the elimination of cancer cells and to remove solid tumors. Along with this, these MXenes were applicable in CT imaging and photoacoustic imaging for *in vitro* tumor representation. The outstanding biocompatibility, tissue toxicity, and minimal hematotoxicity of Mo_2C nanosheets was examined on a solid tumor by magnetic resonance imaging and B-mode ultrasonography [46].

Another group also explored the Mo_2C MXenes for photonic tumor hyperthermia in the year 2019. They explained the structure of both bulk Mo_2Ga_2C ceramic and Mo_2C MXenes and studied their computational simulation for photonic performance prediction. The Mo_2C MXenes also have intense near-infrared (NIR) absorption, which covered the first and second biological transparency window to display superior degradability and biocompatibility. This MXene expanded the nanomedicine applications and enhanced high therapeutic performance [47].

7.2.5 TWO-DIMENSIONAL VANADIUM CARBIDE (MXENE) THERANOSTIC APPROACH TOWARDS CANCER

MXenes i.e., two-dimensional vanadium carbide (V_2C) also have significant photothermal behavior towards photothermal therapy. Due to high-quality structural reliability and remarkably elevated absorption in the near-infrared region, these MXenes could be considered well-organized PTAs for magnetic resonance imaging-guided PTT and photoacoustic imaging of cancer cell lines. Also, this group of MXenes has attractive properties for biomedical applications [48].

7.3 FUTURE PERSPECTIVES OF MXENES

The recent scientific world has searched for an ideal nanomaterial with various applications in the field of biomedical chemistry towards cancer theranostic approaches [49]. In this phase of research, MXenes have emerged as interesting nanosheets in cancer theranostic applications [50]. Mainly, the photothermal behavior and exceptional magnetic and electrical properties of MXenes make them different from other candidates [51]. With the help of surface engineering, a wide platform was given to MXenes [52]. The different types of MXenes participated in cell culture and were investigated for the biological importance of nanoparticles [53]. To obtain basic knowledge about their cancer therapeutic applications, MXenes were studied *in vitro*, which highlights the preliminary thoughts about the movement against cancer cells, the nature of biodegradation, and apoptosis pathways in normal healthy cells [54]. On the other hand, MXenes have also been reported as antibacterial agents via both ROS-independent and -dependent toxicity mechanisms [55]. Along with cell culture studies, these MXene nanoparticles were also examined against animal models to demonstrate the biocompatibility in living organisms [56]. As animal replicas are extremely essential for the investigation of the biocompatibility of MXenes with living cells, it explained its unusual destructive effects that can be activated in the body after administration [57]. The major side effects of MXenes included reproductive, embryonic, neurological, and long-term toxicity in the body [58]. Furthermore, these MXenes provided valuable information regarding the theranostic approach of MXenes to the cancer cell lines with multi-modality imaging and synergistic therapy [59]. But it is worth mentioning that the harmful toxic effect of MXenes was ruled out when they underwent biodegradability and excretion pathways [57]. After *in vitro* and *in vivo* studies, the final and most vital approach is performing clinical tests to investigate the anticancer and toxicity features of MXenes [60]. To date, cancer management is very challenging in the healthcare division, so appropriate clinical experiments can lay concrete light on the diverse effects of MXenes.

7.4 CONCLUSION

MXenes have a broad range of physiological mechanisms and photothermal behaviors that enhance the conventional theranostic approach towards cancer cells. Here, different types of MXenes were discussed along with novel surface engineering. Along with this, the impact of nanotechnology on cancer is also discussed. By the use of these types of two-dimensional nanoparticles, various novel biomedical devices were developed for drug delivery vehicles, sensing probes, and imaging. Hence, MXenes combined with imaging agents can provide precious results in cancer theranostics which will give noteworthy dives into oncology research.

REFERENCES

1. Xiaomei, M. and Herbert, Y. Global Burden of Cancer. *Yale J. Biol. Med.*, 2006, *79(3–4)*, 85–94.
2. Nagai, H. and Kim, Y. H. Cancer Prevention from the Perspective of Global Cancer Burden Patterns. *J. Thorac. Dis.*, 2017 Mar, *9(3)*, 448–451.

3. Carrera, P. M., Kantarjian, H. M. and Blinder, V. S. The Financial Burden and Distress of Patients with Cancer: Understanding and Stepping-Up Action on the Financial Toxicity of Cancer Treatment. *CA: Cancer J. Clin.*, 2018 Mar, *68(2)*, 153–165.

4. Remen, T., Pintos, J., Abrahamowicz, M. and Siemiatycki, J. Risk of Lung Cancer in Relation to Various Metrics of Smoking History: A Case-Control Study in Montreal. *BMC Cancer*, 2018, *18*, 1275.

5. Wargo, J. A., Reuben, A., Cooper, Z. A., Oh, K. S. and Sullivan, R. J. Immune Effects of Chemotherapy, Radiation, and Targeted Therapy and Opportunities for Combination With Immunotherapy. *Semin. Oncol.*, 2015 Aug, *42(4)*, 601–616.

6. Baskar, R., Lee, K. A., Yeo, R. and Yeoh, K.-W. Cancer and Radiation Therapy: Current Advances and Future Directions. *Int. J. Med. Sci.*, 2012, *9(3)*, 193–199.

7. Patra, J. K., Das, G., Fraceto, L. F., Campos, E. V. R., Rodriguez-Torres, M. D. P., Acosta-Torres, L. S., Diaz-Torres, L. A., Grillo, R., Swamy, M. K., Sharma, S., Habtemariam, S. and Shin, H.-S. Nano Based Drug Delivery Systems: Recent Developments and Future Prospects. *J. Nanobiotechnology*, 2018, *16*, 71.

8. Bae, K. H., Chung, H. J. and Park, T. G. Nanomaterials for Cancer Therapy and Imaging. *Mol. Cell*, 2011 Apr, *31(4)*, 295–302.

9. Din, F. U., Aman, W., Ullah, I., Qureshi, O. S., Mustapha, O., Shafique, S. and Zeb, A. Effective Use of Nanocarriers as Drug Delivery Systems for the Treatment of Selected Tumors. *Int. J. Nanomed.*, 2017, *12*, 7291–7309.

10. Li, H., Fan, R., Zou, B., Yan, J., Shi, Q. and Guo, G. Roles of MXenes in Biomedical Applications: Recent Developments and Prospects. *J. Nanobiotechnology*, 2023, *21*, 73.

11. Siwal, S. S., Kaur, H., Chauhan, G. and Thakur, V. K. MXene-Based Nanomaterials for Biomedical Applications: Healthier Substitute Materials for the Future. *Advanced Nanobiomed Research*, 2023, *3(1)*, 2200123.

12. Lu, B., Zhu, Z., Ma, B., Wang, W., Zhu, R. and Zhang, J. 2D MXene Nanomaterials for Versatile Biomedical Applications: Current Trends and Future Prospects. *Nano. Micro Small*, 2021 Nov, *17(46)*, 2100946.

13. Chen, L., Dai, X., Feng, W. and Chen, Y. Biomedical Applications of MXenes: From Nanomedicine to Biomaterials. *Acc. Mater. Res.*, 2022, *3(8)*, 785–798.

14. Solangi, N. H., Mazari, S. A., Mubarak, N. M., Karri, R. R., Rajamohan, N., Vo, D.-V. N. Recent Trends in MXene-Based Material for Biomedical Applications. *Environ. Res.*, 2023 Apr, *222*, 115337.

15. Deysher, G., Shuck, C. E., Hantanasirisakul, K., Frey, N. C., Foucher, A. C., Maleski, K., Sarycheva, A., Shenoy, V. B., Eric, A., Stach, E. A., Anasori, B. and Gogotsi, Y. Synthesis of Mo_4VAlC_4 MAX Phase and Two-Dimensional Mo_4VC_4 MXene with Five Atomic Layers of Transition Metals. *ACS Nano*, 2020, *14(1)*, 204–217.

16. Hussein, E. A., Zagho, M. M., Rizeq, B. R., Younes, N. N., Pintus, G., Mahmoud, K. A., Nasrallah, G. K. and Elzatahry, A. A. Plasmonic MXene-Based Nanocomposites Exhibiting Photothermal Therapeutic Effects with Lower Acute Toxicity Than Pure MXene. *Int. J. Nanomed.*, 2019, *14*, 4529–4539.

17. Gazzi, A., Fusco, L., Khan, A., Bedognetti, D., Zavan, B. and Vitale, F. Photodynamic Therapy Based on Graphene and MXene in Cancer Theranostics. *Front. Bioeng. Biotechnol.*, 2019, *7*, 295.

18. Shukla, V., Jena, N. K., Naqvi, S. R., Luo, W. and Ahuja, R. Modelling High-Performing Batteries with Mxenes: The Case of S-Functionalized Two-Dimensional Nitride Mxene Electrode. *Nano Energy*, 2019 Apr, *58*, 877–885.

19. Ahmaruzzaman, M. MXenes Based Advanced Next Generation Materials for Sequestration of Metals and Radionuclides from Aqueous Stream. *J. Environ. Chem. Eng.*, 2022 Oct, *10(5)*, 108371.

20. Barsoum, M. W. The MN+1AXN Phases: A New Class of Solids: Thermodynamically Stable Nanolaminates. *Prog. Solid State Chem.*, 2000, *28*, 201–281.

21. (a) Malaki, M., Maleki, A., and Varma, R. S. Mxenes and Ultrasonication. *J. Mater. Chem. A*, 2019, *7*, 10843–10857. (b) Sivasankarapillai, V. S., Somakumar, A. K., Joseph, J., Nikazar, S., Rahdar, A., and Kyzas, G. Z. Cancer Theranostic Applications of MXene Nanomaterials: Recent Updates. *Nano-Struct. Nano-Objects*, 2020, *22*, 100457.

22. Lin, H., Chen, Y., Shi, J. Insights into 2D Mxenes for Versatile Biomedical Applications: Current Advances and Challenges Ahead. *Adva. Sci.*, 2018, *5*, 1800518.

23. Barsoum, M. W. and El-Raghy, T. The MAX Phases: Unique New Carbide and Nitride Materials: Ternary Ceramics Turn Out to Be Surprisingly Soft and Machinable, Yet Also Heat-Tolerant, Strong and Lightweight. *Am. Sci.*, 2001, *89*, 334–343.

24. Wu, L., Lu, X., Dhanjai, Wu, Z.-S., Dong, Y., Wang, X., Zheng, S. and Chen, J. 2D Transition Metal Carbide MXene as a Robust Biosensing Platform for Enzyme Immobilization and Ultrasensitive Detection of Phenol. *Biosens. Bioelectron.*, 2018, *107*, 69–75.

25. Lin, H., Wang, X., Yu, L., Chen, Y. and Shi, J. Two-Dimensional Ultrathin MXene Ceramic Nanosheets for Photothermal Conversion. *Nano Lett.*, 2017, *17*, 384–391.

26. Yang, B., Chen, Y. and Shi, J. Material Chemistry of Two-Dimensional Inorganic Nanosheets in Cancer Theranostics. *Chem.*, 2018, *4*, 1284–1313.

27. Dai, C., Lin, H., Xu, G., Liu, Z., Wu, R. and Chen, Y. Biocompatible 2D Titanium Carbide (MXenes) Composite Nanosheets for pH-Responsive MRI-Guided Tumor Hyperthermia. *Chem. Mater.*, 2017, *29*, 8637–8652.

28. Yu, Z., Gao, L., Chen, K., Zhang, W., Zhang, Q., Li, Q. and Hu, K. Nanoparticles: A New Approach to Upgrade Cancer Diagnosis and Treatment. *Nanoscale Res. Lett.*, 2021, *16*, 88.

29. Khan, I., Saeed, K. and Khan, I. Nanoparticles: Properties, Applications and Toxicities. *Arab. J. Chem*, 2019, *12(7)*, 908–931.

30. Zhu, W., Li, H. and Luo, P. Emerging 2D Nanomaterials for Multimodel Theranostics of Cancer. *Front Bioeng. Biotechnol.*, 2021, *9*, 769178.

31. Shekhirev, M., Shuck, C. E., Sarycheva, A. and Gogotsi, Y. Characterization of MXenes at Every Step, from Their Precursors to Single Flakes and Assembled Films. *Prog. Mater. Sci.*, 2021, *120*, 100757.

32. Lin, H., Chen, Y. and Shi, J. Insights into 2D MXenes for Versatile Biomedical Applications: Current Advances and Challenges Ahead. *Adv. Sci.* (Weinh), 2018, *5(10)*, 1800518.

33. Wu, J., Yu, Y. and Su, G. Safety Assessment of 2D MXenes: In Vitro and In Vivo. *Nanomaterials*, 2022, *12(5)*, 828.

34. Shao, J., Zhang, J., Jiang, C., Lin, J. and Huang, P. Biodegradable Titanium Nitride MXene Quantum Dots for Cancer Phototheranostics in NIR-I/II Biowindows. *Chem. Eng. J.*, 2020, *400*, 126009.

35. Jiang, Z., Li, T., Cheng, H., Zhang, F., Yang, X., Wang, S., Zhou, J. and Ding, Y. Nanomedicine Potentiates Mild Photothermal Therapy for Tumor Ablation. *Asian J. Pharm. Sci.*, 2021, *16(6)*, 738–761.

36. Dai, C., Chen, Y., Jing, X., Xiang, L., Yang, D., Lin, H., Liu, Z., Han, X. and Wu, R. Two-Dimensional Tantalum Carbide (MXenes) Composite Nanosheets for Multiple Imaging-Guided Photothermal Tumor Ablation. *ACS Nano*, 2017, *11*, 12696–12712.

37. Lin, H., Wang, Y., Gao, S., Chen, Y. and Shi, J. Theranostic 2D Tantalum Carbide (MXene). *Adv. Mater.*, 2018, *30*, 1703284.

38. Liu, Z., Lin, H., Zhao, M., Dai, C., Zhang, S., Peng, W. and Chen, Y. 2D Superparamagnetic Tantalum Carbide Composite MXenes for Efficient Breast-Cancer Theranostics. *Theranostics*, 2018, *8(6)*, 1648–1664.

39. Dai, D., Lin, H., Xu, G., Liu, Z., Wu, R. and Chen, Y. Biocompatible 2D Titanium Carbide (MXenes) Composite Nanosheets for pH-Responsive MRI-Guided Tumor Hyperthermia. *Chem. Mater.*, 2017, *29*, 8637–8652.

40. Yu, X., Cai, X., Cui, H., Lee, S.-W., Yu, X.-F. and Liu, B. Biocompatible 2D Titanium Carbide (MXenes) Composite Nanosheets for pH-Responsive MRI-Guided Tumor Hyperthermia. *Nanoscale*, 2017, *9*, 17859.

41. Liu, Z., Zhao, M., Lin, H., Dai, C., Ren, C., Zhang, S., Peng, W. and Chen, Y. 2D Magnetic Titanium Carbide MXene for Cancer Theranostics. *J. Mater. Chem. B*, 2018, *6*, 3541.

42. Zong, L., Wu, H., Lin, H. and Chen, Y. A Polyoxometalate-Functionalized Two-Dimensional Titanium Carbide Composite MXene for Effective Cancer Theranostics. *Nano Res.*, 2018, *11(8)*, 4149–4168.

43. Pan, J., Zhang, M., Fu, G., Zhang, L., Yu, H., Yan, X., Liu, F., Sun, P., Jia, X., Liu, X. and Lu, G. Ti3C2 MXene Nanosheets Functionalized with NaErF4:0.5%Tm@NaLuF4 Nanoparticles for Dual-Modal Near-Infrared IIb/Magnetic Resonance Imaging-Guided Tumor Hyperthermia. *ACS Appl. Nano Mater.*, 2022, *5*, 8142–8153.

44. Zhu, Y., Wang, Z., Zhao, R., Zhou, Y., Feng, L., Gai, S. and Yang, P. Pt Decorated Ti3C2Tx MXene with NIR-II Light Amplified Nanozyme Catalytic Activity for Efficient Phototheranostics. *ACS Nano*, 2022, *16*, 3105–3118.

45. Han, X., Jing, X., Yang, D., Lin, H., Wang, Z., Ran, H., Li, P. and Chen, Y. Therapeutic Mesopore Construction on 2D Nb2C MXenes for Targeted and Enhanced Chemo-Photothermal Cancer Therapy in NIR-II Biowindow. *Theranostics*, 2018, *8(16)*, 4491–4508.

46. Zhang, Q., Huang, W., Yang, C., Wang, F., Song, C., Gao, Y., Qiu, Y., Yan, M., Yang, B. and Guo, C. The Theranostic Nanoagent Mo2C for Multi-Modal Imaging-Guided Cancer Synergistic Phototherapy. *Biomater. Sci.*, 2019, *7*, 2729.

47. Feng, W., Wang, R., Zhou, Y., Ding, L., Gao, X., Zhou, B., Hu, P. and Chen, Y. Ultrathin Molybdenum Carbide MXene with Fast Biodegradability for Highly Efficient Theory-Oriented Photonic Tumor Hyperthermia. *Adv. Funct. Mater.*, 2019, *29*, 1901942.

48. Zada, S., Dai, W., Kai, Z., Lu, H., Meng, X., Zhang, Y., Cheng, Y., Yan, F., Fu, P., Zhang, X. and Dong, H. Algae Extraction Controllable Delamination of Vanadium Carbide Nanosheets with Enhanced Near-Infrared Photothermal Performance. *Angew. Chem. Int. Ed.*, 2020, *59*, 6601–6606.

49. Khursheed, R., Dua, K., Vishwas, S., Gulati, M., Jha, N. K., Aldhafeeri, G. M., Alanazi, F. G., Goh, B. H., Gupta, G., Paudel, K. R., Hansbro, P. M., Chellappan, D. K. and Singh, S. K. Biomedical Applications of Metallic Nanoparticles in Cancer: Current Status and Future Perspectives. *Biomed. Pharmacother.*, 2022, *150*, 112951.

50. Iravani, S. and Varma, R. S. MXenes in Cancer Nanotheranostics. *Nanomaterials*, 2022, *12(19)*, 3360.

51. Shukla, V. The Tunable Electric and Magnetic Properties of 2D MXenes and Their Potential Applications. *Mater. Adv.*, 2020, *1*, 3104–3121.

52. Li, N., Peng, J., Ong, W.-J., Ma, T., Arramel, Zhang, P., Jiang, J., Yuan, X., Zhang, C. J. MXenes: An Emerging Platform for Wearable Electronics and Looking Beyond. *Matter*, 2021, *4(2)*, 377–407.

53. Zamhuri, A., Lim, G. P., Ma, N. L., Tee, K. S. and Soon, C. F. MXene in the Lens of Biomedical Engineering: Synthesis, Applications and Future Outlook. *BioMed. Eng. OnLine*, 2021, *20*, 33.

54. Liu, S., Wei, W., Wang, J. and Chen, T. Theranostic Applications of Selenium Nanomedicines against Lung Cancer. *J. Nanobiotechnology*, 2023, *21*, 96.

55. Iravani, S. and Varma, R. S. MXene-Based Composites against Antibiotic-Resistant Bacteria: Current Trends and Future Perspectives. *RSC Adv.*, 2023 Mar, *13(14)*, 9665–9677.

56. Lim, G. P., Soon, C. F., Ma, N. L., Morsin, M., Nayan, N., Ahmad, M. K. and Tee, K. S. Cytotoxicity of MXene-Based Nanomaterials for Biomedical Applications: A Mini Review. *Environ. Res.*, 2021 Oct, *201*, 111592.

57. Vasyukova, I. A., Zakharova, O. V., Kuznetsov, D. V. and Gusev, A. A. Synthesis, Toxicity Assessment, Environmental and Biomedical Applications of MXenes: A Review. *Nanomaterials(basel)*, 2022 Jun, *12(11)*, 1797.

58. Gogotsi, Y. and Anasori, B. The Rise of MXenes. *ACS Nano*, 2019, *13(8)*, 8491–8494.

59. Han, X., Jing, X., Yang, D., Lin, H., Wang, Z., Ran, H., Li, P. and Chen, Y. Therapeutic Mesopore Construction on 2D Nb_2C MXenes for Targeted and Enhanced Chemo-Photothermal Cancer Therapy in NIR-II Biowindow. *Theranostics*, 2018, *8(16)*, 4491–4508.

60. Ranjbari, S., Darroudi, M., Hatamluyi, B., Arefinia, R., Aghaee-Bakhtiari, S. H., Rezayi, M. and Khazaei, M. Application of MXene in the Diagnosis and Treatment of Breast Cancer: A Critical Overview. *Front. Bioeng. Biotechnol.*, 2022, *10*, 984336.

8 Innovations in MXenes for Photocatalysis

Sigamani Saravanan and M.V. Someswararao

8.1 INTRODUCTION

Photocatalysis has been an environmentally friendly technology in development for the past few decades. The promising photocatalysis performance has drawn ever-growing technologies with MXene-based two-dimensional nanomaterials because of their unique optical, electronic, high feasibility, easy fabrication and physicochemical properties with larger surface-to-volume ratios. The wider range of 2D nanomaterials has been playing a crucial role, like graphene-based carbon nitride (g-C_3N_4), metal oxide (WO_3 and TiO_2), transition metal dichalcogenides (TMDs) and metals (BiWO4, $BiMoO_6$), as shown in Figure 8.1. Recently, the significance and rapid growth of MXene-based two-dimensional (2D) nanomaterials used in various applications such as batteries, capacitors, oxygen reduction, CO_2 reduction, hydrogen evolution method, etc., has been observed [1]. Generally, MXenes are originated from the MAX phase of ternary ceramic nanomaterials. These phases form larger families of ternary ceramic material substances with the broad formula as follows:

$$M_{n+1}AX_n$$

Here, n = 1–3, M is a transition metal, A is a group IIIA–IVA element and X is C, N or B [2]. These materials are considered the primary source of MXenes.

Further, MXene materials become highly desirable materials for different environmental remediation and energy harvesting or conversion applications, for example as co-catalysts in photocatalytic activities [3]. When light illuminates the MXene materials it can be absorbed by the materials. This phenomenon or process is also well known as photocatalysis. During this process, the valence orbital electrons are excited to the conduction band because the holes' (+ve) charges are formed in the primary positions [4]. The selection of a photocatalyst produces significant hydrogen fuel using sunlight for various applications. The growth and development of density functional theory (DFT) can help in the investigation of photocatalysts. Recently, two-dimensional nanostructures with transition metal carbide, nitrides and carbon nitrides have attracted researchers due to their outstanding performance. Materials such as graphene, hexagonal boron nitride, metal (transition) dichalcogenides and black phosphorous have

 DOI: 10.1201/9781003366225-8

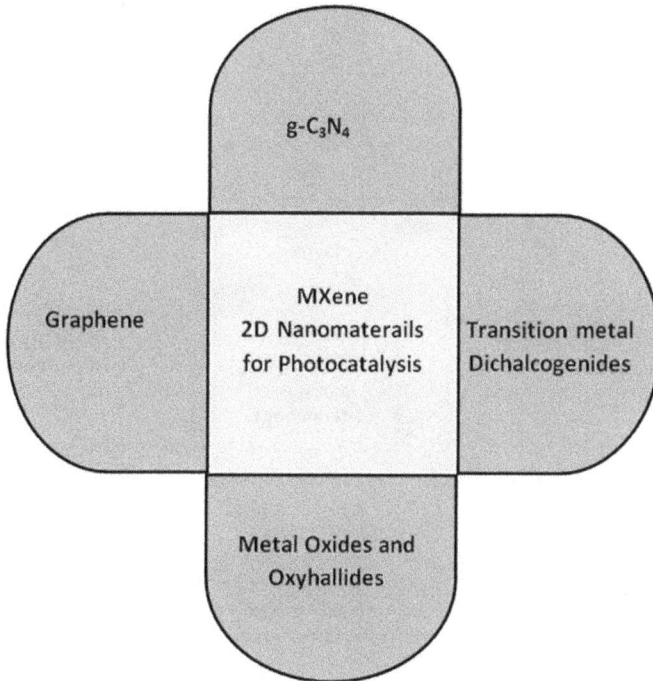

FIGURE 8.1 Block diagram of 2D nanomaterials for photocatalysis.

garnered great attention because of ultrahigh surface areas, efficient charge–carrier transfer, optimal electronic properties and excellent ionic conductivity with high Young's moduli at nanoscale level [5]. Significantly, the work function and carrier transport of MXenes can be manipulated with their terminated groups surface like -F, -OH and -O. MXene transition metal carbides and nitrides were first reported at 2011 [6]. Nowadays, 'N' number of two-dimensional MXene materials has been newly used and fabricated, which is generating novel and efficient photocatalytic performance, for example MoS_2, VS_2, WX_y (X = C, N, S), WC_{1-x}, etc. [7–12].

Figure 8.2 shows the schematic diagrams for the production of different 2D hybrid nanosheets. Nowadays, a major effort has been made by developing highly efficient and strong photocatalysts with different carbon-based nanomaterials because the carbon-related 2D-materials are the strongest organic structures at the nanoscale level, as reported by various scientific community members, for example nanotubes, graphene oxide, metal–organic frameworks and MXenes [14–19]. In this section, we are focusing on an introduction to MXenes from recent work along with the synthesis approaches and fabrication of 2D nanomaterials for photocatalysis.

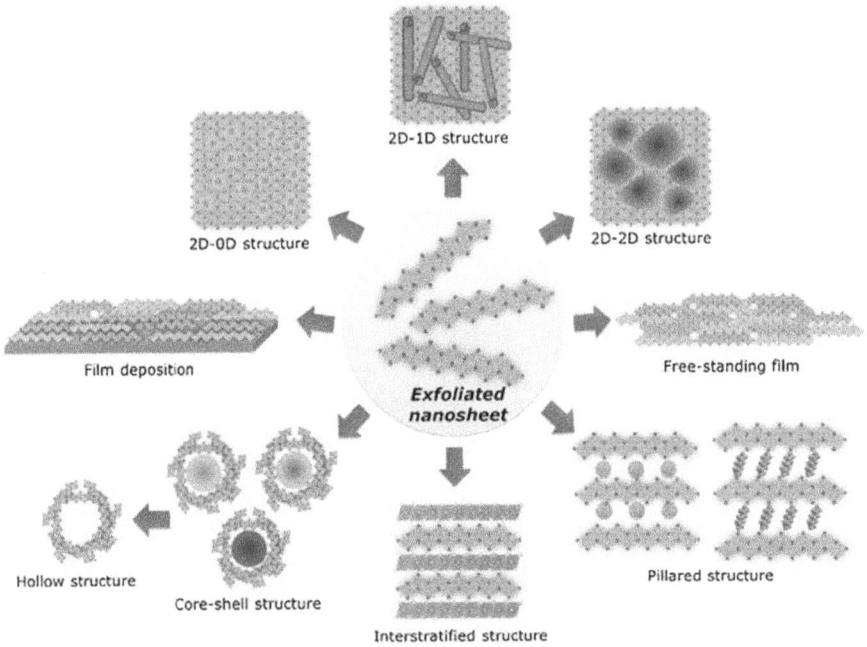

FIGURE 8.2 The schematic representation of various 2D nanostructures [13].

Copyright 2022, Elsevier.

8.2 SYNTHESIS TECHNIQUES

MXene-related photocatalysts are fabricated by replacing metal co-catalysts. Majorly, the fabrication of photocatalysts or related composites is used the following techniques such as mechanical mixing, self-assembly, in situ decoration/oxidation, etc. [19]. Also, it is growing the attention towards intellectual engineering and synthesis of MXenes as catalysts in photocatalytic applications. These structured materials are gaining desirable results due to their importance in photocatalysis. Presently, various methods for the fabrication of MXenes include wet etching, self-assembly, in situ oxidation, ultrasonic exfoliation and mechanical mixing (Table 8.1).

The 2D MXenes emerged as hot photocatalytic materials. The MXenes are synthesized by using selectively etching with atomic layers from various parent MAX phase. Figure 8.3 shows the different atomic layers in the unit cell and typical structure of M_2XT_x, $M_3X_2T_x$ and $M_4X_3T_x$ respectively.

8.2.1 HF ETCHING

Two-dimensional heterojunction was used and demonstrated the synthetic process of 2D heterojunction of ultrathin Ti_3C_2/Bi_2WO_6 nanosheets [29]. Initially, the bulk Ti_3AlC_2 etched according to the shape structure using HF etching and then

TABLE 8.1

The Various Mxene Material and Fabrication Methods

S.N.	MXene materials	Method	Ref.
1	MXene/NiFe$_2$O$_4$	Hydrothermal	[20]
2	Nickel-cobalt layered double hydroxide/MXene	Heterojunction surface	[21]
3	Ti$_3$C$_2$T$_x$	Controlled oxidation action	[22]
4	Ti$_2$CT$_x$	3D spheroid type cultures	[23]
5	Ti$_3$C$_2$T$_z$	Acid etching	[24]
6	Cr$_2$CT$_x$	Etching	[25]
7	TiO$_2$@Ti$_3$C$_2$	In situ Transformation	[26]
8	Mo$_2$CT$_x$	Etching	[27]

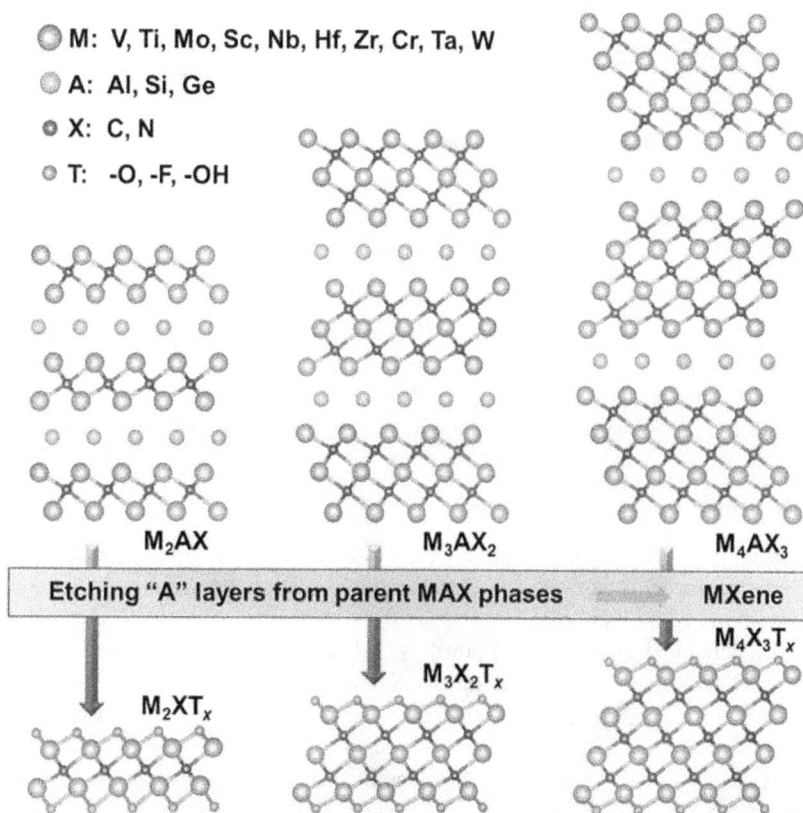

FIGURE 8.3   The MXene parent phases (M$_2$AX, M$_3$AX$_2$ and M$_4$AX$_3$) and crystal structure (M$_2$XT$_x$, M$_3$X$_2$T$_x$, M$_4$X$_3$T$_x$) [28].

dimethyl sulfoxide assisted ultrasonic exfoliation was employed for the few-layer Ti_3C_2 nanosheets.

8.2.2 WET ETCHING

The HF-mediated wet etching is the most frequently used synthesis technique for the preparation of MXenes. This etching processing is appropriate for the preparation of multilayered MXenes with parent Ti_3AlC_2 with a hydrofluoride (HF) solution. Furthermore, an ultrasonication process was carried out in methanol (CH_3OH) solution. In general, the HF etching chemical reactions for the synthesis of MXenes ($M_{n+1}X_nT_x$) are as follows,

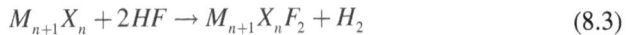

$$M_{n+1}AlX_n + 3HF \rightarrow M_{n+1}X_n + AlF_3 + 1.5H_2 \tag{8.1}$$

$$M_{n+1}X_n + 2H_2O \rightarrow M_{n+1}X_n(OH)_2 + H_2 \tag{8.2}$$

$$M_{n+1}X_n + 2HF \rightarrow M_{n+1}X_nF_2 + H_2 \tag{8.3}$$

Here, the first equation represents the aluminum (Al) layer placed in the MAX phase (Ti_3AlC_2) and reacting with hydrofluoride (HF) for the generation of AlF_3 and H_2 gas using exfoliation on the thin MXene layers [13]. Further, multiple-layer MXenes are formed using O, OH and F terminations by using all species.

8.2.3 SELF-ASSEMBLY

The self-assembling method and preparation of the MXene-based nanomaterials could be improved highly as compared to other induced techniques where photocatalysts and co-catalysts are prepared and reported by researchers [30–31]. The composite structures were obtained by mechanical mixing, and final samples were obtained by the self-assembly method. This method has advantages such as more intimate contact between the layers and highly uniform dispersion.

8.2.4 MECHANICAL MIXING

This mechanical mixing is a relatively easy method for the fabrication of composite catalyst structures. It has various advantages by saving energy and being economical, low-cost and convenient.

The mechanical grinding solid powders and substances in solutions were used to coat the MXene surface as photocatalysts. Sun et al. reported Ti_3C_2 on a g-C_3N_4 photocatalyst by using easy grinding mechanisms. They further conducted mechanical mixing of g-C_3N_4 and Ti_3C_2 powders, followed by calcination to yield g-C_3N_4/$Ti_3C_2T_x$x composite structures. Further, researchers noticed an improved charge carrier separation and increased photocatalytic H_2 emission performance [32]. Next, Ti_3C_2 (MXene) along with sacrificial alkalinization P25 was demonstrated by Ye et al.

Next, an improvement in photocatalytic CO_2 reduction was observed. Before, Ti_3C_2 and P25 were mixed in H_2O (water) using magnetic stirring. The result exhibited an excellent photocatalytic performance up to superior conductivity, carbon

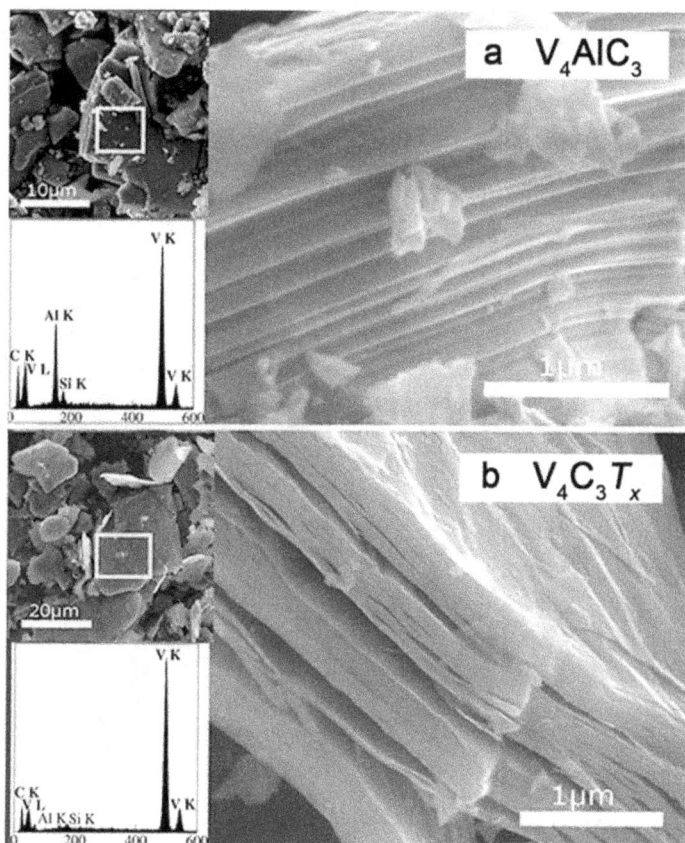

FIGURE 8.4 The SEM micro image and EDX analyses of V_4AlC_3 (a) and $V_4C_3T_x$ [2].
Copyright 2021, Elsevier.

dioxide adsorption and potential activation of Ti_3C_2-OH [33]. Figure 8.4 depicts an overview of top and cross-sectional morphological micro images from scanning electron microscopy (SEM). The inserted image shows the elemental spectrum by energy dispersive X-ray spectroscopy (EDX). These 2D MXene sheets or flakes were prepared by wet chemical methods and enhanced the physical and chemical properties which are useful for various desirable applications [34–39]. The EDX spectrum revealed the successful elemental analyses and confirmed the presence of the corresponding samples like V, Al, C and T. Figure 8.5 shows morphologies of MXenes which provided post-processing images. Figure 8.4a displays a top-view SEM image of Ti_3C_2 nanosheets with three-dimensional architectures. The MXene sheet shows crumpled shapes of various ridges and a rough surface by using a capillary forced assembling strategy [34]. Figure 8.5b demonstrates the wrinkled morphological image of $Ti_3C_2T_x$ flakes along with NaOH treated. Also, Figure 8.5c–d shows the cross-sectional SEM micrograph of compact $Ti_3C_2T_x$ film and $Ti_3C_2T_x$/CNT composites [40].

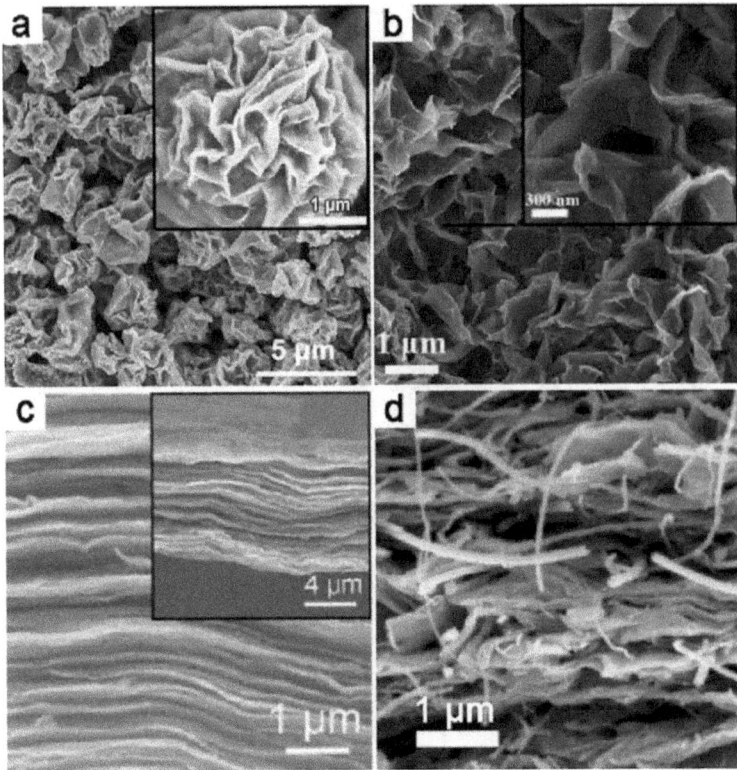

FIGURE 8.5 SEM micrograph of top-view and cross-sectional MXenes (a) $Ti_3C_2T_x$, (b) Na-c-$Ti_3C_2T_x$, (c) $Ti_3C_2T_x$ and (d) porous $Ti_3C_2T_x$/carbon nanotube composite [2].

Copyright 2021, Elsevier.

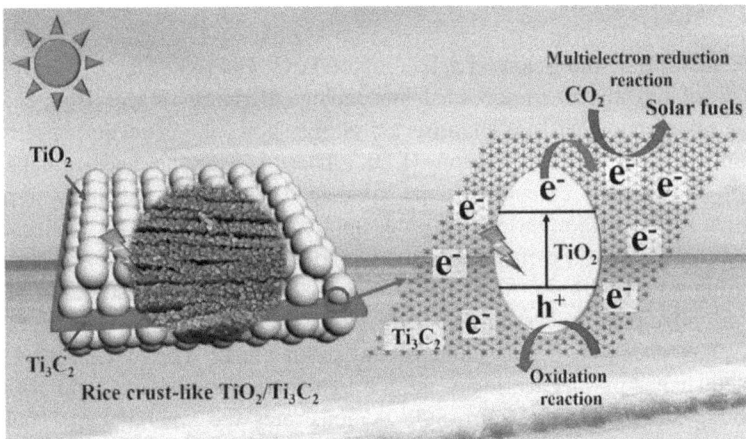

FIGURE 8.6 Schematic diagram of the photocatalytic CO_2 reduction on a TiO_2/Ti_3C_2 composite nanostructure [27].

Copyright 2020, Elsevier.

8.3   RESULTS AND DISCUSSION

8.3.1   MXENE/TiO$_2$-RELATED COMPOSITE STRUCTURES

The rapid expansion of the human population and industries has vastly consumed the fossil fuels and caused rising environmental problems for several decades. For that, artificial photosynthesis with CO_2 and H_2O has attracted great attention by converting greenhouse gases. In this section, different semiconductors (CDs, TiO_2, g-C_3N_4, BiOBr, Fe_3O_4, etc.) extensively used and investigated by researchers are discussed [34–35].

Recently, titanium dioxide nanoparticles (NPs) in situ grew on an extremely conductive Ti_3C_2 MXene prepared by the calcination (annealing) method. Particularly, a crust-like structure was achieved by uniform TiO_2 NP distribution on the Ti_3C_2 MXene. The samples (TiO_2/Ti_3C_2 (TT)) were calcinated at different temperatures such as 0, 550 or 650°C. Further, the optimal TiO_2/MXene (Ti_3C_2) composite revealed a higher photocatalytic carbon dioxide reduction performance under CH_4^- production than the P25 TiO_2 nanoparticles (commercially purchased), and the schematic diagram of the processing steps is depicted in Figure 8.6. Based on the results and discussion, a method for the enhancement of the photocatalytic mechanism is discussed [34]. Many strategies, instances, surface modifications (nanoscale level), co-catalyst loading and doping elements of impurity with heterojunction constructions were proposed by various researchers and showed enhanced photocatalytic performance compared to that of TiO_2. In nanoscale engineering, these MXene 2D nanomaterials attributed their maximum electrical conductivity, large specific area to volume ratio and hydrophilicity. MXenes have been extensively investigated and applied in various applications like photocatalysis. Various researchers explored and indicated that the photocatalytic hydrogen production performance of cadmium selenide (CDs) enhanced their performance with the addition a MXene (Ti_3C_2) co-catalyst and improved the high conductivity [21].

8.3.2   MXENE-BASED PHOTOCATALYST PERFORMANCE

MXenes are emerging as better alternatives to enhance photocatalytic performance for various environmental remediation- and renewable energy-related applications due to their high surface area, tuned chemistry and also noticeably adjustable elemental compositions. Currently, around 30 members of the MXene family are available with different elemental compositions and utilized as catalysts. These materials successfully functioned as photocatalysts for nitrogen (N_2) fixation, photochemical degradation, CO_2 reduction, etc. The structure of MXenes and hydrophilic functional groups on the surface generate excellent photocatalytic hydrogen evolution. Also, the surface defects provide major carbon dioxide adsorption sites, and their improved efficient catalytic oxidation properties showed excellent 2D nanomaterials with fast electron transport channels. This MXene material depends on various parameters such as the preparation structure and various synthesis techniques. In photocatalytic applications, MXenes have several benefits such as hydrogen evolution, nitrogen fixation, carbon dioxide reduction, degradation of pollutants and recyclability. Also, researchers critically studied the MXene-based heterostructure and composites in the photocatalyst synthesis process for their best performance for organic pollutant degradation [41].

8.3.3 MXENES AND OTHER MATERIAL-BASED COMPOSITES

The family of 2D transition metal carbides, nitrides and graphene-related nitrides have grabbed considerable attention in photocatalysis. MXenes form composites with other materials like polymers, carbon nanotubes (CNT) and metal oxides. The properties of MXenes are more helpful in various applications. MXenes and MXene composites play two significant roles and show great potential for environmental applications like electro- and photocatalytic water splitting and photocatalytic reduction. These MXene-related composite materials have the highest conductivity, biocompatibility and reductivity. The synthesis methodology and properties of MXenes and MXene-related composites are highlighted in the current advanced technological applications [42].

8.3.4 CO_2 REDUCTION, H_2 EVALUATION AND N_2 FIXATION

8.3.4.1 CO_2 Reduction

In the recent era, the reduction and conversion of carbon dioxide (CO_2) into other fuels like carbon monoxide (CO), methane (CH_4), formaldehyde (HCHO), methanol CH_3OH and formic acid (HCOOH) by using numerous diversified photocatalysts has become more popular to address the problems of energy deficiency and global warming. Because of the low thermodynamic stability of CO_2 molecules, low carrier utilization rate and the poor adsorption and activation of CO_2 molecules, CO_2 reduction using photocatalysts is still challenging. This creates the need for development of highly efficient photocatalysts for the reduction of CO_2 [43–45]. Due to the remarkable properties of MXenes, combining them with other photocatalysts such as perovskite and metal oxides results in different structures like heterostructures and ternary composites, which may boost the photoreduction of CO_2. One such combination is with the two-dimensional g-C_3N_4, which was reported by many researchers. Despite the superior properties of g-C_3N_4, its photocatalytic performance was limited by fast recombination of light-induced electron hole pairs. With superior conductivity, MXenes resolved this issue. Similarly, MXenes enhanced the photocatalytic performance of perovskite by inducing more light absorption, which resolved the major problem of large excitation binding energy that limited the photocatalytic reaction by reducing the charge separation that was faced by most perovskite materials like $Cs_2AgBiBr_6$. Similarly, for metal oxides such as CeO_2, photocatalytic activity can be improved by coupling them with MXenes. This is an effective strategy for the separation of photo-generated carriers by accelerating the transfer of electrons.

8.3.4.2 H_2 Evolution

MXenes have been garnering great attention from the scientific community in the past few decades as the noble metal-free photocatalysts for the evolution of H_2 [27] because of their excellent properties such as hydrophilicity, tunable functional groups, excellent electron transfer efficiency, good chemical stability, large interlayer spacing, suitable Gibbs adsorption free energies and outstanding thermal/electrical conductivity. They can boost the photocatalytic performance of other catalyst materials in various ways. So, in general, they have been combined with other photocatalytic materials for improving the H_2 generation efficiency. To study the effects of $Ti_3C_2T_x$ on the photocatalytic performance using titanium dioxide, photocatalytic H_2 reactions were performed by various $TiO_2/Ti_3C_2T_x$ nanocomposites with different amounts of $Ti_3C_2T_x$ under light irradiation while CH_3OH was used as a hole scavenger. The obtained results identified that the $TiO_2/Ti_3C_2T_x$ nanocomposite performed well over pure TiO_2. An improved photocatalytic performance was attributed to the enhanced light-induced charge carrier separation in TiO_2 by $Ti_3C_2T_x$, which was found by using photoluminescence (PL) and passing photo-current responses for electrochemical impedance (EI) spectrum measurements. Further, it was also identified that the one with a 5% monolayer among various amounts (0–6%) of $Ti_3C_2T_x$ produced the maximum amount of H_2 (2.7 mmol g^{-1} h^{-1}), which was nearly nine times more than that of pure TiO_2 particles (0.2 mmol g^{-1} h^{-1}). Besides the advantages, there are some disadvantages with the monolayer $Ti_3C_2T_x$ such as complex preparation, low structural stability, easy oxidation and difficulty in the manipulation of monolayers. In another study, a ternary $Ni_2P/TiO_2/Ti_3C_2$ hybrid composite was prepared and evaluated for photocatalytic H_2 generation against TiO_2, TiO_2/Ti_3C_2 and Ni_2P/TiO_2. It was found that the MXene-incorporated composite under UV–visible light irradiation achieved the highest H_2 evolution rate of 9425 ppm g^{-1} h^{-1}, which is about 8.28, 4.81 and 2.77 times more than that of the TiO_2, TiO_2/Ti_3C_2 and Ni_2P/TiO_2, respectively. For other MXenes, in situ oxidation worked as a strategy. For instance, a 0D/1D/2D nanohybrid was prepared through the photo-deposition of Ag nanoparticles over the Nb_2O_5 nanorod arrays grown via an in situ process on the Nb_2CT_x. Under the solar simulator, the prepared hierarchical photocatalyst demonstrated an excellent H_2 production rate by the existence of methanol (CH_3OH) and glycerol ($C_3H_8O_3$) with the rate of ~682.2 and ~824.2 µmol g^{-1} h^{-1}, respectively.

8.3.4.3 N_2 Fixation

Nitrogen is a naturally abundant gas and plays pivotal role by preserving life on the earth. At present, nitrogen-based ammonia is an important raw material for the large-scale production of different chemical products for fertilizers. Naturally available nitrogen is unusable due to its stable non-polar energetic triple bond. At high temperature and pressure, the Haber–Bosch process generated a landmark by utilizing atmospheric nitrogen and hydrogen converted to ammonia (NH_3). The traditional synthetic process has hard conditions and causes extensive energy consumption. But the synthesis of ammonia by using the photocatalytic reduction of nitrogen under mild conditions has shown better significance. According to literature surveys, MXene-based photocatalytic nitrogen fixation showed excellent performance and included a Ti_3C_2 and $AgInS_2$ MXene, a RuO_2-loaded MXene hybrid and a microporous MIL-100 (Fe)/Ti_3C_2 MXene composite [44, 46–47]. Also, the density functional theory calculations confirmed that the MXene is a potential catalyst for N_2 reduction [45]. Furthermore, analyzed MXene-based materials are of relative significance for photocatalytic nitrogen fixation. Nowadays, these materials are gaining greater attention in both experimental and theoretical research in the photocatalytic field.

8.3.4.4 Stability

Stability is an important parameter that determines the future of any photocatalyst [44], as well as the efficiency and overall cost of the application [31]. In general, MXenes have low stability [45], and they are prone to oxidation due to larger amounts of metal atoms exposed on the surface [48]. For instance, when Ti_3AlC_2 treated with HF is converted into $Ti_3C_2T_x$, it produces sp^2 π-conjugated graphene quantum dots due to the reaction between unsaturated carbon bonds resulting from the removal of the atoms. On the other hand, during the preparation of $La_2Ti_2O_7$/$Ti_3C_2T_x$ composite, $Ti_3C_2T_x$ is totally oxidized to CO_x-modified TiO_x species, causing the loss of electrical conductivity and the role of co-catalyst. A similar phenomenon has been identified during the synthesis of other MXene-based composites as photocatalysts. Further, to verify the reusability of MXene composites, a cycling test performed using Ag_2WO_4/Ti_3C_2 in terms of antibiotics and observed a clear reduction trend in cyclical production after three cycles. This resulted in approximately 9% and 22% lower removal efficiencies for tetracycline hydrochloride (HCl) and sulfadimidine, respectively. The cause for the declination might be because of photolysis, photocorrosion or photocatalyst damage during reuse. Primarily, in the fabrication process of the Ag_2WO_4/Ti_3C_2 composite, minor quantities of Ag were expected due to the exhibition of stable oxidation and reduction reactivity at the terminal metallic sites of the Ti_3C_2 MXene. Surprisingly, there is no change in the Ag_2WO_4/Ti_3C_2 X-ray diffraction (XRD) patterns before and after the reaction, which revealed the constant MXene maintains better stability [49].

8.4 CONCLUSION AND FUTURE

This chapter summarized the recent progress of MXenes and MXene-based composites as co-catalysts in photocatalysis and discussed various synthesis methods. The synthesized MXene-based nanostructures and their possible mechanism are discussed. The introduction of basic MXenes has shown improved performance of photocatalysis at various sizes and shapes at the nanoscale level. For the past few years, more than 20 types of MXenes have been reported by various researchers through particular etching and exfoliation of metal carbide and carbonitride layers. Also, the synthesis process leads to the surface modification of MXenes using different functional groups (O, F and OH), which depends on the deposition time (t) and temperature (°C). Because of this unique surface chemistry, MXenes have shown fascinating optical, electronic, magnetic, mechanic and electrochemical properties. Further, the half metallicity and topological insulative ability were noticed for the few MXenes and are appealing for two-dimensional (2D) spintronics thanks to their flexibility. These MXene-based metal oxides and other composites improve the specific capacity and generate more electron-conductive pathways by promoting electrolyte ion transportation, which gives better photocatalytic performances of MXenes.

REFERENCES

1. K. Kannan, S. Kishor Kumar, A.M. Abdullah, and B. Kumar, "Current trends in MXene-based nanomaterials for energy storage and conversion system: A mini review", *Catalysts*, 10(5), 495 (2020). https://doi.org/10.3390/catal10050495

2. H. Alnoor, A. Elsukova, J. Palisaitis, I. Persson, E.N. Tseng, J. Lu, L. Hultman, and P.O.A. Persson, "Exploring MXenes and their MAX phase precursors by electron microscopy", *Mater. Today Adv.*, 9, 100123 (2021). https://doi.org/10.1016/j.mtadv.2020.100123

3. Q. Zhong, L. Yuan, and G. Zhang, "Two-dimensional MXene-based and MXene derived photocatalysts: Recent developments and perspectives", *Chem. Eng. J.*, (2020). https://doi.org/10.1016/j.cej.2020.128099

4. X. Li, Y. Bai, X. Shi, N. Su, G. Nie, R. Zhang, H. Nie, and L. Ye, "Applications of MXene (Ti3C2Tx) in photocatalysis: A review", *Mater. Adv.*, 2, 1570 (2021). https://doi.org/10.1039/D0MA00938E

5. P. Miro, M. Audiffred, and T. Heine, "An atlas of two-dimensional materials", *Chem. Soc. Rev.*, 43(18), 6537–6554 (2014). https://doi.org/10.1039/C4CS00102H

6. G.R. Bhimanapati, Z. Lin, V. Meunier, Y. Jung, J. Cha, S. Das, D. Xiao, Y. Son, M.S. Strano, V.R. Cooper, L. Liang, S.G. Louie, E. Ringe, W. Zhou, S.S. Kim, R.R. Naik, B.G. Sumpter, H. Terrones, F. Xia, Y. Wang, J. Zhu, D. Akinwande, N. Alem, J.A. Schuller, R.E. Schaak, M. Terrones, and J.A. Robinson, "Recent advances in two-dimensional materials beyond graphene", *ACS Nano*, 9(12), 11509–11539 (2015). https://doi.org/10.1021/acsnano.5b05556

7. R. Tong, K.W. Ng, X. Wang, S. Wang, X. Wang, and H. Pan, "Two-dimensional materials as novel co-catalysts for efficient solar-driven hydrogen production", *J. Mater. Chem. A*, 8(44), 23202–23230 (2020). https://doi.org/10.1039/D0TA08045D

8. H. Pan, "Principles on design and fabrication of nanomaterials as photocatalysts for water-splitting", *Renew. Sustain. Energy Rev.*, 57, 584–601 (2016). https://doi.org/10.1016/j.rser.2015.12.117

9. M. Shao, Y. Shao, S. Ding, R. Tong, X. Zhong, L. Yao, W.F. Ip, B. Xu, X.-Q. Shi, Y.-Y. Sun, X. Wang, and H. Pan, "Carbonized MoS2: Super-active co-catalyst for highly efficient water splitting on CdS", *ACS Sustain. Chem. Eng.*, 7(4), 4220–4229 (2019). https://doi.org/10.1021/acssuschemeng.8b05917

10. M. Shao, Y. Shao, S. Ding, J. Wang, J. Xu, Y. Qu, X. Zhong, X. Chen, W.F. Ip, N. Wang, B. Xu, X. Shi, X. Wang, and H. Pan, "Vanadium disulfide decorated graphitic carbon nitride for super-efficient solar-driven hydrogen evolution", *Appl. Catal.*, B 237, 295–301 (2018). https://doi.org/10.1016/j.apcatb.2018.05.084

11. R. Tong, Z. Sun, X. Wang, S. Wang, and H. Pan, "Ultrafine WC1-x nanocrystals: An efficient cocatalyst for the significant enhancement of photocatalytic hydrogen evolution on g-C3N4", *J. Phys. Chem.*, C 123(43), 26136–26144 (2019). https://doi.org/10.1021/acs.jpcc.9b07922

12. M. Shao, W. Chen, S. Ding, K.H. Lo, X. Zhong, L. Yao, W.F. Ip, B. Xu, X. Wang, and H. Pan, "WXy/g-C3N4 (X = C, N & S) composites for highly efficient photocatalytic water splitting", *ChemSusChem*, 12(14), 3355–3362 (2019). https://doi.org/10.1002/cssc.201900844

13. A. Raza, A. Rafiq, U. Qumar, and J.Z. Hassan, "2D hybrid photocatalysts for solar energy harvesting", *Sustainable Materials and Technologies*, 33, e00469–1–43 (2022). https://doi.org/10.1016/j.susmat.2022.e00469

14. Y.K. Jo, J.M. Lee, S. Son, and S.-J. Hwang, "2D inorganic nanosheet-based hybrid photocatalysts: Design, applications, and perspectives", *J. Photochem. Photobiol. C: Photochem. Rev.*, 40, 150–190 (2019). https://doi.org/10.1016/j.jphotochemrev.2018.03.002

15. J. Di, S.X. Li, Z.F. Zhao, Y.C. Huang, Y. Jia, and H.J. Zheng, "Biomimetic CNT@TiO2 composite with enhanced photocatalytic properties", *Chem. Eng. J.*, 281, 60e68 (2015). https://doi.org/10.1016/j.cej.2015.06.067

16. A. Bafaqeer, M. Tahir, and N.A.S. Amin, "Synergistic effects of 2D/2D ZnV2O6/RGO nanosheets heterojunction for stable and high performance photo-induced CO2 reduction to solar fuels", *Chem. Eng. J.*, 334, 2142e2153 (2018). https://doi.org/10.1016/j.cej.2017.11.111

17. H.Z. Luo, Z.T. Zeng, G.M. Zeng, C. Zhang, R. Xiao, D.L. Huang, C. Lai, M. Cheng, W.J. Wang, W.P. Xiong, Y. Yang, L. Qin, C.Y. Zhou, H. Wang, Y. Zhou, and S.H. Tian, "Recent progress on metal-organic frameworks based- and derived photocatalysts for water splitting", *Chem. Eng. J.*, 383, 123196 (2020). https://doi.org/10.1016/j.cej.2019.123196

18. J.X. Low, L.Y. Zhang, T. Tong, B.J. Shen, and J.G. Yu, "TiO2/MXene Ti3C2 composite with excellent photocatalytic CO2 reduction activity", *J. Catal.*, 361, 255e266 (2018). https://doi.org/10.1016/j.jcat.2018.03.009

19. J.K. Im, E.J. Sohn, S. Kim, M. Jang, A. Son, K.-D. Zoh, and Y. Yoon, "Review of MXene-based nanocomposites for photocatalysis", *Chemosphere*, 270, 129478 (2021). https://doi.org/10.1016/j.chemosphere.2020.129478

20. F. Qiu, Z. Wang, M. Liu, Z. Wang, and S. Ding, "Synthesis, characterization and microwave absorption of MXene/NiFe2O4 composites", *Cream. Int.*, 47, 24713–24720 (2021). https://doi.org/10.1016/j.ceramint.2021.05.194

21. X. Gao, Z. Jia, B. Wang, X. Wu, T. Sun, X. Liu, Q. Chi, and G. Wu, "Synthesis of NiCo-LDH/MXene hybrids with abundant heterojunction surfaces as a lightweight electromagnetic wave absorber", *Chem. Eng.*, 419, 130019 (2021). https://doi.org/10.1016/j.cej.2021.130019

22. A. Shahzad, K. Rasool, J. Iqbal, J. Jang, Y. Lim, B. Kim, J.-M. Oh, and D.S. Lee, "MXsorption of mercury: Exceptional reductive behavior of titanium carbide/carbonitrides MXenes", *Environ. Res.*, 205, 112532 (2022). https://doi.org/10.1016/j.envres.2021.112532

23. G.P. Lim, C.F. Soon, A.M. Jastrzebska, N.L. Ma, A.R. Wojciechpowska, A. Szuplewska, W.I. Wan Omar, M. Morsin, N. Nayan, and K.S. Tee, "Synthesis, characterization and biophysical evaluation of the 2D Ti2CTx MXene using 3D spheroid-type cultures", *Cream. Int.*, 47, 22567–22577 (2021). https://doi.org/10.1016/j.ceramint.2021.04.268

24. S. Jolly, M.P. Paranthaman, and M. Naguib, "Synthesis of Ti3C2Tz MXene from low-cost and environmentally friendly precursors", *Mater. Today Adv.*, 10, 100139 (2021). https://doi.org/10.1016/j.mtadv.2021.100139

25. X. Zou, H. Liu, X. Wu, X. Han, J. Kang, and K. Madhav Reddy, "A simple approach to synthesis Cr2CTx MXene for efficient hydrogen evolution reaction", *Ceram. Int.*, 20, 100668 (2021). https://doi.org/10.1016/j.mtener.2021.100668

26. V.T. Quyen, L.R. Grummitt, L. Birrell, L. Stapinski, E.L. Barrett, J. Boyle, M. Teesson, and N.C. Newton, "Advanced synthesis of MXene-derived nanoflower-shaped TiO2@Ti3C2 heterojunction to enhance photocatalytic degradation of rhodamine B", *Environ. Technol. Innov.*, 21, 101286 (2021). https://doi.org/10.1016/j.eti.2020.101286

27. Y. Guo, S. Jin, L. Wang, P. He, Q. Hu, L.-Z. Fan, and A. Zhou, "Synthesis of two-dimensional carbide Mo2CTx MXene by hydrothermal etching with fluorides and its thermal stability", *Cream. Int.*, 46, 19550–19556 (2020). https://doi.org/10.1016/j.ceramint.2020.05.008

28. P. Kuang, J. Low, B. Cheng, J. Yu, and J. Fan, "MXene-based photocatalysts", *J. Mater. Sci. Technol.*, 56, 18–44 (2020). https://doi.org/10.1016/j.jmst.2020.02.037

29. S. Cao, B. Shen, T. Tong, J. Fu, and J. Yu, "2D/2D heterojunction of ultrathin MXene/ Bi2WO6 nanosheets for improved photocatalytic CO2 reduction", *Adv. Funct. Mater.*, 28(21), 1800136 (2018). https://doi.org/10.1002/adfm.201800136

30. N. Liu, N. Lu, Y. Su, P. Wang, and X. Quan, "Fabrication of g-C3N4/Ti3C2 composite and its visible-light photocatalytic capability for ciprofloxacin degradation", *Sep. Purif. Technol.*, 211, 782–789 (2019). https://doi.org/10.1016/j.seppur.2018.10.027

31. Y.L. Sun, X. Meng, Y. Dall'Agnese, C. Dall'Agnese, S.N. Duan, Y. Gao, G. Chen, and X.F. Wang, "2D MXenes as co-catalysts in photocatalysis: Synthetic methods", *Nano-Micro Lett.*, 11(1), 79 (2019). https://doi.org/10.1007/s40820-019-0309-6

32. Y.L. Sun, D. Jin, Y. Sun, X. Meng, Y. Gao, Y. Dall'Agnese, G. Chen, and X.F. Wang, "g-C3N4/Ti3C2Tx (MXenes) composite with oxidized surface groups for efficient photocatalytic hydrogen evolution", *J. Mater. Chem. A*, 6, 9124–9131 (2018). https://doi.org/10.1039/C8TA02706D

33. M.H. Ye, X. Wang, E.Z. Liu, J.H. Ye, and D.F. Wang, "Boosting the photocatalytic activity of P25 for carbon dioxide reduction by using a surface-alkalized titanium carbide MXene as cocatalyst", *ChemSusChem*, 11, 1606–1611 (2018). https://doi.org/10.1002/cssc.201800083

34. L. Xiu, Z. Wang, M. Yu, X. Wu, and J. Qiu, "Aggregation-resistant 3D MXene-based architecture as efficient bifunctional electrocatalyst for overall water splitting", *ACS Nano*, 12(8), 8017e8028 (Aug. 2018). https://doi.org/10.1021/acsnano.8b02849

35. D. Zhao, M. Clites, G. Ying, S. Kota, J. Wang, V. Natu, X. Wang, E. Pomerantseva, M. Cao, and M.W. Barsoum, "Alkali-induced crumpling of Ti3C2TX (MXene) to form 3D porous networks for sodium ion storage", *Chem. Commun.*, 54(36), 4533e4536 (2018). https://doi.org/10.1039/C8CC00649K

36. X. Wang, Q. Fu, J. Wen, X. Ma, C. Zhu, X. Zhang, and D. Qi, "3D Ti_3C_2TX aerogels with enhanced surface area for high performance supercapacitors", *Nanoscale*, 10(44), 20828e20835 (2018). https://doi.org/10.1039/C8NR06014B

37. V. Natu, M. Clites, E. Pomerantseva, and M.W. Barsoum, "Mesoporous Mxene powders synthesized by acid induced crumpling and their use as Na-ion battery anodes", *Mater. Res. Lett.*, 6(4), 230e235 (2018). https://doi.org/10.1080/21663831.2018.1434249

38. R. Thakur, A. Vahid Mohammadi, J. Moncada, W.R. Adams, M. Chi, B. Tatarchuk, M. Beidaghi, and C.A. Carrero, "Insights into the thermal and chemical stability of multilayered V_2CT_X MXene", *Nanoscale*, 11(22), 10716e10726 (2019). https://doi.org/10.1039/C9NR03020D

39. M. Seredych, C.E. Shuck, D. Pinto, M. Alhabeb, E. Precetti, G. Deysher, B. Anasori, N. Kurra, and Y. Gogotsi, "High-temperature behavior and surface chemistry of carbide MXenes studied by thermal analysis", *Chem. Mater.*, 31(9), 3324e3332 (2019). https://doi.org/10.1021/acs.chemmater.9b00397

40. X. Xie, M.Q. Zhao, B. Anasori, K. Maleski, C.E. Ren, J. Li, B.W. Byles, E. Pomerantseva, G. Wang, and Y. Gogotsi, "Porous heterostructured MXene/carbon nanotube composite paper with high volumetric capacity for sodium-based energy storage devices", *Nanomater. Energy*, 26, 513e523 (Aug. 2016). https://doi.org/10.1016/j.nanoen.2016.06.005

41. T. Aneef, K. Rasool, J. Iqbal, R. Nawaz, M. Raza Ul Mustafa, K. Mohmoud, T. Sarkar, and A. Shahzad, "Recent progress in two dimensional MXenes for photocatalysis: A critical review", *2D Materials*, 10, 012001 (2023). https://doi.org/10.1088/2053-1583/ac9e66

42. X. Zhan, C. Si, J. Zhou, and Z. Sun, "MXene and MXene-based composites: Synthesis, properties and environment-related applications", *Nanoscale Horiz.*, 5, 235–258 (2020). https://doi.org/10.1039/C9NH00571D

43. J. Low, L. Zhang, T. Tong, B. Shen, and J. Ju, "TiO$_2$/MXene Ti$_3$C$_2$ composite with excellent photocatalytic CO$_2$ reduction activity", *J. Catal.*, 361, 255–266 (2018). https://doi.org/10.1016/j.jcat.2018.03.009

44. J.R. Ran, G.P. Gao, F.T. Li, T.Y. Ma, A.J. Du, and S.Z. Qiao, "Ti3C2 MXene co-catalyst on metal sulfide photo-absorbers for enhanced visible-light photocatalytic hydrogen production," *Nat. Commun.*, 8, 13907 (2017). https://doi.org/10.1038/ncomms13907

45. Z. Guo, J. Zhou, L. Zhu, and Z. Sun, "MXene: A promising photocatalyst for water splitting", *J. Mater. Chem.*, 4(29), 2016. https://doi.org/10.1039/C6TA04414J

46. C.Y. Hao, Y. Liao, Y. Wu, Y.J. An, J.N. Lin, Z.F. Gu, M.H. Jiang, S. Hu, and X.T. Wang, "RuO$_2$-loaded TiO$_2$-MXene as a high performance photocatalyst for nitrogen fixation", *J. Phys. Chem. Solids*, 136, 109141 (2020). https://doi.org/10.1016/j.jpcs.2019.109141

47. H.M. Wang, R. Zhao, J.Q. Qin, H.X. Hu, X.W. Fan, X. Cao, and D. Wang, "MIL-100 (Fe)/Ti3C2 MXene as a Schottky Catalyst with enhanced photocatalytic oxidation for nitrogen fixation activities", *ACS Appl. Mater. Interfaces*, 11, 44249–44262 (2019). https://doi.org/10.1021/acsami.9b14793

48. J.Z. Qin, X. Hu, X.Y. Li, Z.F. Yin, B.J. Liu, and K. Lam, "0D/2D AgInS2/MXene Z-scheme heterojunction nanosheets for improved ammonia photosynthesis of N$_2$", *Nano Energy*, 61, 27–35 (2019). https://doi.org/10.1016/j.nanoen.2019.04.028

49. J.H. Peng, X.Z. Chen, W.J. Ong, X.J. Zhao, and N. Li, "Surface and heterointerface engineering of 2D MXenes and their nanocomposites: Insights into electro- and photo-catalysis", *Chem*, 5(1), 18–50 (2019). https://doi.org/10.1016/j.chempr.2018.08.037

9 MXene-Based Materials for Antibacterial Activity

*Usha Kiran Rout, Chita Ranjan Sahoo,
and Debdutta Bhattacharya*

9.1 INTRODUCTION

A broad set of early transition metal carbides is newly added to the family of 2D materials. MXenes are an intriguing class of two-dimensional (2D) materials that are made out of transition metals, carbon, and nitrogen. MXenes were initially synthesized in 2011 by Gogotsi and Barsoum from the University of Drexel in Philadelphia. MXenes are a sort of layered material, and their structure looks like a sandwich. The transition metal layer in the middle is flanked by carbon and nitrogen, forming the outer layers. MXenes are exceptionally conductive, which makes them a great possibility for use in energy capacity gadgets like batteries and supercapacitors. Moreover, MXenes have high mechanical strength and can be effortlessly synthesized in enormous amounts, making them ideal for use in many applications [1].

MXenes have been viewed as reasonable for different purposes including energy stockpiling, water cleaning, catalysis, and sensing. Research is in progress to investigate their true capacity in different fields like biomedicine, natural remediation, and electrochemical detecting. The discovery of MXenes has opened up another domain of possibilities in materials for science and engineering. With their exceptional properties and colossal applications, they are ready to fundamentally affect various mechanical fields from now on [2].

Presently, antimicrobial resistance (AMR) is a worldwide developing threat. The WHO has proclaimed that AMR is one of the top 10 worldwide public health threats confronting humankind. The excessive use and misuse of antibiotics has caused increasing cases of antibiotic resistance. Bacterial resistance to more than one class of antibiotics is very common nowadays. This increasing antibiotic resistance has led to the development of novel approaches to possessing antibacterial activity against multidrug-resistant bacterial infection. In this prospect, nanomaterials have a considerably greater chance of withstanding and escaping bacterial resistance than antibiotics. This is because of the high membrane permeability and bio-/cyto-compatibility nanomaterials. The antimicrobial activity of the nanomaterial depends upon the kind of microbes and the physicochemical features of nanoparticles. MXenes, a novel family of two-dimensional nanomaterials, have sparked considerable attention in biological applications and provide promising results to overcome antimicrobial resistance. These MXenes are expressed as $M_{n+1}AX_n$, where

DOI: 10.1201/9781003366225-9

M denotes early transition metal elements, i.e. Ti, Ni, Mo, V, and so on; X denotes nitrogen and/or carbon; A denotes the primary group of SP elements (Al, Ga, Si, etc.); and n = 1, 2, 3.

The suffix 'ene' indicates that these 2D-layered MXenes are analogous to graphene. The layered hexagonal MAX phases are made up of tightly packed M layers and octahedral sites inhabited by X atoms, with the layers of M and X bound together by A atoms. The M–X bond shows a combination of ionic/metallic/covalent properties in the MAX phase, whereas the M–A bond has a metallic character in most circumstances. These MXenes have a hexagonally close-packed (HCP) crystal structure. The M–X bond in MAX is metallic and difficult to break with simple mechanical shearing.

9.2   PREPARATION OF MXENES

MXenes are synthesized using the MAX ternary phase [3–5]. The following sections describe a general preparation method for MXenes.

9.2.1   Selecting MAX Phase Materials

The first step in MXene synthesis involves the selection of the appropriate MAX phase precursor, which is typically chosen based on its availability, desired MXene composition, and properties. These M, A, and X elements from the periodic table react together and form MAX phases (Figure 9.1). These MAX phases are hexagonally closely packed ternary carbides, nitrides, or carbonitrides consisting of M, A, and X layers.

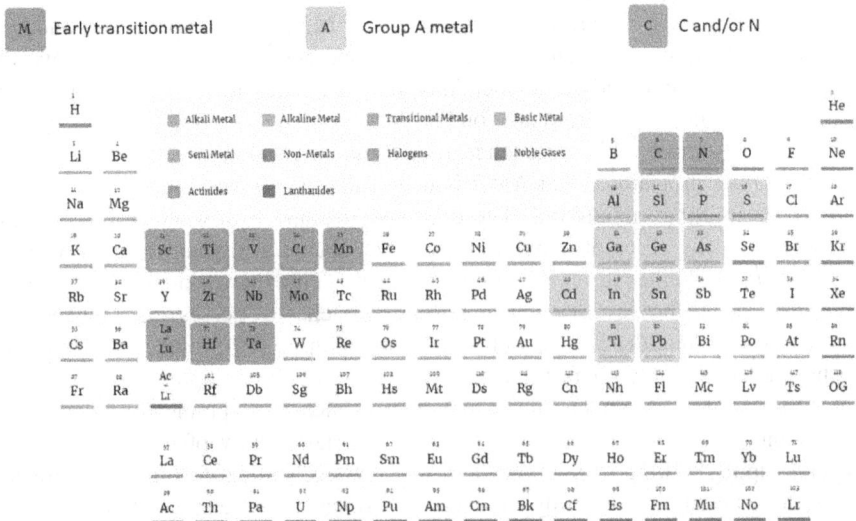

FIGURE 9.1   Graphical representation of the MAX phase designed by Slidechef template.

9.2.2 ETCHING

These MAX phases are selectively etched with acid containing aqueous fluoride which eliminates the A layer from the MAX phase. This is etched using a solution of hydrofluoric acid (HF) and/or a combination of HF and another etchant such as lithium fluoride (LiF) or sodium chloride (NaCl). The etching solution removes the A layer, leaving behind the MXene. In this method, the MAX phase powders are first layered and are mixed with aqueous hydrofluoric (HF) acid at room temperature for a certain time. As a consequence, the A layers are selectively etched from the MAX phase. The metallic linkages between the MX layers are replaced on the surface of the multi-layered MXene by weak bonds of surface terminations such as fluoride, hydroxyl, or oxygen.

9.2.3 WASHING AND FILTERING

The resulting material is then washed several times to remove any residual acid and dried to obtain a partially exfoliated material that has a layered structure with MXene layers interleaved with water. The process is repeated until the pH of the entire mixture remains between 4 and 6. Further lowering the pH of the MXene solution might cause the crumpling of MXene flakes. This multi-layered MXene is reduced to a few-layered MXene consisting of fewer than five layers. The multi-layered MXene is further subjected to sonication, which results in the formation of delaminated singular MXene flakes. Centrifugation and filtration separate the supernatant from the solid to obtain a pure MXene product. These MXene flakes are dried in a vacuum oven. The general process is given in the Figure 9.2. To date, more than

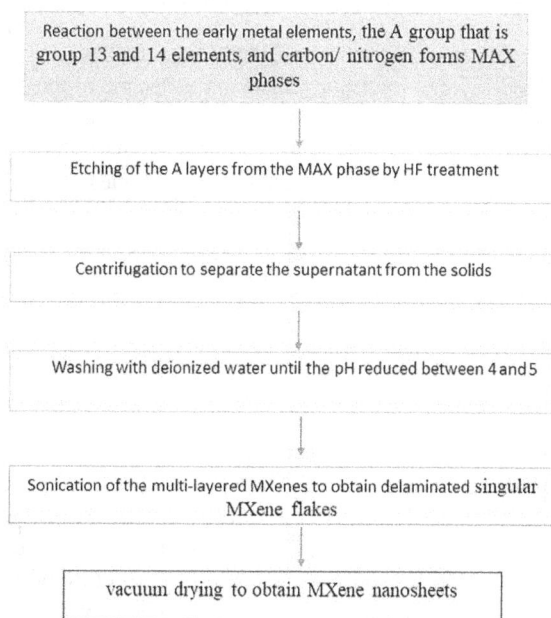

FIGURE 9.2 Schematic illustration of the MXene synthesis procedure.

70 MAX phases have been identified; 37 of them have been experimentally pre-dicted and effectively exfoliated into 2D MXenes.

M–X bonds are more powerful than M–A bonds. Other techniques for produc-ing an A element with selected characteristics includes heating MAX phases under vacuum or by immersion in molten salts or molten metals at high temperatures. Other method to etch the A element for the synthesis of MXene nanosheets includes hydrothermal treatment and chlorination at high temperature. However, HF etching is most commonly used.

9.3 PROCEDURE OF $Ti_3C_2T_x$ MXENES

i. In a beaker, mix the 10mL of 9M of HCl solution with 0.8 g of LiF and leave 30 minutes at 300rpm until the LiF completely dissolved.

ii. The process is exothermic reaction. So, to avoid bubbling and overheating, add 0.5g of Ti_3AlC_2 MAX phase gently to the etchant solution. The entire mixture is then subjected to agitation at room temperature for 24 hours.

iii. Wash the solution with deionized water by centrifugation at 3500rpm for 5min multiple times until the pH reached between 4 and 5. After centrifu-gation, discard the supernatant and again wash with fresh deionized water.

iv. Upon reaching pH >5, a dark green supernatant and $Ti_3C_2T_x$ sediment can be observed. It can be seen that the $Ti_3C_2T_x$ slurry sediment will swell up to twice the volume after removing the supernatant.

v. Carefully remove the slurry and wash the solids three to four times with DI water.

vi. Again, rinse the product four more times using an ultrahigh-speed centri-fuge at 3500rpm for 10 minutes.

vii. To acquire delaminated solitary Ti_3C_2MXene flakes, sonicate the multi-layered Ti_3C_2MXene for 1 hour using an ultrasonic probe sonicator.

viii. Dry the delaminated MXene flakes in a vacuum oven at 120°C for 24 hours. The series of chemical reactions aiming to remove the A layers from MAX phases which are ternary carbides or nitrides can be described as follows:

MXenes involve a series of chemical reactions that are:

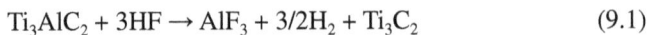

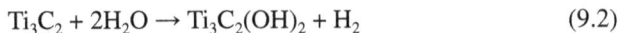

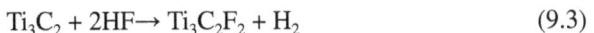

$$Ti_3AlC_2 + 3HF \rightarrow AlF_3 + 3/2H_2 + Ti_3C_2 \qquad (9.1)$$

$$Ti_3C_2 + 2H_2O \rightarrow Ti_3C_2(OH)_2 + H_2 \qquad (9.2)$$

$$Ti_3C_2 + 2HF \rightarrow Ti_3C_2F_2 + H_2 \qquad (9.3)$$

The surface morphology can be characterized by scanning electronic microscopy (SEM) of Ti_3AlC_2 powders and $Ti_3C_2T_x$. Ti_3AlC_2 particles exhibited layered struc-tures held together firmly. The precursor Ti_3AlC_2 powders are very dense due to the metallic Ti–Al bond in their structure and are difficult to delaminate with sonication. The $Ti_3C_2T_x$ MXene powders show an accordion-like structure with loosely stacked layers and can be quickly produced as MXene nanosheets by sonication. The use of HF poses a risk to both the environment and human health. Alternatives to the use of HF include chlorination and hydrothermal procedures.

In the chlorination method, the MXene powder is put into a tube furnace, heated to a favourable temperature under a flow of argon, and then chlorinated with Cl_2 gas. After the chlorination process, the samples are cooled under argon or ammonia flow to remove residual metal chlorides trapped in the pores. The synthesized MXenes have a layered structure with an enormous specific surface area.

In the hydrothermal process, synthesis involves a multilayer $Ti_3C_2T_x$ in a NaOH–water solution (27.5M) at 72°C under an argon atmosphere to prevent oxidation of the sample. MXenes synthesized with a fluorine-free method have a compact layered structure similar to $Ti_3C_2T_x$ prepared with 5%wt HF and have a high purity of about 92%wt. Reducing the temperature to 100–220°C at any concentration of the NaOH solution, no MXene will form. So, temperature is the main factor in the production of $Ti_3C_2T_x$ by hydrothermal process, and engagement of NaOH only affects purity.

9.4   STUDIES DONE TO CHECK MXENE ACTIVITY SO FAR

S. No.	Composition	Mechanism	Activity checked against	Results available	Reference
1	$Ti_3C_2T_x$ MXene	MXene caused irreparable damage to the bacterial cell membrane, resulting in growth inhibition.	*E. coli* and *B. subtilis*	Flow cytometric assay for bacterial viability analysis, colony forming counts, cell morphology observation with SEM and AFM	[1]
2	$Ti_3C_2T_x$ MXene	Direct-contact killing, cell membrane damage resulting in prevention of nutrient intake, oxidative stress	*E. coli* and *B. subtilis*	SEM and TEM images, as well as glutathione oxidation assays	[3]
3	$Ti_3C_2T_x$ within chitosan (CS) nanofibres	Reduces growth rate	*E. coli* and *S. aureus*	SEM and culture based	[4]
4	MXene/tungsten oxide (WO_3)	Photocatalytic degradation, electrostatic interaction	*S. aureus, E. coli, K. pneumonia, P. vulgaris*	Culture based	[5]
5	MXene + diacetoxyl thioketal (TK) linkage + doxorubicin coated with polydopamine (MXene-TK-DOX@PDA)	Interact with the bacterial secretory proteins and their metabolism. Also, PDA possesses photothermal antibacterial activity.	*E. coli, B. subtilis*	Culture based	[6]

(*Continued*)

9.4 CONTINUED

S. No.	Composition	Mechanism	Activity checked against	Results available	Reference
6	MXene/cobalt ferrite (MXene/CoFe$_2$O$_4$)	Phototherapy performance inducing membrane permeability and intracellular protein leakage. MXene/ CoFe$_2$O$_4$ acts as a peroxidase-mimicking catalyst that disintegrates the bacterial antioxidant defence system.	E. coli and S. aureus	Spread plate method, fluorescent microscopy, SEM, bacterial membrane permeability test, ROS level, GSH oxidase-mimicking activity, peroxidase-mimicking property, CAT-mimicking activity, hemocompatibility test, cytotoxicity study, in vivo animal model	[7]
7	MXene– functionalized graphene (FG) nanocomposites	A combination of physical (direct contact) and chemical (oxidative stress) mechanisms	E. coli and S. aureus (MRSA)	SEM, fluorescence microscopy, total ROS assessment, cytotoxicity study	[8]
8	Ruthenium (II) complex grafted Ti$_3$C$_2$T$_x$	Physical cutting of bacterial cell membranes, hyperthermia, ROS production, germicidal effect by adhering to bacteria with negative potential	E. coli and S. aureus	Plate-counting method, live/dead staining, and TEM	[9]
9	Titanium dioxide (TiO$_2$)/MXene	Photocatalytic activity and ROS generation	E. coli and B. subtilis	Plate culture based	[10]
10	Ti$_3$C$_2$T$_x$/TiO$_2$	Photocatalytic activity and ROS generation	E. coli O111	Colony-counting method	[11]
11	Ag/Ti$_3$C$_2$T$_x$	Photothermal and intrinsic activity	E. coli and S. aureus	Growth curve and colony-counting method	[12]
12	Colloidal Ti$_3$C$_2$T$_x$	Oxidative stress and fast membrane depolarization	Staphylococcus aureus and Pseudomonas aeruginosa	Flow cytometry and fluorescence imaging (FI)	[13]
13	Bi$_2$S$_3$/Ti$_3$C$_2$T$_x$	Photothermal effect, ROS generation, and hyperthermia	E. coli and S. aureus	Fluorescence imaging and SEM	[14]

S. No.	Composition	Mechanism	Activity checked against	Results available	Reference
14	$Ti_3C_2T_x$ immobilized lysozyme	Light-enhanced enzymatic inactivation	methicillin-resistant *Staphylococcus aureus*	Live/dead bacterial staining assay, SEM, in vitro and in vivo biosafety analysis, i.e. haemolysis assay and cytotoxicity test	[15]
15	2D MXene sheets incorporated into the silane film on metal surface	Inhibits growth	*E. coli*	Plate-counting method, propidium iodide (PI) staining, and morphology check by SEM	[16]
16	$Nb_2CT_x/Nb_4C_3T_x$	Mechanical damage, oxidative stress	*E. coli* and *S. aureus*	In vitro and in vivo assays	

9.5 ANTIBACTERIAL PROPERTIES

MXene-based amalgams offer diverse benefits due to exhibiting broad-range antibacterial potential, effectively decreasing bacterial resistance, as well as non-invasive strategies for combating antibiotic-resistant bacteria. These materials have exceptional physical, chemical, and mechanical properties which enhance therapeutic properties, drug delivery, and tissue engineering systems. Their outstanding optical, thermal, and electrical capabilities make them a profound agent for electromagnetic shielding, electrochemical capacitors, photocatalysts, and biosensors. MXene nanosheets are metal-based bactericides that cause physical/mechanical destruction to bacterial cell. MXene nanomaterials have been reported to possess activity against Gram-positive bacteria like *Staphylococcus aureus* and *Bacillus subtilis* and Gram-negative bacteria like *Escherichia coli*, *Staphylococcus aureus*, *Pseudomonas aeruginosa*, and *Salmonella typhimurium*. Several experiments on MXenes' antibacterial activity have been performed, which have shown they mainly act on bacterial cell membranes.

The sharp edges of MXenes interact with the bacterial cell membrane by penetrating the bacterial surface, thus causing disruption of the cell membrane and leakage of intracellular components along with chemical damage upon the generation of oxidative stress and charge transfer. These mechanisms slow down the bacterial growth, ultimately causing bacterial cell death. Due to their hydrophilicity, negatively charged MXene nanosheets cause the microbial inactivation with direct-contact killing mechanisms. The hydrogen bonding between MXene nanomaterials and the polysaccharide membrane disables the bacterial cells for food intake, thus inhibiting the bacterial cells. Another innovative mode of action of MXenes is photothermal therapy (PTT). PTT employing MXene amalgams is a non-invasive technique to control bacterial multiplication and a good alternative to antibiotics. In this method, the PTT agents cause hyperthermia that interferes with the DNA structure, inactivates the proteins, and thus damages the cell membrane. This allows the antimicrobial compounds to penetrate and eliminate biofilm structures.

Photothermal therapy (PTT) in combination with photodynamic therapy (PTT/PDT) or chemo treatment (chemo/PTT) has shown synergistic effects that can significantly improve antibacterial efficiency, particularly against antibiotic-resistant bacteria. The heat caused by PTT increases the cell permeability, promoting the free movement of reactive oxygen species (ROS) or metal ions within the cells. Furthermore, these combinational strategies reduce the dose requirements both for photothermal agents and antibiotics, resulting in better elimination of pathogens, reducing potential side effects, and improving therapeutic efficiency [2].

Research has been conducted with graphene and graphene oxide (GO) to check bacterial removal activity. The researchers concluded that adding graphene to membrane filters improves their antimicrobial properties. The antibacterial properties of a $Ti_3C_2T_x$ MXene were studied for two bacteria, *Bacillus subtilis* (*B. subtilis*) and *Escherichia coli* (*E. coli*), and then analyzed by the colony-counting method. $Ti_3C_2T_x$ nanosheet dispersion showed growth inhibition of more than 98% for both bacteria after 4hr treatment with 100g/mL concentration of nanomaterial. The antibacterial activity of $Ti_3C_2T_x$ was compared with GO using *E. coli* and *B. subtilis* under the same conditions and different concentrations varying from 2 to 200g/mL. At the concentration of 2g/mL, the survival rate of *E. coli* and *B. subtilis* were 92.53% and 93.96%, respectively. As the MXene concentration increased up to 100g/mL, more than 96% of bacterial cells started losing their viability [17].

Another study introduced the antibacterial properties of $Ti_3C_2T_x$ within chitosan (CS) nanofibres on Gram-negative *E. coli* and Gram-positive *S. aureus*. Treatment with 0.75%wt $Ti_3C_2T_x$ up to 4 hours resulted in a reduction of 95% in *E. coli* cells and 62% in *S. aureus* cells. The reduced effectiveness in *S. aureus* was observed due to the thicker cell structure of Gram-positive bacteria. $Ti_3C_2T_x$/CS composite nanofibres efficiently decreased *E. coli* and *S. aureus* at low $Ti_3C_2T_x$ concentrations (0.75%wt. $Ti_3C_2T_x$/CS) [4].

The antibacterial activity of MXenes in combination with cuprous oxide (Cu_2O) has also been studied against *P. aeruginosa* and *S. aureus*. The Cu_2O nanospheres anchored on the MXene surface damage cell membranes and kill bacteria by releasing Cu^{2+} and generating reactive oxygen species when making contact with the bacterial membrane. The antibacterial activity against *P. aeruginosa* was about 83.21% and that against *S. aureus* was 81.56%. A Cu_2O/MXene combination strategy showed the loss of cell viability in *P. aeruginosa* of 97.04% and *S. aureus* of 95.59%. Cu_2O/MXene had higher bactericidal activity due to the MXene nanosheets increasing the antibacterial properties of Cu_2O nanospheres [18].

In another work, an SEM study was done with *P. aeruginosa* and *S. aureus* cell morphology. Before treatment with Cu_2O/MXene, *P. aeruginosa* and *S. aureus* were rod shaped and sphere shaped, respectively. After treatment with the Cu_2O/MXene, the shape of the cells changed. The above research concluded that MXenes have better antibacterial activity against *E. coli* and *B. subtilis* bacteria than against *P. aeruginosa* and *S. aureus*, as well as a more significant reduction in cell viability.

The combination of MXenes with cellulose acetate (CA) has shown significant antibacterial activity. These cellulose acetates are cost friendly, hydrophilic, tough, biocompatible, and have great film-forming capability. Due to their fouling potential, CA membranes were incorporated with 2D nanomaterials to improve their antifouling properties. A phase inversion method was used to prepare MXene/CA cross-linked nanoporous membranes with formaldehyde crosslinks in different weight percentages of MXene (2, 6, 8 and, 10%wt). The antibacterial activity was evaluated against *E. coli* and *B. subtilis*. The result showed that an increased concentration of MXene decreased the cell viability. The composite consisting of 10% MXene/CA showed about 98% and 96% cell death for *E. coli* and *B. subtilis*, respectively. The antibacterial activity was highly dependent on the concentration of $Ti_3C_2T_x$ in the composite membrane [19].

MXene/CA has excellent properties for water treatment. The antibacterial activity of 10% MXene/CA is greater than the activity of a crosslinked cellulose acetate (CCA) membrane. The bacterial cells on the CCA membrane were found to be healthy; however, very few or no viable cells were observed on the 10% MXene/CA composite. The $Ti_3C_2T_x$ prepared on poly(vinylidene) fluoride (PVDF) membrane was studied against *E. coli* and *B. subtilis*. The MXene with PVDF membrane treated for 24 hours caused a reduction in bacterial colonies. The growth of bacterial colonies on fresh $Ti_3C_2T_x$/PVDF membranes showed 67% inhibition of growth for *E. coli* and 73% for *B. subtilis*. The *E. coli* and *B. subtilis* cells were cultured on fresh as well as aged $Ti_3C_2T_x$ MXene-coated PVDF membranes for 24 hours. The result showed the aged membrane to have better antibacterial properties. This may be due to the oxidation of the $Ti_3C_2T_x$ surface. More than 90% of bacterial growth is inhibited on the aged $Ti_3C_2T_x$/PVDF membranes [1].

The antibacterial behaviour of MXene, 21% Ag/MXene, and PVDF composite membranes were also compared toward *E. coli* bacteria at 35°C. *E. coli* growth inhibition for 21% Ag/MXene was 99%, while only 60% growth inhibition was observed for the MXene membrane against the PVDF-based membrane, indicating that the 21% Ag/MXene membrane is more effective than the other two membranes [20]. The difference in antibacterial activity against both bacteria can be due to the difference in the cell wall of both bacteria; *B. subtilis* has a peptide glycan layer between 20 to 80 nm within the inner and outer cell wall membranes, but the layer of peptidoglycan for *E. coli* is thinner (2–3 nm) and as a result has less resistance to $Ti_3C_2T_x$ film.

9.6 CONCLUSION

MXene nanomaterials have emerged as promising candidates for various biomedical applications, especially as a novel class of antibacterial agents, due to their broad-spectrum antibacterial activity, nanoscale size, and high surface area-to-volume ratio. They offer great potential for the treatment of bacterial infections, including antibiotic-resistant bacterial infections. In addition to their antibacterial properties, MXenes have also been found to be non-toxic to human cells, making them a safe and promising material for use in medical applications. However, more research is needed to fully understand the potential of MXenes as antibacterial agents and determine their potential applications in various fields. Further studies are needed to ascertain the exact mechanism of their antibacterial activity and to optimize their properties for biomedical applications. This chapter explored MXene nanomaterials and their potential to serve as novel antibacterial agents.

REFERENCES

[1] K. Rasool, K. A. Mahmoud, D. J. Johnson, M. Helal, G. R. Berdiyorov, and Y. Gogotsi, "Efficient Antibacterial Membrane based on Two-Dimensional Ti3C2Tx (MXene) Nanosheets," *Sci Rep*, vol. 7, no. 1, p. 1598, May 2017, doi: 10.1038/s41598-017-01714-3.

[2] S. Iravani and R. S. Varma, "MXene-based composites against antibiotic-resistant bacteria: Current trends and future perspectives," *RSC Adv*, vol. 13, no. 14, pp. 9665–9677, 2023, doi: 10.1039/D3RA01276J.

[3] K. Rasool, M. Helal, A. Ali, C. E. Ren, Y. Gogotsi, and K. A. Mahmoud, "Antibacterial activity of $Ti_3C_2T_x$ MXene," *ACS Nano*, vol. 10, no. 3, pp. 3674–3684, Mar. 2016, doi: 10.1021/acsnano.6b00181.

[4] E. A. Mayerberger, R. M. Street, R. M. McDaniel, M. W. Barsoum, and C. L. Schauer, "Antibacterial properties of electrospun $Ti_3C_2T_z$ (MXene)/chitosan nanofibers," *RSC Adv*, vol. 8, no. 62, pp. 35386–35394, 2018, doi: 10.1039/C8RA06274A.

[5] A.-Z. Warsi, F. Aziz, S. Zulfiqar, S. Haider, I. Shakir, and P. O. Agboola, "Synthesis, characterization, photocatalysis, and antibacterial study of WO3, MXene and WO3/MXene nanocomposite," *Nanomaterials*, vol. 12, no. 4, p. 713, Feb. 2022, doi: 10.3390/nano12040713.

[6] W.-J. Zhang et al., "ROS- and pH-responsive polydopamine functionalized Ti3C2Tx MXene-based nanoparticles as drug delivery nanocarriers with high antibacterial activity," *Nanomaterials*, vol. 12, no. 24, p. 4392, Dec. 2022, doi: 10.3390/nano12244392.

[7] J. Shi et al., "Multi-activity cobalt ferrite/MXenenanoenzymes for drug-free phototherapy in bacterial infection treatment," *RSC Adv*, vol. 12, no. 18, pp. 11090–11099, 2022, doi: 10.1039/D2RA01133F.

[8] M. S. Salmi, U. Ahmed, N. Aslfattahi, S. Rahman, J. G. Hardy, and A. Anwar, "Potent antibacterial activity of MXene—Functionalized graphene nanocomposites," *RSC Adv*, vol. 12, no. 51, pp. 33142–33155, 2022, doi: 10.1039/D2RA04944A.

[9] X. Liu et al., "Ru(II) complex grafted Ti3C2Tx MXene nano sheet with photothermal/photodynamic synergistic antibacterial activity," *Nanomaterials*, vol. 13, no. 6, p. 958, Mar. 2023, doi: 10.3390/nano13060958.

[10] S. Lu, G. Meng, C. Wang, and H. Chen, "Photocatalytic inactivation of airborne bacteria in a polyurethane foam reactor loaded with a hybrid of MXene and anatase TiO2 exposing {0 0 1} facets," *Chem Eng J*, vol. 404, p. 126526, Jan. 2021, doi: 10.1016/j.cej.2020.126526.

[11] K. Rajavel, S. Shen, T. Ke, and D. Lin, "Photocatalytic and bactericidal properties of MXene-derived graphitic carbon-supported TiO2 nanoparticles," *Appl Surf Sci*, vol. 538, p. 148083, Feb. 2021, doi: 10.1016/j.apsusc.2020.148083.

[12] X. Zhu et al., "A near-infrared light-mediated antimicrobial based on $Ag/Ti_3C_2T_x$ for effective synergetic antibacterial applications," *Nanoscale*, vol. 12, no. 37, pp. 19129–19141, 2020, doi: 10.1039/D0NR04925E.

[13] A. Arabi Shamsabadi, M. Sharifian, B. Anasori, and M. Soroush, "Antimicrobial mode-of-action of colloidal Ti3C2Tx MXene nanosheets," *ACS Sustain Chem Eng*, vol. 6, no. 12, pp. 16586–16596, Dec. 2018, doi: 10.1021/acssuschemeng.8b03823.

[14] J. Li et al., "Interfacial engineering of Bi2S3/Ti3C2Tx MXene based on work function for rapid photo-excited bacteria-killing," *Nat Commun*, vol. 12, no. 1, p. 1224, Feb. 2021, doi: 10.1038/s41467-021-21435-6.

[15] D. Zhang, L. Huang, D.-W. Sun, H. Pu, and Q. Wei, "Bio-interface engineering of MXene nanosheets with immobilized lysozyme for light-enhanced enzymatic inactivation of methicillin-resistant Staphylococcus aureus," *Chemical Engineering Journal*, vol. 452, p. 139078, Jan. 2023, doi: 10.1016/j.cej.2022.139078.

[16] Y. Nie *et al.*, "MXene-hybridized silane films for metal anticorrosion and antibacterial applications," *Appl Surf Sci*, vol. 527, p. 146915, Oct. 2020, doi: 10.1016/j.apsusc.2020.146915.

[17] F. Abbasi, N. Hajilary, and M. Rezakazemi, "Antibacterial properties of MXene-based nanomaterials: A review," *Materials Express*, vol. 12, no. 1, pp. 34–48, Jan. 2022, doi: 10.1166/mex.2022.2138.

[18] W. Wang *et al.*, "A photo catalyst of cuprous oxide anchored MXene nanosheet for dramatic enhancement of synergistic antibacterial ability," *Chemical Engineering Journal*, vol. 386, p. 124116, Apr. 2020, doi: 10.1016/j.cej.2020.124116.

[19] S. Shen, H. Chen, R. Wang, W. Ji, Y. Zhang, and R. Bai, "Preparation of antifouling cellulose acetate membranes with good hydrophilic and oleophobic surface properties," *Mater Lett*, vol. 252, pp. 1–4, Oct. 2019, doi: 10.1016/j.matlet.2019.03.089.

[20] R. P. Pandey, K. Rasool, V. E. Madhavan, B. Aïssa, Y. Gogotsi, and K. A. Mahmoud, "Ultrahigh-flux and fouling-resistant membranes based on layered silver/MXene ($Ti_3 C_2 T_x$) nanosheets," *J Mater Chem A Mater*, vol. 6, no. 8, pp. 3522–3533, 2018, doi: 10.1039/C7TA10888E.

10 Applications of MXenes in Electrochemical Sensors

Prakash Chandra, E. Shakerzadeh, Tarun Yadav, and Subhendu Chakroborty

10.1 INTRODUCTION

Recently, electrochemical sensors have fitted out new prospects for chemists to conduct detailed research for deeper understanding in analytical science and technology [1]. Electrochemical sensors are further sub-categorized into biosensors and non-biosensors. Electrocatalyst-modified electrode electrochemical non-biosensors often have low selectivity. Electrochemical biosensors combine electroanalytical sensitivity with inherent bio-selectivity. Furthermore, these electrochemical biosensors are sub-classified into two major sub-groups depending on the type of biological recognition methods, that is, biocatalytic and affinity sensors [2]. In order to recognize the anticipated analyte, biocatalytic sensors integrate biological entities like enzymes, whole cells, and fragments of tissue, creating active species or other noticeable consequences. Whereas affinity sensors mostly use membrane receptors, antibodies, and oligonucleotides, which are bio-related molecules, for strong and precise binding to afford a computable electrical signal in response to a specific analyte. Electrochemiluminescence and photoelectrochemical sensors have made significant advancements in addition to conventional electrochemical non-biosensors or biosensors [3–9]. Figure 10.1 represents the applications of MXene-based electrode materials in sensor application.

Chemiluminescence that is driven by electrochemical processes is known as electrochemiluminescence, sometimes known as electrogenerated chemiluminescence [10]. ECL uses high-energy electron transfer processes to create excited states in species produced at electrodes, which ultimately emit light [11]. ECL today is a very potent analytical technique that is frequently employed in water, food testing, immunoassays, and the identification of biowarfare agents. In the basic mechanism of the PEC process, photons are converted into electrical energy because of the charge separation followed by the transfer of charges after the absorption of photons during the illumination process. In contrast to conventional electrochemical sensing, in PEC sensing, light is used as an excitation source that results in electrical signal output. PEC sensors have higher sensing ability than traditional sensors based on electrochemical mechanisms as well as chemiluminescent mechanisms. This is

140

DOI: 10.1201/9781003366225-10

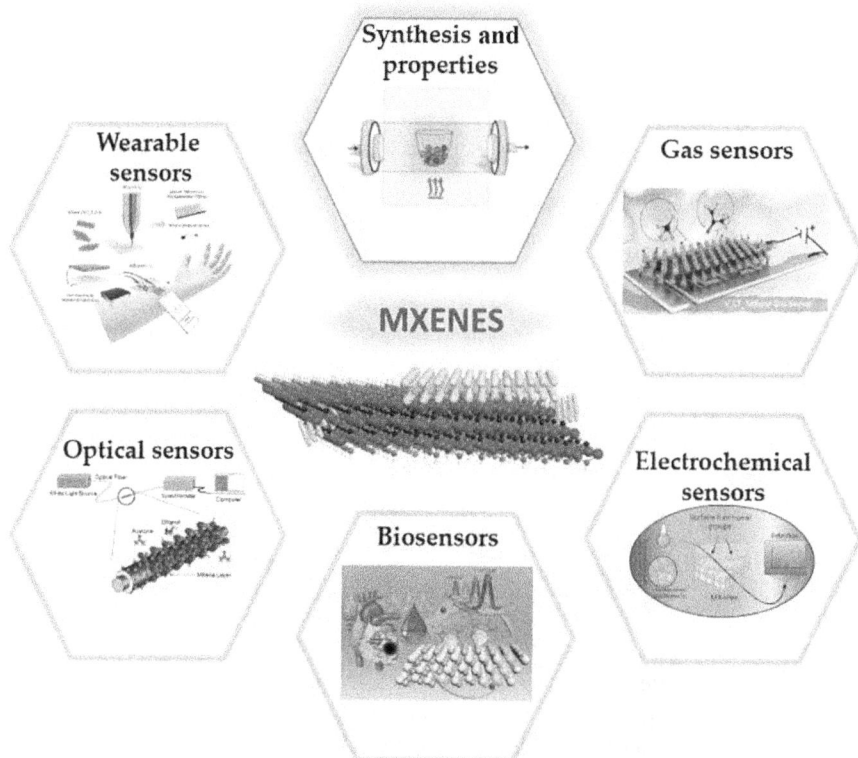

FIGURE 10.1 Applications of MXene-based electrode material for miscellaneous sensing applications [17].

Copyright 2015, American Chemical Society.

because of the difference in energy in the excitation source and the detection signal [12]. Moreover, electrochemical signal patterns give PEC biosensors attributes like higher sensitivity, resilience, simpler instrumentation, reduced cost, and ease of miniaturization when compared to numerous optical approaches using expensive and specialized equipment. Due to the reduced noise caused by the entire separation along with various energy forms of the excitation and detection signals, ECL and PEC sensing techniques naturally provide advantages of better sensitivity, simplicity in equipment and cost effectiveness [3]. In the current chapter, several recent advancements in ECL- and PEC-based sensors, specifically those for bioanalytical chemistry, are presented [13–16].

Several nanomaterials have been concentrated for electrochemical sensors as a result of outstanding advancements in nanotechnology and nanoscience [17]. Among several types of nanomaterials, MXenes are a type of two-dimensional (2D) layered inorganic nanomaterials that have garnered a lot of attention in recent years [18]. Up until now, MXenes, which refers to nitrides, transition metal carbides, and

M in synthesized MXenes   X C, N

M only in theoretical MXenes   T Surface terminations

Periodic table:

H, He
Li Be, B C N O F Ne
Na Mg, Al Si P S Cl Ar
K Ca Sc Ti V Cr Mn Fe Co Ni Cu Zn Ga Ge As Se Br Kr
Rb Sr Y Zr Nb Mo Tc Ru Rh Pd Ag Cd In Sn Sb Te I Xe
Cs Ba 57-71 Hf Ta W Re Os Ir Pt Au Hg Tl Pb Bi Po At Rn
Fr Ra 89-103 Rf Db Sg Bh Hs Mt Ds Rg Cn Nh Fl Mc Lv Ts Og

M_2XT_x $M_3X_2T_x$ $M_4X_3T_x$ $M_5X_4T_x$

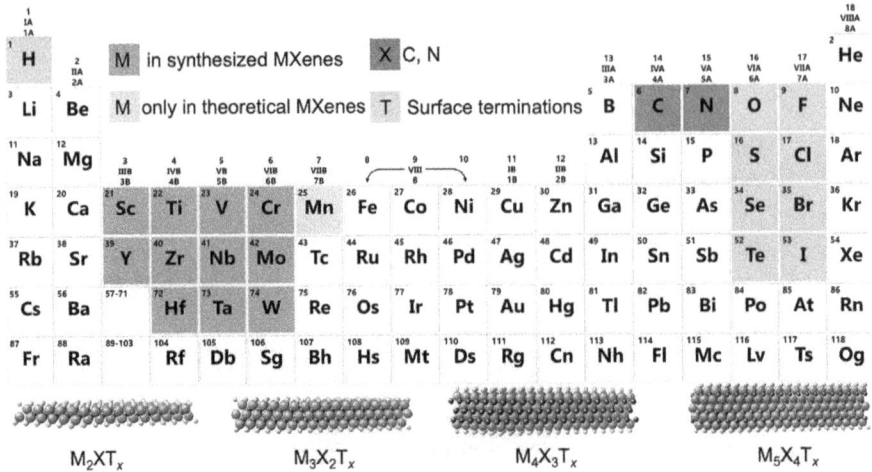

FIGURE 10.2   Schematic presenting constituents in MXenes.

carbonitrides, have swiftly grown into a vast family, as shown in Figure 10.2 [19]. In 2004, mechanically exfoliated graphene was discovered, which produced an international impression [20, 21]. Experimental investigations on graphene convincingly show that layered materials will manifest novel features distinct from their bulk counterparts when thinned to their physical limitations (having a width of just few layers, two layers, or even one layer) [20, 22]. As a result, when the first MXene, Ti_3C_2, was found in 2011, material scientists paid close courtesy to this novel 2D nanomaterial. MXenes are often created by selective removal of an A element, which is typically a group 13 or 14 member, which is from MAX phases like Ti_2AlC, Ti_3AlC_2, and Ta_4AlC_3. M is an early transition metal, X is carbon and/or nitrogen, and T_x denotes surface-terminated functional groups such as hydroxyl groups, oxygen, or fluorine groups obtained from various types of the synthesis methods. The chemical formula of MXenes is $M_{n+1}X_nT_x$ (n = 13) [23]. The term "MXene" was chosen to recognize the parent MAX phases as well as the parallels between the family of 2D material and graphene [24, 25]. A variety of single/few-layered MXenes can be generated by using various synthesis techniques, which are often categorized into bottom-up approaches and top-down approaches [26]. Several hundreds of MXenes with stable phases have so far been identified or anticipated and are mainly from the combinations of several transition metals, including titanium, molybdenum, vanadium, chromium, and their carbon- and nitrogen-based alloys [27]. The exceptional properties of MXenes, including their lower resistivity, good hydrophilicity, higher stability, high specific surface areas, surface functionalities, ease of synthesis in large quantities in water, environmentally friendly traits, and non-toxic nature, makes them attractive materials for analytical chemistry [28–33]. Using $Ti_3C_2T_x$ MXene nanosheets embellished with quantum dots of iron phthalocyanine, Faming and coworkers created a nanocarrier of the complementary DNA to miRNA-155 and built an impedimetric aptasensor device for miRNA-155 with an LOD of 4.3 aM. The use of MXenes in ECL biosensors is comparable [34]. However, MXenes are typically used as conductive films to improve electronic properties and catalytic activity for PEC sensors and flexible sensors [35–36]. In a nutshell, in present chapter emphasizes applications of MXenes for sensing applications.

10.2 EXCLUSIVE CHARACTERISTICS OF MXENES

Ti_3C_2 is a common form of MXene and has distinct advantages over other 2D materials that are driving its application in the development of electrochemical sensors. These distinct qualities can be summed up as follows.

1. MXenes have a clear benefit over other 2D materials in terms of high electrical conductivity, which is essential for efficient electron transport in electrochemical sensors. Such impressive electrical conductivity of MXenes is due to their metallic nature, as they are derived from a family of transition metal carbides and nitrides.

   MXenes have greater electrical conductivity than graphene and TMDs. MXenes have 10^2 to 10^4 S/cm electrical conductivity, several orders of magnitude greater than graphene and TMDs. MXenes are ideal for electrochemical sensor electrodes due to their strong electrical conductivity. Electrochemical sensors depend on effective electron transport between the analyte and the electrode surface. The high electrical conductivity of MXenes ensures that this electron transfer occurs efficiently, resulting in better sensor performance. Furthermore, the high surface area of MXenes, coupled with their high electrical conductivity, makes them a suitable material for sensing applications. The large surface area provides more active sites for the immobilization of sensing molecules or catalysts, while the high electrical conductivity enables rapid electron transfer, resulting in high sensitivity and selectivity in sensing.
2. MXenes are uncomplicated chemicals to manufacture and have acceptable stability and dispersibility in solutions. The preparation of modified electrodes using the most popular method, drop-casting, which calls for preparation prior to the application of the well-dispersed coating solution, is necessary for the creation of electrochemical sensors.
3. MXenes have also been explored as promising materials for printing technologies due to their good electrical conductivity, mechanical strength, and flexibility. MXene-based inks can be used for printing conductive patterns and circuits on various substrates such as paper, plastics, and textiles. This opens up the possibility of creating flexible and wearable electronic devices with customized designs. Additionally, MXene inks can also be used for 3D printing, which enables the fabrication of complex structures with high resolution and precision [37]. The use of MXenes in printing and coating techniques can enhance the performance and functionality of electronic devices, while also improving the effectiveness and sustainability of the manufacturing process. MXene-based inks can be easily printed or coated

onto various substrates using techniques such as inkjet printing, screen printing, or spray coating. Moreover, pre/post-patterned coating of MXenes can also be used to create complex and multifunctional structures, such as electrodes, sensors, and energy storage devices. This makes MXenes promising materials for the industrial-scale production of electronic devices with enhanced performance and reduced environmental impact.

4. MXenes have outstanding mechanical properties, together with high flexibility and stretchability, which are responsible for making them ideal materials for the formation of elastic and wearable electrochemical sensors. The biocompatibility of MXenes also makes them suitable for use in biomedical applications such as non-invasive body fluid monitoring. By using MXenes as a substrate, electrochemical sensors can be designed to conform to the contours of the human body, providing comfortable and non-invasive monitoring of vital signs and other health parameters. MXene-based flexible sensors can also be integrated into clothing, making them convenient and unobtrusive [38].

5. MXenes are potential materials that can be coupled with several nanomaterials/biomolecules for several analytical applications because of their 2D structures and surfaces with numerous functional groups. Various MXene-based nanostructures demonstrate divergent and intriguing properties [39] that are suitable to develop electrochemical sensors with diverse functions, particularly for the class of ECL or PEC sensors.

6. MXenes are great carriers in bio-related applications such as in biosensors or the biomedical area because of their non-toxic nature and biocompatibility with molecules like proteins, nucleic acids, and enzymes.

7. The exclusive photothermal transformation capability of MXenes is responsible for a dual-model recognition approach feasible in the fabrication of electrochemical sensors [40] that extend the signal tactics of electrochemical sensors.

10.3 APPLICATIONS OF MXENES IN ELECTROCHEMICAL SENSORS

MXenes are attractive candidates for electrochemical sensors because of their distinct properties. And given the increased interest in MXenes, we can conclude that their use in electrochemical sensors is on the rise. Several important applications of MXenes in the sensing domain are discussed in the upcoming section.

10.3.1 MXENES AND MXENE COMPOSITES FOR ELECTROCHEMICAL BIOSENSORS

Due to their high electrical conductivity, huge surface area, and unique surface chemistry, MXenes and their composites have been intensively studied for electrochemical biosensors. MXenes' high surface area allows for biomolecule loading, and their strong electrical conductivity improves electron transport between immobilized biomolecules and the electrode surface, boosting biosensor sensitivity and selectivity.

10.3.1.1. MXenes and Their Composite Materials for Enzyme-Based Biosensing Applications

Building electrochemical biosensors requires the transfer of electrons directly between enzyme and electrode [41]. The aforementioned electron transfer process is favored by MXenes due to their unique characteristics like high specific surface area and outstanding electrical conductivity [41]. The first Ti_3C_2 MXene was used to fabricate the first enzyme-based electrochemical sensor via the incorporation of hemoglobin (Hb) for hydrogen peroxide detection [42]. This Hb enzyme-immobilized Ti_3C_2 MXene exhibited excellent capability for the immobilization of the enzyme and created a perfect microenvironment for stability and activity of protein. Additionally, the Ti_3C_2 MXene provided a platform for the successful immobilization of enzyme-based biosensors without the application of mediators. After successful immobilization of the enzyme Hb, several other enzymes like acetylcholinesterase and tyrosinase were successfully immobilized over the MXene (Ti_3C_2) surface [43–44]. These results clearly demonstrate that Ti_3C_2 MXenes can be utilized as excellent candidates for efficient enzyme immobilization to generate enzyme-based biosensors possessing high specific surface area, outstanding biocompatibility, hydrophilic surfaces, and great electrical conductivity.

Mixing Ti_3C_2 MXenes with other materials can improve the properties and performance of enzyme-based biosensors. By incorporating different nanomaterials such as TiO_2 nanoparticles or other types of nanoparticles into Ti_3C_2 MXene composites, the active surface area can be increased, and the enzymatic stability and activity can be maintained or enhanced. This can result in improved sensitivity, selectivity, and stability of the biosensor, making it a more reliable and effective tool for a variety of applications, such as in medical diagnosis, environmental monitoring, and food safety testing [45]. The TiO_2 NPs increase the active surface area accessible for protein adsorption, leading to improved sensitivity and a lower limit of detection. Similarly, Au/Ti_3C_2 MXene nanocomposites produced a GOx-based biosensor for the detection of glucose. By incorporating gold nanoparticles into the Ti_3C_2 MXene composite, the active surface area of the biosensor was further increased, leading to improved sensitivity and selectivity [46]. The addition of Au nanoparticles into the Ti_3C_2 MXene composite creates a unique nanocomposite with synergistic electrocatalytic properties, which can enhance the performance of enzyme-based biosensors. A $GOx/AuNPs/Ti_3C_2/Nafion/GCE$ biosensor demonstrated a low limit of detection of 5.9 M for glucose, indicating its high sensitivity for glucose detection. Additionally, the biosensor showed a high amperometric sensitivity of 4.2 aM/M/cm², which means it can detect glucose with high accuracy and precision, making it a promising tool for glucose monitoring in various applications. The $Ag@Ti_3C_2$ nanocomposite as a nanocarrier for AChE can improve the performance of biosensors for the detection of malathion, an organophosphate pesticide. AChE is an enzyme that is commonly used as a biorecognition element in biosensors for the detection of organophosphate compounds. Incorporating AChE into $Ag@Ti_3C_2$ nanocomposites can enhance their stability, activity, and specificity, resulting in a more sensitive and selective biosensor for the detection of malathion. The $Ag@Ti_3C_2/AChE$ biosensor can detect malathion with high sensitivity and accuracy, making it a promising tool for environmental monitoring and pesticide residue analysis [47]. AChE-based biosensors are mostly used for the organophosphorus pesticides detection, including methamidophos. A 3D MnO_2/Mn_3O_4/MXene/AuNPs composite was used to generate a biosensor for the detection of methamidophos. The use of the 3D structure can augment the surface area and upsurge the active sites for reactions, leading to improved sensitivity and specificity of the biosensor. The biosensor demonstrated a low limit of detection of 0.134 pM for methamidophos, indicating its high sensitivity for the pesticide [32]. The researchers used a Ti_3C_2 MXene, Au@Pt nanoflowers, 5'-nucleotidase, and xanthine oxidase to create a biosensor for inosine monophosphate detection. The biosensor identified beef inosine monophosphate at 2.73 ng/mL. In real-world glucose biosensing systems, hydrogen peroxide from enzymatic glucose oxidation often inhibits GOx performance [48–49]. To solve this, Wu and colleagues built a combination Ti_3C_2/poly-L-lysine (PLL)/glucose oxidase (GOx) nanoreactor. This nanoreactor catalyzed glucose oxidation and hydrogen peroxide breakdown. The Ti_3C_2 MXene catalyzed the hydrogen peroxide process, and when paired with loaded GOx, it generated a glucose cascade reaction. Ti_3C_2/PLL/GOx nanoreactors with good catalytic performance were deposited on glassy carbon electrodes to make a 2.6 M glucose biosensor. MXenes' biosensing uses may be hindered by their HF etching synthesis [49–50].

Hydrofluoric acid (HF) used in the MXene nanosheets can affect the liberation of HF during the biosensing process, which can harm normal cells in vivo or decrease enzyme activity. Additionally, HF-etched MXenes can be vulnerable to oxidation, leading to reduced stability and efficiency in biosensing applications. Therefore, it is important to develop fluoride-free synthesis techniques to produce biocompatible and stable nanosheets of MXenes. An electrochemical etching and exfoliation method is used to create 2D fluoride-free Nb_2C MXene nanosheets. The fluoride-free Nb_2C MXene had improved low cytotoxicity, chemical stability, high porosity, and increased conductivity, which helped to stabilize enzyme activity and encourage electrochemical activity. The Nb_2C/AChE-based biosensor created by Song et al. achieved a low LOD of 0.046 ng/mL for phosmet, indicating its high sensitivity for the pesticide. Moreover, the use of Nb_2C as a biosensing platform can result in improved biosensor performance due to its superior metallic characteristics and close to zero energy bandgap. Nb_2C-based biosensors have been shown to perform better than those based on V_2C and Ti_3C_2 [50]. Nanocomposites of Ti_3C_2/Au-Pd were used as the functional platform for a disposable electrochemical biosensor (AChE based) to detect paraoxonin. The researchers synthesized a nanocomposite of Ti_3C_2 sheets with gold and gold–palladium bimetallic nanoparticles by self-reducing. The application of the AChE-based biosensor for paraoxon detection in samples of pears and cucumbers proved effective. In addition to a glassy carbon electrode (GCE), Ti_3C_2 MXenes can also be used as conductive substrates in various electrode systems [51]. The combination of Ti_3C_2 and graphene oxide (GO) results in a material that possesses excellent electrical conductivity, high surface area, and strong mechanical properties. An inkjet-printed hydrogen peroxide biosensor was fabricated using Ti_3C_2-GO by dispersing the Ti_3C_2-GO nanocomposite in a suitable solvent, such as water or ethanol, to form a stable ink that can be used for printing. In the subsequent step, the ink was loaded into an inkjet printer and printed onto a substrate, such as a flexible or rigid substrate. The printing process was controlled to deposit a precise amount of ink at specific locations on the substrate, resulting in the formation of a patterned biosensor. To create a hydrogen peroxide biosensor, the printed pattern was modified with an enzyme such as horseradish peroxidase (HRP), which reacts specifically with hydrogen peroxide. The enzyme was immobilized onto the surface of the Ti_3C_2-GO nanocomposite using a variety of methods, such as physical adsorption or covalent bonding. Once the enzyme is immobilized onto the Ti_3C_2-GO nanocomposite, it can be used to detect the presence and concentration of hydrogen peroxide. This can be achieved by measuring the changes in electrical conductivity or optical properties of the Ti_3C_2-GO nanocomposite as a result of the enzymatic reaction [52]. The screen-printed carbon electrode (SPCE) is a low-cost and portable electrode commonly employed in electrochemical sensing. To create the biosensor, the SPCE was modified with a nanocomposite of polyaniline, platinum particles, and a Ti_3C_2 MXene (Pt/PANI/MXene). This modification allowed for the detection of both hydrogen peroxide and lactate, which are important biomarkers in various applications, such as clinical diagnosis and sports science. The platinum particles provide catalytic activity for the electrochemical detection of hydrogen peroxide, while polyaniline and the Ti_3C_2 MXene enhance the sensitivity and selectivity of the biosensor. The use of the Ti_3C_2 MXene in the nanocomposite also contributes to the stability and durability of the biosensor [53]. The limit of detection (LOD) for hydrogen peroxide of 1.0 M in the prepared SPCE and the LOD for lactate of 5.0 M after immobilizing lactate oxidase indicate the sensitivity of the biosensor. An LOD of 1.0 M for hydrogen peroxide suggests that the biosensor is capable of detecting very low concentrations of this analyte. An LOD of 5.0 M for lactate indicates that the biosensor is also capable of detecting lactate with good sensitivity. The use of amperometry for lactate assays allows for the quantification of lactate based on the current produced by the enzymatic reaction. The consistent and accurate measurement of lactate in milk samples indicates the potential of the biosensor for various applications, such as food safety and quality control.

10.3.1.2 MXenes and Their Nanocomposites for Immunosensors

The first MXene-based PSA immunosensor was disclosed in 2019. The sensor uses a Ti_3C_2 MXene, a 2D nanomaterial with excellent surface area and electrical conductivity. The immunosensor detects the prostate cancer biomarker PSA. Early diagnosis and therapy require PSA detection. The MXene-based immunosensor employed a sandwich immunoassay to collect PSA with an immobilized antibody on the sensor surface and detect it with a gold nanoparticle-labeled antibody. The immunosensor's sensitivity was increased by the Ti_3C_2 MXene's huge surface area and electrical conductivity, which supported antibody immobilization. An immunosensor with excellent PSA specificity and selectivity was created by immobilizing antibodies on the Ti_3C_2 MXene. The MXene-based immunosensor detected PSA from 0.1 to 100 ng/mL with an LOD of 0.02. The LOD reflects the sensor's sensitivity and the lowest analyte concentration that can be reliably detected. Figure 10.3 shows a schematic of an electrochemical immunosensor.

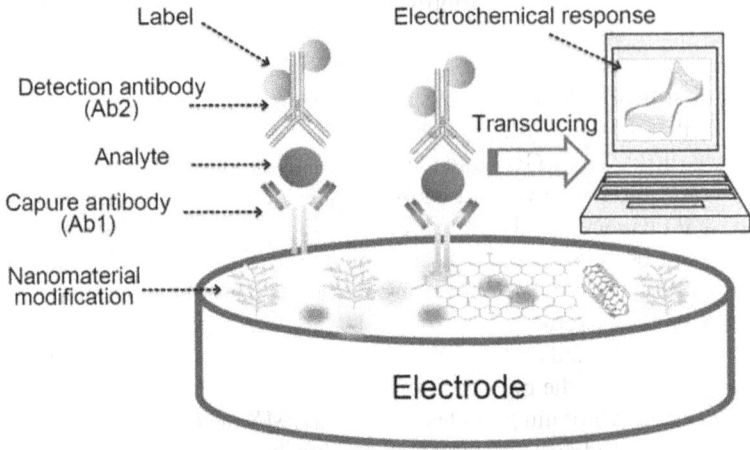

FIGURE 10.3 Schematic depicting an electrochemical immunosensor.

The MXene-based immunosensor detected PSA with great sensitivity and selectivity, making it a viable clinical candidate [54]. The sandwich immunosensor (CuPtRh/NH$_2$-Ti$_3$C$_2$) detects cardiac troponin I. The nanocomposite had many layers of ultrathin ammoniated Ti$_3$C$_2$ (NH$_2$-Ti$_3$C$_2$) and trimetallic hollow CuPtRh cubic nanoboxes (CuPtRh CNBs). The CuPtRh CNBs implanted in NH$_2$-Ti$_3$C$_2$ were a connection to install extra secondary antibodies (Ab2) through stable Pt–N and Rh–N bonds and a spacer to avoid irreversible restacking. CuPtRh CNBs and secondary antibodies (Ab2) were backboned with NH$_2$-Ti$_3$C$_2$. The CuPtRh CNBs/ NH$_2$-Ti$_3$C$_2$ were used as an Ab2 label and had a strong catalytic activity for reducing hydrogen peroxide, which greatly increased the signal from the immunosensor. The LOD for the developed immunosensor was 8.3 fg/mL, which is a very low concentration and point to high sensitivity. Cardiac troponin I is a biomarker for cardiac injury, and the detection of this biomarker is important for the diagnosis and management of cardiovascular diseases. The developed immunosensor has the potential to be used for the early detection of cardiac troponin I in patient samples, which could improve patient outcomes. In summary, a nanocomposite (CuPtRh/NH$_2$-Ti$_3$C$_2$) was developed and used to create a sandwich kind of immunosensor to detect cardiac troponin I. To increase immunosensor stability and sensitivity, CuPtRh CNBs implanted in NH$_2$-Ti$_3$C$_2$ were connectors and spacers. The immunosensor's LOD was 8.3 fg/mL, showing great sensitivity. These findings highlight the potential of MXene-based nanocomposites for clinical biosensors with great sensitivity and selectivity [33]. A self-assembled delaminated MXene and gold nanoparticle (d-Ti$_3$C$_2$T$_x$ MXene@AuNPs) biosensor for prostate-specific antigen (PSA) detection uses PSA Ab2 as a signal amplification. The self-assembled d-Ti$_3$C$_2$T$_x$ MXene@ AuNPs were stable in various solvents. The biosensor detected PSA by immobilizing PSA Ab2 on d-Ti$_3$C$_2$T$_x$ MXene@AuNPs. The biosensor's limit of detection

(LOD) was 3.0 fg/mL, better than a prior study's 31 pg/mL. The d-$Ti_3C_2T_x$ MXene@ AuNPs biosensor signal amplification platform is very sensitive. PSA, a prostate cancer biomarker, must be detected early to diagnose and treat the illness. The biosensor may detect PSA in patient samples quickly, improving patient outcomes. In conclusion, gold nanoparticles and a self-assembled delaminated MXene (d-$Ti_3C_2T_x$ MXene@AuNPs) were employed to designate PSA Ab2 as a signal amplification in a biosensor for PSA detection. The biosensor had a better LOD than in prior research at 3.0 fg/mL. The aforementioned work highlights the potential of MXene-based nanocomposites to produce clinically sensitive biosensors [55]. A sandwich electrochemical immunosensor was utilized to detect the sepsis biomarker procalcitonin (PCT). The immunosensor detected PCT using two signal amplification techniques. In the first method, carboxylated graphitic carbon nitride (c-g-C_3N_4) labeled PCT Ab2 and amplified it. Due to its large surface area, conductivity, and biomolecule interaction, c-g-C_3N_4 was exploited as a signal amplification platform. The second technique immobilized PCT Ab2 on a delaminated sulfur-doped MXene (d-S-Ti_3C_2 MXene)-modified glassy carbon electrode (GCE) with AuNPs. Due to its large surface area, conductivity, and biomolecule interaction, d-S-Ti_3C_2 MXene was used to immobilize PCT Ab2. The AuNPs' large surface area and conductivity made them signal amplification platforms. The immunosensor had an LOD of 0.008 pg/mL and a linear response to PCT concentrations from 0.01 to 100 ng/mL. The immunosensor detected PCT in human serum samples with good selectivity and accuracy. Using two signal amplification techniques, a sandwiched electrochemical immunosensor detected procalcitonin (PCT). Carboxylated graphitic carbon nitride (c-g-C_3N_4) and delaminated sulfur-doped MXene (d-S-Ti_3C_2 MXene) signal amplification platforms with AuNPs produced a very sensitive and selective immunosensor for PCT detection [56].

10.3.1.3   MXene-Based Nanostructured Materials for Aptasensors

Single-stranded DNA or RNA aptamers may precisely and selectively attach to target molecules. This method requires repeated selection, amplification, and enrichment of target-binding aptamer sequences. Aptamers that attach to tiny molecules, proteins, peptides, bacteria, viruses, and living cells have been created over time. These aptamers can be used in a variety of applications, such as biosensors, therapeutics, and imaging agents. For example, aptamers that bind to small molecules, such as adenosine, cocaine, and theophylline, have been developed for use in biosensors and drug delivery systems. Aptamers that bind to bacteria, such as *Staphylococcus aureus* and *Escherichia coli*, have been developed for use in rapid diagnostic tests and antimicrobial therapies. Aptamers that bind to live cells, such as cancer cells, have been developed for use in imaging and targeted drug delivery. The diversity of aptamers that can be developed using the SELEX technique has made them a valuable tool for a wide range of applications in the biomedical field [57–59]. Electrochemical aptasensors are a type of biosensor that uses aptamers as the recognition element and electrochemical transduction to detect the target molecule. These electrochemical aptasensors have advantages over other types of biosensors, including simplicity, quick response, low cost, recyclability, and miniaturization. In an electrochemical aptasensor, the aptamer is immobilized on a transducer surface, such as a screen-printed electrode or a gold electrode, and the target molecule binds specifically to the aptamer. This binding event causes a change in the electrochemical signal, which can be measured and quantified to determine the concentration of the target molecule in the sample. They have been used in a variety of applications, such as clinical diagnostics, environmental monitoring, and food safety testing. The advantages of electrochemical aptasensors, such as their simplicity, low cost, and miniaturization, make them particularly attractive for point-of-care testing and field-deployable applications. As a result, there is currently extensive research being conducted in this area, with a focus on improving the sensitivity, specificity, and stability of electrochemical aptasensors for a wide range of applications [60]. MXenes and their nanocomposites have been recently recognized as valuable materials for aptasensor applications. The PPy@Ti_3C_2/PMo_{12} composite was chosen due to its high surface area and good conductivity, which makes it an ideal material for biosensing applications. The OPN-specific DNA aptamer (OPN aptamer is a short single-stranded DNA or RNA molecule that is capable of binding specifically to osteopontin (OPN), a protein that is involved in various physiological processes such as bone remodeling, immune response, and cancer progression) was immobilized onto the surface of the composite using a simple adsorption method. The aptamer-functionalized PPy@Ti_3C_2/PMo_{12} composite exhibited a linear response to OPN concentrations in the range of 0.1 ng/mL to 10 ng/mL, with a detection limit of 0.03 ng/mL. The aptasensor also showed high selectivity for OPN over other interfering proteins [61]. OPN can interfere with the ability of the OPN aptamer strand immobilized over the PPy@Ti_3C_2/PMo_{12} hybrid to reach the electrode surface by forming a G-quadruplex with OPN. Moreover, an impedimetric aptasensor has been investigated that sensitively detects miRNA-155 based on a unique zero dimensional (0D)/2D hybrid of Ti_3C_2 sheets embellished with quantum dots of Fe phthalocyanines (FePc QDs). The complementary DNA (cDNA)-targeted miRNA-155 was anchored using the Ti_3C_2/FePcQDs nanohybrid as a nanocarrier. The Ti_3C_2/FePcQDs-based aptasensor exhibited outstanding sensing properties for the detection of miRNA-155, with a detection limit of 4.3 aM for miRNA-155 concentrations between 0.01 fM and 10 pM without using a labeled probe or other electrochemical indicators. The design of an electrochemical sensor could potentially make use of the change in conductance [34]. The use of a Ti_3C_2 MXene as a signal amplifying transduction material in a conductometric electrochemical sensor for *Mycobacterium tuberculosis* detection is a promising application of MXenes in biosensing. The PNA-AuNPs nanogap network electrode surface must hybridize the probe peptide nucleic acid (PNA) with a 16S rDNA fragment to detect *Mycobacterium tuberculosis*. This procedure selectively captures target pieces and links them to the conductive Ti_3C_2 MXene. A zirconium-cross-linked Ti_3C_2 MXene and target fragment phosphate groups effectively bind the fragments. This linkage creates a conductive connection between the electrodes, changing conductance and detecting *Mycobacterium tuberculosis*. This change in conductance is then used as a signal to detect the presence of *Mycobacterium tuberculosis*. The LOD of the proposed method was reported to be 20 CFU/mL, which is a relatively low concentration and indicates a high sensitivity of the sensor. This sensitivity is attributed to the amplification effect of the Ti_3C_2 MXene as a signal transduction material, which enhances the detection signal and improves the accuracy and precision of the sensor [62].

A competitive electrochemical aptasensor utilizing a cDNA-Fc/MXene probe detected breast cancer marker mucin 1 (MUC1). Au nanoparticles (AuNPs), a cDNA-Fc/MXene probe, and a MUC1-specific aptamer were added to the sensing electrode. cDNA-Fc/MXene and MUC1 competed during detection. Competition separated the cDNA-Fc/MXene probe from the detecting electrode and lowered the electrical signal. MUC1 concentration was measured by electrical signal reduction. The proposed electrochemical aptasensor had a low LOD of 0.33 pM, which indicates high sensitivity, and could detect MUC1 at very low concentrations. The linear range of the sensor was reported to be broad, ranging from 1.0 pM to 10 mM, which means that the sensor could detect MUC1 over a wide concentration range [63]. A $MoS_2@Ti_3C_2T_x$ MXene hybrid-based electrochemical aptasensor was developed for the sensitive and rapid detection of thyroxine (T4). According to the investigations, MoS_2 was used as a signal amplification strategy, while $Ti_3C_2T_x$ MXene was used as a platform to immobilize the T4 aptamer. The $MoS_2@Ti_3C_2T_x$ MXene-based aptasensor showed a detection limit of 0.005 nM for T4 and a linear range of 0.01–10 nM. The aptasensor also demonstrated good selectivity, stability, and reproducibility for T4 detection [64]. A three-dimensional nanocomposite based on AuNPs and Ti_3C_2 MXene has been developed for the sensitive detection of miRNA-155. The aptamer specific to miRNA-155 was immobilized onto the AuNPs/Ti_3C_2 MXene nanocomposite and used for electrochemical detection. The nanocomposite was found to enhance the sensitivity of the aptasensor by providing a large surface area for the immobilization of the aptamer and promoting electron transfer between the electrode and the target. The developed aptasensor demonstrated high sensitivity and specificity towards miRNA-155, with a limit of detection of 0.85 Fm [65]. A PB/Ti_3C_2 hybrid-based sensor was developed for the detection of exosomes, which are small vesicles secreted by cells that contain important biomolecules and can serve as potential diagnostic markers for various diseases. The sensor was able to detect exosomes with high sensitivity and selectivity and could potentially be useful for the early diagnosis of diseases such as cancer [66].

10.3.2 MXENE-BASED NANOSTRUCTURED MATERIALS AS ELECTROCHEMICAL NON-BIOSENSORS

MXenes can be effectively used to create electrochemical non-biosensors without enzyme immobilization. The very first electrochemical non-biosensor with a pure Ti_3C_2 MXene was developed and investigated in 2018 for BrO_3^-, which is known as a common drinking water contaminant [67]. As a reducing agent and signal enhancer matrix, the Ti_3C_2 MXene demonstrated exceptional electrocatalytic capabilities for efficient BrO_3^- reduction. Thereafter, two identical Ti_3C_2 MXene-modified GCE electrochemical sensors were developed to detect the insecticide carbendazim [68] and the neurotransmitter dopamine [69]. With the Ti_3C_2 MXene, carbendazim's redox was obtained with fewer overpotentials than with the graphene-based sensor [70]. The latter was effective in detecting dopamine in actual samples with a robust sensitivity and an LOD of 3 nM. Adding MXenes to other electrode systems, such as a paste electrode made of composite graphite (GCPE) [51] and a screen-printed electrode (SPE) [71], has also been investigated.

A MXene/GCPE electrochemical sensor was effectively used to detect adrenaline. The constructed sensor attained an LOD of 9.5 nM and had 99.2–100.8% recovery when used to detect adrenaline in pharmaceutical samples. Acetaminophen (ACOP) and isoniazid were determined by voltammetry using the MXene/SPE electrochemical sensor (INZ). As compared to bare SPE in 0.1 M H_2SO_4, the Ti_3C_2 MXene has superior electrocatalytic activity for the oxidation of ACOP and INZ, where the separated oxidation peak potentials help in the simultaneous detection of the targets. For ACOP and INZ, respectively, the constructed sensor attained LODs of 0.048 M and 0.064 mM. Moreover, the Ti_3C_2 MXene was also incorporated in a variety of nanocomposites with other substances or molecules for several distinct analytical uses, for example a NiO/Ti_3C_2 hybrid for non-enzymatic hydrogen peroxide sensing [72], a 3D porous layered material of nickel–cobalt double hydroxide for non-enzymatic glucose sensing [73], nitrile sensing using a $AuNPs/Ti_3C_2$ composite [74], l-cysteine fast determination using a $Pd@Ti_3C_2$ nanocomposite [75–77], and a Ti_3C_2 composite for ratiometric electrochemical detection of piroxicam using methylene blue (MB)/CuNPs [78]. In all the examples mentioned, the Ti_3C_2 MXene demonstrated excellent performance in these efforts for the creation and use of electrochemical sensors.

Due to their propensity to restack and the hydrogen bonding and van der Waals interactions that exist between their sheets, MXenes' electrochemical performance may be further constrained [79, 80]. Carbon nanohorns (CNHs) were used as spacers in Ti_3C_2/CNHs nanocomposites to fix these issues. Layered MXene/CNHs offered excellent conductivity, catalytic activity, and ion transport channels. Nanocomposites were used to construct a 1.0 nM carbendazim electrochemical sensor [80]. To prevent Ti_3C_2 MXene sheets from restacking, porous nitrogen-doped carbon (N-PC) from MOF-5-NH_2 was used as a spacer in the alk-Ti_3C_2/N-PC electrochemical sensor. The sensor detected catechol (CT) and hydroquinone (HQ) in industrial effluent [79]. Molecularly imprinted polymers can increase the selectivity of electrochemical sensors in tracking studies because to their recognition property, low cost, and fast production time [81]. A hierarchical porous Ti_3C_2 MXene/amino carbon nanotube (MXene/NH_2-CNTs) composite and MIP were produced as electrochemical sensors for selective and sensitive fisetin detection. Positive and negative NH_2-CNTs and Ti_3C_2 flakes self-assembled to form the composite. Amino-functionalized CNTs can prevent Ti_3C_2 MXenes from aggregating as interlayer spacers because of their high conductivity and positive charges. The synthetic MIP/Ti_3C_2 MXene/NH_2-CNTs/GCE measured fisetin well at 1.0 nM [82–85].

MXene-based electrochemical non-biosensors also advanced. An electrochemical biosensor based on Ti_3C_2 nanosheets (PB NPs/Ti_3C_2) with intercalated PB nanoparticles had a 0.20 M detection limit for real-time and in situ H_2O_2 detection from living cells. PB/Ti_3C_2 displayed no cytotoxicity to normal fibroblast cells at any period or concentration, indicating its use in live cell fields [86–94]. Ti_3C_2/BN nanocomposites produced an electrochemical catalytic sensor for sulfadiazine detection with a 3.0 nM detection limit [95].

10.3.2.1 Electrochemiluminescence Sensors

In 2018, Fang, Y. F., et al. [96] created the first Ti_3C_2 MXene-based ECL sensors. The negatively charged Ti_3C_2 MXene surface enabled electrostatic immobilization of $Ru(bpy)_3^{2+}$ due to its fluorine (F) and hydroxyl (OH/=O) surface functional groups. The Ti_3C_2 MXene improved $Ru(bpy)_3^{2+}$ adsorption and nanocomposite conductivity. After then, an aptamer-modified Ti_3C_2 MXene ECL nanoprobe was employed to build a sensitive exosome biosensor [97]. This study used $Ru(bpy)_2(mcpbpy)^{2+}$ as an intercalation molecule and an ECL luminophore to functionalize the surface and delaminated the multilayered Ti_3C_2. Ru@MXene ECL sensors measured MUC1 with a low LOD of 26.9 ag/mL [98]. Electrochemiluminescence sensing, like other electroanalysis methods, is vulnerable to environmental errors such co-reactant concentration, pH, temperature, etc. ECL ratiometric tests have been suggested to control the external environment. This immobilizes Ab1 and Ti_3C_2/Au nanoparticle hybrids to load Ab2 [99]. Luminol showed ECL cathodic and ECL anodic cyclic potential scanning in GR-IL-Pt nanocomposites and Ti_3C_2/Au nanoparticle hybrids. As CEA concentration rose, ECL anodic/ECL cathodic increased, allowing for the sensitive ratiometric detection of CEA. The novel ECL sensor detected CEA at 34.58 fg/mL with sensitivity.

Photothermal amplification for ECL sensors was developed using MXenes' increased photothermal conversion efficiency [100]. An ECL and photothermal dual-mode biosensor for exosome detection was made using a Ti_3C_2 MXene and BPQD composite probe. The Ti_3C_2 MXene supported BPQD and $Ru(dcbpy)_3^{2+}$ immobilization. The Ti_3C_2 MXene and BPQDs were used to build the ECL and temperature dual-mode sensor because of their photothermal properties [40]. For photothermal detection, the logarithm of exosome concentration increased linearly with temperature from 1.1102 to 1.1107 particles/L, whereas for ECL detection, the sensor had an LOD of 37.0 particles/L. This demonstrated how photothermal amplification may boost ECL sensor sensitivity [100].

10.3.4 PHOTOELECTROCHEMICAL SENSORS

MXene-based PEC sensors detected glucose using a Ti_3C_2 MXene/Cu_2O heterostructure. The Ti_3C_2 MXene/Cu_2O composite has good PEC performance and a reasonable glucose photoelectric response at 0.17 nM. A glutathione-detecting composite film of TiO_2 and Ti_3C_2 quantum dots was then produced [101–102]. The TiO_2 IOPCs/Ti_3C_2 composite film may display photocurrent effects from 280 to 900 nm, and its internal power conversion efficiency may reach 26% at 350 nm. These results suggest that the Ti_3C_2 MXene might be a PEC sensor. Controlled oxidation can form transition metal oxide nanoparticles on MXene sheets, changing their surface properties [103]. This heterojunction features an optimum interfacial design for photocatalytic redox activity with low recombination. This heterojunction was used to create a PEC biosensor to detect soluble CD44 protein at 0.000014 ng/mL. This novel heterojunction may be suitable for PEC biosensors, which use light to transform metabolic reactions into electrical signals [104]. Oxidizing Ti_3C_2 with hydrogen peroxide and heating MXene sheets makes Ti_3C_2–TiO_2 composites in one process. Ti_3C_2–TiO_2 composites were used to study gallic acid, proanthocyanin, epigallocatechin gallate, vitamin E, and ascorbic acid (AA) surface interactions. The findings imply that AA and other antioxidants work synergistically. AA's low redox potential helps lower other antioxidants' radicals and regenerate them. The PEC sensor's $Ti3*C_2$-TiO_2 combination may enable sensitive and precise assessment of these interactions, which might impact nutrition and health. The PEC sensor measured the antioxidant capacity of a food product using grape seed and vitamins C and E [103]. A Z-scheme $TiO_2/Ti3C_2/Cu_2O$ heterostructure photocathode was used to build self-powered PEC glucose sensors. The artificial indirect "Z-scheme" system used two semiconductors, two photosystems (PS), and electron mediators. The thermal treatment of a Ti_3C_2 MXene in ethanol generated TiO_2 NPs on Ti_3C_2 nanosheets in situ, and hybridization with Cu_2O created a Z-scheme $TiO_2/Ti_3C_2/Cu_2O$ heterojunction. The Z-scheme $TiO_2/Ti_3C_2/Cu_2O$ heterojunction increased PEC performance over TiO_2/Ti_3C_2 and Cu_2O, resulting in a glucose PEC sensor with a remarkably low LOD of 33.75 nM and a broad range of 100 nM to 10 M [35].

In situ heterojunctions were made from oxidized Ti_3C_2 sheets and photoactive $NiWO_4$ nanoparticles. The $TiO_2/Ti_3C_2/NiWO_4$ PEC platform detected PSA at 0.15 fg/mL [105]. Yuan, C., et al. developed a photoelectrochemical (PEC) sensor to detect ciprofloxacin (CFX) with an LOD of 0.13 nM and a dynamic range of 0.4 to 1000 nM [106]. The heterojunction between the Ti_3C_2 MXene and $BiVO_4$ semiconductor was spin coated with a single layer of Ti_3C_2 to build a PEC sensor for Hg^{2+} detection. The PEC sensor showed a broad linear range of 1 pM to 2 nM with a detection limit of 1 pM using Yangtze River water. It was accurate and reproducible [107].

10.3.5 APPLICATIONS OF MXENE-BASED NANOCOMPOSITE MATERIALS AS FLEXIBLE ELECTROCHEMICAL SENSORS

When using paper substrate to build innovative implanted biological devices and miniature lap-on-a-chip systems for the testing of real samples, flexibility is essential [108]. As compared to conventional rigid sensing platform/materials, flexible sensing platforms/material forms more compatible interfaces with soft biological materials, such as human skin. This enhances biological signal molecule identification using pressure-based electrochemical flexible sensors [109–112]. Materials used for interfaces have a significant impact on flexible sensors [113]. MXenes' introduction has proven to be a successful strategy [114–116].

Metal nanoparticles imitate biological tissue in a flexible Ti_3C_2 MXene paper. Because of their electrocatalytic capabilities, biocompatibility, stability, and flexibility, Ti_3C_2 MXene hybrid sheets were employed as high-performance extracellular biosensing substrates to sensitively monitor superoxide (O^{2-}) released by live cells. After 50 continuous bending cycles, the Ti_3C_2 MXene paper retained 80.1% of its initial condition [108]. A flexible electrochemical sensor constructed of $Ti_3C_2/Mn_3(PO_4)_2$ detects O_2^- released by human lung cancer cells in real time and quantitatively. Researchers constructed a biomimetic enzyme using 2D $Mn_3(PO_4)_2$ and Ti_3C_2 MXene nanosheets in a 2D/2D heterostructure. The 2D components and strong interfacial connection allowed the heterostructure's amazing flexibility. The heterostructure maintained 100% reactivity up to 180° and 96% after 100 bending/relaxing cycles [109]. Wearable biosensors monitored biomarkers in bodily fluids such

FIGURE 10.4 Application of MXenes as wearable sensors [118].

sweat, tears, and saliva to assess body health [117]. A MXene-based wearable sensor concept is shown in Figure 10.4. An electrochemical sensor patch system that is sensitive, battery-free, adaptable, wireless, and fully integrated (no external analysis equipment) was developed to monitor human sweat potassium ion (K^+) concentration on-site. The flexible electrochemical sensor and low-power near-field communication (NFC) wireless patch device detected potassium content in human sweat on-site and wirelessly sent measurements to a smartphone. Smartphones were used as RF power sources to activate NFC electronic equipment and receive sensor data. The Android Studio app displayed K^+ concentration values. The electrochemical sensor patch was combined with a microfluidic channel to collect perspiration from human skin and prevent sensor surface contamination. For detecting human sweat K^+ concentration, the sensor has a high linear range of 1 to 32 mM and sensitivity of 173 mV/dec [117]. Li et al. produced a highly integrated sensing (HIS) paper with a signal processing system to detect lactate and glucose in real time utilizing Ti_3C_2 as the active material and foldable all-paper substrates as sweat analysis patches. The HIS paper functional zones were folded into three-dimensional structures. This used paper substrates' hydrophilic qualities and the well-constructed 3D diffusion channel. Human–device interactions can swiftly gather and disseminate vertical sweat. Based on the HIS work, they developed a dual-channel electrochemical sensor that could simultaneously detect glucose and lactate at 2.4 and 0.49 nA/mM, respectively. HIS paper is inexpensive and practical for wearable bioelectronic devices and non-invasive electrochemical sensors [118]. Electrochemical sensing devices are challenging to make because nanomaterials must be placed on an electrode with regulated thickness, shape, and architecture [119].

The DIDE sensor platform can identify many biomarkers in a single sample, which might benefit clinical applications. Ti_3C_2-AFBPB film and label-free detection provide a sensitive and selective immunosensor. The dual immunosensor's LODs for Apo-A1 and NMP 22 (0.3 and 0.7 pg/mL, respectively) show its potential for early and accurate bladder cancer diagnosis [119]. A 96-well adaptive sensor system evaluated cancer plasma samples. For point-of-care (POC) use, this device may quantify more analytes [120].

10.4   CONCLUSIONS AND OUTLOOK

MXenes' strong conductivity, surface area, and surface chemistry make them promising materials for electrochemical sensing devices. In biosensors, MXenes transport biomolecules, and direct chemical interaction between biomolecules and MXenes can improve biocomposite consistency. Thus, MXene surface amination or carboxylation may be a viable study area. Electrochemical etching (E-etching) exfoliation is an intriguing approach for making MXene-based electrodes with controllable thickness and shape. Additionally, it is also worth investigating the use of MXenes in the progress of multifunctional sensors that can simultaneously distinguish multiple analytes, as well as the incorporation of MXene-based sensors with other technologies such as microfluidics and machine learning for enhanced sensing performance and data analysis [50]. The simultaneous electrochemical functionalization and exfoliation of MXenes affords an appropriate and well-organized technique to acquire MXenes with definite surface functionalities for numerous applications, together with electrochemical sensing. Moreover, the photothermal properties of MXenes can also be utilized for sensitivity enhancement in electrochemical sensing. For instance, MXene-based nanocomposites have been shown to exhibit enhanced electrochemical performance in various applications, such as the detection of heavy metal ions and small molecules. The incorporation of MXenes into electrochemical sensors can improve their sensitivity, selectivity, and stability due to the exclusive properties of MXenes, such as their high surface area, excellent conductivity, and photothermal conversion efficiency. Additionally, this sensitivity enhancement approach can potentially be applied to other types of electrochemical sensors beyond electrochemiluminescence (ECL) sensors. For example, MXene-based nanocomposites can also be incorporated into amperometric and potentiometric sensors to improve their performance. The simultaneous electrochemical functionalization and exfoliation of MXenes, combined with their photothermal properties, offer a promising approach for boosting the performance of electrochemical sensors and can be applied in plentiful sensing applications beyond ECL sensors [40,42]. Photoelectrochemical and flexible sensors depend on heterojunction structure and conductive and sensing film composition. MXenes' strong conductivity, adjustable surface chemistry, and photothermal conversion efficiency make them ideal for high-performance sensor materials. New hetero-junctions and composite films for sensing may be created by combining MXenes with metal oxides, polymers, and nanocarbons. Photoelectrochemical sensors can employ MXene-based heterojunctions for charge separation and high sensitivity. MXenes may also be utilized as conductive and sensing films for flexible sensors, which can be readily incorporated into wearable devices for non-invasive and continuous biomarker monitoring. MXene's unique features allow flexible sensors to be sensitive, flexible, and durable. MXenes, as a strong tool, will enable more complex electrochemical sensors for healthcare, environmental monitoring, and energy conversion applications.

REFERENCES

[1] J.M. Díaz-Cruz, N. Serrano, C. Pérez-Ràfols, C. Ariño, M. Esteban, Electroanalysis from the past to the twenty-first century: Challenges and perspectives, Journal of Solid State Electrochemistry, 24 (2020) 2653–2661.

[2] N.J. Ronkainen, H.B. Halsall, W.R. Heineman, Electrochemical biosensors, Chemical Society Reviews, 39 (2010) 1747–1763.

[3] W.-W. Zhao, J.-J. Xu, H.-Y. Chen, Photoelectrochemical immunoassays, Analytical Chemistry, 90 (2018) 615–627.

[4] W.-W. Zhao, J.-J. Xu, H.-Y. Chen, Photoelectrochemical bioanalysis: The state of the art, Chemical Society Reviews, 44 (2015) 729–741.

[5] Z. Liu, W. Qi, G. Xu, Recent advances in electrochemiluminescence, Chemical Society Reviews, 44 (2015) 3117–3142.

[6] W.-W. Zhao, J.-J. Xu, H.-Y. Chen, Photoelectrochemical DNA biosensors, Chemical Reviews, 114 (2014) 7421–7441.

[7] A. Devadoss, P. Sudhagar, C. Terashima, K. Nakata, A. Fujishima, Photoelectrochemical biosensors: New insights into promising photoelectrodes and signal amplification strategies, Journal of Photochemistry and Photobiology C: Photochemistry Reviews, 24 (2015) 43–63.

[8] K. Mao, H. Zhang, Z. Wang, H. Cao, K. Zhang, X. Li, Z. Yang, Nanomaterial-based aptamer sensors for arsenic detection, Biosensors and Bioelectronics, 148 (2020) 111785.

[9] S. Tajik, Z. Dourandish, F.G. Nejad, H. Beitollahi, P.M. Jahani, A. Di Bartolomeo, Transition metal dichalcogenides: Synthesis and use in the development of electro-chemical sensors and biosensors, Biosensors and Bioelectronics, (2022) 114674.

[10] L. Hu, G. Xu, Applications and trends in electrochemiluminescence, Chemical Society Reviews, 39 (2010) 3275–3304.

[11] W. Miao, Electrogenerated chemiluminescence and its biorelated applications, Chemical Reviews, 108 (2008) 2506–2553.

[12] Q. Zhou, D. Tang, Recent advances in photoelectrochemical biosensors for analysis of mycotoxins in food, TrAC Trends in Analytical Chemistry, 124 (2020) 115814.

[13] C. Chen, M. La, Recent developments in electrochemical, electrochemiluminescent, photoelectrochemical methods for the detection of caspase-3 activity, International Journal of Electrochemical Science, 15 (2020) 6852–6862.

[14] M. La, Electrochemical, electrochemiluminescent and photoelectrochemical immuno-sensors for procalcitonin detection: A review, International Journal of Electrochemical Science, 15 (2020) 6436–6447.

[15] Z. Wang, R. Yu, H. Zeng, X. Wang, S. Luo, W. Li, X. Luo, T. Yang, Nucleic acid-based ratiometric electrochemiluminescent, electrochemical and photoelectrochemical bio-sensors: A review, Microchimica Acta, 186 (2019) 1–19.

[16] Y. Zhou, H. Yin, W.-W. Zhao, S. Ai, Electrochemical, electrochemiluminescent and photoelectrochemical bioanalysis of epigenetic modifiers: A comprehensive review, Coordination Chemistry Reviews, 424 (2020) 213519.

[17] C. Zhu, G. Yang, H. Li, D. Du, Y. Lin, Electrochemical sensors and biosensors based on nanomaterials and nanostructures, Analytical Chemistry, 87 (2015) 230–249.

[18] P.K. Kannan, D.J. Late, H. Morgan, C.S. Rout, Recent developments in 2D layered inorganic nanomaterials for sensing, Nanoscale, 7 (2015) 13293–13312.

[19] B. Anasori, M.R. Lukatskaya, Y. Gogotsi, 2D metal carbides and nitrides (MXenes) for energy storage, Nature Reviews Materials, 2 (2017) 1–17.

[20] G.R. Bhimanapati, Z. Lin, V. Meunier, Y. Jung, J. Cha, S. Das, D. Xiao, Y. Son, M.S. Strano, V.R. Cooper, Recent advances in two-dimensional materials beyond graphene, ACS Nano, 9 (2015) 11509–11539.

[21] K.S. Novoselov, A.K. Geim, S.V. Morozov, D.-E. Jiang, Y. Zhang, S.V. Dubonos, I.V. Grigorieva, A.A. Firsov, Electric field effect in atomically thin carbon films, Science, 306 (2004) 666–669.

[22] M. Naguib, V.N. Mochalin, M.W. Barsoum, Y. Gogotsi, 25th anniversary article: MXenes: A new family of two-dimensional materials, Advanced Materials, 26 (2014) 992–1005.

[23] M. Naguib, O. Mashtalir, J. Carle, V. Presser, J. Lu, L. Hultman, Y. Gogotsi, M.W. Barsoum, Two-dimensional transition metal carbides, ACS Nano, 6 (2012) 1322–1331.

[24] B. Anasori, Y. Xie, M. Beidaghi, J. Lu, B.C. Hosler, L. Hultman, P.R. Kent, Y. Gogotsi, M.W. Barsoum, Two-dimensional, ordered, double transition metals carbides (MXenes), ACS Nano, 9 (2015) 9507–9516.

[25] M. Alhabeb, K. Maleski, B. Anasori, P. Lelyukh, L. Clark, S. Sin, Y. Gogotsi, Guidelines for synthesis and processing of two-dimensional titanium carbide (Ti3C2T x MXene), Chemistry of Materials, 29 (2017) 7633–7644.

[26] L. Verger, C. Xu, V. Natu, H.-M. Cheng, W. Ren, M.W. Barsoum, Overview of the syn-thesis of MXenes and other ultrathin 2D transition metal carbides and nitrides, Current Opinion in Solid State and Materials Science, 23 (2019) 149–163.

[27] J. Pang, R.G. Mendes, A. Bachmatiuk, L. Zhao, H.Q. Ta, T. Gemming, H. Liu, Z. Liu, M.H. Rummeli, Applications of 2D MXenes in energy conversion and storage systems, Chemical Society Reviews, 48 (2019) 72–133.

[28] M. Sajid, MXenes: Are they emerging materials for analytical chemistry applications?— A review, Analytica Chimica Acta, 1143 (2021) 267–280.

[29] L. Lu, X. Han, J. Lin, Y. Zhang, M. Qiu, Y. Chen, M. Li, D. Tang, Ultrasensitive fluo-rometric biosensor based on Ti 3 C 2 MXenes with Hg 2+-triggered exonuclease III-assisted recycling amplification, Analyst, 146 (2021) 2664–2669.

[30] R. Zeng, W. Wang, M. Chen, Q. Wan, C. Wang, D. Knopp, D. Tang, CRISPR-Cas12a-driven MXene-PEDOT: PSS piezoresistive wireless biosensor, Nano Energy, 82 (2021) 105711.

[31] G. Cai, Z. Yu, P. Tong, D. Tang, Ti 3 C 2 MXene quantum dot-encapsulated liposomes for photothermal immunoassays using a portable near-infrared imaging camera on a smartphone, Nanoscale, 11 (2019) 15659–15667.

[32] D. Song, X. Jiang, Y. Li, X. Lu, S. Luan, Y. Wang, Y. Li, F. Gao, Metal-organic frameworks-derived MnO2/Mn3O4 microcuboids with hierarchically ordered nanosheets and Ti3C2 MXene/Au NPs composites for electrochemical pesticide detection, Journal of Hazardous Materials, 373 (2019) 367–376.

[33] H. Dong, L. Cao, Z. Tan, Q. Liu, J. Zhou, P. Zhao, P. Wang, Y. Li, W. Ma, Y. Dong, A signal amplification strategy of CuPtRh CNB-embedded ammoniated Ti3C2 MXene for detecting cardiac troponin I by a sandwich-type electrochemical immunosensor, ACS Applied Bio Materials, 3 (2019) 377–384.

[34] F. Duan, C. Guo, M. Hu, Y. Song, M. Wang, L. He, Z. Zhang, R. Pettinari, L. Zhou, Construction of the 0D/2D heterojunction of Ti3C2Tx MXene nanosheets and iron phthalocyanine quantum dots for the impedimetric aptasensing of microRNA-155, Sensors and Actuators B: Chemical, 310 (2020) 127844.

[35] G. Chen, H. Wang, X. Wei, Y. Wu, W. Gu, L. Hu, D. Xu, C. Zhu, Efficient Z-scheme heterostructure based on TiO2/Ti3C2Tx/Cu2O to boost photoelectrochemical response for ultrasensitive biosensing, Sensors and Actuators B: Chemical, 312 (2020) 127951.

[36] X. Hui, M. Sharifuzzaman, S. Sharma, X. Xuan, S. Zhang, S.G. Ko, S.H. Yoon, J.Y. Park, High-performance flexible electrochemical heavy metal sensor based on layer-by-layer assembly of Ti3C2T x/MWNTs nanocomposites for noninvasive detection of copper and zinc ions in human biofluids, ACS Applied Materials & Interfaces, 12 (2020) 48928–48937.

[37] Y.Z. Zhang, Y. Wang, Q. Jiang, J.K. El-Demellawi, H. Kim, H.N. Alshareef, MXene printing and patterned coating for device applications, Advanced Materials, 32 (2020) 1908486.

[38] M. Mathew, S. Radhakrishnan, A. Vaidyanathan, B. Chakraborty, C.S. Rout, Flexible and wearable electrochemical biosensors based on two-dimensional materials: Recent developments, Analytical and Bioanalytical Chemistry, 413 (2021) 727–762.

[39] W. Huang, L. Hu, Y. Tang, Z. Xie, H. Zhang, Recent advances in functional 2D MXene-based nanostructures for next-generation devices, Advanced Functional Materials, 30 (2020) 2005223.

[40] D. Fang, D. Zhao, S. Zhang, Y. Huang, H. Dai, Y. Lin, Black phosphorus quantum dots functionalized MXenes as the enhanced dual-mode probe for exosomes sensing, Sensors and Actuators B: Chemical, 305 (2020) 127544.

[41] H.L. Chia, C.C. Mayorga-Martinez, N. Antonatos, Z.K. Sofer, J.J. Gonzalez-Julian, R.D. Webster, M. Pumera, MXene titanium carbide-based biosensor: Strong dependence of exfoliation method on performance, Analytical Chemistry, 92 (2020) 2452–2459.

[42] F. Wang, C. Yang, C. Duan, D. Xiao, Y. Tang, J. Zhu, An organ-like titanium carbide material (MXene) with multilayer structure encapsulating hemoglobin for a mediator-free biosensor, Journal of the Electrochemical Society, 162 (2014) B16.

[43] L. Zhou, X. Zhang, L. Ma, J. Gao, Y. Jiang, Acetylcholinesterase/chitosan-transition metal carbides nanocomposites-based biosensor for the organophosphate pesticides detection, Biochemical Engineering Journal, 128 (2017) 243–249.

[44] L. Wu, X. Lu, Z.-S. Wu, Y. Dong, X. Wang, S. Zheng, J. Chen, 2D transition metal carbide MXene as a robust biosensing platform for enzyme immobilization and ultra-sensitive detection of phenol, Biosensors and Bioelectronics, 107 (2018) 69–75.

[45] F. Wang, C. Yang, M. Duan, Y. Tang, J. Zhu, TiO2 nanoparticle modified organ-like Ti3C2 MXene nanocomposite encapsulating hemoglobin for a mediator-free biosensor with excellent performances, Biosensors and Bioelectronics, 74 (2015) 1022–1028.

[46] R. Rakhi, P. Nayak, C. Xia, H.N. Alshareef, Novel amperometric glucose biosensor based on MXene nanocomposite, Scientific Reports, 6 (2016) 1–10.

[47] Y. Jiang, X. Zhang, L. Pei, S. Yue, L. Ma, L. Zhou, Z. Huang, Y. He, J. Gao, Silver nanoparticles modified two-dimensional transition metal carbides as nanocarriers to fabricate acetycholinesterase-based electrochemical biosensor, Chemical Engineering Journal, 339 (2018) 547–556.

[48] G. Wang, J. Sun, Y. Yao, X. An, H. Zhang, G. Chu, S. Jiang, Y. Guo, X. Sun, Y. Liu, Detection of Inosine Monophosphate (IMP) in meat using double-enzyme sensor, Food Analytical Methods, 13 (2020) 420–432.

[49] M. Wu, Q. Zhang, Y. Fang, C. Deng, F. Zhou, Y. Zhang, X. Wang, Y. Tang, Y. Wang, Polylysine-modified MXene nanosheets with highly loaded glucose oxidase as cascade nanoreactor for glucose decomposition and electrochemical sensing, Journal of Colloid and Interface Science, 586 (2021) 20–29.

[50] M. Song, S.Y. Pang, F. Guo, M.C. Wong, J. Hao, Fluoride-free 2D niobium carbide MXenes as stable and biocompatible nanoplatforms for electrochemical biosensors with ultrahigh sensitivity, Advanced Science, 7 (2020) 2001546.

[51] S.S. Shankar, R.M. Shereema, R. Rakhi, Electrochemical determination of adrenaline using MXene/graphite composite paste electrodes, ACS Applied Materials & Interfaces, 10 (2018) 43343–43351.

[52] J. Zheng, J. Diao, Y. Jin, A. Ding, B. Wang, L. Wu, B. Weng, J. Chen, An inkjet printed Ti3C2-GO electrode for the electrochemical sensing of hydrogen peroxide, Journal of the Electrochemical Society, 165 (2018) B227.

[53] S. Neampet, N. Ruecha, J. Qin, W. Wonsawat, O. Chailapakul, N. Rodthongkum, A nanocomposite prepared from platinum particles, polyaniline and a Ti 3 C 2 MXene for amperometric sensing of hydrogen peroxide and lactate, Microchimica Acta, 186 (2019) 1–8.

[54] J. Chen, P. Tong, L. Huang, Z. Yu, D. Tang, Ti3C2 MXene nanosheet-based capacitance immunoassay with tyramine-enzyme repeats to detect prostate-specific antigen on interdigitated micro-comb electrode, Electrochimica Acta, 319 (2019) 375–381.

[55] H. Medetalibeyoglu, G. Kotan, N. Atar, M.L. Yola, A novel and ultrasensitive sandwich-type electrochemical immunosensor based on delaminated MXene@ AuNPs as signal amplification for prostate specific antigen (PSA) detection and immunosensor validation, Talanta, 220 (2020) 121403.

[56] H. Medetalibeyoglu, M. Beytur, O. Akyıldırım, N. Atar, M.L. Yola, Validated electrochemical immunosensor for ultra-sensitive procalcitonin detection: Carbon electrode modified with gold nanoparticles functionalized sulfur doped MXene as sensor platform and carboxylated graphitic carbon nitride as signal amplification, Sensors and Actuators B: Chemical, 319 (2020) 128195.

[57] S. Ranallo, A. Porchetta, F. Ricci, DNA-based scaffolds for sensing applications, Analytical Chemistry, 91 (2018) 44–59.

[58] S. Lv, K. Zhang, L. Zhu, D. Tang, ZIF-8-assisted NaYF4: Yb, Tm@ ZnO converter with exonuclease III-powered DNA walker for near-infrared light responsive biosensor, Analytical Chemistry, 92 (2019) 1470–1476.

[59] K. Zhang, S. Lv, Q. Zhou, D. Tang, CoOOH nanosheets-coated g-C3N4/CuInS2 nanohybrids for photoelectrochemical biosensor of carcinoembryonic antigen coupling hybridization chain reaction with etching reaction, Sensors and Actuators B: Chemical, 307 (2020) 127631.

[60] A. Abi, Z. Mohammadpour, X. Zuo, A. Safavi, Nucleic acid-based electrochemical nanobiosensors, Biosensors and Bioelectronics, 102 (2018) 479–489.

[61] S. Zhou, C. Gu, Z. Li, L. Yang, L. He, M. Wang, X. Huang, N. Zhou, Z. Zhang, Ti3C2Tx MXene and polyoxometalate nanohybrid embedded with polypyrrole: Ultra-sensitive platform for the detection of osteopontin, Applied Surface Science, 498 (2019) 143889.

[62] J. Zhang, Y. Li, S. Duan, F. He, Highly electrically conductive two-dimensional Ti3C2 Mxenes-based 16S rDNA electrochemical sensor for detecting Mycobacterium tuberculosis, Analytica Chimica Acta, 1123 (2020) 9–17.

[63] H. Wang, J. Sun, L. Lu, X. Yang, J. Xia, F. Zhang, Z. Wang, Competitive electrochemical aptasensor based on a cDNA-ferrocene/MXene probe for detection of breast cancer marker Mucin1, Analytica Chimica Acta, 1094 (2020) 18–25.

[64] L. Kashefi-Kheyrabadi, A. Koyappayil, T. Kim, Y.-P. Cheon, M.-H. Lee, A MoS2@ Ti3C2Tx MXene hybrid-based electrochemical aptasensor (MEA) for sensitive and rapid detection of thyroxine, Bioelectrochemistry, 137 (2021) 107674.

[65] X. Yang, M. Feng, J. Xia, F. Zhang, Z. Wang, An electrochemical biosensor based on AuNPs/Ti3C2 MXene three-dimensional nanocomposite for microRNA-155 detection by exonuclease III-aided cascade target recycling, Journal of Electroanalytical Chemistry, 878 (2020) 114669.

[66] H. Zhang, Z. Wang, F. Wang, Y. Zhang, H. Wang, Y. Liu, Ti3C2 MXene mediated Prussian blue in situ hybridization and electrochemical signal amplification for the detection of exosomes, Talanta, 224 (2021) 121879.

[67] P.A. Rasheed, R.P. Pandey, K. Rasool, K.A. Mahmoud, Ultra-sensitive electrocatalytic detection of bromate in drinking water based on Nafion/Ti3C2Tx (MXene) modified glassy carbon electrode, Sensors and Actuators B: Chemical, 265 (2018) 652–659.

[68] D. Wu, M. Wu, J. Yang, H. Zhang, K. Xie, C.-T. Lin, A. Yu, J. Yu, L. Fu, Delaminated Ti3C2Tx (MXene) for electrochemical carbendazim sensing, Materials Letters, 236 (2019) 412–415.

[69] F. Shahzad, A. Iqbal, S.A. Zaidi, S.-W. Hwang, C.M. Koo, Nafion-stabilized two-dimensional transition metal carbide (Ti3C2Tx MXene) as a high-performance electrochemical sensor for neurotransmitter, Journal of Industrial and Engineering Chemistry, 79 (2019) 338–344.

[70] T.S.H. Pham, L. Fu, P. Mahon, G. Lai, A. Yu, Fabrication of β-cyclodextrin-functionalized reduced graphene oxide and its application for electrocatalytic detection of carbendazim, Electrocatalysis, 7 (2016) 411–419.

[71] Y. Zhang, X. Jiang, J. Zhang, H. Zhang, Y. Li, Simultaneous voltammetric determination of acetaminophen and isoniazid using MXene modified screen-printed electrode, Biosensors and Bioelectronics, 130 (2019) 315–321.

[72] R. Ramachandran, C. Zhao, M. Rajkumar, K. Rajavel, P. Zhu, W. Xuan, Z.-X. Xu, F. Wang, Porous nickel oxide microsphere and Ti3C2Tx hybrid derived from metal-organic framework for battery-type supercapacitor electrode and non-enzymatic H2O2 sensor, Electrochimica Acta, 322 (2019) 134771.

[73] M. Li, L. Fang, H. Zhou, F. Wu, Y. Lu, H. Luo, Y. Zhang, B. Hu, Three-dimensional porous MXene/NiCo-LDH composite for high performance non-enzymatic glucose sensor, Applied Surface Science, 495 (2019) 143554.

[74] H. Zou, F. Zhang, H. Wang, J. Xia, L. Gao, Z. Wang, Au nanoparticles supported on functionalized two-dimensional titanium carbide for the sensitive detection of nitrite, New Journal of Chemistry, 43 (2019) 2464–2470.

[75] P.A. Rasheed, R.P. Pandey, K.A. Jabbar, J. Ponraj, K.A. Mahmoud, Sensitive electrochemical detection of l-cysteine based on a highly stable Pd@ Ti 3 C 2 T x (MXene) nanocomposite modified glassy carbon electrode, Analytical Methods, 11 (2019) 3851–3856.

[76] J. Zheng, B. Wang, Y. Jin, B. Weng, J. Chen, Nanostructured MXene-based biomimetic enzymes for amperometric detection of superoxide anions from HepG2 cells, Microchimica Acta, 186 (2019) 1–9.

[77] R. Huang, S. Chen, J. Yu, X. Jiang, Self-assembled Ti3C2/MWCNTs nanocomposites modified glassy carbon electrode for electrochemical simultaneous detection of hydroquinone and catechol, Ecotoxicology and Environmental Safety, 184 (2019) 109619.

[78] R. Zhang, J. Liu, Y. Li, MXene with great adsorption ability toward organic dye: An excellent material for constructing a ratiometric electrochemical sensing platform, ACS Sensors, 4 (2019) 2058–2064.

[79] R. Huang, D. Liao, S. Chen, J. Yu, X. Jiang, A strategy for effective electrochemical detection of hydroquinone and catechol: Decoration of alkalization-intercalated Ti3C2 with MOF-derived N-doped porous carbon, Sensors and Actuators B: Chemical, 320 (2020) 128386.

[80] X. Tu, F. Gao, X. Ma, J. Zou, Y. Yu, M. Li, F. Qu, X. Huang, L. Lu, Mxene/carbon nanohorn/β-cyclodextrin-Metal-organic frameworks as high-performance electrochemical sensing platform for sensitive detection of carbendazim pesticide, Journal of Hazardous Materials, 396 (2020) 122776.

[81] X. Ma, X. Tu, F. Gao, Y. Xie, X. Huang, C. Fernandez, F. Qu, G. Liu, L. Lu, Y. Yu, Hierarchical porous MXene/amino carbon nanotubes-based molecular imprinting sensor for highly sensitive and selective sensing of fisetin, Sensors and Actuators B: Chemical, 309 (2020) 127815.

[82] Y. Dang, X. Guan, Y. Zhou, C. Hao, Y. Zhang, S. Chen, Y. Ma, Y. Bai, Y. Gong, Y. Gao, Biocompatible PB/Ti3C2 hybrid nanocomposites for the non-enzymatic electrochemical detection of H2O2 released from living cells, Sensors and Actuators B: Chemical, 319 (2020) 128259.

[83] T. Kokulnathan, E. Ashok Kumar, T.-J. Wang, Design and in situ synthesis of titanium carbide/boron nitride nanocomposite: Investigation of electrocatalytic activity for the sulfadiazine sensor, ACS Sustainable Chemistry & Engineering, 8 (2020) 12471–12481.

[84] P.K. Kalambate, A. Sinha, Y. Li, Y. Shen, Y. Huang, An electrochemical sensor for ifosfamide, acetaminophen, domperidone, and sumatriptan based on self-assembled MXene/ MWCNT/chitosan nanocomposite thin film, Microchimica Acta, 187 (2020) 1–12.

[85] S. Elumalai, V. Mani, N. Jeromiyas, V.K. Ponnusamy, M. Yoshimura, A composite film prepared from titanium carbide Ti 3 C 2 T x (MXene) and gold nanoparticles for voltammetric determination of uric acid and folic acid, Microchimica Acta, 187 (2020) 1–10.

[86] X. Wang, M. Li, S. Yang, J. Shan, A novel electrochemical sensor based on TiO2-Ti3C2TX/CTAB/chitosan composite for the detection of nitrite, Electrochimica Acta, 359 (2020) 136938.

[87] Y. Wang, Z. Zeng, J. Qiao, S. Dong, Q. Liang, S. Shao, Ultrasensitive determination of nitrite based on electrochemical platform of AuNPs deposited on PDDA-modified MXene nanosheets, Talanta, 221 (2021) 121605.

[88] E.A. Kumar, T. Kokulnathan, T.-J. Wang, A.J. Anthuvan, Y.-H. Chang, Two-dimensional titanium carbide (MXene) nanosheets as an efficient electrocatalyst for 4-nitroquinoline N-oxide detection, Journal of Molecular Liquids, 312 (2020) 113354.

[89] Y. He, L. Ma, L. Zhou, G. Liu, Y. Jiang, J. Gao, Preparation and application of bismuth/ MXene nano-composite as electrochemical sensor for heavy metal ions detection, Nanomaterials, 10 (2020) 866.

[90] X. Lv, F. Pei, S. Feng, Y. Wu, S.-M. Chen, Q. Hao, W. Lei, Facile synthesis of protonated carbon nitride/Ti3C2Tx nanocomposite for simultaneous detection of Pb2+ and Cd2+, Journal of the Electrochemical Society, 167 (2020) 067509.

[91] P.A. Rasheed, R.P. Pandey, T. Gomez, M. Naguib, K.A. Mahmoud, Large interlayer spacing Nb 4 C 3 T x (MXene) promotes the ultrasensitive electrochemical detection of Pb 2+ on glassy carbon electrodes, RSC Advances, 10 (2020) 24697–24704.

[92] Y. Yao, X. Han, X. Yang, J. Zhao, C. Chai, Detection of hydrazine at MXene/ZIF-8 nanocomposite modified electrode, Chinese Journal of Chemistry, 39 (2021) 330–336.

[93] N. Arif, S. Gul, M. Sohail, S. Rizwan, M. Iqbal, Synthesis and characterization of layered Nb2C MXene/ZnS nanocomposites for highly selective electrochemical sensing of dopamine, Ceramics International, 47 (2021) 2388–2396.

[94] P.A. Rasheed, R.P. Pandey, T. Gomez, K.A. Jabbar, K. Prenger, M. Naguib, B. Aïssa, K.A. Mahmoud, Nb-based MXenes for efficient electrochemical sensing of small biomolecules in the anodic potential, Electrochemistry Communications, 119 (2020) 106811.

[95] T. Kokulnathan, T.-J. Wang, Vanadium carbide-entrapped graphitic carbon nitride nanocomposites: Synthesis and electrochemical platforms for accurate detection of furazolidone, ACS Applied Nano Materials, 3 (2020) 2554–2561.

[96] Y. Fang, X. Yang, T. Chen, G. Xu, M. Liu, J. Liu, Y. Xu, Two-dimensional titanium carbide (MXene)-based solid-state electrochemiluminescent sensor for label-free single-nucleotide mismatch discrimination in human urine, Sensors and Actuators B: Chemical, 263 (2018) 400–407.

[97] H. Zhang, Z. Wang, Q. Zhang, F. Wang, Y. Liu, Ti3C2 MXenes nanosheets catalyzed highly efficient electrogenerated chemiluminescence biosensor for the detection of exosomes, Biosensors and Bioelectronics, 124 (2019) 184–190.

[98] W. Huang, Y. Wang, W.-B. Liang, G.-B. Hu, L.-Y. Yao, Y. Yang, K. Zhou, R. Yuan, D.-R. Xiao, Two birds with one stone: Surface functionalization and delamination of multilayered Ti3C2T x MXene by grafting a ruthenium (II) complex to achieve conductivity-enhanced electrochemiluminescence, Analytical Chemistry, 93 (2021) 1834–1841.

[99] L. Shang, X. Wang, W. Zhang, L.-P. Jia, R.-N. Ma, W.-L. Jia, H.-S. Wang, A dual-potential electrochemiluminescence sensor for ratiometric detection of carcinoembryonic antigen based on single luminophor, Sensors and Actuators B: Chemical, 325 (2020) 128776.

[100] D. Fang, H. Ren, Y. Huang, H. Dai, D. Huang, Y. Lin, Photothermal amplified cathodic ZnO quantum dots/Ru (bpy) 32+/S2O82-ternary system for ultrasensitive electrochemiluminescence detection of thyroglobulin, Sensors and Actuators B: Chemical, 312 (2020) 127950.

[101] M. Li, H. Wang, X. Wang, Q. Lu, H. Li, Y. Zhang, S. Yao, Ti3C2/Cu2O heterostructure based signal-off photoelectrochemical sensor for high sensitivity detection of glucose, Biosensors and Bioelectronics, 142 (2019) 111535.

[102] X. Chen, J. Li, G. Pan, W. Xu, J. Zhu, D. Zhou, D. Li, C. Chen, G. Lu, H. Song, Ti3C2 MXene quantum dots/TiO2 inverse opal heterojunction electrode platform for superior photoelectrochemical biosensing, Sensors and Actuators B: Chemical, 289 (2019) 131–137.

[103] F. Han, Z. Song, J. Xu, M. Dai, S. Luo, D. Han, L. Niu, Z. Wang, Oxidized titanium carbide MXene-enabled photoelectrochemical sensor for quantifying synergistic interaction of ascorbic acid based antioxidants system, Biosensors and Bioelectronics, 177 (2021) 112978.

[104] R.A. Soomro, S. Jawaid, N.H. Kalawar, M. Tunesi, S. Karakuş, A. Kilislioğlu, M. Willander, In-situ engineered MXene-TiO2/BiVO4 hybrid as an efficient photoelectrochemical platform for sensitive detection of soluble CD44 proteins, Biosensors and Bioelectronics, 166 (2020) 112439.

[105] R.A. Soomro, S. Jawaid, P. Zhang, X. Han, K.R. Hallam, S. Karakuş, A. Kilislioğlu, B. Xu, M. Willander, NiWO4-induced partial oxidation of MXene for photo-electrochemical detection of prostate-specific antigen, Sensors and Actuators B: Chemical, 328 (2021) 129074.

[106] C. Yuan, Z. He, Q. Chen, X. Wang, C. Zhai, M. Zhu, Selective and efficacious photoelectrochemical detection of ciprofloxacin based on the self-assembly of 2D/2D g-C3N4/Ti3C2 composites, Applied Surface Science, 539 (2021) 148241.

[107] Q. Jiang, H. Wang, X. Wei, Y. Wu, W. Gu, L. Hu, C. Zhu, Efficient BiVO4 photoanode decorated with Ti3C2TX MXene for enhanced photoelectrochemical sensing of Hg (II) ion, Analytica Chimica Acta, 1119 (2020) 11–17.

[108] Y. Yao, L. Lan, X. Liu, Y. Ying, J. Ping, Spontaneous growth and regulation of noble metal nanoparticles on flexible biomimetic MXene paper for bioelectronics, Biosensors and Bioelectronics, 148 (2020) 111799.

[109] S.F. Zhao, F.X. Hu, Z.Z. Shi, J.J. Fu, Y. Chen, F.Y. Dai, C.X. Guo, C.M. Li, 2-D/2-D heterostructured biomimetic enzyme by interfacial assembling Mn 3 (PO 4) 2 and MXene as a flexible platform for realtime sensitive sensing cell superoxide, Nano Research, 14 (2021) 879–886.

[110] Z. Yu, G. Cai, X. Liu, D. Tang, Platinum nanozyme-triggered pressure-based immunoassay using a three-dimensional polypyrrole foam-based flexible pressure sensor, ACS Applied Materials & Interfaces, 12 (2020) 40133–40140.

[111] Z. Yu, Y. Tang, G. Cai, R. Ren, D. Tang, Paper electrode-based flexible pressure sensor for point-of-care immunoassay with digital multimeter, Analytical Chemistry, 91 (2018) 1222–1226.

[112] Z. Yu, G. Cai, P. Tong, D. Tang, Saw-toothed microstructure-based flexible pressure sensor as the signal readout for point-of-care immunoassay, ACS Sensors, 4 (2019) 2272–2276.

[113] L. Wang, K. Jiang, G. Shen, A perspective on flexible sensors in developing diagnostic devices, Applied Physics Letters, 119 (2021) 150501.

[114] S. Zhao, W. Ran, L. Wang, G. Shen, Interlocked MXene/rGO aerogel with excellent mechanical stability for a health-monitoring device, Journal of Semiconductors, 43 (2022) 082601.

[115] X. Fu, L. Zhao, Z. Yuan, Y. Zheng, V. Shulga, W. Han, L. Wang, Hierarchical MXene@ ZIF-67 film based high performance tactile sensor with large sensing range from motion monitoring to sound wave detection, Advanced Materials Technologies, 7 (2022) 2101511.

[116] L. Zhao, X. Fu, H. Xu, Y. Zheng, W. Han, L. Wang, Tissue-like sodium alginate-coated 2D MXene-based flexible temperature sensors for full-range temperature monitoring, Advanced Materials Technologies, 7 (2022) 2101740.

[117] S. Zhang, M.A. Zahed, M. Sharifuzzaman, S. Yoon, X. Hui, S. Chandra Barman, S. Sharma, H.S. Yoon, C. Park, J.Y. Park, A wearable battery-free wireless and skin-interfaced microfluidics integrated electrochemical sensing patch for on-site biomarkers monitoring in human perspiration, Biosens Bioelectron, 175 (2021) 112844.

[118] M. Li, L. Wang, R. Liu, J. Li, Q. Zhang, G. Shi, Y. Li, C. Hou, H. Wang, A highly integrated sensing paper for wearable electrochemical sweat analysis, Biosensors and Bioelectronics, 174 (2021) 112828.

[119] M. Sharifuzzaman, S.C. Barman, M.A. Zahed, S. Sharma, H. Yoon, J.S. Nah, H. Kim, J.Y. Park, An electrodeposited MXene-Ti3C2Tx nanosheets functionalized by task-specific ionic liquid for simultaneous and multiplexed detection of bladder cancer biomarkers, Small, 16 (2020) 2002517.

[120] M. Mohammadniaei, A. Koyappayil, Y. Sun, J. Min, M.-H. Lee, Gold nanoparticle/ MXene for multiple and sensitive detection of oncomiRs based on synergetic signal amplification, Biosensors and Bioelectronics, 159 (2020) 112208.

11 Insights into MXenes for Versatile Biomedical Applications

*Monika Ahuja, Prem P. Sharma,
and Pratik Kumar Jagtap*

11.1 INTRODUCTION

The rapid advancement of nanobiotechnology and biomedicine has encouraged the invention of various new inorganic nanosystems as potential alternatives in fighting different diseases by multimodal imaging and synergistic treatment [1–11]. Currently, multidisciplinary research on the biomedical applications of 2D nanomaterials has been of great interest to material researchers because of their ultrathin layer-structured topology as well as unique optical, electronic and mechanical properties [12–17]. MXenes (pronounced 'maxenes') have emerged as multifaceted 2D nanomaterials and are currently replacing traditional 2D materials, bringing disruptive technology by transforming many aspects of human life [18]. The first ever MXene was discovered by Professor Gogotsi, Professor Barsoum, and coworkers at Drexel University, Philadelphia, in 2011, where 2D titanium carbide (Ti_3C_2) layers were produced when they exfoliated 3D titanium aluminum carbide (Ti_3AlC_2) using hydrofluoric acid (HF) [19–21].

MXenes characteristically have the following three different formulae: M_2X, M_3X_2, and M_4X_3. MXenes ($M_{n+1}X_n$ layers) are generally fashioned by the selective etching of 'A' layers from layered ternary carbides of MAX phases *via* a top-down approach. The 3D precursors for MXenes such as $M_{n+1}AX_n$ phases (n = 1–3) are made of transition metal carbides/nitrides interconnected by an 'A' element with strong sigma and pi bonds, where M is an early transition metal, A is mostly a main group IIIA or IVA element, and X is C/N. M–X bonds are stronger than M–A bonds. The 'A' layer can be selectively removed by an etching process using a strong acid (hydrofluoric acid) to fabricate $M_{n+1}X_n$ layers followed by sonication [22]. The surfaces of MXenes are generally terminated with functional groups such as fluorine (-F), hydroxide (-OH), and oxygen (-O) groups [23–25], which modify their chemical formula to $M_{n+1}X_nT_x$ (T_x = surface functional groups), as shown in Figure 11.1. So far, Ti_3C_2 and Ti_2C groups are the most commonly used groups in MXenes for biomedical applications, as new arrangements of M and A group elements are possible besides Ti and C.

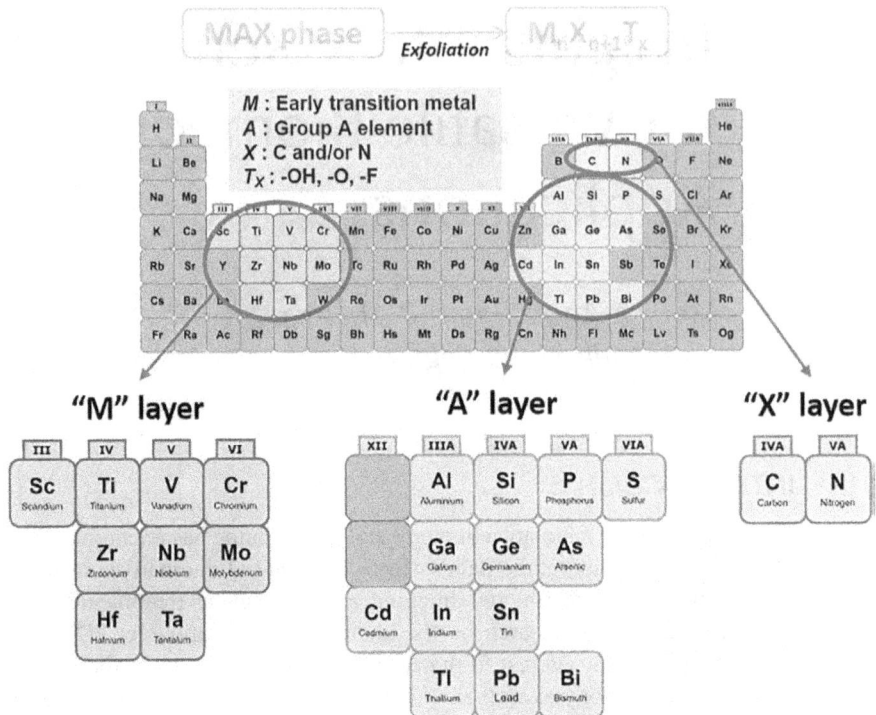

FIGURE 11.1 General elemental composition of a MAX phase and MXene.

The nanoscale lateral sizes of MXenes have been investigated for various biomedical applications in the last few years [23]. MXenes satisfy the strict demands in variety of fields such as in nanomedicine owing to their captivating physiochemical properties such as semiconductive, electrical, magnetic, and mechanical properties. MXenes possess large specific surface area, modifiable surface functional groups, biocompatibility, atomic layer thickness, a hydrophilic nature, and metallic conductivity [24,25]. MXene nanosheets are suitable for drug loading/delivery because of their biocompatibility with living cell/tissues, which is also enhanced due to their due to surface functionalization and hydrophilicity [26,27]. The existence of metals within the atomic layers makes MXenes feasible for proficient diagnosis and treatment of different diseases. They show strong absorption at near-infrared regions, which enables them to be used for photothermal therapy [28,29]. Their surface plasmon resonance effect and tunable bandgap makes MXenes more capable of ultrasonic irradiation for nanodynamic therapy [30]. The reduced sizes of MXenes with numerous defects have been found to possess luminescence properties and can be utilized as biosensors [31]. MXenes containing paramagnetic transition metals element with high atomic numbers makes them more suitable for CT scans and MRI techniques [32]. Furthermore, the redox activities of the metal atoms

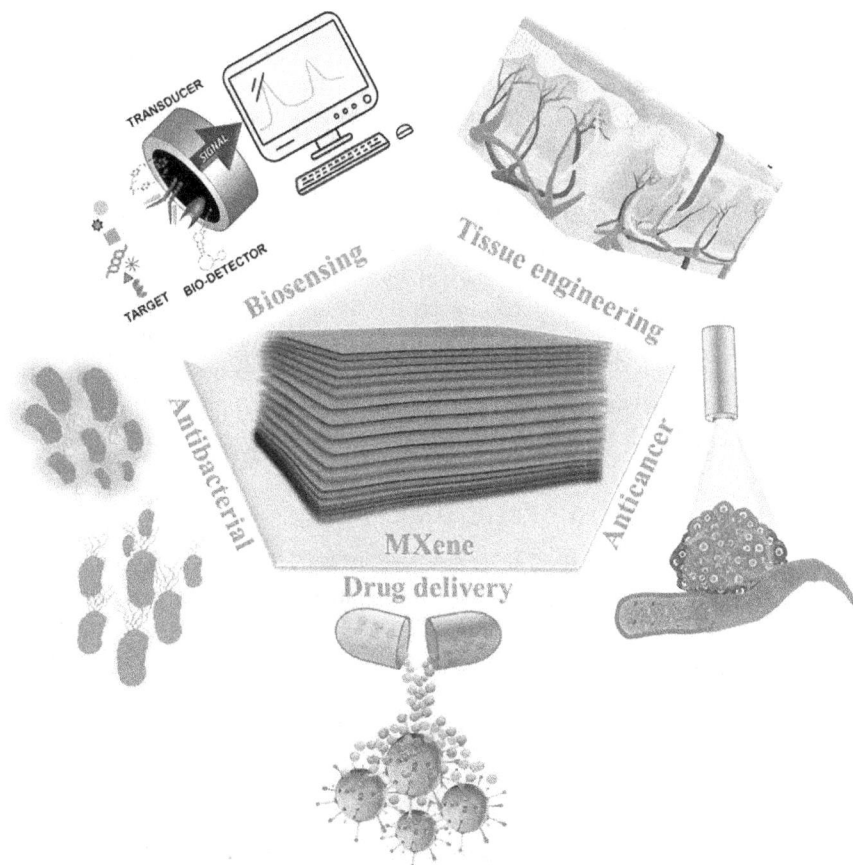

FIGURE 11.2 An overview of applications of MXenes in the field of biomedicine.
Copyright with permission from Ref [30]. Open access under a CC BY 4.0 license.

provide an advantage towards catalytic multienzyme activities. Having these properties, MXenes have been proven to be suitable candidates for nanomedicine in cancer therapy, bioimaging, biosensing, tissue engineering, drug delivery, and antibacterial applications (Figure 11.2).

The synthesis and engineering of MXene-based materials is of supreme importance for biomedical applications. MXenes can be imparted with different physical, mechanical, and chemical properties so as to make them compatible with the physiological environment. These need to have sufficient mechanical strength, degradability, and the ability to overcome biological rejection when used in living cells and tissues. Recently, intensive research efforts on MXenes have been made to improve their performance and open a new horizon for the functional modifications suitable for their many applications. The surfaces of MXenes can be functionalized by combining MXenes with polymeric materials from 0D to 3D. Moreover, MXenes can be modified with heteroatoms (N, P, and S) to produce diverse functional MXenes. Apart from this, in order to achieve enhanced properties of MXenes, they can be doped with boron, platinum, niobium, silicon, germanium, vanadium, and alkali/ alkaline earth metal cations. These doped MXenes and their composites have better compatibility and are proven to be non-toxic to living organisms, like a biodegradable niobium carbide MXene used in mice [28].

In this chapter, we have summarized the important applications of 2D MXenes, their composites, and derivatives in a variety of diverse fields such as in drug delivery [33], antimicrobial applications [34–35], biosensing [36–39], cancer theranostics, and tissue engineering [40–46]. Moreover, the biological behavior of 2D MXenes is yet to be explored and can provide a robust nanoplatform, facilitating promising applications along with clinical translation potential to benefit human health. Furthermore, current challenges and limitations in the developments of 2D MXene-based biomaterials will be elaborated on in the conclusion.

11.2 ANTIMICROBIAL APPLICATIONS

MXenes and their composites show promising antibacterial activities against *Bacillus subtilis*, *Escherichia coli*, *Pseudomonas aeruginosa*, *Staphylococcus aureus*, and *Shigella* which pose serious health issues. These enhanced antimicrobial properties of MXenes are because of their sharp edges, hydrophilicity, semiconductor properties, good electrical conductivity, hydrogen bonding with the cell membrane lipopolysaccharide molecules, oxygen-containing surface functionalities, and optical properties.

MXenes such as $Ti_3C_2T_x$ and $TiVCT_x$ have shown inherent antimicrobial properties, even better than graphene oxide. Kashif et al. reported that $Ti_3C_2T_x$ can inhibit the growth of Gram-negative *E. coli* and Gram-positive *Bacillus subtilis*, depending on the dose of MXenes. Due to the differences in the peptidoglycan thicknesses in Gram-negative *E. coli* and Gram-positive *B. subtilis* bacteria, a subsequent difference was observed in their resistance towards the MXene. The antimicrobial mechanism of $Ti_3C_2T_x$ nanosheets can be associated with their reactive surface, strong reducing activity, and smaller size. The interaction of the $Ti_3C_2T_x$ MXene with the microbial cell membrane and cytoplasm through direct physical penetration led to the destruction of cell structure and microorganism death [47]. Another reason is anionic properties of the negatively charged Ti_3C_2 nanosheets due to their conductive properties, which promote the electron transfer from bacterial components to the outside environment, resulting in cell death. Moreover, environmental conditions also assist in the aging of the membrane and surface oxidation of $Ti_3C_2T_x$. The TiO_2-catalyzed free radical increases the antibacterial properties of $Ti_3C_2T_x$ by creating oxidative stress on the lipid bilayer.

11.3 BIOSENSING

MXenes play a key role as biosensors to selectively detect some specific molecules such as proteins, amino acids, RNAs, etc., in the human body due to their extensive physical, chemical, and surface properties. Using biosensors, diseases can be easily detected at earlier stages of their proliferation. Biosensors can be based on electrical signals or optical signals. Surface plasmon resonance biosensors have been used in pathology labs for real-time analysis, as they require only a few mg of samples with fast detection and high sensitivity. MXenes have been applied for sensing various gases through surface alteration, which ultimately leads to conductivity changes in MXenes. Yu et al. explored the potential of MXenes in the sensing of ammonia gas selectively by Ti_2C monolayers with oxygen termination due to charge transfer [48]. This study was later experimentally confirmed by Kim and coworkers [49]. MXene-based sensors have been also used to detect H_2O_2 and other small molecules like glucose and phenols and also for metal ions such as Cd^{2+}, Pb^{2+}, Cu^{2+}, and Hg^{2+} with enhanced selectivity and sensitivity. They have also been used in pH sensors (Song et al.), as their adsorption intensity decreases by 10% with an increase in pH value from 5 to 9 for Ti_3C_2 MXene dots [50]. Currently Au-, Ag-, and Pd-based MXene hybrid probes are applied as Raman sensing probes, which opens gates for biosensing applications. Low-cost piezoresistive sensors have been used in the electronic devices used in skin treatments and other portable healthcare monitors for better screen resolution. MXene aerogels have been reported to be used as probes in voice monitoring and pulse beating applications. MXenes coupled with sponge piezosensitive sensors have been used to monitor other human physiological signals such as those during respiration, joint movement, and heart pulses specially used in SARS-COV-19 cases. MXenes coupled with SPR biosensors have been used because of their wide applications range and ability to track and monitor the dosage of the drug. This method has excellent reproducibility and high throughput with lower LOD values.

11.4   BIOIMAGING

Creatively-designed MXene-based fluorescent agents are a main focus and have recently been explored in order to initiate a green approach and to overcome various limitations of fluorescent probes such as their instability, non-biodegradability, biological toxicity, and inadequate fluorescent signal output. Xue et al. applied a hydrothermal technique using zinc ion sensors with advanced properties (biocompatible multicolored cellular imaging and better photoluminescence) based on MXenes coupled with advanced Ti_3C_2QDs [51]. Later, in another approach, Xu et al. demonstrated the use of S- and N-doped Nb_2C QDs with enhanced properties such as better quantum yield, enhanced fluorescence, photo-stability, and high pH [52]. Thus, with further advances, these fluorescent probes can be suitable candidates for efficient three-dimensional (3D) brain organoid labeling with better fluorescent properties and contrast.

Liu et al. explained the mechanism for the use of smart Ti_3C_2 QDs designed for the detection of hypochlorite (ClO^-) and curcumin. Basically, here, the curcumin is converted to quinones *via* the oxidization of the methoxy and phenolic groups. The linear uncovering range was ~0.05–10 µM for curcumin, and the linear detection limit for ClO^- was reported to be ~25–150 µM; 150–275 µM. The LOD was reported as ~20 nM and 5 µM for curcumin and ClO^-, respectively [53]

11.5   CONTRAST-ENHANCED BIOIMAGING

Bioimaging itself can be considered a powerful technique for the sensing and diagnosing of diseases. Having superior photothermal effects, MXenes are promising agents with contrast-enhanced bioimaging ability. Lin et al. developed 2D Ta_4C_3 nanosheets used for instantaneous PTT and PA imaging of 4T1 tumors [54].

11.6   DRUG DELIVERY

Controlled drug delivery to cutaneous and subcutaneous wounds with reduced toxic effects has been achieved using MXene-wrapped 2D $Ti_3C_2T_x$ magnetic colloids. Ultrathin Ti_3C_2 nanosheets have been promising candidates for the efficient eradication of tumor cells, as their surface properties enable efficient coating of the anticancer drug doxorubicin. Moreover, the efficacy of the drug is enhanced due to the smaller sizes of these nanoparticles (1–100 nm), which enable them to be easily transported through the blood vessels. This, on the other hand, indirectly decreases the drug dosages without compromising with their therapeutic levels, thus reducing the toxicity of normal cells and tissues and also minimizing their side effects. They have been reported to have improved coating and higher drug-releasing properties even in acidic environments. Bae et al. experimentally demonstrated that drug release was enhanced by 23.5% even in an acidic environment and upon irradiation at 808 nm using near-infrared (NIR) [55,56].

11.7   TISSUE ENGINEERING

The MXene-based sensors used in tissue engineering have wide pressure-sensing ranges (0–400 KPa). Moreover, MXene-based photothermal therapy converts light energy to heat energy, causing the elimination of tumor cells, and has been proven to be better than radiation and chemotherapy, as MXenes' provide vast surface area acts as an anchoring site for the accumulation of toxins in tumor cells during irradiation. Cellulose- and Ti_3C_2-based MXene composite hydrogels act as anticancer agents for faster drug delivery. Moreover, they help not only in tumor elimination but also in further suppression of tumor recurrence. Thus, these MXene-based carriers allow active targeting and better biocompatibility. Recently Hussein et al. reported titanium carbide ($Ti_3C_2T_x$)-based 2D plasmonic nanocomposites (Au/MXene and Au/Fe_3O_4/MXene) with corresponding anticancer and photothermal capabilities [58]. These modified MXenes have lower toxicity than traditional MXenes, as shown by lower mortality rates in zebrafish embryos. Combining a fluorescent substance with MXenes can increase their surface brightness and is thus used in bioimaging. Tantalum-rich MXenes like Ta_4C_3 are suitable for use in CT scans as contrast agents due to their high biocompatibility. Gadolinium- and manganese-based MXenes have been used as MRI contrast agents.

11.8   REGENERATION

MXene-based composites have been used in bone regeneration in oral and orthopedic treatments. Titanium-based polylactic membrane MXene nanosheets are also reported to be used in guided bone regeneration. All these applications are possible because of the hydrophobic surface of MXene films. MXene nanoparticles are used for bone tissue regeneration because of their enhanced mechanical and biocompatibility properties. The ultrathin structure and properties of MXene nanoparticles help in regenerating alveolar bone defects at impaired sites. Moreover, some surface-modified MXenes have been found to stimulate osteogenesis activity in both *in vitro* and *in vivo* applications by triggering the non-osteogenic tissues from interfering with bone healing and guiding bone cell attachment.

MXene-based polymeric fibrous nanocomposites have been found to be compatible with different cell lines such as NIH-373 and MC-373-E1 cells due to their wettability properties, biomineralization, and protein adsorption properties, allowing them to be exploited in wound healing, bone tissue engineering, and cancer therapy. Graphene and TiC MXene nanocomposites are being used as adsorbents for the suspension of cytokine driven inflammatory stimulus, indicating factors like integrated biological interactions and surface properties [25,28].

11.9   CANCER THERANOSTICS

MXene-based nanocomposites have been used for drug delivery for anticancer agents due to their low toxicity, high biocompatibility, and enhanced drug bonding capacity. The adjusted quantum confinement and edge effect and localized surface plasma resonance effects can be exploited in the core of their applications in cancer therapy for fluorescent imaging, computed tomography imaging, and photothermal therapy. Core shell MXenes/mesoporous silica have been reported for drug loading and also in functional group modifications with PEG and other active materials at the site of tumor development for controlled drug release, facilitating chemotherapy in terms of its efficacy.

Chemotherapy and radiotherapy suffer from disadvantages such as low selectivity; however, Ta_4C_3 MXene-based sheets showed better photothermal conversion efficiency and were found to be suitable under physiological media. They have longer blood circulation lifetimes and lower accumulation levels in tumors cells due to differences in nanoparticle size and enhanced permeability and retention effects.

Further, when fluorescent species are attached to the surface of MXenes, they can be used for cancer treatment, such as those obtained by using Ti_3C_2 loaded with doxorubicin (DOX) and hyaluronic acid. All oxyanion-terminated MXenes and cationic DOX can also be used in anticancer therapy. Tantalum-rich Ta_4C_3 MXenes are excellent candidates for CT imaging contrast agents because of their ecocompatible synthesis, biocompatibility, and suitable nanoparticle size. Even Ta_4C_3 can be used for bioimaging tumors. Ta_4C_3 MXene nanocomposites have been reported to be used in MRIs.

11.10   SPINAL CORD INJURY REPAIR

A polyvinyl pyrrolidone/phytic acid/MXene hydrogel was developed for spinal cord injury repair. An appropriate composition of these three helped in a fast and facile gelatin process of spinal cord injury repair. This hybrid material helps to fill the cavity because of their self-healing property due to the distribution of abundant active groups in their hydrogel structure. These hybrid matrices help in the angiogenesis, recyclization, and axon regeneration and also in calcium channel activation, which accelerates the regeneration of spinal cord tissues and hence repairs spinal cord injury.

11.11 ENZYME-MIMICKING CAPABILITY

Enzymes, or bio-catalysts, not only catalyze biochemical reactions but also maintain metabolic equilibrium in biological systems. Zhang et al. in 2021 proposed the concept of a novel modified nanomaterial called 'nanozymes' [57] with inherent enzyme-mimicking activities of MXene-based nanomaterials. W. Feng et al. later modified them to 'MXenzymes' by coupling the enzyme-mimicking activities of transition metals such as V_2C MXene into 'nanozymes' to serve its dual purposes in biomedical applications [58].

11.12 BIODEGRADABILITY

The evaluation of biosafety, which is an important index for a material to be used for biomedical applications, can be dominantly affected by the biodegradability of the materials. In the work proposed by Lin, H., et al., it was found that a Nb_2C MXene undergoes decay within 24 hours in the presence of human myeloperoxidase (hMPO) and H_2O_2 [59]. Small MXenes were found to degrade *via* oxidation reaction. Moreover, it was found that the rate of their degradation was enhanced under ionic conditions and overflowing enzymes in the physiological environment.

11.13 ANTIOXIDANT PROPERTIES

MXene quantum dots (QDs) demonstrate adequate antioxidant property; however, they undergo oxidation in the course of their hydrothermal production, and perhaps a few defects may happen during their synthesis, limiting their antioxidant capability [60]. One such work suggested that using ethylenediamine precursor can cause N-element protection in Ti_2C MXene QDs. The fabricated N-doped Ti_2C MXene QDs showed strong antioxidant behavior with excellent scavenging of •OH radicals, $KMnO_4$ reduction, and dye protection. As a result of the N doping, the electrochemical interface between the free radicals and QDs was found to be enhanced. Furthermore, density functional theory simulations revealed the hydroxyl radical quenching procedure and corroborated the stimulatory effects of doped N element.

11.14 CONCLUSIONS AND FUTURE OUTLOOKS

This chapter highlights the current developments in the biomedical relevance of functionalized 2D MXenes and their composites, particularly emphasizing the exceptional applications in bioimaging probes, antibacterial materials, biosensors, anticancer agents, drug delivery, tissue engineering, and theranostics. MXenes possess tunable optical, electronic, and magnetic properties, and by selective preparation approaches, MXenes with desirable properties can be achieved for distinctive biomedical applications. It is noteworthy that the research and applications of 2D MXenes in the field of nanomedicine is yet in its early stage, and there are still many issues with MXenes to be solved.

More effort should be devoted to designing and developing new surface modification strategies to meet the future requirements of MXenes and their composites in various biomedical applications. In addition, more efforts are expected in the examination of biocompatibility properties, toxicity, and pharmacokinetics applications of MXenes. Finally, multidisciplinary efforts among chemistry, physics, biology, and engineering should be explored to understand the mechanistic details of MXene-based nanosystems and applications in unique biomedical fields such as acoustic dynamics therapy.

REFERENCES

1. Ashley, C. E.; Carnes, E. C.; Phillips, G. K.; Padilla, D.; Durfee, P. N.; Brown, P. A.; Hanna, T. N.; Liu, J.; Phillips, B.; Carter, M. B.; Carroll, N. J.; Jiang, X.; Dunphy, D. R.; Willman, C. L.; Petsev, D. N.; Evans, D. G.; Parikh, A. N.; Chackerian, B.; Wharton, W.; Peabody, D. S.; Brinker, C. J. The Targeted Delivery of Multicomponent Cargos to Cancer Cells by Nanoporous Particle-Supported Lipid Bilayers. *Nat. Mater.* 2011, 10, 389.

2. Suryavanshi, V. S.; Maharana, T.; Jagtap, P. K. Microencapsulation of Cassia Fistula Flower Extract with Chitosan and Its Antibacterial Studies. *Curr. Drug Deliv.* 2022, 19 (9), 980–990.

3. Kotagiri, N.; Sudlow, G. P.; Akers, W. J.; Achilefu, S. Breaking the Depth Dependency of Phototherapy with Cerenkov Radiation and Low-radiance-Responsive Nanophotosensitizers. *Nat. Nanotechnol.* 2015, 10, 370.

4. Mura, S.; Nicolas, J.; Couvreur, P. Stimuli-Responsive Nanocarriers for Drug Delivery. *Nat. Mater.* 2013, 12, 991.

5. Ahuja, M.; Das, S.; Sharma, P.; Kumar, A.; Srivastava, A.; Samanta, S. Facile Access to Furo[2′,3′:4,5]Pyrido[3,2,1-jk]Carbazol-5-Ones as Blue Emitters: Photophysical, Electrochemical, Thermal and DFT Studies. *J. Mol. Struct.* (Elsevier). 2021, 1233, 130044.

6. Ahuja, M.; Majee, D.; Sharma, P.; Kumar, A.; Mobin, S. M.; Samanta, S. Metal-Free Based Domino Approach to Pyrano-Fused-Pyrido[3,2,1-jk]Carbazolones: Antibacterial and Molecular Docking Studies. *ChemistrySelect* (Wiley). 2019, 4, 9096–9101.

7. Panda, P.; Pal, K.; Chakroborty, S. Smart Advancements of Key Challenges in Graphene-Assembly Glucose Sensor Technologies: A Mini Review. *Mater. Lett.* 2021, 130508.

8. Chakroborty, S.; Bharadwaj, V.; Sahoo, S. K. Sensing and Biosensing with Optically Active Nanomaterials; Sensing and Biosensing with 2D Nanosheets beyond Graphene. *Elsevier.* 2021, 119–141. ISBN: 9780323902458.

9. Ahuja, M.; Biswas, S.; Sharma, P.; Samanta, S. Synthesis and Photophysical Properties of α-Pyrone-Fused-Pyrido[3,2,1-jk]Carbazolone Derivatives: DFT/TD-DFT Insights. *ChemistrySelect.* 2018, 3, 4354–4360.

10. Jagtap, P. K.; Tapadia, K. Development of Novel Dynamic Drop-to-Drop Microextraction Coupled with GC-MS for the Trace Level Screening of Phenothiazine Drug in Human Biological Samples. *Indian J. Pharm. Educ. Res.* 2018, 52 (4), 303–308.

11. Jagtap, P. K.; Tapadia, K. Green Chemistry: An Eco-Compatible Approach for Nanodrop Spectrophotometric Determination of Amitriptyline in Pure and Pharmaceutical Dosage Forms. *Rasayan J. Chem.* 2016, 9 (4), 582–587.

12. Soren, S.; Chakroborty, S.; Pradhan, L.; Chandra, P.; Sahu, J.; Parhi, P. Hydrothermal Synthesis of Graphene Modified SnO Nanocomposite for Oxygen Reduction Reaction. *Mater. Today: Proc.* 2022, 57 (1), 72–76.

13. Soren, S.; Chakroborty, S.; Pal, K. Enhanced in Tunning of Photochemical and Electrochemical Responses of Inorganic Bi-Metal Oxide Nanoparticles via rGO Frameworks (MO/rGO): A Comprehensive Review, Elsevier. *Mater. Sci. Eng. B.* 2022, 278, 115632.

14. Chakroborty, S.; Panda, P. *Nanovaccinology against Infectious Diseases: Nanovaccinology Outbreak as Targeted Therapeutics.* Weinheim, Germany: Wiley-VCH Verlag GmbH, 2022.

15. Tan, C.; Zhang, H. Two-Dimensional Transition Metal Dichalcogenide Nanosheet-Based Composites. *Chem. Soc. Rev.* 2015, 44, 2713.

16. Kalantar-Zadeh, K.; Ou, J. Z.; Daeneke, T.; Mitchell, A.; Sasaki, T.; Fuhrer, M. S. Two Dimensional and Layered Transition Metal Oxides. *Appl. Mater. Today.* 2016, 5, 73.

17. Jagtap, P. K.; Meher, R. K.; Biswal, M. M. Facile Eco-Compactable Design for the Synthesis and Characterization of Silver Nanoparticles. *Nanosci. Nanotechnol.-Asia.* 2021, 11 (4), 88–97.

18. Barsoum, M. W. The $M_{N+1}AX_N$ Phases: A New Class of Solids: Thermodynamically Stable Nanolaminates. *Prog. Solid State Chem.* 2000, 28, 201.

19. Naguib, M.; Kurtoglu, M.; Presser, V.; Lu, J.; Niu, J.; Heon, M.; Hultman, L.; Gogotsi, Y.; Barsoum, M. W. Two-Dimensional Nanocrystals Produced by Exfoliation of Ti_3AlC_2. *Adv. Mater.* 2011, 23, 4248.

20. Naguib, M.; Mashtalir, O.; Carle, J.; Presser, V.; Lu, J.; Hultman, L.; Gogotsi, Y.; Barsoum, M. W. Two-Dimensional Transition Metal Carbides. *ACS Nano.* 2012, 6, 1322.

21. Naguib, M.; Mochalin, V. N.; Barsoum, M. W.; Gogotsi, Y. MXenes: A New Family of Two-dimensional Materials. *Adv. Mater.* 2014, 26, 992.

22. Kannan, K.; Sadasivuni, K. K.; Abdullah, A. M.; Kumar, B. Current Trends in MXene-Based Nanomaterials for Energy Storage and Conversion System: A Mini Review. *Catalysts.* 2020, 10, 1–28.

23. Gogotsi, Y.; Anasori, B. The Rise of MXenes. *ACS Nano.* 2019, 13, 8491–8494.

24. Ihsanullah, I. MXenes as Next-Generation Materials for the Photocatalytic Degradation of Pharmaceuticals in Water. *J. Environ. Chem. Eng.* 2022, 10, 107381.

25. Huang, K.; Li, Z.; Lin, J.; Han, G.; Huang, P. Two-Dimensional Transition Metal Carbides and Nitrides (MXenes) for Biomedical Applications. *Chem. Soc. Rev.* 2018, 47, 5109–5124.

26. Wang, Y.; Feng, W.; Chen, Y. Chemistry of Two-Dimensional MXene Nanosheets in Theranostic Nanomedicine. *ChinChem. Lett.* 2020, 31, 937–946.

27. Huang, H.; Jiang, R.; Feng, Y.; Ouyang, H.; Zhou, N.; Zhang, X.; Wei, Y. Recent Development and Prospects of Surface Modification and Biomedical Applications of MXenes. *Nanoscale.* 2020, 12, 1325–1338.

28. Huang, K.; Li, Z. J.; Lin, J.; Han, G.; Huang, P. Two-Dimensional Transition Metal Carbides and Nitrides (MXenes) for Biomedical Applications. *Chem. Soc. Rev.* 2018, 47, 5109–5124.

29. Xuan, J. N.; Wang, Z. Q.; Chen, Y. Y.; Liang, D. J.; Cheng, L.; Yang, X. J.; Liu, Z.; Ma, R. Z.; Sasaki, T.; Geng, F. X. Organic-Base Driven Intercalation and Delamination for the Production of Functionalized Titanium Carbide Nanosheets with Superior Photothermal Therapeutic Performance. *Angew. Chem. Int. Ed.* 2016, 55, 14569–14574.

30. Li, G.; Zhong, X.; Wang, X.; Gong, F.; Lei, H.; Zhou, Y.; Li, C.; Xiao, Z.; Ren, G.; Cheng, L. Titanium Carbide Nanosheets with Defect Structure for Photothermal-Enhanced Sonodynamic Therapy. *Bioact. Mater.* 2022, 8, 409–419.

31. Lu, S. Y.; Sui, L. Z.; Liu, Y.; Yong, X.; Xiao, G. J.; Yuan, K. J.; Liu, Z. Y.; Liu, B. Z.; Zou, B.; Yang, B. White Photoluminescent Ti3C2 MXene Quantum Dots with Two-Photon Fluorescence. *Adv. Sci.* 2019, 6, 1801470.

32. Cao, Y.; Wu, T. T.; Zhang, K.; Meng, X. D.; Dai, W. H.; Wang, D. D.; Dong, H. F.; Zhang, X. J. Engineered Exosome-Mediated NearInfrared-II Region V2C Quantum Dot Delivery for Nucleus-Target Low-Temperature Photothermal Therapy. *ACS Nano.* 2019, 13 (2), 1499–1510.

33. Mura, S.; Nicolas, J.; Couvreur, P. Stimuli-Responsive Nanocarriers for Drug Delivery. *Nat. Mater.* 2013, 12, 991–1003.

34. Rasool, K.; Helal, M.; Ali, A.; Ren, C. E.; Gogotsi, Y.; Mahmoud, K. A. Antibacterial Activity of Ti3C2Tx MXene. *ACS Nano.* 2016, 10, 3674.

35. Rasool, K.; Mahmoud, K. A.; Johnson, D. J.; Helal, M.; Berdiyorov, G. R.; Gogotsi, Y. Efficient Antibacterial Membrane Based on Two-Dimensional Ti3C2Tx (MXene) Nanosheets. *Sci. Rep.* 2017, 7, 1598.

36. Liu, H.; Duan, C.; Yang, C.; Shen, W.; Wang, F.; Zhu, Z. A Novel Nitrite Biosensor Based on the Direct Electrochemistry of Hemoglobin Immobilized on MXene-Ti3C2. *Sens. Actuators, B.* 2015, 218, 60.

37. Rakhi, R.; Nayak, P.; Xia, C.; Alshareef, H. N. Novel Amperometric Glucose Biosensor Based on MXene Nanocomposite. *Sci. Rep.* 2016, 6, 36422.

38. Wang, F.; Yang, C. H.; Duan, C. Y.; Xiao, D.; Tang, Y.; Zhu, J. F. An Organ-Like Titanium Carbide Material (MXene) with Multilayer Structure Encapsulating Hemoglobin for a Mediator-Free Biosensor. *J. Electrochem. Soc.* 2015, 162, B16.

39. Xu, B.; Zhu, M.; Zhang, W.; Zhen, X.; Pei, Z.; Xue, Q.; Zhi, C.; Shi, P. Ultrathin MXene-Micropattern-Based Field-Effect Transistor for Probing Neural Activity. *Adv. Mater.* 2016, 28, 3333.

40. Qian, X.; Gu, Z.; Chen, Y. Two-Dimensional Black Phosphorus Nanosheets for Theranostic Nanomedicine. *Mater. Horiz.* 2017, 4, 800.

41. Mao, H. Y.; Laurent, S.; Chen, W.; Akhavan, O.; Imani, M.; Ashkarran, A. A.; Mahmoudi, M. Graphene: Promises, Facts, Opportunities, and Challenges in Nanomedicine. *Chem. Rev.* 2013, 113, 3407.

42. Kurapati, R.; Kostarelos, K.; Prato, M.; Bianco, A. Biomedical Uses for 2D Materials Beyond Graphene: Current Advances and Challenges Ahead. *Adv. Mater.* 2016, 28, 6052.

43. Feng, L.; Wu, L.; Qu, X. New Horizons for Diagnostics and Therapeutic Applications of Graphene and Graphene Oxide. *Adv. Mater.* 2013, 25, 168.

44. Chimene, D.; Alge, D. L.; Gaharwar, A. K. Two-Dimensional Nanomaterials for Biomedical Applications: Emerging Trends and Future Prospects. *Adv. Mater.* 2015, 27, 7261.

45. Chen, Y.; Wu, Y.; Sun, B.; Liu, S.; Liu, H. Two-Dimensional Nanomaterials for Cancer Nanotheranostics. *Small.* 2017, 13, 1603446.

46. Chen, Y.; Tan, C.; Zhang, H.; Wang, L. Two-Dimensional Nanomaterials for Cancer Nanotheranostics. *Chem. Soc. Rev.* 2015, 44, 2681.

47. Rasool, K.; Helal, M.; Ali, A.; Ren, C. E.; Gogotsi, Y.; Mahmoud, K. A. Antibacterial Activity of Ti3C2Tx MXene. *ACS Nano.* 2016, 10, 3674.

48. Yu, X.; Li, Y.; Cheng, J.; Liu, Z.; Li, Q.; Li, W.; et al. Monolayer Ti2CO2: A Promising Candidate for NH3 Sensor or Capturer with High Sensitivity and Selectivity. *ACS Appl. Mater. Interfaces.* 2015, 7, 13707–13713.

49. Kim, S. J.; Koh, H.-J.; Ren, C. E.; Kwon, O.; Maleski, K.; Cho, S.-Y. Metallic Ti3C2Tx MXene Gas Sensors with Ultrahigh Signal-to-Noise ratio. *ACS Nano*. 2018, 12, 986–993.

50. Chen, X.; Sun, X.; Xu, W.; Pan, G.; Zhou, D.; Zhu, J.; Song, H. Ratiometric Photoluminescence Sensing Based on Ti 3 C 2 MXene Quantum Dots as an Intracellular pH Sensor. *Nanoscale*. 2018, 10 (3), 1111–1118.

51. Xue, Q.; Zhang, H.; Zhu, M.; Pei, Z.; Li, H.; Wang, Z.; Huang, Y.; Huang, Y.; Deng, Q.; Zhou, J.; et al. Photoluminescent Ti_3C_2 MXene Quantum Dots for Multicolor Cellular Imaging. *Adv. Mater.* 2017, 29, 1604847.

52. Xu, Q.; Ma, J.; Khan, W.; Zeng, X.; Li, N.; Cao, Y.; Zhao, X.; Xu, M. Highly Green Fluorescent Nb_2C MXene Quantum Dots. *Chem. Commun.* 2020, 56, 6648–6651.

53. Liu, M.; Bai, Y.; He, Y.; Zhou, J.; Ge, Y.; Zhou, J.; Song, G. Facile Microwave-Assisted Synthesis of Ti_3C_2 MXene Quantum Dots for Ratiometric Fluorescence Detection of Hypochlorite. *Microchim. Acta*. 2021, 188, 15.

54. Zhou, B. G.; Yin, H. H.; Dong, C. H.; Sun, L. P.; Feng, W.; Pu, Y. Y.; Han, X. X.; Li, X. L.; Du, D.; Xu, H. X.; Chen, Y. Biodegradable and Excretable 2D W1.33C i-MXene with Vacancy Ordering for Theory-Oriented Cancer Nanotheranostics in Near-Infrared Biowindow. *Adv. Sci.* 2021, 8, 2101043.

55. Lin, H.; Wang, X.; Yu, L.; Chen, Y.; Shi, J. Two-Dimensional Ultrathin MXene Ceramic Nanosheets for Photothermal Conversion. *Nano Lett.* 2017, 17, 384–391.

56. Bae, Y.; Nishiyama, N.; Fukushima, S.; Koyama, H.; Yasuhiro, M.; Kataoka, K. Preparation and Biological Characterization of Polymeric Micelle Drug Carriers with Intracellular pH-Triggered Drug Release Property: Tumor Permeability, Controlled Subcellular Drug Distribution, and Enhanced in Vivo Antitumor Efficacy. *Bioconjug. Chem.* 2005, 16, 122–130.

57. Zhang, R. F.; Yan, X. Y.; Fan, K. L. Nanozymes Inspired by Natural Enzymes. *Acc. Mater. Res.* 2021, 2, 534–547.

58. Feng, W.; Han, X. G.; Hu, H.; Chang, M. Q.; Ding, L.; Xiang, H. J.; Chen, Y.; Li, Y. H. 2D Vanadium Carbide MXenzyme to Alleviate ROS-Mediated Inflammatory and Neurodegenerative Diseases. *Nat. Commun.* 2021, 12, 2203.

59. Lin, H.; Gao, S. S.; Dai, C.; Chen, Y.; Shi, J. L. A Two-Dimensional Biodegradable Niobium Carbide (MXene) for Photothermal Tumor Eradication in NIR-I and NIR-II Biowindows. *J. Am. Chem. Soc.* 2017, 139, 16235–16247.

60. Gou, J.; Zhao, L.; Li, Y.; Zhang, J. Nitrogen-Doped Ti_2C MXene Quantum Dots as Antioxidants. *ACS Appl. Nano Mater.* 2021, 4, 12308–12315.

12 The Potential of MXenes in Water Desalination

Vandana Sharma, Anchit Modi,
Poonam Deshmukh, Rashmi Chourasia,
Sonali Saha, and N.K. Gaur

12.1 INTRODUCTION

Water scarcity is a critical global issue exacerbated by population growth, urbanization, and climate change. Access to freshwater is vital for sustaining life and supporting economic development, yet many regions worldwide struggle with limited water resources, necessitating alternative sources. Water desalination has emerged as a crucial solution to address the growing demand for freshwater [1–3]. Desalination removes salt and impurities from seawater or brackish water, making it suitable for various applications like drinking, agriculture, and industry. However, traditional desalination methods face significant challenges, including high energy consumption, costliness, environmental concerns, and limited scalability. Desalination holds immense potential to provide a sustainable and reliable freshwater supply, especially in coastal areas with abundant seawater [3–6]. By employing desalination technologies, nations can tackle water scarcity, enhance water security, and drive socio-economic growth.

Nevertheless, conventional methods like reverse osmosis (RO) and multi-stage flash (MSF) distillation have long-standing limitations. RO, the predominant technique, requires high operating pressures and consumes substantial energy, increasing costs. Furthermore, RO membranes are susceptible to fouling, reducing efficiency and necessitating frequent maintenance. MSF distillation, while effective, is energy-intensive and requires extensive infrastructure, limiting its applicability in decentralized settings [7–10].

MXenes, a class of two-dimensional (2D) materials, hold immense promise for transforming water desalination [11,12]. Derived from layered parent materials known as MAX phases, MXenes exhibit remarkable properties, including high electrical conductivity, mechanical strength, and tunable surface chemistry [13]. These attributes make MXenes exceptionally well-suited for various desalination applications, mainly capacitive deionization (CDI) and membrane-based processes. MXenes' large surface area, ion selectivity, and fast ion transport enable efficient desalination while minimizing energy consumption and enhancing overall performance. The unique properties of MXenes position them as highly favorable materials for addressing the challenges associated with traditional desalination methods

[14, 15]. By harnessing MXenes in water desalination technologies, it is possible to achieve significant advancements in the field, opening doors to more efficient, cost-effective, and sustainable solutions for water scarcity and ensuring access to clean water for various societal needs.

The objective of this chapter is to comprehensively explore the potential of MXenes in water desalination and to elucidate the unique properties that make them highly suitable for this application. The chapter will overview traditional desalination methods, emphasizing their inherent challenges. Subsequently, it will delve into an in-depth analysis of the properties and synthesis of MXenes, with a specific focus on their applicability in desalination processes. Furthermore, the chapter will critically discuss recent advancements, existing limitations, and prospects of MXene-based desalination technologies. To conclude, it will concisely summarize the essential findings and offer valuable recommendations for further research and development in this dynamic field. By thoroughly examining the potential of MXenes in water desalination, this chapter aims to contribute significantly to advancing desalination technologies and mitigating water scarcity. It offers valuable insights into the unique properties of MXenes that enable efficient and sustainable desalination processes. Moreover, it sheds light on the challenges and prospects associated with MXene-based desalination technologies, providing a roadmap for future research endeavors. Ultimately, harnessing the immense potential of MXenes in water desalination can pave the way for a more secure and sustainable water future.

12.2 MXENES: PROPERTIES AND SYNTHESIS

12.2.1 Overview of MXenes and Their Layered Structure

MXenes have garnered significant attention as a class of two-dimensional (2D) materials with unique properties and potential applications, including water desalination. Derived from layered parent materials called MAX phases, MXenes feature a composition where "M" represents a transition metal, "A" means aluminum or silicon, and "X" denotes carbon and nitrogen. The layered structure of MXenes consists of transition metal carbide, nitride, or carbonitride layers separated by intercalated species or surface terminations. This distinctive layered structure endows MXenes with remarkable properties and enables their synthesis and functionalization [11–13,16].

12.2.2 Methods of MXene Synthesis

MXenes can be synthesized through a selective etching process of the MAX phase precursor. This involves the removal of the "A" layer using powerful etchants such as hydrofluoric acid (HF) and an intercalating agent like lithium or sodium. The etching process leads to the exfoliation of the MAX phase, resulting in the formation of MXene layers. Various techniques, including liquid and solid-state etching methods, have been explored for MXene synthesis. Liquid exfoliation methods employ acid etchants, while solid-state methods rely on high-temperature treatments to promote intercalation and exfoliation [17].

12.2.3 KEY PROPERTIES OF MXENES RELEVANT TO WATER DESALINATION

MXenes exhibit fundamental properties that make them highly suitable for water desalination applications:

- **High Surface Area:** MXenes possess a large surface area, providing ample space for ion adsorption and interaction. This property is particularly advantageous in capacitive deionization (CDI), where ions are adsorbed onto MXene surfaces [12].
- **High Electrical Conductivity:** MXenes exhibit high electrical conductivity, facilitating efficient charge transfer and ion transport during desalination. This conductivity is beneficial in CDI, enabling rapid and reversible adsorption and desorption of ions [15].
- **Selective Ion Adsorption:** MXenes can selectively adsorb specific ions based on their surface chemistry and functionalization. This capability allows for the targeted removal of particular ions from water, enhancing desalination efficiency and selectivity [11].
- **Remarkable Mechanical Properties:** MXenes possess exceptional mechanical strength, combining the power of ceramics with the flexibility of graphene. This mechanical resilience ensures the stability and durability of MXene-based membranes and electrodes during water desalination processes [13].

12.2.4 SURFACE FUNCTIONALIZATION AND ITS IMPACT ON DESALINATION PERFORMANCE

Surface functionalization is crucial in optimizing MXene-based materials for water desalination, leading to enhanced ion adsorption, improved stability, and superior performance. MXene chemistry, hydrophilicity, and ion selectivity can be tailored by introducing various functional groups or molecules onto the MXene surface [18].

Chemical modification through functionalization enables MXenes to be customized for specific desalination applications, optimizing their properties accordingly, as shown in Figure 12.1. MXenes exhibit a unique layered structure derived from MAX phases, providing exceptional properties for various applications, including water desalination. Their synthesis methods involve selective etching of MAX phases, and their properties such as high surface area, electrical conductivity, selective ion adsorption, and mechanical strength, make them highly suitable for efficient desalination processes [19]. Additionally, surface functionalization allows MXenes to be tailored for specific desalination applications, further enhancing their performance and versatility. Continued research and development in MXene-based desalination technologies hold immense promise for addressing water scarcity challenges and achieving a more secure and sustainable water future.

(a)

H C T ● M(Transition metal)
● C(Carbon) ● T(Oxygen, Fluorine)

Hollow site
Carbon-top
Transition metal-top

(1) HH   (2) CC   (3) TT

(4) HC   (5) CT

(b)

Insulator / Semimetal
Metal / Depending on methods

Ferro / Depending on methods
Antiferro
Nonmag

$Sc_{3d^14s^2}$	$Ti_{3d^24s^2}$	$V_{3d^34s^2}$	M_2CO_2 M^{+4}
Sc_2CO_2 d⁰(HC)	Ti_2CO_2 d¹(HH)	V_2CO_2 d¹(HH)	M_2CF_2 M^{+3}
Sc_2CF_2 d⁰(HH)	Ti_2CF_2 d¹(HH)	V_2CF_2 d²(HH)	

$Y_{4d^15s^2}$	$Zr_{4d^25s^2}$	$Nb_{4d^45s^1}$	$Mo_{4d^55s^1}$
Y_2CO_2 d⁰(HC)	Zr_2CO_2 d¹(HH)	Nb_2CO_2 d¹(HH)	Mo_2CO_2 d³(CC)
Y_2CF_2 d⁰(HH)	Zr_2CF_2 d¹(HH)	Nb_2CF_2 d¹(HH)	Mo_2CF_2 d³(HH)

$La_{5d^16s^2}$	$Hf_{5d^26s^2}$	$Ta_{5d^36s^2}$	$W_{5d^46s^2}$
La_2CO_2 d⁰(HC)	Hf_2CO_2 d¹(HH)	Ta_2CO_2 d¹(HH)	W_2CO_2 d³(CC)
La_2CF_2 d⁰(HH)	Hf_2CF_2 d¹(HC)	Ta_2CF_2 d²(CC)	W_2CF_2 d³(HH)

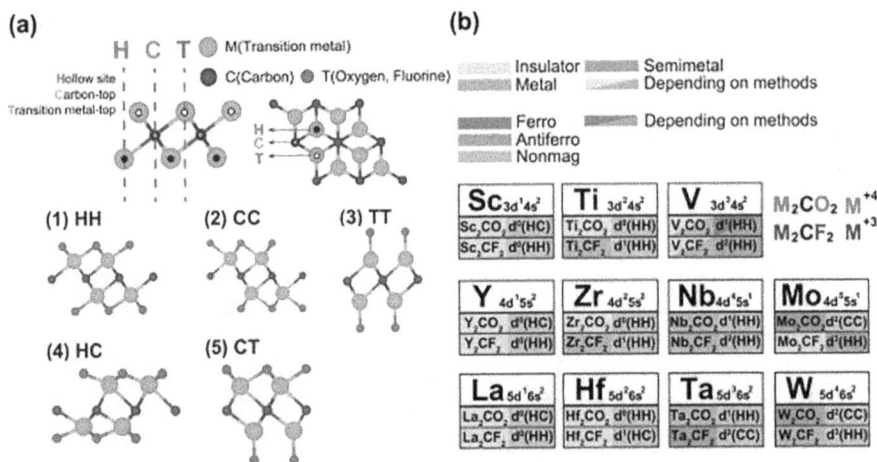

FIGURE 12.1 (a) Geometric structures of the studied functionalized MXenes. (b) An illustrative description of the structure, electronic, and magnetic properties of carbide MXenes. For each MXene, information on electronic properties (insulator, metal, or semimetal) and magnetic properties (non-magnetic, ferromagnetic, and antiferromagnetic), as well as the ground-state structure and the formal dn configuration, are given.

Copyright with permission from Ref [19]. Open access under a CC BY 4.0 license.

12.3 MXENES IN WATER DESALINATION

12.3.1 CAPACITIVE DEIONIZATION (CDI) AND MXENES

Capacitive deionization (CDI) is an emerging water desalination technique that utilizes the electrostatic attraction between charged electrodes and ions in water to remove salt and other impurities. MXenes have shown great potential in enhancing the performance of CDI systems. With their high surface area and electrical conductivity, MXenes have emerged as promising electrodes for CDI. The large surface area of MXenes provides more sites for ion adsorption, resulting in higher ion removal capacity. Furthermore, the high electrical conductivity of MXenes enables efficient charge transfer during the adsorption and desorption processes, facilitating effective ion removal [20]. The unique properties of MXenes, including their high surface area and rapid ion transport, contribute to improved ion adsorption and desorption kinetics in CDI. MXene electrodes have demonstrated fast ion adsorption rates, enhancing desalination efficiency. The rapid ion transport within the MXene layers ensures swift ion desorption, enabling quicker regeneration of the electrodes and shorter desalination cycles.

12.3.2 MXENES IN MEMBRANE-BASED DESALINATION PROCESSES

In addition to CDI, MXenes have demonstrated potential in membrane-based desalination processes such as reverse osmosis (RO) and polymeric membranes. MXenes act as nanoscale reinforcements in the polymer matrix, enhancing the membrane's performance and stability [12]. To improve their performance, MXenes can be incorporated as active layers in RO membranes. The high surface area and ion selectivity of MXenes facilitate the efficient removal of salts and contaminants, resulting in improved desalination efficiency and reduced fouling. MXenes also enhance the mechanical strength of RO membranes, rendering them more pressure resistant and providing increased durability. Another approach is to utilize MXenes as fillers in polymeric membranes. Adding MXenes improves the membrane's water permeability, salt rejection, and mechanical properties. This approach holds promise for developing high-performance and robust membranes for efficient desalination processes.

12.3.3 Selective Ion Removal Using MXenes

MXenes' tunable surface chemistry and functionalization allow for selective ion removal, making them attractive for targeted pollutant removal in water desalination. By modifying the surface functional groups of MXenes, their ion selectivity can be tailored to remove specific ions or contaminants. Introducing specific functional groups enhances the affinity of MXenes toward certain ions, enabling selective removal and the purification of water. MXenes' particular ion removal capability opens up potential applications in targeted pollutant removal [13,14]. They can be functionalized to target specific contaminants such as heavy metals or organic pollutants, offering a versatile and efficient water treatment and remediation approach. Integrating MXenes in water desalination processes, whether in CDI, membrane-based techniques, or selective ion removal, showcases the significant potential of MXene-based materials for addressing water scarcity challenges [20]. Continued research and development in MXene-based desalination technologies will contribute to the advancement of efficient and sustainable water treatment solutions, promoting a more secure and environmentally friendly water future.

12.4 MXENE-BASED MEMBRANES FOR WATER DESALINATION

12.4.1 MXenes in Reverse Osmosis (RO) Membranes

MXenes have demonstrated significant potential in enhancing the performance of reverse osmosis (RO) membranes, a widely adopted desalination technology. Incorporating MXenes into RO membranes can enhance water permeability while maintaining efficient salt rejection. MXenes' high surface area and unique nanoscale morphology create additional water channels, increasing water flux through the membrane. Furthermore, MXenes' tunable surface chemistry enables selective ion transport, improving the membrane's salt rejection capability. Fouling, a significant challenge in RO systems, can be mitigated by MXenes, which provide antifouling properties [21]. The hydrophilic nature of MXenes promotes water transport while inhibiting foulant and contaminant adhesion, reducing membrane fouling and enhancing the long-term performance and durability of RO membranes [12].

12.4.2 MXenes in Nanofiltration (NF) Membranes

Nanofiltration membranes, operating between the range of RO and ultrafiltration, selectively remove ions and micropollutants. MXenes offer distinct advantages in improving the performance of NF membranes [22,23]. MXenes' tunable surface chemistry enables selective ion removal and micropollutant adsorption. By modifying the surface functional groups of MXenes, membranes can be customized to target specific ions or contaminants, offering precise water purification capabilities. Incorporating MXenes into NF membranes enhances water permeability while maintaining high selectivity. MXenes' nanoscale morphology and interconnected structure create additional water pathways, facilitating efficient water transport through the membrane without compromising separation efficiency.

12.4.3 Challenges and Future Prospects of MXene-Based Membranes

Despite the potential of MXene-based membranes for water desalination, several challenges and future research directions must be addressed. Developing cost-effective and scalable production methods while preserving MXenes' desired properties is critical for their practical implementation. Scalable synthesis of MXenes and their integration into membrane fabrication processes is a challenge. The long-term stability and fouling behavior of MXene-based membranes under actual operating conditions need to be understood. Research is required to investigate the membrane's performance, durability, and fouling resistance over extended periods, considering different water sources [24]. Prospects for MXene-based membranes include exploring advanced functionalization techniques to enhance selectivity, investigating hybrid membranes combining MXenes with other materials for synergistic properties, and optimizing membrane structure and architecture for improved performance. MXene-based membranes hold significant promise for enhancing the performance of RO and NF membranes for water desalination. Their ability to improve permeability, selectivity, fouling resistance, and antifouling properties make them attractive for next-generation membrane technologies. However, further research must address synthesis, stability, and fouling mitigation challenges [25,26].

12.5 MXENE-BASED ELECTRODES FOR CAPACITIVE DEIONIZATION

Capacitive deionization (CDI) is an electrochemical desalination technique that relies on the electrical double layer (EDL) formed between electrodes and an electrolyte solution. CDI offers numerous advantages over traditional desalination methods, including low energy consumption, environmental friendliness, and the capability to remove both ions and micropollutants [20].

MXenes have emerged as promising materials for CDI electrodes due to their unique properties and structural characteristics. With a high surface area typically ranging from hundreds to thousands of square meters per gram, MXenes provide abundant adsorption sites for ions, resulting in a greater ion removal capacity. Surface functional groups in MXenes enhance ion adsorption through electrostatic interactions and specific ion affinity. Efficient regeneration is crucial for the continuous operation of CDI systems. MXene electrodes excel with their fast ion desorption kinetics, allowing for efficient revival and shorter desalination cycles. The rapid ion transport within the MXene layers facilitates the quick release of ions from the EDL, facilitating the electrode's regeneration process [27]. Comparing the performance of MXene-based CDI electrodes with that of traditional carbon-based electrodes, MXenes have demonstrated superior performance. They offer higher specific capacitance, leading to enhanced ion adsorption capacity. The faster ion transport kinetics of MXenes contribute to improved desalination efficiency and shorter operation times. MXenes exhibit better stability and durability, maintaining their electrochemical performance over extended cycles [28].

Scalability and cost considerations are vital for implementing MXene-based CDI systems. MXenes have shown tremendous potential as high-performance electrodes in CDI systems, thanks to their high surface area, enhanced ion adsorption capacity, fast ion desorption kinetics, and superior performance compared to traditional carbon-based electrodes. MXenes can be synthesized in large quantities, making them suitable for large-scale production. The cost of MXene production may vary depending on the synthesis method and precursor materials used [29,30]. However, considering the potential benefits of MXene-based CDI systems, such as energy efficiency and prolonged electrode lifespan, a comprehensive evaluation of their overall cost-effectiveness is necessary. Addressing scalability and cost considerations will pave the way for the practical implementation of MXene-based CDI systems.

12.6 CHALLENGES AND FUTURE DIRECTIONS

12.6.1 SCALABILITY AND STABILITY OF MXENE SYNTHESIS

One of the primary challenges in utilizing MXenes for water desalination is achieving scalability in their synthesis. MXenes are primarily synthesized in small, laboratory-scale quantities, limiting their practical applications in large-scale desalination processes. Developing scalable synthesis methods that can produce MXenes in large quantities while maintaining their high quality and properties is crucial for widespread adoption. Additionally, ensuring the stability of MXenes during synthesis and subsequent processing steps is essential. MXenes are susceptible to oxidation and hydrolysis, which can degrade their structural integrity and impact their performance in water desalination applications. Research efforts should focus on optimizing synthesis conditions, exploring protective coatings or encapsulation techniques, and developing strategies to enhance MXenes' stability for long-term operation [31].

12.6.2 LONG-TERM STABILITY OF MXENE-BASED MEMBRANES AND ELECTRODES

The long-term stability of MXene-based membranes and electrodes represents another significant challenge that needs to be addressed. MXenes may undergo degradation or structural changes over time, which can affect their performance and efficiency in water desalination. Factors such as exposure to harsh operating conditions, fouling, and mechanical stress can contribute to the degradation of MXene-based materials. Understanding the degradation mechanisms and developing strategies to mitigate or prevent degradation is vital for ensuring the long-term stability of MXene-based membranes and electrodes. This involves studying the effects of various operating parameters such as temperature, pH, and pressure on MXenes' stability. Additionally, developing surface modification techniques or protective coatings can enhance the strength and durability of MXene-based materials in water desalination applications [32].

12.6.3 Cost-Effectiveness and Commercialization Potential

The cost-effectiveness of MXene-based technologies is a critical factor for their commercialization and widespread adoption. Currently, the synthesis of MXenes involves complex, energy-intensive, costly processes. Research efforts should focus on optimizing MXene synthesis methods to reduce production costs while maintaining the desired properties. Furthermore, evaluating the overall cost-effectiveness of MXene-based desalination systems should consider the material costs, energy consumption, maintenance requirements, and lifespan of the membranes and electrodes. Exploring cost-effective fabrication techniques such as roll-to-roll processing or additive manufacturing can contribute to reducing overall costs and enhancing the commercialization potential of MXene-based water desalination technologies [33].

12.6.4 Emerging Research Areas and Potential Applications

In addition to water desalination, MXenes hold potential for various emerging research areas and applications. Tailoring the surface functional groups of MXenes can enhance their ion selectivity, enabling targeted pollutant removal in water treatment processes. Moreover, integrating MXenes with other materials, such as polymers or nanoparticles, can create hybrid membranes or composite electrodes with synergistic properties, improving desalination performance. Exploring the potential of MXenes in advanced desalination technologies beyond reverse osmosis and capacitive deionization is another exciting research direction [34]. For example, investigating the use of MXenes in membrane distillation or forward osmosis can open up new possibilities for energy-efficient desalination processes. Furthermore, MXenes have shown promise in catalysis and pollutant degradation due to their unique surface chemistry and high conductivity. Exploring their potential in these areas can provide insights into their broader environmental remediation and sustainable water treatment applications.

Addressing the challenges of scalability and stability in MXenes synthesis, ensuring the long-term stability of MXene-based membranes and electrodes, evaluating the cost-effectiveness and commercialization potential, and exploring emerging research areas are critical for the successful implementation of MXenes in water desalination and related applications. Continued research efforts in these areas will contribute to unlocking the full potential of MXenes and advancing the field.

12.7 CONCLUSION

Throughout this chapter, we have explored the potential of MXenes in water desalination and highlighted the unique properties that make them highly promising for this application. MXenes, with their layered structure and excellent conductivity, have emerged as attractive candidates for various desalination processes, including capacitive deionization (CDI) and membrane-based desalination. They offer advantages such as high ion adsorption capacity, selective ion removal, improved membrane permeability and selectivity, and enhanced fouling resistance. MXenes also provide opportunities for tailoring their properties through surface functionalization, paving the way for advanced water treatment solutions. Our comprehensive analysis has revealed several key findings regarding the potential of MXenes in water desalination. Firstly, MXenes have demonstrated exceptional performance as electrodes in capacitive deionization systems. Their large surface area and rapid ion adsorption/desorption kinetics significantly enhance ion removal efficiency and regeneration capability in CDI processes. Incorporating MXenes as electrodes improves the overall desalination performance, leading to efficient and sustainable water treatment solutions.

Furthermore, MXenes show promise in membrane-based desalination processes. When used as active layers in reverse osmosis (RO) membranes, MXenes contribute to improved permeability, selectivity, and fouling resistance. Integrating MXenes into polymeric membranes as fillers enhances ion removal, water permeability, and antifouling properties. Incorporating MXenes in membrane-based desalination technologies offers an opportunity to overcome the limitations of conventional membranes and achieve more efficient and sustainable water desalination. The surface functionalization of MXenes plays a crucial role in tailoring their properties and achieving selective ion removal. Introducing specific functional groups onto the MXene surface enhances their ion selectivity, enabling targeted pollutant removal in water treatment processes. This capability holds promise for addressing challenges associated with removing specific ions or micropollutants from water sources, contributing to developing more tailored and effective water treatment technologies.

There are critical areas for future research and development in MXenes for water desalination. The scalability and stability of MXene synthesis should be further investigated to enable large-scale production without compromising quality and properties. Ensuring the long-term stability of MXene-based membranes and electrodes is crucial, requiring an understanding of degradation mechanisms and the development of strategies to enhance their durability under actual operating conditions. Techno-economic analyses should be conducted to evaluate the cost-effectiveness and commercialization potential of MXene-based desalination systems. Optimizing synthesis methods and exploring cost-effective fabrication techniques will benefit their commercial feasibility. Exploring emerging research areas, such as MXenes in membrane distillation and forward osmosis and their potential in catalysis and pollutant degradation, will unlock new possibilities for MXenes in water treatment beyond desalination. In conclusion, MXenes hold significant promise for water desalination, and addressing the challenges and recommendations outlined in this chapter will advance their implementation in practical applications. Continued research efforts and interdisciplinary collaborations are essential to fully harness the potential of MXenes and propel the field of water desalination toward more efficient and sustainable solutions.

REFERENCES

[1]. I. Ihsanullah, M. Bilal, Potential of MXene-based membranes in water treatment and desalination: A critical review, Chemosphere. 303 (3) (2022) 135234, https://doi.org/10.1016/j.chemosphere.2022.135234.

[2]. F. A. Janjhi, I. Ihsanullah, M. Bilal, R. Castro-Munoz, G. Boczkaj, F. Gallucci, MXene-based materials for removal of antibiotics and heavy metals from wastewater—A review, Water Resour. Ind. 29 (2023) 100202, https://doi.org/10.1016/j.wri.2023.100202.

[3]. S. Lustenberger, R. Castro-Munoz, Advanced biomaterials and alternatives tailored as membranes for water treatment and the latest innovative European water remediation projects: A review, Case Stud. Chem. Environ. Eng. 5 (2022) 100205, https://doi.org/10.1016/j.cscee.2022.100205.

[4]. A. Azimi, A. Azari, M. Rezakazemi, M. Ansarpour, Removal of heavy metals from industrial wastewaters: A review, Chem. Bio. Eng. Rev. 4 (2017) 37–59, https://doi.org/10.1002/cben.201600010.

[5]. F. Fu, Q. Wang, Removal of heavy metal ions from wastewaters: A review, J. Environ. Manag. 92 (2011) 407–418, https://doi.org/10.1016/j.jenvman.2010.11.011.

[6]. J. K. Im, E. J. Sohn, S. Kim, M. Jang, A. Son, K. D. Zoh, Y. Yoon, Review of MXene-based nanocomposites for photocatalysis, Chemosphere. 270 (2021) 129478, https://doi.org/10.1016/j.chemosphere.2020.129478.

[7]. C. Fritzmann, J. Löwenberg, T. Wintgens, T. Melin, State-of-the-art of reverse osmosis desalination, Desalination. 216 (2007) 1–76. https://doi.org/10.1016/j.desal.2006.12.009.

[8]. M. F. A. Goosen, S. S. Sablani, H. Al-Hinai, S. Al-Obeidani, R. Al-Belushi, D. Jackson, Fouling of reverse osmosis and ultrafiltration membranes: A critical review, Sep. Sci. Technol. 39(10) (2005) 2261–2297, https://doi.org/10.1081/SS-120039343.

[9]. M. Qasim, M. Badrelzaman, N. N. Darwish, N. A. Darwish, N. Hilal, Reverse osmosis desalination: A state-of-the-art review, Desalination. 459 (2019) 59–104, https://doi.org/10.1016/j.desal.2019.02.008.

[10]. S. Burn, S. Gray, Efficient desalination by reverse osmosis: A guide to RO practice, IWA Publishing, London, UNITED KINGDOM, 2015, http://ebookcentral.proquest.com/lib/aus-ebooks/detail.action?docID=4354916.

[11]. T. Rasheed, 3D MXenes as promising alternatives for potential electrocatalysis applications: Opportunities and challenges, J. Mater. Chem. C. 10 (2022) 9669–9690, https://doi.org/10.1039/D2TC01542K.

[12]. Y. A. J. Al-Hamadani, B. Jun, M. Yoon, N. Taheri-Qazvini, S. A. Snyder, M. Jang, J. Heo, Y. Yoon, Chemosphere. 254 (2020) 126821, https://doi.org/10.1016/j.chemosphere.2020.126821.

[13]. B. M. Jun, C. M. Park, J. Heo, Y. Yoon, Adsorption of Ba^{2+} and Sr^{2+} from model fracking wastewater by $Ti_3C_2T_x$ MXene. J. Environ. Manag. 256 (2020) 109940, https://doi.org/10.1016/j.jenvman.2019.109940.

[14]. B. Anasori, M. R. Lukatskaya, Y. Gogotsi, 2D metal carbides and nitrides (MXenes) for energy storage. Nat. Rev. Mater. 2 (2) (2017) 16098, https://doi.org/10.1038/natrevmats.2016.98.

[15]. M. Naguib, V. N. Mochalin, M. W. Barsoum, Y. Gogotsi, 25th anniversary article: MXenes: A new family of two-dimensional materials. Adv. Mater. 26 (7) (2014) 992–1005, https://doi.org/10.1002/adma.201304138

[16]. J. Ren, Z. Zhu, Y. Qiu, F. Yu, T. Zhou, J. Ma, J. Zhao, Enhanced adsorption performance of alginate/MXene/CoFe$_2$O$_4$ for antibiotic and heavy metal under rotating magnetic field, Chemosphere. 284 (2021), 131284, https://doi.org/10.1016/j.chemosphere.2021.131284.

[17]. M. Khatami, S. Iravani, M. Khatami, Comments on inorganic chemistry a journal of critical discussion of the current literature MXenes and MXene-based materials for the removal of water pollutants: Challenges and opportunities MXenes and MXene-based materials for the removal of water pollut, Comments Mod. Chem. 41 (2021) 213–248, https://doi.org/10.1080/02603594.2021.1922396.

[18]. J. Zhao, Y. Yang, C. Yang, Y. Tian, Y. Han, J. Liu, X. Yin, W. Que, A hydrophobic surface enabled salt-blocking 2D Ti$_3$C$_2$ MXene membrane for efficient and stable solar desalination, J. Mater. Chem. A. 6 (2018) 16196–16204, https://doi.org/10.1039/C8TA05569F.

[19]. S. Bae, Y.-G. Kang, M. Khazaei, K. Ohno, Y.-H. Kim, M. J. Han, K. J. Chang, H. Raebiger, Electronic and magnetic properties of carbide MXenes–the role of electron correlations, Mater. Today Adv. 9 (2021) 100118, https://doi.org/10.1016/j.mtadv.2020.100118.

[20]. B. Zhang, A. Boretti, S. Castelletto, Mxene pseudocapacitive electrode material for capacitive deionization, Chem. Eng. J. 435 (2022) 134959, https://doi.org/10.1016/j.cej.2022.134959.

[21]. J. L. Fajardo-Diaz, A. Gomez, R. C. Silva, A. Matsumoto, Y. Ueno, N. Takeuchi, K. Kitamura, H. Miyakawa, S. Tejima, K. Takeuchi, K. Tsuzuki, M. Endo, Antifouling performance of spiral wound type module made of carbon nanotubes/polyamide composite RO membrane for seawater desalination, Desalination. 523 (2022) 115445, https://doi.org/10.1016/j.desal.2021.115445.

[22]. Q. Xue, K. Zhang, MXene nanocomposite nanofiltration membrane for low carbon and long-lasting desalination, J. Membr. Sci. 640 (2021) 119808, https://doi.org/10.1016/j.memsci.2021.119808.

[23]. Y. Zhang, S. Li, R. Huang, J. He, Y. Sun, Y. Qin, L. Shen, Stabilizing MXene-based nanofiltration membrane by forming analogous semi-interpenetrating network architecture using flexible poly(acrylic acid) for effective wastewater treatment, J. Membr. Sci. 648 (2021) 120360, https://doi.org/10.1016/j.memsci.2022.120360.

[24]. I. Ihsanullah, Potential of MXenes in water desalination: Current status and perspectives, Nano-Micro Lett. 12 (2020) 72, https://doi.org/10.1007/s40820-020-0411-9.

[25]. R. Malik, Maxing out water desalination with MXenes, Joule. 2 (2018) 591–593, https://doi.org/10.1016/j.joule.2018.04.001.

[26]. J.-C. Lei, X. Zhang, Z. Zhou, Recent advances in MXene: Preparation, properties, and applications, Front. Phys. 10 (2015) 276–286, https://doi.org/10.1007/s11467-015-0493-x.

[27]. S. Porada, R. Zhao, A. van der Wal, V. Presser, P. M. Biesheuvel, Review on the science and technology of water desalination by capacitive deionization, Prog. Mater. Sci. 58 (8) (2013) 1388–1442, https://doi.org/10.1016/j.pmatsci.2013.03.005.

[28]. G. Zhang, L. Wang, R. Sa, C. Xu, Z. Li, L. Wang, Interconnected N-doped MXene spherical shells for highly efficient capacitive deionization, Environ. Sci.: Nano. 9 (2022) 204–213, https://doi.org/10.1039/D1EN00821H.

[29]. X. Shen, Y. Xiong, R. Hai, F. Yu, J. Ma, All-MXene-based integrated membrane electrode constructed using Ti$_3$C$_2$T$_x$ as an intercalating agent for high-performance desalination, Environ. Sci. Technol. 54 (7) (2020) 4554–4563, https://doi.org/10.1021/acs.est.9b05759.

[30]. H. Qie, M. Liu, X. Fu, X. Tan, Y. Zhang, M. Ren, Y. Zhang, Y. Pei, A. Lin, B. Xi, J. Cui. Interfacial charge-modulated multifunctional MoS/Ti$_3$C$_2$T$_x$ penetrating electrode for high-efficiency freshwater production. ACS Nano. 16 (11) (2022) 18898–18909, https://doi.org/10.1021/acsnano.2c07810.

[31]. A. Iqbal, J. Hong, T. Y. Ko, C. M. Koo, Improving oxidation stability of 2D MXenes: Synthesis, storage media, and conditions, Nano Converg. 8 (2021) 9, https://doi.org/10.1186/s40580-021-00259-6.

[32]. B. Anasori, Y. Xie, M. Beidaghi, J. Lu, B. C. Hosler, L. Hultman, P. R. C. Kent, Y. Gogotsi, M. W. Barsoum, Two-dimensional, ordered, double transition metals carbides (MXenes). ACS Nano. 9 (10) (2015) 9507–9516, https://doi.org/10.1021/acsnano.5b03591.

[33]. R. A. Soomro, P. Zhang, B. Fan, Y. Wei, B. Xu, Progression in the oxidation stability of MXenes, Nano-Micro Lett. 15 (2023) 108, https://doi.org/10.1007/s40820-023-01069-7.

[34]. G. Murali, J. K. Reddy Modigunta, Y. H. Park, J.-H. Lee, J. Rawal et al., A review on MXene synthesis, stability, and photocatalytic applications. ACS Nano. 16 (9) (2022) 13429, https://doi.org/10.1021/acsnano.2c04750.

13 Electrocatalytic Applications of MXenes

Deepak Ranjan Mishra, Asish Dasmohapatra,
Pravati Panda, and Priyanka Chandra

13.1 INTRODUCTION

Research and development of innovative clean energy conversion technologies based on renewable sources, for example, the hydrogen economy, is in great demand [1,2]. A chain system of energy flow such as the conversion of electrical energy into molecular hydrogen via water electrolysis and, further, the utilization of hydrogen as a fuel in fuel cell vehicles, are some of the most anticipated aims of the hydrogen economy [1,2]. However, the efficiency of this energy flow is significantly restricted due to the slow kinetics of the hydrogen evolution reaction (HER; $2H_2O + 2e^- \rightarrow H_2 + 2OH^-$) and oxygen reduction reaction (ORR; $O_2 + 4H^+ + 4e^- \rightarrow 2H_2O$) in water electrolyzers and fuel cells, respectively [3–6]. This inefficiency mainly arises from the high activation energy barrier of multiple electron transfer steps on an electrode surface. Therefore, designing highly active, cost-effective electrode materials for HER and ORR is crucial for the development of water electrolyzers and fuel cells [3–6].

Two-dimensional transition metal carbides or nitrides (MXenes), which exhibit high catalytic activity in various electrocatalytic and photocatalytic devices, have attracted much attention in recent years as frontier materials for a wide range of applications. Their hydrophilicity is enhanced by characteristics like metal sites exposed at the terminals of MXenes and a significant number of -OH and -O groups on the surface. This characteristic is thought to make electrocatalysis in an aqueous media easier to perform by ensuring that MXenes have sufficient interaction with water molecules. Therefore, they will be essential in the development of future energy, electronic, and optoelectronic devices [7–9]. MXene-based materials are currently gaining a lot of research interest as potential platinum (Pt) alternatives for HER and ORR reactions. Numerous experimental and theoretical investigations have been carried out to comprehend and enhance the HER and ORR activity of MXenes [7–9].

Herein, we summarize the progress of MXene-based electrocatalysts for the HER, oxygen evolution reaction, and oxygen reduction reaction, including regulated pristine MXenes and modified hybrid MXenes, from both theoretical and experimental perspectives. A brief overview of MXene synthesis is presented first, accompanied by a discussion on the relationship between electrocatalytic properties and M, X, T, vacancies, and morphologies. After reviewing strategies in terms of atom substitution, functional modification, defect engineering, and morphology control,

DOI: 10.1201/9781003366225-13

we emphasize the construction of heterojunctions between MXenes and other nanostructures such as metal nanoparticles, oxides, hydroxides, sulfides, and phosphides. We finally discuss prospects for the future development of MXene-based electrocatalysts.

13.2 APPLICATIONS IN ELECTROCATALYSIS

MXenes, which are composed of two-dimensional (2D) Ti metal carbides, have been chosen as prospective electrocatalysts for hydrogen evolution reaction (HER) applications owing to their metal conductivity, rich surface chemistry, and atomic thickness with highly exposed active sites. Furthermore, the density functional theory (DFT) calculation of the Gibbs free energy for hydrogen adsorption (ΔGH^*) on Ti_3C_2 with O-terminal groups revealed a nearly zero value of $|G_{H^*}| = 0.00283$ eV at its optimal H* coverage ($\theta = 1/2$), which is lower than that of Pt ($\Delta GH^* = -0.009$ eV). Jiang's group, in 2018, designed an oxygen-functionalized Ti_3C_2MXene ($Ti_3C_2O_x$) as an electrocatalyst for HER in acid media *via* simple post-processing of the OH-terminated Ti_3C_2MXene. The resultant $Ti_3C_2O_x$ with O terminations on the basal plane enhanced the surface area and the conductivity of the Ti_3C_2MXene, exhibiting improved catalytic activity for HER with an overpotential of 190 mV at a current density of 10 mA cm^{-2} accompanied by long-term stability. Therefore, an improved HER performance was displayed by oxygen-functionalized Ti_3C_2MXenes ($Ti_3C_2O_x$) compared to the analogous E-$Ti_3C_2T_x$ and E$Ti_3C_2(OH)_x$ [10].

Investigating the effect of MXenes on the OER activity of MOF, a number of CoNi-MOFNs@MX nanocomposites were designed by the Cheng group in 2021. The OER activity of CoNi-MOFNs was discovered to be diminished after combining with MXenes, despite the fact that the nanocomposites exhibited higher electronic conductivity. The TOF calculation disclosed that the inclusion of the $Ti_3C_2T_x$MXene reduced the average OER activity of Co/Ni atoms in CoNi-MOFNs. The decreased OER activity of the CoNi-MOFNs@MX nanocomposites may be caused by the electron-donating property of the $Ti_3C_2T_x$ MXene, as seen by the reduced oxidation of Co^{2+} and Ni^{2+} active species in CoNi-MOFNs, despite the almost unchanged valence state and chemical bonding environment. $Ti_3C_2T_x$ MXenes have been found to have a deleterious impact on CoNi-MOFN OER activity, highlighting the necessity for careful consideration of MXene usage in electrocatalytic applications [11].

Owing to the acute toxicity of HF acid, conventional preparation of MXenes by hydrofluoric (HF) acid etching hinders the large-scale fabrication of MXenes and their widespread use in applications pertaining to energy. Consequently, the development of a new innocuous protocol for MXene synthesis is necessarily desirable. Pang's group in 2019 disclosed the synthesis of MXenes (e.g., Ti_2CT_x, Cr_2CT_x, and V_2CT_x) utilizing a collective strategy based on a thermal-assisted electrochemical etching route. Additionally, Co^{3+} ion doped MXenes displayed multifunctional catalytic activity with noticeably enhanced competency of the oxygen evolution reaction (OER), with $\eta@Js = 425$ mV, and the hydrogen evolution reaction (HER), with $\eta@Js = 404$ mV. Furthermore, a Co^{3+}-MXene used as a cathode was evaluated as a switchable battery mode, exhibiting significant energy conversion and storage capabilities. In aqueous ZIB systems, the energy storage of corresponding optimized E-etched MXenes showed a specific capacity of 100 mA g^{-1}. This work proves the usefulness of the materials and paves a nontoxic and HF-free route to produce diverse MXenes [12].

According to 2016 research by Seh and colleagues, both theory and experiments have shown that MXene materials can operate as stable, active catalysts for the HER in acid. Computational screening of a 2D layered M_2XT_x (M = metal; X = (C, N); and T_x = surface functional groups) revealed that Mo_2CT_x was a promising system

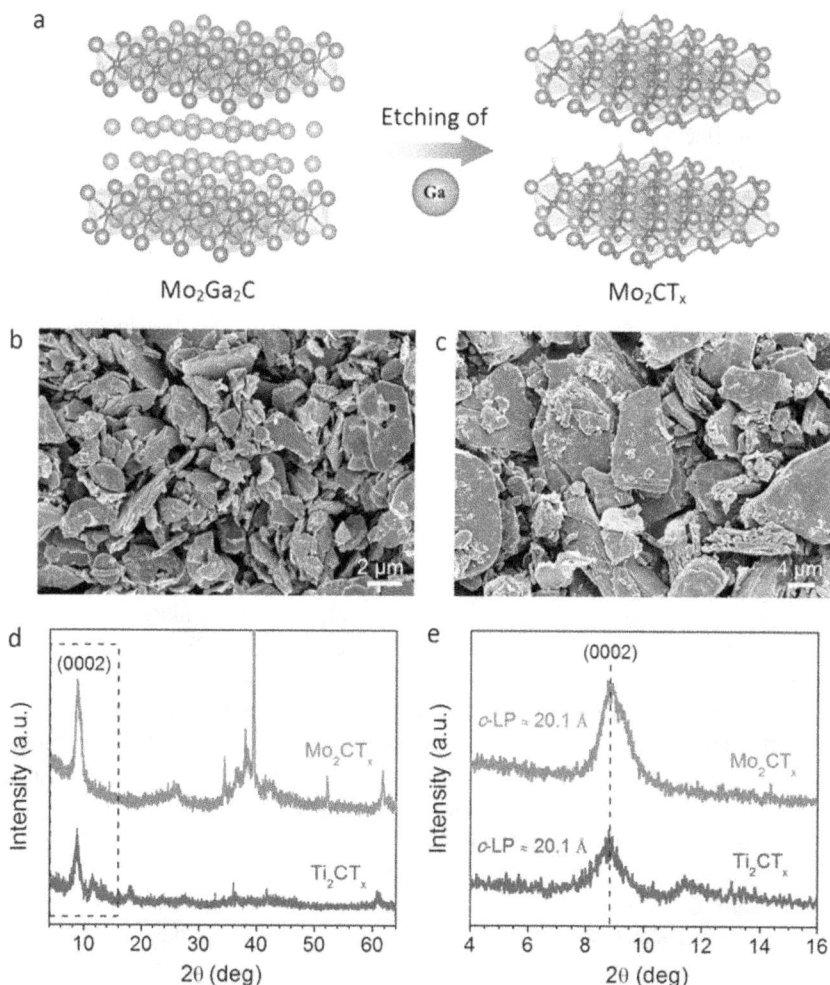

FIGURE 13.1 Characterization of MXenes. (a) Schematic of the synthesis process of Mo_2CT_x by selective etching the two Ga layers from Mo_2Ga_2C. Purple, green, brown, red, and white balls represent Mo, Ga, C, O, and H atoms, respectively. (b, c) SEM images of (b) Mo_2CT_x and (c) Ti_2CT_x. (d) XRD patterns of Mo_2CT_x and Ti_2CT_x (dried films) with (e) particular focus on the (0002) peak.

Reprinted with permission from Ref. [13].

for this reaction, exhibiting far higher HER activity than Ti_2CT_x. The basal plane may be active in this reaction, unlike MoS_2, in which only the edge sites of the 2H phase are active, negating the requirement for careful and intricate material design to increase the density of the edge sites exposed. With the help of this work, prospective 2D layered materials with active basal planes can be created in the future and used in a variety of additional significant clean energy reactions, such as carbon dioxide reduction, oxygen evolution, and oxygen reduction, among others as shown in Figure 13.1 [13].

The foundation of capturing and using renewable energy is the development of effective non-noble metal-based electrocatalysts for electrochemical water splitting. Electrocatalysis is a promising method for producing sustainable hydrogen. In a recent study, a remarkable electrocatalyst with multi-heterostructure interfaces and a three-dimensional porous structure was created by Dang's group, and the processes of increased electrocatalytic activity using multi-characterizations and density functional calculations were clarified. In particular, the developed $Co_2P/N@Ti_3C_2T_x@NF$ (abbreviated as CPN@TC) demonstrated an extremely low overpotential of 15 mV to reach a current density of 10 mA cm^{-2}, long-term durability, and a tiny Tafel slope of 30 mV dec^{-1} in 1 M KOH, which even favorably compares with noble metal catalysts. Multi-hetero interfaces for adsorbing H_2O and H^*, fine conductivity for electronic transmission, and a well-designed structure for quick ion and gas transit are all credited with exceptional HER activity. It is conceivable that the production of transition metal-based phosphides for improved catalytic performance can be accomplished using the CPN@TC synthetic approach [14].

In contrast to graphene, layered two-dimensional titanium carbide (MXene) materials have a better ability to self-assemble some unique MXene derivatives with fascinating chemical/physical properties. An analogous urchin-shaped MXene-$Ag_{0.9}Ti_{0.1}$ bimetallic nanowire composite exhibited unexpected electrocatalytic activity for the oxygen reduction reaction. These new MXene–Ag composites were created by mixing $AgNO_3$ directly with an alkalization-intercalated MXene (alk-MXene, $Ti_3C_2(OH/ONa)_2$) solution at room temperature. A PVP solution was added to cause the growth of five-fold nano twin Ag seeds, which developed into bimetallic Ag/Ti ($Ag_{0.9}Ti_{0.1}$) nanowires. By providing a large number of oxygen adsorption sites and decreasing the diffusion path of adsorbed oxygen, the distinctive bimetallic nanowires favored a four-electron transfer process and displayed high current density and good stability. The outcomes reflect a new development in the use of MXene materials for electrocatalysis and inspire enthusiasm for the discovery of fresh MXene derivations as shown in Figure 13.2 [15].

A straightforward and incredibly sensitive electrochemical sensor based on lamellar $Ti_3C_2T_x$ sheets was generated by Rasheed and co-workers for the detection of BrO^{3-} ions in water. The sensor, which was employed for the quantitative detection of BrO^{3-} in water, was based on the electrocatalytic reduction capability of selective $Ti_3C_2T_x$ sheets towards the reduction of BrO^{3-} ions. With a linear response for the BrO^{3-} concentration from 50 nM to 5 M, the developed sensor demonstrated great selectivity and stability. Its detection limit was 41 nM. By using various spectroscopic techniques, the chemical changes in $Ti_3C_2T_x$ during the BrO^{3-} reduction were further investigated. It was discovered that the $Ti_3C_2T_x$ was largely oxidized at the reaction's surface edges. The proposed sensor had a linear response to the

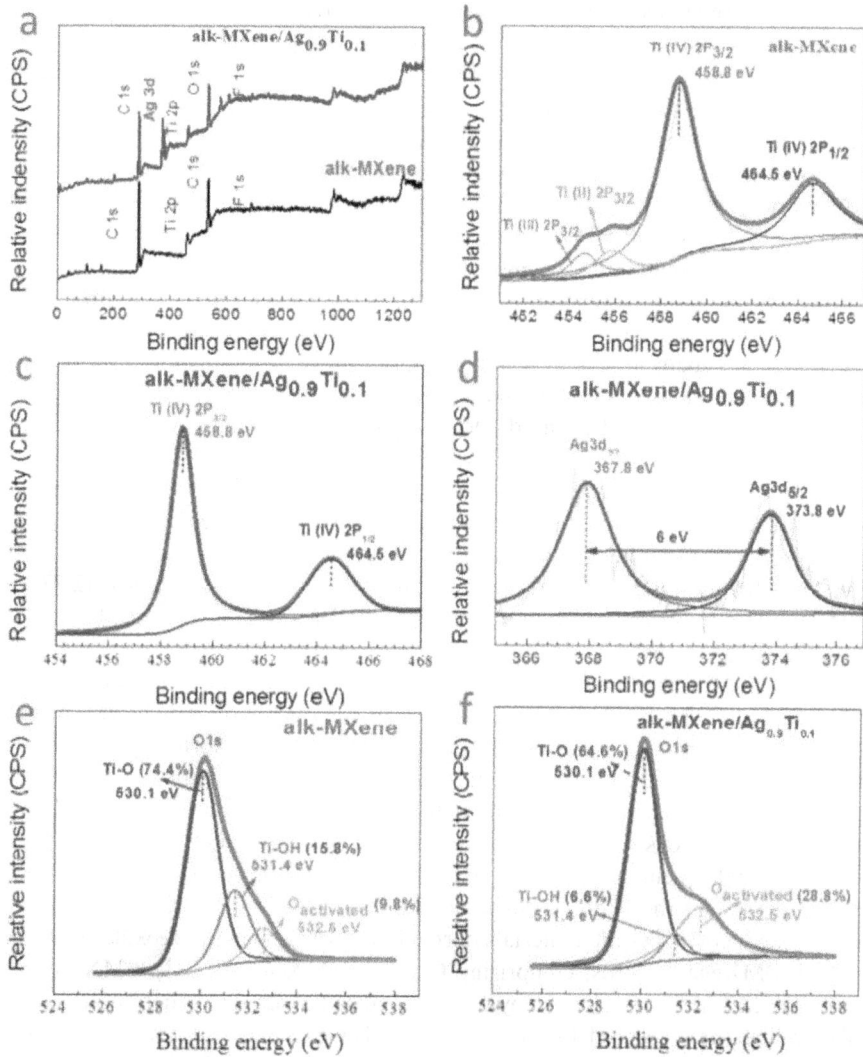

FIGURE 13.2 (a) High-resolution XPS spectra of the alk-MXene and MXene/NW-$Ag_{0.9}Ti_{0.1}$. (b) Ti 2p spectra of the alk-MXene. (c) Ti 2p spectra of the MXene/NW-$Ag_{0.9}Ti_{0.1}$. (d) Ag 3d spectra of the MXene/NW-$Ag_{0.9}Ti_{0.1}$. (e) O 1s of the alk-MXene. (f) O 1s of the MXene/NW-$Ag_{0.9}Ti_{0.1}$.

Reprinted with permission from Ref. [15].

BrO_3^- concentration from 50 nM to 5 M with a detection limit of 41 nM, and the MXene showed unique electrocatalytic characteristics towards efficient BrO_3^- reduction. A rise in peak current and a change in potential were indicators that BrO_3^- was being reduced catalytically by the MXene. Scanning electron microscopy (SEM), transmission electron microscopy (TEM), and X-ray photoelectron spectroscopy

(XPS) were used to examine surface alterations in MXenes following the electrocatalytic reduction of BrO^{3-}. The redox reaction between BrO^{3-} and MXenes has been validated by the production of trace TiO_2 crystals at the surface of $Ti_3C_2T_x$ during the electrochemical reduction of BrO^{3-}. Along with other interfering ions, the devised sensor demonstrated good selectivity for BrO^{3-}. The electrochemical detection of BrO^{3-} in actual drinking water samples was shown to be comparable to conventional IC measurement values. The practical applicability of the sensor was also assessed. It is clear that MXenes can offer a viable framework for the construction of electrocatalytic sensors and adsorbents for water pollutants [16].

The conversion of nitrogen into ammonia is crucial for human activities. Since the environmentally friendly electrochemical synthesis of ammonia from nitrogen and water has many potential applications, considerable effort has been put into enhancing the catalytic activity and selectivity. In this instance, an in situ-grown Ti_3C_2MXene-supported Co-based metal–organic framework (MOF), or zeolitic imidazolate framework-67 (ZIF-67), was generated. The MOF's high porosity and substantial active surface area and the Ti_3C_2MXene's higher conductivity enabled an effective electrochemical synthesis of ammonia using the composite. The produced ZIF-67@Ti_3C_2 catalyst, in particular, demonstrated an outstanding NH_3 yield (6.52 mol h^1 cm^2), which was much greater than that of Ti_3C_2 and ZIF-67 (2.77 and 1.61 mol, respectively). Particularly, the prepared ZIF-67@Ti_3C_2 catalyst displayed a good Faraday efficiency (20.2%) at 0.4 V (vs. the reversible hydrogen electrode) and an excellent NH_3 yield (6.52 mol h^1 cm^2), both of which were significantly higher than those attained by Ti_3C_2 and ZIF-67 alone (2.77 and 1.61 mol h^1 cm^2, respectively). These findings provide insights for developing high-performance electrocatalysts for the electrochemical nitrogen reduction reaction in addition to broadening the applications of the MXene family in this process [17].

Metal hydroxides and oxides have recently garnered attention as intriguing materials and essential structures in electrocatalysis; however, their exploration in the context of the hydrogen evolution reaction (HER) has been limited. Here, we present novel hybrid materials termed transition-metal hydroxides@MXene (TMHs@MXene) hybrids, comprising Co(OH)2@MXene, Ni(OH)$_2$@MXene, and FeOOH@MXene, characterized by well-defined components and hierarchical sheet-like architectures tailored for alkaline HER. The distinctive structure and strong interfacial interactions between transition-metal hydroxides (TMHs) and MXene nanosheets endow these nanohybrids with abundant active sites, structural integrity, and enhanced electrochemical kinetics, leading to superior catalytic performance. Theoretical calculations and electrochemical assessments corroborate that interfacial electronic coupling between the two components optimizes water and hydrogen adsorption energies, resulting in Pt-like catalytic behavior, evidenced by a low Tafel slope (31.7 mV dec^{-1}), minimal overpotential (21.0 mV@10 mA cm^{-2}), and exceptional stability, particularly demonstrated by Co(OH)2@MXene. Furthermore, we demonstrate the utility of Co(OH)2@MXene as a cathode in an alkaline water electrolyzer, achieving a current density of 10 mA cm^{-2} at 1.46 V with remarkable stability over 100 hours. This finding underscores the significant potential of interfacial electronic coupling in tailoring advanced electrocatalysts for applications in energy-related domains [18].

In light of the production of metal oxides and the subsequent loss of their intrinsic qualities, particularly grave oxidation of MXenes has turned into a serious issue. Using solvothermal processing, a bimetallic cobalt-manganese organic framework (CMT) was directly grown on a $Ti_3C_2T_x$ MXene sheet to improve electrocatalytic properties for oxygen evolution and reduction reactions in addition to obtaining strong oxidation resistance in an open-structured application. Through Fischer esterification and substitution reaction of fluorine, the carboxyl acids in tetrakis(4-carboxyphenyl)porphyrin, acting as organic linkers, were grafted with the surface terminators of the $Ti_3C_2T_x$ MXene, significantly improving the antioxidation stability. Additionally, through an electron hopping mechanism, the metalloporphyrin structure as-formed and unpaired electrons produced between CMT and the $Ti_3C_2T_x$ MXene during solvothermal treatment enhanced their electrocatalytic activity, durability, and electrical conductivity. As a result, the CMT@MXene exhibited exceptional stability as a bifunctional electrocatalyst over 247 cycles in a lithium–oxygen ($Li-O_2$) battery at a fixed specific capacity of 1000 mAh g^{-1} and a current density of 500 mA g^{-1}. This method offers fresh ideas for MXene and MOF synergistic coupling in the next open-structured applications [19].

Given that 2DTMCs are promising materials for electrochemical applications, it is unusual to come across 2DTMCs with metallicity and active basal planes. In this study, Chen and co-workers built a material library of 79 2DTMCs and proposed a straightforward and efficient method to extract 2DTMCs from non-van der Waals bulk materials. They designated these compounds anti-MXenes, as they are composed up of one M atomic layer sandwiched between two X atomic layers. A total of 24 anti-MXenes were found to be thermodynamically, dynamically, mechanically, and thermally stable by the use of density functional theory calculations. These anti-MXenes have the potential to make good electrode materials, such as electrocatalysts for hydrogen evolution processes, due to their metallicity and active basal plane. CuS can increase HER at all H coverages compared to noble-metal-free anti-MXenes with favorable H-binding, whereas CoSi, FeB, CoB, and CoP demonstrate potential for HER at some particular H coverages. These materials have a relatively high density of active sites because the tetra-coordinating nonmetal atoms at the basal planes serve as active sites. With its extremely high capacity, acceptable open circuit voltage, and low Li diffusion energy barriers, CoB is also a viable anode material for lithium-ion batteries. This work illustrates the use of anti-MXenes in various electrochemical processes and encourages the "computational exfoliation" of 2D materials from non-van der Waals bulks [20].

It is highly desirable to develop low-cost yet effective bifunctional electrocatalysts that can directly split water for the expanding range of uses for hydrogen energy. In this instance, the Tao group introduced a multiphasic 1T/2H $MoSe_2$-on-MXene nanosheet heterostructure (referred to as 1T/2H $MoSe_2$/MXene) that is both effective and economical to use as an electrocatalyst for total water splitting. On the one hand, the 1T/2H $MoSe_2$ has a lot of active sites that have a good amount of electrocatalytic activity. On the other hand, the MXene nanosheets act as highly conductive 2D substrates, enhancing charge transfer and preventing the aggregation of 1T/2H $MoSe_2$. In contrast to its pure 1T/2H $MoSe_2$, MXene, and 2H $MoSe_2$/MXene counterparts, the 1T/2H $MoSe_2$/MXene catalyst in an alkaline medium demonstrated outstanding

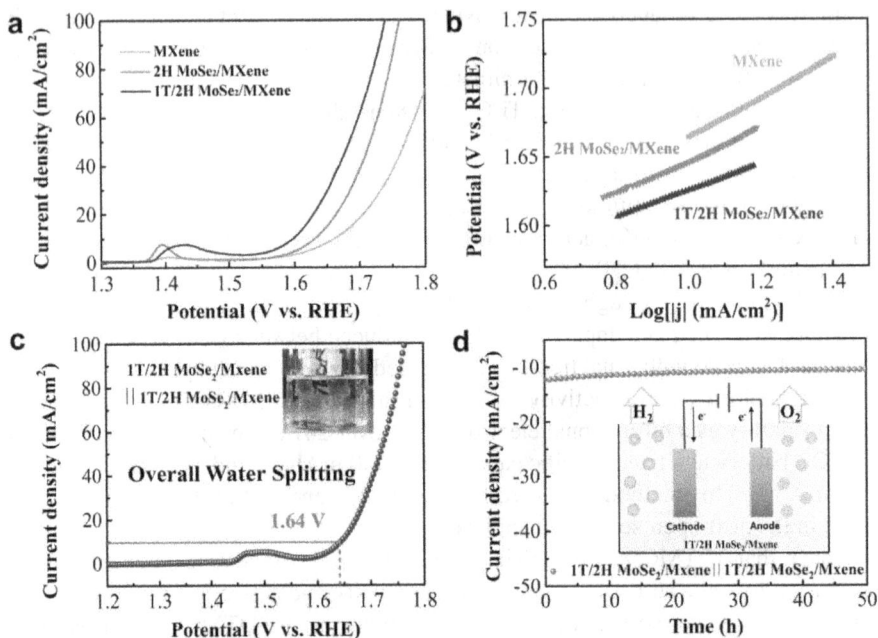

FIGURE 13.3 (a) OER polarization curves and (b) corresponding Tafel plots of catalysts. (c) The LSV curves of 1T/2H $MoSe_2$/MXene k 1T/2H $MoSe_2$/MXene. Inset in (c) shows the photograph of overall water splitting. (d) Stability test of the 1T/2H $MoSe_2$/MXene k 1T/2H $MoSe_2$/MXene for 50 h. Inset in (d) shows a schematic description of the 1T/2H $MoSe_2$/MXene for overall water splitting.

Reprinted with permission from Ref. [21].

synergy in both hydrogen evolution reaction (HER) and oxygen evolution reaction (OER) activity. For instance, the 1T/2H $MoSe_2$/MXene displayed an overpotential (95 mV) and Tafel slope (91 mV dec^l) for HER performance and an overpotential (340 mV) and Tafel slope (90 mV dec^l) for OER performance as shown in Figure 13.3. To achieve a current density of 10 mA cm^2, especially when the 1T/2H $MoSe^2$/MXene was acting as a bifunctional electrocatalyst for total water splitting, just 1.64 V was required. In the total water splitting system, the 1T/2H $MoSe_2$/MXene heterostructure also demonstrated good endurance [21].

Given that electrocatalytic nitrate reduction to ammonia (ENRA) is a useful method for addressing the environmental and energy crises, it still faces significant difficulties in achieving high activity and stability in a way that is practical for use in a fluid environment. Owing to its benefits in terms of low cost, lightweight, environmental friendliness, simple and scalable manufacture, extensive structural stability, and electrocatalytic reliability, the flexible film electrode may be able to resolve the aforementioned issue of practical catalytic application. A 2D flexible CuBDC@ $Ti_3C_2T_x$ electrode for ENRA was prepared by seamlessly growing copper 1,4-benzene di-carboxylate (CuBDC) on electronegative MXene nanosheets ($Ti_3C_2T_x$) via a 2D hybridization process. Corresponding to previously reported nanomaterials towards ENRA, the flexible electrode simultaneously displayed strong Faradaic efficiency (86.5%) and outstanding stability for NH_3 production. After bending, twisting, folding, and crumpling tests, the flexible electrode in particular maintained exceptional $FENH_3$ towards ENRA, indicating excellent electroconductibility, high stability, and durability. These findings present the practical uses of the flexible electrode with effective environmental adaptability in resolving the worldwide environmental contamination and energy crises by effective ENRA. They also offer a mild permeation-mediated method to produce a flexible electrode [22].

Two-dimensional transition metal-based carbides (or nitrides), also known as MXenes, which can be produced from the three-dimensional MAX phases, have received a lot of interest lately. They are potential prospects for a wide range of applications, including sensors, electrodes, and catalysts, because of their unique structure as well as their hydrophilic and metallic nature. It goes without saying that the chemical makeup, stoichiometry, and surface structure of the MXenes have a significant impact on the corresponding chemical and physical properties. Based on the chemical exfoliation of the 413 MAX phase V_4AlC_3 by treatment with aqueous hydrofluoric acid, the Birkel group introduced a novel member of the MXene family, $V_4C_3T_x$ (T indicating the surface groups). Scale-bridging electron microscopy research and X-ray powder diffraction data demonstrated that aluminum was successfully removed from the MAX phase structure. This novel MXene was tested for 100 cycles in an acidic solution for its electrocatalytic activity in the hydrogen evolution reaction. The overpotential needed to achieve a current density of 10 mA cm^{-2} decreased by almost 200 mV over time, which is an interesting improvement in the catalytic performance that the authors attribute to the removal of an oxide species from the MXene surface, as demonstrated by XPS measurements. The electrocatalytic activity of MXenes was experimentally determined in this study, along with the evolution of their surface structure, which is important for other transition metal-based MXenes in the context of future possible applications [23].

The primary obstacle to the adoption and development of direct methanol fuel cells (DMFCs) is the slow methanol oxidation reaction kinetics with present electrocatalysts. In a recent study, a polyaniline/palladium/$Ti_3C_2T_x$ (PANI/Pd/MXene) nanocomposite electrocatalyst for the methanol electro-oxidation process (MEOR) was developed. Using a pre-anodized screen-printed electrode (SPE) and an acidic electrolyte solution that contained $Ti_3C_2T_x$, aniline, and palladium chloride as precursors, the PANI/Pd/MXene nanocomposite was created. The authors used FESEM, TEM, FT-IR, XPS, and cyclic voltammetry to analyze the PANI/Pd/MXene nanocomposite. When compared to Pd/MXene, which had a peak current density of 106 mA cm^2, the electrochemical response of the PANI/Pd/MXene nanocomposite demonstrated an enhanced electrocatalytic response towards the oxidation of methanol. Additionally, it remained stable for 100 cycles. MXene nanosheets incorporated the electrochemically active PANI/Pd sites, enabling a more effective MEOR. The strong metal–support interactions between PANI/Pd and MXene nanosheets, which maximized methanol adsorption on the electrode surface for effective electrocatalytic oxidation, were responsible for this encouraging outcome. The metal electrocatalyst interface and surface features were thereby tailored by the $Ti_3C_2T_x$ support, leading to increased electrocatalytic performance. The design of MXene-supported noble metal electrocatalysts for MEOR in direct methanol fuel cells was made simple in this study [24].

Numerous techniques for storing and converting renewable energy use the oxygen evolution reaction (OER), which is a crucial reaction. Growing interest has been shown in creating effective non-precious metal-based electrocatalysts for OER. An approach to creating hierarchical cobalt borate/$Ti_3C_2T_x$ MXene (Co-Bi/$Ti_3C_2T_x$) hybrids using quick chemical reactions at room temperature is described in a recent study. The Co-Bi/$Ti_3C_2T_x$ hybrid's intriguing hierarchical structure is useful for exposing more active sites, enhancing mass diffusion, and optimizing charge transfer paths for electrochemical reactions. Furthermore, a robust interaction between Co-Bi and $Ti_3C_2T_x$ ensures effective charge transfer and makes it easier for more anionic intermediates to be attracted electrostatically for a quick redox process. Given this, the hierarchical Co-Bi/$Ti_3C_2T_x$ hybrid exhibits exceptional OER catalytic activity to produce a current density of 10 mA cm^{-2} at a 250 mV overpotential and a Tafel slope of about 53 mV dec^{-1}. Cobalt borate (Co-Bi), a two-dimensional $Ti_3C_2T_x$-supported compound, has the potential to be a catalyst for the oxygen evolution reaction in an alkaline media. The Co-Bi/$Ti_3C_2T_x$ hybrid has a hierarchical structure that allows for better exposure of the active surface sites, mass diffusion, and charge transfer properties. As a result, at a low overpotential of 250 mV, the hybrid material can deliver a stable current density of 10 mA cm^{-2} [25].

Spinel oxides composed of cobalt are excellent candidates for the oxygen evolution process (OER), but further development could be hampered by their poor electrical conductivity and unstable crystal structure. In a recent study, a hybrid electrocatalyst composed of $CuCo_2O_4$ nanoparticles, $Ti_3C_2T_x$, and nickel foam (NF) was introduced as a reliable electrode for OER. The presence of the $Ti_3C_2T_x$ MXene structure improves the electrocatalysts' electrical conductivity. The electrocatalytic performance could be effectively improved by adding more electrochemically active sites to the ultrathin hybrid structure's huge surface area. According to experimental results, $CuCo_2O_4$/$Ti_3C_2T_x$ hybrid structure on NF exhibited increased OER electrocatalytic activity over $Ti_3C_2T_x$ in its purest form. As a result, the hybrid electrocatalyst displayed excellent long-term durability in addition to a low overpotential of 1.67 V at 100 mA.cm^2 and a tiny Tafel slope of 49 mV dec^{-1} [26].

As a part of the MXene family, two-dimensional vanadium carbide (V2CTx, known as V-MXene) has primarily been explored theoretically for its potential in the electrocatalytic hydrogen evolution reaction (HER), with limited experimental validation. This study presents a combined theoretical and experimental investigation into the inherent structural and electrochemical properties of a V-MXene-based hybrid (NiS2/V-MXene) to assess its suitability as an HER electrocatalyst. Additionally, we examine the widely studied titanium carbide (Ti3C2Tx, or Ti-MXene) for comparison. In this research, we design, synthesize, and experimentally evaluate NiS2/V-MXene and NiS2/Ti-MXene as proof of concept. As anticipated, NiS2/V-MXene demonstrates superior electrocatalytic performance for HER compared to NiS2/Ti-MXene, owing to its enhanced electronic transfer and favorable interactions. The close packing of NiS2 nanoparticles not only prevents V-MXene from stacking but also exposes additional active sites, facilitating faster electrolysis. This unique sandwich-like architecture enhances the contact interface and facilitates electron migration and mass diffusion during electrolysis. Furthermore, the predicted electron transfer from V-MXene to NiS2 is experimentally validated using X-ray photoelectron spectroscopy. These insights contribute to the design and advancement of V-MXene-based electrocatalysts for HER, broadening their potential applications. [27].

In a recent investigation, Huang and co-workers exploited conductive Ti_3C_2 MXenes as a promoter to quicken MoS_2's charge transfer and achieve extremely effective HER electrocatalysis. It is shown that in situ growth of MoS_2 nanosheets vertically perched on planar Ti_3C_2 nanosheets can successfully produce hierarchical heterostructures using a simple hydrothermal method. The resulting MoS_2/Ti_3C_2 heterostructures produce a massive increase in HER activity compared to pristine MoS_2 nanosheets thanks to the opened layer structures and high interfacial coupling effect. More particular, at an overpotential of about 400 mV, the catalytic current density caused by the MoS_2/Ti_3C_2 heterostructure is almost 6.2 times higher than that of the pure MoS_2 nanosheets. This research shows that Ti_3C_2 nanosheets are excellent building blocks for extremely effective electrocatalysts for water splitting [28].

The electrocatalytic performance of a fuel cell system can be enhanced by the use of a new material (MXene). MXenes haven't yet been found in the field of electrocatalysis for a fuel cell, however. Therefore, the purpose of a recent study was to improve the electrocatalyst performance of direct methanol fuel cells (DMFCs) by utilizing a bimetallic PtRu and MXene mixture. Utilizing response surface methodology (RSM), optimization is performed. The optimization analysis uses the composition of MXene, Nafion content, and methanol concentration as parameters (input), and current density is employed as a response (output). Cyclic voltammetry (CV) is used to measure the current density. RSM produces the best factors when the MXene is composed at 78.90 weight percent, Nafion at 19.71 weight percent, and methanol at 2.82 M. The optimal response is anticipated to be 186.59 mA/mg$_{PtRu}$. It is established through the validation test that the average current density is 187.05 mA/mg$_{PtRu}$. When compared to PtRu/C commercial electrocatalyst, the PtRu/MXene electrocatalyst generates a current density that is 2.34 times higher. This suggests that MXenes have excellent promise as nanocatalysts for the fuel cell's production of greener energy [29].

Developing efficient non-precious electrocatalysts for hydrogen evolution reaction (HER) in alkaline media is enticing yet problematic in the field of renewable energies. In a recent study, researchers present an experimental and theoretical investigation to demonstrate how binary MXene/Ni_3S_2 nanosheets can form in situ over three-dimensional Ni foam (NF) in a target, acting as a self-supported and incredibly potent electrocatalyst for alkaline HER. The HER test on $Ti_3C_2T_x/Ni_3S_2/NF$ in 1 M KOH yields a low overpotential (72 mV) to produce a current density of 10 mA cm^{-2} thanks to the excellent electrical conductivity, superior hydrophilic interface for gas release, and synergistic coupling of $Ti_3C_2T_x$ and Ni_3S_2. The Tafel slope as a result is 45 mV dec^{-1}, one of the lowest values ever recorded for both modern MXene-based electrocatalysts and self-supported Ni-based electrocatalysts in alkaline conditions. With no change in current density for at least 12 hours, the catalyst is extremely stable. Calculations using density functional theory further show that the hybridization of $Ti_3C_2T_x$ and Ni_3S_2 results in the lowest possible hydrogen adsorption-free energy on the catalyst surface. This research could lead to the development of self-sustaining and affordable electrocatalysts for use in next-generation energy applications [30].

Strategies to develop an ingenious and effective electrocatalyst that combines the properties of abundant surface deficiency, good dispersibility, high conductivity, and large surface specific area (SSA) in a straightforward manner is highly essential in order to achieve energy-effective ammonia (NH_3) production via the ambient-condition electrochemical N_2 reduction reaction (NRR). The ethanol-thermal treatment of the $Ti_3C_2T_x$ MXene is used to create oxygen-vacancy-rich TiO_2 nanoparticles (NPs) in situ grown on the $Ti_3C_2T_x$ nanosheets ($TiO_2/Ti_3C_2T_x$), which is motivated by the fact that the MXene contains thermodynamically metastable marginal transition metal atoms. The primary active sites for the production of NH_3 are oxygen vacancies. Untreated, highly conductive $Ti_3C_2T_x$ nanosheets in the interior could prevent the self-aggregation of the TiO_2 NPs while also facilitating electron transport. In the meantime, the TiO_2 NPs formation might, in turn, improve the SSA of the $Ti_3C_2T_x$. As a result, the as-prepared electrocatalyst displays a remarkable Faradaic efficiency of 16.07% at 0.45 V vs RHE in 0.1 M HCl and an impressive NH_3 production of 32.17 g h^{-1} mg^{-1}cat. at 0.55 V versus reversible hydrogen electrode (RHE), making it one of the most promising NRR electrocatalysts. Furthermore, compared to $Ti_3C_2T_x$ or TiO_2 (101) alone, the TiO_2 (101)/$Ti_3C_2T_x$ combination has a lower NRR energy barrier (0.40 eV), according to calculations using density functional theory [31].

A key area of research for industrial-scale hydrogen production is the creation of low-cost, highly effective electrocatalysts for the hydrogen evolution process (HER). The P-Mo_2C/Ti_3C_2@NC nanohybrid described by Que's group, which can act as a high-performance nonprecious-metal electrocatalyst for HER, is composed of phosphorus-doped molybdenum carbide nanodots supported on Ti_3C_2 flakes enclosed by nitrogen-doped carbon. The P-doped Mo_2C nanodots and the conductive Ti_3C_2 matrix are coupled in an optimized interfacial manner thanks to the intrinsic anchoring sites of the conductive Ti_3C_2 matrix, which also aids in the excellent distribution and limited development of the electrocatalytically active P-doped Mo_2C nanodots. Meanwhile, nitrogen-doped porous carbon shells can stifle spontaneous oxidation of the Ti_3C_2 flakes. Consequently, a synergistic effect of cooperative catalytic interfaces between Ti_3C_2 and extremely tiny P-Mo_2C nanodots enclosed in nitrogen-doped porous carbon and a suitable introduction of a Ti_3C_2 substance afford overall improvement in electrical conductivity and exposure of reactive sites. As a result, the P-Mo_2C/Ti_3C_2@NC nanohybrid catalyst exhibits remarkable HER activity with an overpotential of 177 mV at 10 mA cm^{-2}, fast reaction kinetics of 57.3 mV dec^{-1}, and long-term stability over 60 h in the acidic electrolyte, which exceeds the P-Mo_2C@NC, Ti_3C_2@NC, P-Mo_2C/CNT@NC, and P-Mo_2C/rGO@NC nanohybrids. This research opens the door to the creation of sophisticated MXene-based electrocatalysts for HER and encourages the investigation of a brand-new class of MXene-based nanohybrids for use in renewable energy applications [32].

The widespread adoption of the hydrogen economy depends on the creation of long-lasting, extremely effective catalysts for the HER (hydrogen evolution process). Recently, it has been discovered that members of the 2D MXenes family, particularly Mo_2CT_x, are promising HER catalysts. Their widespread use is hampered by their inherent oxidative instability in air and aqueous electrolyte solutions. Recently, researchers outlined a quick and scalable technique for creating a $Mo_2CT_x/2H$-MoS_2 nanohybrid by sulfidating Mo_2CT_x MXenes in place to avoid accidental oxidation. Superior HER activities were enabled by the close epitaxial coupling at the $Mo_2CT_x/2H$-MoS_2 nanohybrid interface, which required only 119 or 182 mV overpotential to produce current densities of 10 or 100 mA cm^{-2}, respectively. Calculations using density functional theory show that the nanohybrid structure has stronger interfacial adhesion than the nanohybrid that has been physisorbed and that the HER overpotential may be tuned by varying the degree of MXene sulfidation. Importantly, the presence of $2H$-MoS_2 prevents further oxidation of the MXene layer, allowing the nanohybrid to sustain exceptionally long-lasting industrially relevant current densities of over 450 mA cm^{-2}. After 10 continuous days of electrolysis at a constant 10 mA cm$^{-2}_{geom}$ current density or 100,000 successive cycles of cyclic voltammetry, less than 30 mV overpotential degradation was seen. The $Mo_2CT_x/2H$-MoS_2 nanohybrid's excellent HER endurance represents a significant advancement in the practical deployment of MXenes as noble metal-free catalysts for a variety of applications in energy conversion and water splitting [33].

To address the energy issue, it is crucial to investigate effective electrocatalysts devoid of noble metals for the HER. MXenes, a class of two-dimensional transition metal carbides, have received relatively little study in relation to their exceptional potential as HER electrocatalysts, including the reactivity and number of active sites. To accomplish the synergistic modulation of both reactivity and the number of active sites, Zhang's group presented a model of atomically thin Ti_2CT_x nanosheets with rich surface fluorine termination groups (where T_x denotes the surface termination groups; -F, -O, and -OH). Theoretical studies and electrochemical measurements demonstrate that the Ti_2CT_x nanosheets' rich surface F terminations can enhance the proton adsorption kinetics and lower their charge transfer resistance, which increases the activity of the active sites and favors electrode kinetics. Additionally, the Ti_2CT_x nanosheets' ultrathin thickness provides HER with high-density active sites. Rich F-terminated Ti_2CT_x nanosheets display a substantial exchange current density of 0.41 mA cm^{-2} and a tiny onset overpotential of 75 mV, as expected. This discovery increases the MXene-based materials' already-expanding potential for use in hydrogen evolution reactions [34].

For the electro-oxidation of methanol, wet impregnation of nickel onto MoS_2/MXene composites (NiMoS_2/MXene) serves as the anode electrode material. The synthesis of MoS_2/MXene was verified by X-ray diffraction, X-ray photoelectron spectra, and scanning electron microscopy with energy-dispersive X-ray spectroscopy techniques. Analyses of N_2 adsorption–desorption yielded information on the textural characteristics of catalysts. NiMoS_2/MXene catalysts had improved electrocatalytic activity, as shown by electrochemical tests in 0.1 M KOH. The low onset potential provided by Ni, the high tolerance for CO poisoning provided by MoS_2, the high conductivity and high mechanical stability of the MXene, and the NiMoS_2/MXene system all contributed to the NiMoS_2/MXene system's increased electrocatalytic activity for methanol oxidation. High current density, electrochemically active surface area, long-term stability, and a low Rct value were all characteristics of NiMoS$_{2-3}$/MXene catalysts. According to the electrochemical findings, the catalyst NiMoS$_2$/MXene is a very electroactive anode material. Consequently, it can be applied to fuel cell systems like direct methanol fuel cells (DMFCs) [35].

Electrocatalysts for the hydrogen evolution reaction (HER) have been regarded as one of the key elements in current electrochemical hydrogen generation systems. Hydrogen energy has garnered sustained attention in the exploitation and application of sophisticated power-generating technologies. A recent study creates an easy and affordable bottom-up method for the in situ stereo-assembly of 1D ultrafine cobalt selenide nanowires entangled with 2D Ti$_3$C$_2$T$_x$ MXene nanosheets (CoSe NW/Ti$_3$C$_2$T$_x$). Such an architectural layout provides the hybrid system with several exposed CoSe edge sites in addition to a sizable accessible surface for the quick transportation of reactants, producing significant synergic coupling effects. The as-derived CoSe NW/Ti$_3$C$_2$T$_x$ hybrid exhibits competitive electrocatalytic properties towards the HER with a small onset potential of 84 mV, a low Tafel slope of 56 mV dec^{-1}, and exceptional cycling performance, which are better than those of bare CoSe and Ti$_3$C$_2$T$_x$ materials. It is anticipated that this promising nanoarchitecture may open up new opportunities for the design and construction of precious metal-free electrocatalysts [36].

The quest for clean, green, sustainable, and renewable energy sources has been a focus of research over the past few decades due to the impending, unrelenting demand for energy. With hydrogen energy, the issue of an energy shortage can be avoided, and non-renewable fossil fuels can be perfectly replaced. One of the greenest energy sources is the creation of oxygen and hydrogen from water electrolysis, also known as the hydrogen evolution reaction (HER) and the oxygen evolution reaction (OER), or the half-cell processes. To overcome the aforementioned difficulty, cost-effective, noble metal-free, effective, and robust electrocatalysts are required. For a promising overall water splitting application, Rout and colleagues created a strong interface using orthorhombic CoSe$_2$ nanorods and two-dimensional Ti$_3$C$_2$T$_x$ MXene sheets. Both in HER and OER, the catalyst performance was outstanding in terms of electrocatalytic activity. A value of 230 mV for overpotential and 65 mV/dec of Tafel value was achieved by the amalgamated catalyst for effective HER application, whereas 270 mV of overpotential and 71 mV/dec of Tafel slope were disclosed for OER at a current density of 10 mA/cm^2. In both instances, the catalyst withstood a long-term stability test for more than 12 hours, and the pre-stability polarization and post-stability LSV curves were nearly identical, illustrating the material's durability. Because of its superior charge transfer properties, larger surface area, plenty of electrochemically active sites, high hydrogen binding, and high dissociations that can be attributed to the synergistic effect, we anticipate that this catalyst will open new doors in the field of energy conversion [37].

REFERENCES

1. Gray, H. B., 2009. Powering the planet with solar fuel, Nat. Chem., 1, 7. https://doi.org/10.1038/nchem.141.
2. Markovic, N. M., 2013. Interfacing electrochemistry, Nat. Mater., 12, 101–102. https://doi.org/10.1038/nmat3554.

3. Ibrahim, K. B., Tsai, M. C., Chala, S. A., Berihun, M. K., Kahsay, A. W., Berhe, T. A., Su, W. N., Hwang, B. J., 2019. A review of transition metal-based bifunctional oxygen electrocatalysts, J. Chin. Chem. Soc., 66, 8, 829–865. https://doi.org/10.1002/jccs.201900001.

4. Lee, J., Jeong, B., Ocon, J. D., 2013. Oxygen electrocatalysis in chemical energy conversion and storage technologies, Curr. Appl. Phy., 13, 309–321. https://doi.org/10.1016/j.cap.2012.08.008.

5. Lei, Z., Wang, T., Zhao, B., Cai, W., Liu, Y., Jiao, S., Li, Q., Cao, R., Liu, M., 2020. Recent progress in electrocatalysts for acidic water oxidation, Adv. Energy Mater., 2000478. https://doi.org/10.1002/aenm.202000478.

6. Sun, X., Xu, K., Fleischer, C., Liu, X., Grandcolas, M., Strandbakke, R., Bjørheim, T. S., Norby, T., Chatzitakis, A., 2018. Earth-abundant electrocatalysts in proton exchange membrane electrolyzers, Catalysts, 8, 657. https://doi.org/10.3390/catal8120657.

7. Alhabeb, M., Maleski, K., Anasori, B., Lelyukh, P., Clark, L., Sin, S., Gogotsi, Y. 2017. Guidelines for synthesis and processing of two-dimensional titanium carbide ($Ti_3C_2T_x$MXene), Chem. Mater., 29(18), 7633–7644. https://doi.org/10.1021/acs.chemmater.7b02847.

8. Halim, J., Kota, S., Lukatskaya, M. R., Naguib, M., Zhao, M.-Q., Moon, E. J., Pitock, J., Nanda, J., May, S. J., Gogotsi, Y., Barsoum, M. W., 2016. Synthesis and characterization of 2D molybdenum carbide (MXene). Adv. Funct. Mater., 26(18), 3118–3127. https://doi.org/10.1002/adfm.201505328.

9. Zhang, Z., Li, H., Zou, G., Fernandez, C., Liu, B., Zhang, Q., Hu, J., Peng, Q., 2016. Self-reduction synthesis of new MXene/Ag composites with unexpected electrocatalytic activity, ACS Sustain. Chem. Eng., 4(12), 6763–6771. https://doi.org/10.1021/acssuschemeng.6b01698.

10. Jiang, Y., Sun, T., Xie, X., Jiang, W., Li, J., Tian, B., Su, C., 2019. Oxygen-functionalized ultrathin $Ti_3C_2T_x$ MXene for enhanced electrocatalytic hydrogen evolution. ChemSusChem, 12(7), 1368–1373. https://doi.org/10.1002/cssc.201803032.

11. Du, C.-F., Song, Q., Liang, Q., Zhao, X., Wang, J., Zhi, R., Wang, Y., Yu, H., 2021. The passive effect of MXene on electrocatalysis: A case of $Ti_3C_2T_x$/CoNi–MOF nanosheets for oxygen evolution reaction, ChemNanoMat, 7(5), 539–544. https://doi.org/10.1002/cnma.202100061.

12. Pang, S.-Y., Wong, Y.-T., Yuan, S., Liu, Y., Tsang, M.-K., Yang, Z., Huang, H., Wong, W.-T., Hao, J., 2019. Universal strategy for HF-free facile and rapid synthesis of two-dimensional MXenes as multifunctional energy materials, J. Am. Chem. Soc., 141, 24, 9610–9616. https://doi.org/10.1021/jacs.9b02578.

13. Seh, Z. W., Fredrickson, K. D., Anasori, B., Kibsgaard, J., Strickler, A. L., Lukatskaya, M. R., Gogotsi, Y., Jaramillo, T. F., Vojvodic, A., 2016. Two-dimensional molybdenum carbide (MXene) as an efficient electrocatalyst for hydrogen evolution ACS Energy Lett., 1, 3, 589–594. https://doi.org/10.1021/acsenergylett.6b00247.

14. Lv, Z., Ma, W., Wang, M., Dang, J., Jian, K., Liu, D., Huang, D., 2021. Co-constructing interfaces of multiheterostructure on MXene (Ti3C2Tx)-modified 3D self-supporting electrode for ultraefficient electrocatalytic HER in alkaline media. Adv. Funct. Mater., 31(29), 2102576. https://doi.org/10.1002/adfm.202102576.

15. Zhang, Z., Li, H., Zou, G., Fernandez, C., Liu, B., Zhang, Q., Hu, J., Peng, Q., 2016. Self-reduction synthesis of new MXene/Ag composites with unexpected electrocatalytic activity. ACS Sustain. Chem. Eng., 4(12), 6763–6771. https://doi.org/10.1021/acssuschemeng.6b01698.

16. Rasheed, P. A., Pandey, R. P., Rasool, K., Mahmoud, K. A., 2018. Ultra-sensitive electrocatalytic detection of bromate in drinking water based on Nafion/$Ti_3C_2T_x$ (MXene) modified glassy carbon electrode. Sens. Actuators B: Chem., 265, 652–659. https://doi.org/10.1016/j.snb.2018.03.103.

17. Liang, X., Ren, X., Yang, Q., Gao, L., Gao, M., Yang, Y., Zhu, H., Li, G., Ma, T., Liu, A., 2021. A two-dimensional MXene-supported metal-organic framework for highly selective ambient electrocatalytic nitrogen reduction. Nanoscale, 13, 2843–2848. https://doi.org/10.1039/D0NR08744K.

18. Li, L., Yu, D., Li, P., Huang, H., Xie, D., Lin, C. C., Hu, F., Chen, H. Y., Peng, S., 2021. Interfacial electronic coupling of ultrathin transition-metal hydroxide nanosheets with layered MXenes as a new prototype for platinum-like hydrogen evolution. Energy Environ. Sci., 14, 6419–6427. https://doi.org/10.1039/D1EE02538D.

19. Nam, S., Mahato, M., Matthews, K., Lord, R. W., Lee, Y., Thangasamy, P., Ahn, C. W., Gogotsi, Y., Oh, I. K., 2023. Bimetal organic framework—$Ti_3C_2T_x$ MXene with metalloporphyrin electrocatalyst for lithium—Oxygen batteries. Adv. Funct. Mater., 33(1), 2210702. https://doi.org/10.1002/adfm.202210702.

20. Gu, J., Zhao, Z., Huang, J., Sumpter, B. G., Chen, Z., 2021. MX anti-MXenes from non-Van der Waals bulks for electrochemical applications: The merit of metallicity and active basal plane. ACS Nano, 15(4), 6233–6242. https://doi.org/10.1021/acsnano.0c08429.

21. Li, N., Zhang, Y., Jia, M., Lv, X., Li, X., Li, R., Ding, X., Zheng, Y. Z., Tao, X., 2019. 1T/2H $MoSe_2$-on-MXene heterostructure as bifunctional electrocatalyst for efficient overall water splitting. Electrochimica Acta., 326, 134976. https://doi.org/10.1016/j.electacta.2019.134976.

22. Wang, J., Feng, T., Chen, J., He, J. H., Fang, X., 2022. Flexible 2D cu metal: Organic framework@MXene film electrode with excellent durability for highly selective Electrocatalytic NH3 synthesis. Research. https://doi.org/10.34133/2022/9837012.

23. Tran, M. H., Schäfer, T., Shahraei, A., Dürrschnabel, M., Molina-Luna, L., Kramm, U. I., Birkel, C. S., 2018. Adding a new member to the MXene family: Synthesis, structure, and electrocatalytic activity for the hydrogen evolution reaction of V4C3T x. ACS Appl. Energy Mater., 1(8), 3908–3914. https://doi.org/10.1021/acsaem.8b00652.

24. Elancheziyan, M., Eswaran, M., Shuck, C. E., Senthilkumar, S., Elumalai, S., Dhanusuraman, R., Ponnusamy, V. K., 2021. Facile synthesis of polyaniline/titanium carbide (MXene) nanosheets/palladium nanocomposite for efficient electrocatalytic oxidation of methanol for fuel cell application. Fuel, 303, 121329. https://doi.org/10.1016/j.fuel.2021.121329.

25. Liu, J., Chen, T., Juan, P., Peng, W., Li, Y., Zhang, F., Fan, X., 2018. Hierarchical cobalt borate/MXenes hybrid with extraordinary electrocatalytic performance in oxygen evolution reaction. ChemSusChem, 11(21), 3758–3765. https://doi.org/10.1002/cssc.201802098.

26. Ghorbanzadeh, S., Hosseini, S. A., Alishahi, M., 2022. $CuCo_2O_4/Ti_3C_2T_x$ MXene hybrid electrocatalysts for oxygen evolution reaction of water splitting. J. Alloys Compd, 920, 165811. https://doi.org/10.1016/j.jallcom.2022.165811.

27. Kuang, P., He, M., Zhu, B., Yu, J., Fan, K., Jaroniec, M., 2019. 0D/2D NiS2/V-MXene composite for electrocatalytic H_2 evolution. J. Catal., 375, 8–20. https://doi.org/10.1016/j.jcat.2019.05.019.

28. Huang, L., Ai, L., Wang, M., Jiang, J., Wang, S., 2019. Hierarchical MoS_2 nanosheets integrated Ti_3C_2 MXenes for electrocatalytic hydrogen evolution. Int. J. Hydrogen Energy, 44(2), 965–976. https://doi.org/10.1016/j.ijhydene.2018.11.084.

29. Abdullah, N., Saidur, R., Zainoodin, A. M., Aslfattahi, N., 2020. Optimization of electrocatalyst performance of platinum—Ruthenium induced with MXene by response surface methodology for clean energy application. J. Clean. Prod., 277, 123395. https://doi.org/10.1016/j.jclepro.2020.123395.

30. Tie, L., Li, N., Yu, C., Liu, Y., Yang, S., Chen, H., Dong, S., Sun, J., Dou, S., Sun, J., 2019. Self-supported nonprecious MXene/Ni$_3$S$_2$ electrocatalysts for efficient hydrogen generation in alkaline media. ACS Appl. Energy Mater., 2(9), 6931–6938. https://doi.org/10.1021/acsaem.9b01529.

31. Fang, Y., Liu, Z., Han, J., Jin, Z., Han, Y., Wang, F., Niu, Y., Wu, Y., Xu, Y., 2019. High-performance electrocatalytic conversion of N2 to NH3 using oxygen-vacancy-rich TiO2 in situ grown on Ti3C2Tx MXene. Adv. Energy Mater., 9(16), 1803406. https://doi.org/10.1002/aenm.201803406.

32. Tang, Y., Yang, C., Sheng, M., Yin, X., Que, W., 2020. Synergistically coupling phosphorus-doped molybdenum carbide with Mxene as a highly efficient and stable electrocatalyst for hydrogen evolution reaction. ACS Sustain. Chem. Eng., 8(34), 12990–12998. https://doi.org/10.1021/acssuschemeng.0c03840.

33. Lim, K. R. G., Handoko, A. D., Johnson, L. R., Meng, X., Lin, M., Subramanian, G. S., Anasori, B., Gogotsi, Y., Vojvodic, A., Seh, Z. W., 2020. 2H-Mos$_2$ on Mo$_2$CT$_x$Mxene nanohybrid for efficient and durable electrocatalytic hydrogen evolution. ACS Nano, 14(11), 16140–16155. https://doi.org/10.1021/acsnano.0c08671.

34. Li, S., Tuo, P., Xie, J., Zhang, X., Xu, J., Bao, J., Pan, B., Xie, Y., 2018. Ultrathin MXene nanosheets with rich fluorine termination groups realizing efficient electro-catalytic hydrogen evolution. Nano Energy, 47, 512–518. https://doi.org/10.1016/j.nanoen.2018.03.022.

35. Chandran, M., Raveendran, A., Vinoba, M., Vijayan, B. K., Bhagiyalakshmi, M., 2021. Nickel-decorated MoS$_2$/MXene nanosheets composites for electrocatalytic oxidation of methanol. Ceram. Int., 47(19), 26847–26855. https://doi.org/10.1016/j.ceramint.2021.06.093.

36. Hao, L., He, H., Xu, C., Zhang, M., Feng, H., Yang, L., Jiang, Q., Huang, H., 2022. Ultrafine cobalt selenide nanowires tangled with MXene nanosheets as highly efficient electrocatalysts toward the hydrogen evolution reaction. Dalton Trans., 51(18), 7135–7141. https://doi.org/10.1039/D2DT00238H.

37. Patra, A., Samal, R., Rout, C. S., 2022. Promising water splitting applications of synergistically assembled robust orthorhombic CoSe$_2$ and 2D Ti$_3$C$_2$T$_x$MXene hybrid. Catal. Today. https://doi.org/10.1016/j.cattod.2022.07.021.

14 MXenes for Supercapacitors
Current Status and Future Perspectives

*Amarendra Nayak, Alka Priyadarshini,
and Swarna P. Mantry*

14.1 INTRODUCTION

MXenes, a class of two-dimensional (2D) layered transition metal carbides, nitrides, or carbonitrides discovered by Gogotsi et al., have recently attracted much study interest owing to their extraordinary physical and chemical characteristics [1, 2]. MXenes are produced by etching one or many layers from a corresponding parent precursor known as a MAX ($M_{n+1}AX_n$) phase. For instance, the first reported MXene, Ti_3C_2, was derived by selectively etching the monoatomic Al (A layer) from the MAX phase (Ti_3AlC_2) using HF as an etchant [3, 4]. The MAX phase is layered ternary carbides and nitrides, where M = transition metal (Ti, V, Nb, and Ta, etc.); A = group 13 or 14 elements such as Al, Si; X = C and/or N; and n = 1, 2, 3 [5]. To date, depending upon the value of n, three different types of MAX phases have been identified: M_2AX (n = 1), M_3AX_2 (n = 2), and M_4AX_3 (n = 3) as shown in Figure 14.1 [6]. In the MAX phase, the M–X bond exhibits a combination of ionic, metallic, and covalent characteristics, whereas the M–A bond typically has a metallic character. Therefore, unlike graphite and other transition metal dichalcogenides (TMDs) where weak van der Waals forces hold the layers together, mechanical exfoliation or shearing alone is insufficient to break the strong interlayer bonding between the layers of MXenes [7].

The general formula of a MXene is $M_{n+1}X_nT_x$, arranged alternately in n+1 layers of M and n layers of X with ample surface termination groups (denoted as T_x), such as -OH, -Cl, -F, and -O as shown in Figure 14.2 [8]. These surface-functionalized groups generated due to the chemical etching process provide hydrophilicity to the surface and control electronic and ion transport properties of the corresponding MXene [9, 10]. More than 30 MXenes have been synthesized, and more than 100 structural compositions have been predicted by choosing different M, A, and X atoms, both theoretically and experimentally [11]. The exclusive chemistry associated with 'M' renders MXenes exceptionally suitable for various applications [12, 13]. Similarly, the properties of MXenes can be tuned by changing the 'X'. For instance, transition metal nitrides display higher surface activity than carbides, whereas the latter exhibit a substantial

DOI: 10.1201/9781003366225-14

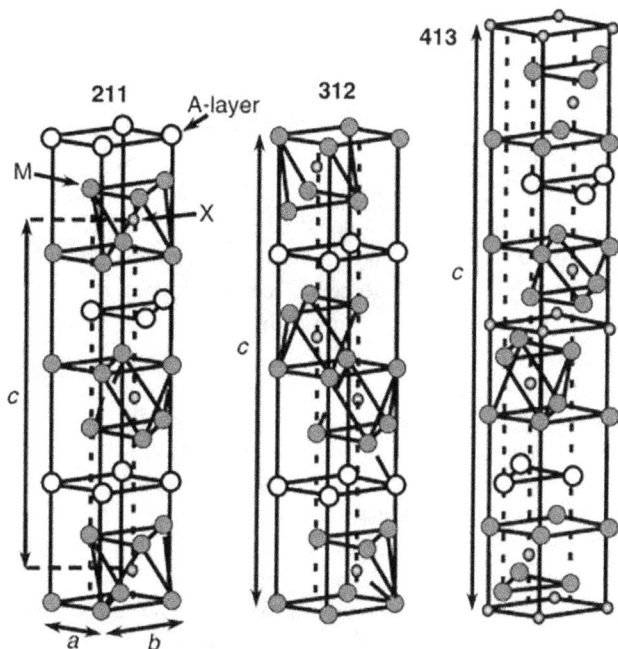

FIGURE 14.1 MAX phase unit cell structures of 211 (n = 1), 312 (n = 2), and 413 (n = 3).
Reproduced with permission from [6]. Copyright 2017, Elsevier.

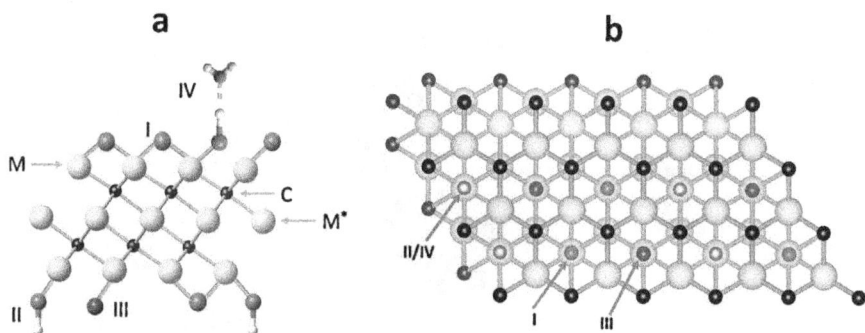

FIGURE 14.2 (a) The side and (b) top view of MXenes representing different surface termination groups; I, II, III, and IV refer to M atoms bonded to O, OH, F, and H_2O surface-functionalized groups, respectively. M* is the M atom that is only bonded to the C atom.
Reproduced with permission from [8]. Copyright 2016, Elsevier.

lattice constant and stability [14]. MXenes have emerged as a potential candidate for supercapacitors (SCs) owing to their distinctive stacking structure, intrinsic electrical conductivity, higher pseudocapacitance, tunable surface, hydrophilicity, and ample opportunity for intercalation [15, 16].

14.2 SYNTHESIS

Generally, MXene synthesis is completed in three stages: (i) preparation of Ti_3AlC_2 MAX phase precursor, (ii) etching the 'A' layer, i.e., Al, from the MAX phase, and (iii) intercalation and exfoliation of the layered structure. Generally, high-temperature sintering and ball milling methods are adopted to prepare the parent MAX phase precursor by altering the M and A sites [17]. Currently, mostly Al-containing MAX phases have been used for MXenes synthesis. However, MAX phases containing different metals, viz. Cu, Mn, and Zn, have been explored, enriching the chemical diversity of MXenes [18–20]. Recently, certain intriguing MAX phases with different chemical ordering, out-of-plane (o-MAX) and in-plane (i-MAX) orders, have been reported [21].

The transformation of a MXene from the MAX phase is usually done by etching the Al (A layer) from the MAX phase. Selective etching is possible because of the different behaviors of the M–X and M–A bonds towards the etchant. The first prepared MXene, $Ti_3C_2T_x$, was etched from its parent MAX phase, i.e., Ti_3AlC_2, by etching the Al layer by using 50% concentrated HF solution [3]. To date, HF etching methods have been extensively used as a common synthetic technique to produce various MXenes with an accordion-like morphology, as shown in Figure 14.3 [22]. However, the efficacy of selective etching of MAX into MXene strongly depends on the chemistry and crystal structure of the parent MAX phase. For instance, 10 wt.% of HF successfully etched the Ti_2AlC but not the Cr_2AlC. On the other hand, Ti_2AlC and Cr_2AlC are completely dissolved in 50 wt.% of HF [23].

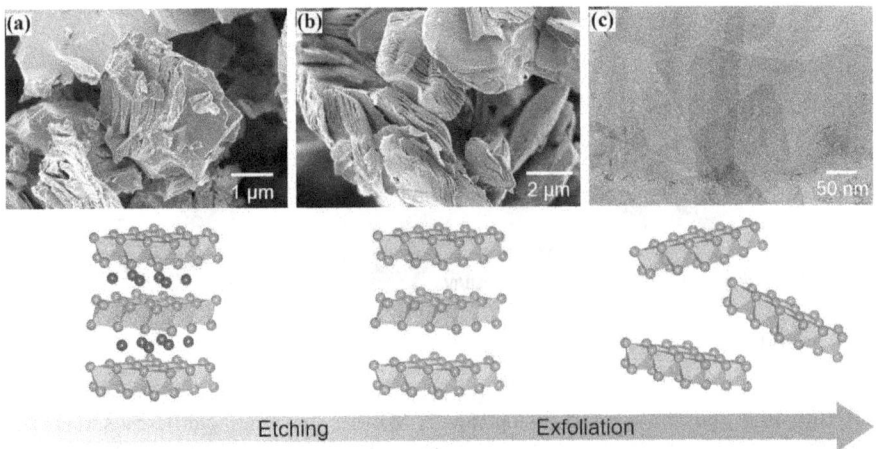

FIGURE 14.3 Pathway to obtaining 2D MXene single flakes by top-down synthesis: first, selective etching of atomic layer(s) from a 3D layered precursor is required; then, exfoliation into single flakes. (a) SEM micrograph of Ti_3AlC_2 particles. (b) SEM micrograph of $Ti_3C_2T_z$ multilayers after etching of the Al layer. (c) TEM micrograph of overlapping $Ti_3C_2T_z$ single layers. Schematics of the MAX-to-multilayer MXene transformation and exfoliation are shown in the bottom row.

Though HF etching techniques have been extensively used as common synthetic techniques to produce various MXenes, the toxicity and hazardous nature associated with the F-containing groups lead to other environmentally friendly synthetic methods. In this context, a mixed solution of HCl and LiF and combinations of acids (H_2SO_4) with F-containing salts (NaF, KF, CaF_2, FeF_3, CsF) have been explored as etchants [24].

Mechanical delamination and exfoliation are two techniques that have been used to obtain single-layer MXenes from as-synthesized multi-layered MXenes. The yield of monolayer MXenes by mechanical delamination is relatively low owing to the high interlayer interactions in multi-layered MXenes. Therefore, intercalation is the most suitable technique to obtain monolayer MXenes. In this method, various intercalants are introduced in the layers of MXene, which increase the interlayer space, thereby weakening the interlayer bonding [3, 25]. Then, the intercalated MXenes undergo sonication or other kinds of agitation techniques to produce single-layer MXenes.

14.3 PROPERTIES

14.3.1 MECHANICAL PROPERTIES

The surface terminations of MXenes have a significant impact on their mechanical properties. The O-terminated MXenes show very high stiffness as compared to F- and OH-terminated MXenes [26]. This may be because the F- or OH-terminated MXenes have larger lattice parameters than O-terminated MXenes [27]. Moreover, compared with pristine MXenes, the surface-functionalized MXenes show more flexibility. Studies suggest that the higher mechanical stability of MXenes is attributed to the electron density created on the MAX surface after the removal of the Al layer [28]. Theoretical studies have revealed that the charge transfer from inner Ti–C bonds to surface Ti–O bonds creates higher stability in O-functionalized MXenes [29].

The mechanical properties of MXenes depend on the number of atomic layers (n). Borysiuk et al. demonstrated that with a lower n value, the strength and hardness of $M_{n+1}X_nT_x$ gradually rise for the functionalized MXenes [30]. It has been observed that the mechanical properties of MXenes are improved by making composites with carbon nanotubes and polymers. The toughness, flexibility, and tensile and compressive strengths of MXenes can be increased by combining different types of polymers to different degrees. For example, a $Ti_3C_2T_x$–polyvinyl alcohol (PVA) composite exhibits superior flexibility and higher tensile and compressive strengths than $Ti_3C_2T_x$ [31]. The critical strain, in-plane stiffness, and Young's modulus can be improved by doping of boron and vanadium at Ti and C sites of Ti_2C [32]. The nitride-based MXenes show superior mechanical properties in terms of a higher Young's modulus and fracture stress than that of carbide-based MXenes [33].

14.3.2 ELECTRONIC PROPERTIES

MXenes display electronic properties owing to their compositional diversity, flexible thickness, and surface functionality [34]. For instance, Ti_3C_2 shows a metallic character with a band gap of 0.05 eV, whereas the F- or OH-functionalized counterparts are semiconductors in nature with a bandgap of 0.1 eV [35]. The band gap of a MXene can also be tuned by varying the elemental composition of the 'M' layer. For example, the replacement of two Ti layers in metallic $Ti_3C_2(OH)_2$ by two Mo layers produced semiconductor $Mo_2Ti-C_2(OH)_2$ with a bandgap of 0.05 eV [36]. Moreover, strain and the external electric field also play a pivotal role in tuning the band gap of MXenes. It has been reported that the bandgap decreases with an increase in strain and approximately 2% critical tensile strain can transit indirect bandgap to direct bandgap [37].

The electrical conductivity of MXenes depends upon the preparation method. The single $Ti_3C_2T_x$ flakes exhibit high field-effect electron mobility of 2.6–0.7 cm^2 V^{-1} s^{-1} and excellent conductivity of up to 4600–1100 S cm^{-1} [38]. The spin-cast and vacuum-filtered $Ti_3C_2T_x$ films possess high electrical conductivity, reaching 6500 S cm^{-1} and 4600 S cm^{-1}, respectively.

14.3.3 CHEMICAL STABILITY

The chemical stability of MXenes depends upon the synthesis condition, elemental composition, and colloidal solution of delaminated MXenes. MXenes prepared with concentrated HF in H_2O degrade more easily than those prepared at a mild concentration [39]. In a recent study, Mathis et al. reported that a MXene ($Ti_3C_2T_x$) etched from the Al-rich MAX phase displayed not only high electronic conductivity (20,000 S/cm) but also high oxidation resistance properties [40].

The production of MXene-based films or coating electrodes relies on the colloidal solutions of delaminated MXene flakes rather than powder form. Therefore, it is highly essential to study the stability of MXene suspensions. Recent studies reported that the decay of MXene ($Ti_3C_2T_x$) solutions is accelerated by dissolved oxygen present in water and high temperature [41]. The proneness of MXenes towards oxidation at ambient conditions is due to the adsorption of oxygen atoms by the under-coordinated Ti atoms of the $Ti_3C_2T_x$ MXene [42].

Therefore, it has been suggested to store the MXenes in an Ar-filled container at a lower temperature due to their susceptibility to oxidation in open air at room temperature. However, the addition of antioxidants, i.e., sodium L-ascorbate, to colloidal MXene solutions prevents oxidation [43]. The stability against oxidation may be credited to the association of L-ascorbate anions to the edges of the MXene nanosheet, the area prone to oxidation. In another study, Zhang et al. demonstrated that freezing MXenes can be a useful technique to prevent the same from oxidation. The authors demonstrate that freezing obstructs the formation of TiO_2 nanoparticles on the edges, thereby retaining the morphologies and elemental composition of the pristine MXene [44]. MXenes are well dispersed in various organic solvents such as ethanol, DMSO, DMF, etc. The solvents not only prepare a stable colloidal suspension but also deactivate the oxidation process [45].

14.3.4 MAGNETIC PROPERTIES

Although most 2D materials are non-magnetic, pristine MXenes such as Ti_2C, Cr_2C, Ti_2N, Mn_2C, and Cr_2N are magnetic; among them, Mn_2C and Cr_2N are antiferromagnetic, whereas Cr_2C, Ti_2C, and Ti_2N are ferromagnetic [46–48]. The magnetic properties of MXenes are due to defects in the monolayer, surface functionalization, and the intrinsic properties of the transition metal (M layer). A spin-polarized DFT investigation revealed that the MXenes are non-magnetic due to their strong covalent bond between transition metal, X, and surface functional groups [49]. However, the magnetic properties can be tuned by applying external strain to [27p]. It has been reported that biaxial tensile strain and compressive strain induce magnetism in V_2C and V_2N MXenes [47].

The magnetic properties of pristine MXenes can be tuned by surface functionalization that endows delocalized d-electrons. For example, Cr_2C functionalized with H, Cl, OH, and F groups displays ferromagnetic behavior from a semi-metal to an antiferromagnetic semiconductor with a large band gap [47, 49]. On the other hand, Cr_2N, which is antiferromagnetic, changes to a ferromagnetic semi-metal after surface functionalization. Studies on Mn-based MXenes revealed the correlation between magnetism with the electronegativity of surface functional groups. MXenes having surface functional groups of charge −1 (F, OH, Cl) exhibit ferromagnetism, whereas functionalization other than −1 (O^{2-}, H^+) shows antiferromagnetism [50]. A recent study reported that the doping of TMs (Cr, Mn, Ti, V) to a MXene (SC_2CT_2) changes its non-magnetic nature to magnetic. Experimental findings demonstrate that the doping of Mn and Cr creates a magnetic moment of 2.1 µB per unit cell due to electron redistribution in the system [51].

14.3.5 Optical Properties

The band structure of transition metals and the electronic transitions between them affect the optical characteristics of MXenes. A MXene ($Ti_3C_2T_x$) exhibited remarkable optical transmittance of 77% at a wavelength of 550 nm [52]. However, transmittance of 90% has been reported by an NH_4HF_2 intercalated MXene, demonstrating the relationship of transmittance with thickness and intercalation [53]. In a study, Lashgari et al. employed the full-potential linearized augmented plane wave method (FP-LAPW) to calculate the reflectivity, absorption, energy loss function, and dielectric function [54]. The authors reported that the MXene displays 100% reflectivity when the applied electric field is along the X-direction and parallel to the surface, whereas it decreases to less than 50% when the field is along the Z-direction [54].

The first principle DFT study done by Berdiyorov et al. revealed that a MXene functionalized with F/OH shows less absorption and reflectivity than one functionalized with O in the visible range, making it a suitable candidate for transparent electrode applications [55]. Moreover, the dielectric properties of functionalized MXenes are two times less than those of the non-functionalized ones. MXenes possess different properties of non-linear absorption, 100% conversion efficiency from light to heat, and high conductivity and absorption in the UV–visible range, making them suitable candidates for optical switching devices, transparent electrodes, photovoltaics, and biomedical applications [56].

14.4 APPLICATIONS OF MXENES FOR SUPERCAPACITORS

14.4.1 MXENES

MXenes have been extensively used in supercapacitors owing to their excellent metallic conductivity, interlayer structure, surface chemistry, unique morphologies, and high hydrophilicity. The symmetric supercapacitor (SSC) device designed by Xie et al. using an equal mass of $Ti_3C_2T_x$ MXene in seawater electrolyte exhibited a volumetric energy density of 1.74×10^{-3} W h cm^{-3} at a power density of 1.53 W cm^{-3} and capacity retention of 93.3% after 5000 cycles, signifying better electrochemical performance than that of the reported same systems [57]. In another study, Zhu et al. investigated the role of etching conditions in terms of Ti_2AlC:HF mass ratio on the electrochemical performance of MXene electrodes. The authors reported that Ti_2CT_x (mass ratio Ti_2AlC:HF is 1:2) exhibits a higher specific capacitance and better stability than $Ti_3C_2T_x$. The Ti_2CT_x/Ti_2CT_x SSC device displayed 100% capacity retention after 3000 cycles at a current density of 20 A g^{-1} [58].

Wu et al. assembled $Ti_3C_2T_x$ microgels with separate $Ti_3C_2T_x$ nanosheets to fabricate a flexible, freestanding, and high-performance electrode with a tunable porous structure [59]. The electrode displays excellent volumetric capacitance of 736 F cm^{-3} at an ultrahigh scan rate of 2000 mV s^{-1}. The $Ti_3C_2T_x/Ti_3C_2T_x$-assembled SSC displayed excellent cycling stability in terms of high capacitance retention of 91.14% at 1000 mV s^{-1} after 20,000 charge–discharge cycles, delivering a high energy density of 40 Wh L^{-1} and 21 Wh L^{-1} at a power density of 0.83 kW L^{-1} and 41.5 kW L^{-1}, respectively, which is the highest reported for an SSC in an aqueous electrolyte.

14.4.2 MXENE-BASED COMPOSITES

While MXenes demonstrate great utility in various electrochemical energy storage devices, poorer mechanical properties and restacking obstruct their further development. These limitations of MXenes can be overcome by forming composites with different materials such as graphene, polymers, and transition metal oxide [15]. The excellent properties of these materials such as high stability, faster redox kinetics, and high conductivity significantly enhance the electrochemical performance of pristine MXenes, motivating researchers to fabricate flexible, portable, and wearable electrochemical devices.

14.4.2.1 MXene/Carbon Composites

Zhu et al. synthesized a ribbon-shaped, highly oriented stacked structure by wet spinning a Ti_3C_2 MXene colloid in a protonated chitosan coagulation bath. The chitosan was further removed by H_2SO_4 [60]. The asymmetric supercapacitor (ASC) was designed by assembling PVA/H_2SO_4 gel electrolyte-coated Ti_3C_2 and reduced graphene oxide (rGO) as the cathode and anode, respectively. The ASC displayed a larger volumetric power density of 7466.0 mW cm^{-3} at 2583.0 W kg^{-1}, an improved volumetric capacitance of 256 F/cm^3 at 5 mV s^{-1}, a volumetric energy density of 58.4 mWh cm^{-3} at 20.2 Wh kg^{-1}, and a capacitance retention of 92.4% over 10,000 cycles at a current density of 10 A/g. In a recent study, Kumar et al.

enhanced the supercapacitor properties by adopting a two-step method of pristine $Ti_3C_2T_x$ MXene. First, they grew graphene on Ni foil by the chemical vapor deposition (CVD) method and used the same as the current collector, and second, they used nanolayered $Ti_3C_2T_x$ MXene structures as the electrode. The $Ti_3C_2T_x$ MXene/graphene/Ni supercapacitor exhibited a high specific capacitance of ~542 F/g at a scan rate of 5 mV s^{-1} [61].

A MXene and graphene spontaneously gelled when exposed to zinc, creating a 3D hydrogel hybrid (MGH) composite with increased resistance to surface oxidation, displaying a higher gravimetric capacitance of 357 F g^{-1} at 10 mV s^{-1} with cyclic stability of 95.6% after 10,000 charge–discharge cycles [62]. The ASC developed using MGH and a hybrid hydrogel of polyaniline–graphene (PGH) as the cathode and anode, respectively, showed a power density of 1.13 kW kg^{-1}, energy density of 30.3 Wh kg^{-1}, and cycling stability of 81% over 10,000 cycles at 10 A g^{-1}. The improved supercapacitor performance of the composite electrode was attributed to the porous structure of the hydrogel, which provides faster ion diffusion within the electrode.

Wang et al. synthesized MXene/CNT composite electrodes by using a layer-by-layer technique [63]. The flexible, compact, and solid-state micro-supercapacitor (mSC) electrode on a paper chip based on MXene/CNT showed a specific capacitance 2.47 times higher than that of pristine $Ti_3C_2T_x$ at a current density of 0.5 mA cm^{-2}. The introduction of CNT into the interlayer of MXene averted restacking, thereby enhancing the cycling stability, specific capacitance, and flexibility.

14.4.2.2 MXene/Polymer Composites

The composites of MXene with polyaniline (PANI) have been synthesized by chemically polymerizing aniline monomers on MXene nanosheets with the help of an oxidant under acidic conditions [64]. The availability of different functional groups such as -OH, -O, and -F on the MXene surface provides ample nucleation sites on the MXene surface for aniline deposition. An ammonium persulfate solution was used as an oxidant and slowly added to aniline to initiate the polymerization process. Porous PANI was formed on the MXene surface, which provided a channel for ion transport resulting in significant improvement in electrochemical activity [65].

Wang et al. fabricated PANI/MXene film electrodes by incorporating PANI (~10 nm) in between the thin layers of MXene [66]. In this study, PANI/MXene ink was prepared by mixing a MXene suspension and PANI nanoparticles dissolved in NMP (N-methyl pyrrolidone) solvent, followed by stirring and filtration. Then, using the blade coating method, the PANI/MXene was used to fabricate a self-supporting PANI/MXene film (Figure 14.4). In this process, MXene nanosheets provide integrated functions for conducting, binding, dispersing, and providing a flexible substrate for PANI nanoparticles. PANI nanoparticles help to reduce the restacking of MXenes and enable ion electron transfer. A fully pseudocapacitive asymmetric supercapacitor (ASC) device was designed using MXene and PANI/MXene as negative and positive electrodes, respectively (Figure 14.5a). As shown in Figure 14.5b, the broad peaks in the 0–1.4 V potential range indicate that the capacitance predominantly originated from the redox pseudocapacitance, as corroborated by the galvanostatic charge/discharge (GCD) profile Figure (14.5c). The well-retained shape of the CV curves even at 100 mV s^{-1} signifies faster ion diffusion and good electrical

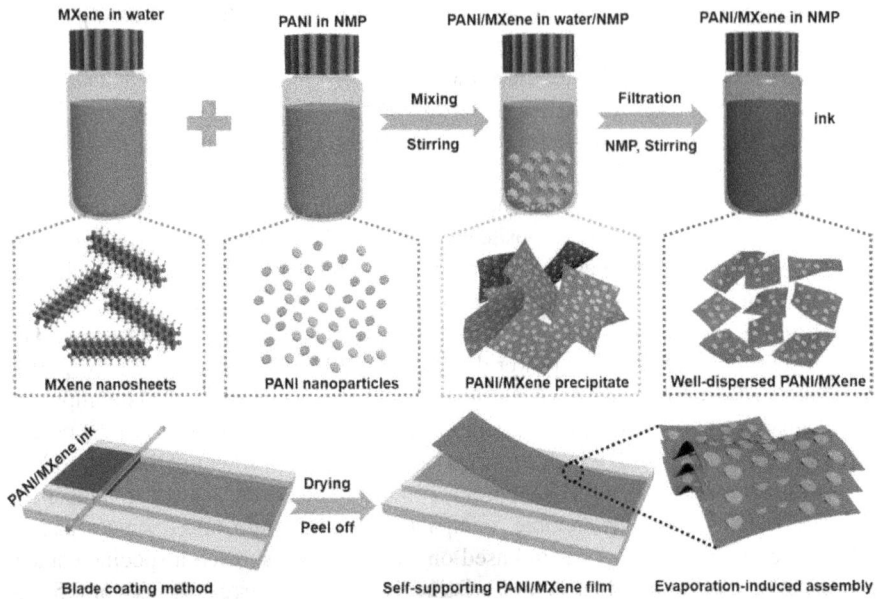

FIGURE 14.4 Schematic presentation of the fabrication process of PANI/MXene inks and self-supporting PANI/MXene flexible films.

Reproduced with permission from [66]. Copyright 2021, Elsevier.

conductivity (Figure 14.5d). The ASC device displays a volumetric capacitance of 231.4F cm^{-3} (82.6F g^{-1}) at 10 mV s^{-1} (Figure 14.5e) and maximum energy density of 65.6 Wh L^{-1} at power density 1687.3 W L^{-1}(Figure 14.5f). As shown in Figure 14.5g, the ASC device displays excellent mechanical stability. As shown in Figure 14.5h, the device exhibits good cycling stability in terms of 87.5% capacitance retention after 5000 charge–discharge cycles at 20 mA cm^{-2}. The actual application demonstration of the device that is shown in the inset of Figure 14.6h can control a red light-emitting diode when two asymmetric devices are linked in series [66].

Beidaghi et al. employed an oxidant-free in situ polymerization method to synthesize the MXene/PANI composite by adding aniline monomer to the MXenes nanosheets. The authors reported that the electrochemical performance of the MXene/PANI electrode strongly depends on the amount of polymer added to MXene sheets [65]. The MXene/PANI electrode with the lowest PANI deposition showed a mass capacitance of 503 F g^{-1} and a volume capacitance of 1682.3 F cm^{-3}.

Wu et al. fabricated organ-like amino–Ti$_3$C$_2$ (N-Ti$_3$C$_2$)/PANI composites via a two-step electrochemical process [67]. The N-Ti$_3$C$_2$ was coated on an FTO substrate by an electrochemical reaction. PANI was polymerized for different durations (240 s, 300 s, 420 s, 600 s) under a constant voltage of 1 V using N-Ti$_3$C$_2$ on FTO as the working electrode, Pt mesh as the counter electrode, and aniline added to 1M H$_2$SO$_4$ (Figure 14.6).

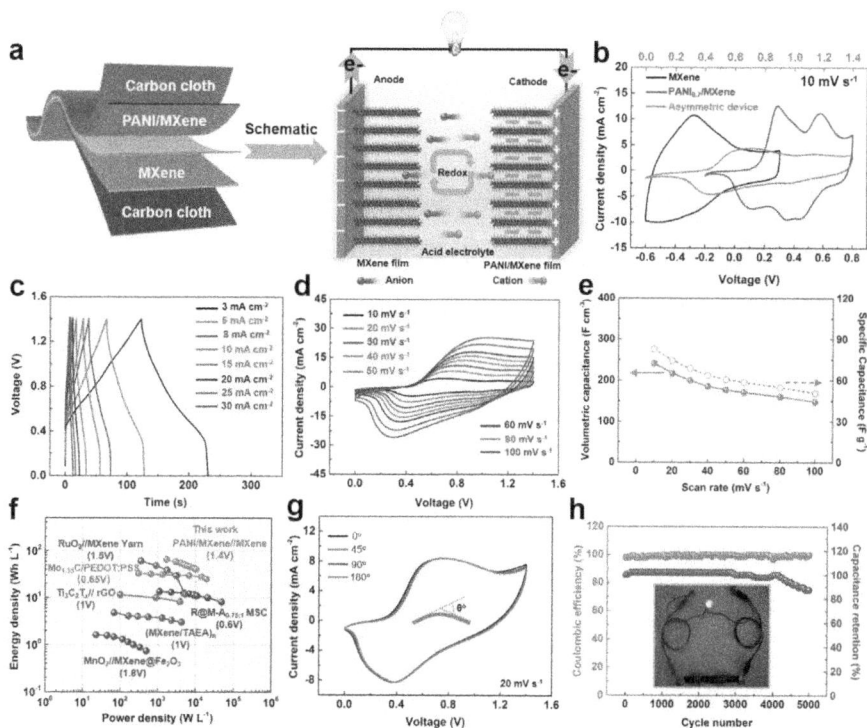

FIGURE 14.5 (a) Schematic representation of a designed all-pseudocapacitive asymmetric device in which PANI$_{0.7}$/MXene and MXene electrodes serve as the cathode and anode, respectively; (b) cyclic voltammogram (CV) curves of PANI/MXene electrode (1 mg cm^{-2}), MXene electrode (1.8 mg cm^{-2}), and the asymmetric supercapacitor (ASC) in 1 M H$_2$SO$_4$ aqueous electrolyte at a scan rate of 10 mV s^{-1}; (c) GCD profiles of the ASC device at various current densities; (d) CV curves of the ASC device at various scan rates; (e) gravimetric and volumetric specific capacitance of the ASC device; (f) Ragone plots displaying energy and power densities of the PANI/Ti$_3$C$_2$T$_x$//Ti$_3$C$_2$T$_x$ ASC in comparison to other state-of-the-art MXene-based supercapacitors; (g) CV curves of the ASC device under different bending states tested at 20 mV s^{-1}; (h) cycling stability and Coulombic efficiency of the aqueous ASC at a current density of 20 mA cm^{-2} for 5000 cycles. Inset shows that two PANI/Ti$_3$C$_2$T$_x$//Ti$_3$C$_2$T$_x$ ASCs connected in series can power up a red light-emitting diode under the discharge state.

As shown in Figure 14.7a, N-Ti$_3$C$_2$/PANI composite electrodes exhibit a greater specific capacitance and higher electrochemical activity than Ti$_3$C$_2$ electrodes. However, the specific capacitance decreases with an increase in scan rate, suggesting that part of the electrode surface is inaccessible at high charge–discharge rates (Figure 14.7b). The GCD profile (Figure 14.7c) of the electrodes demonstrates the reversible nature of an ideal capacitor. As shown in Figure 14.7d, the capacitance

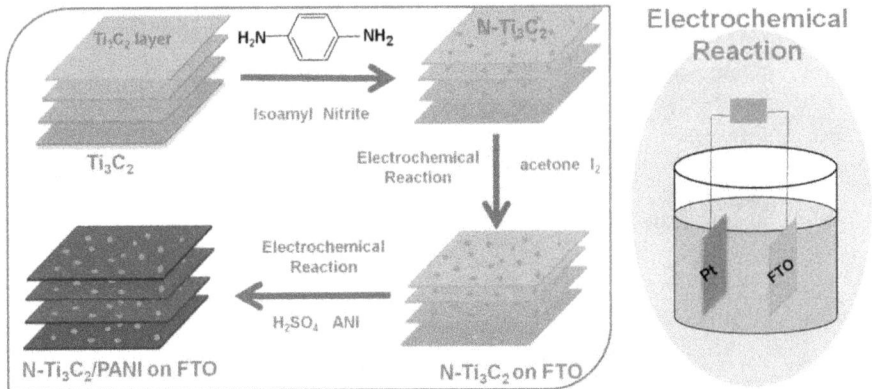

FIGURE 14.6   Schematic presentation for the fabrication of N-Ti$_3$C$_2$/PANI.

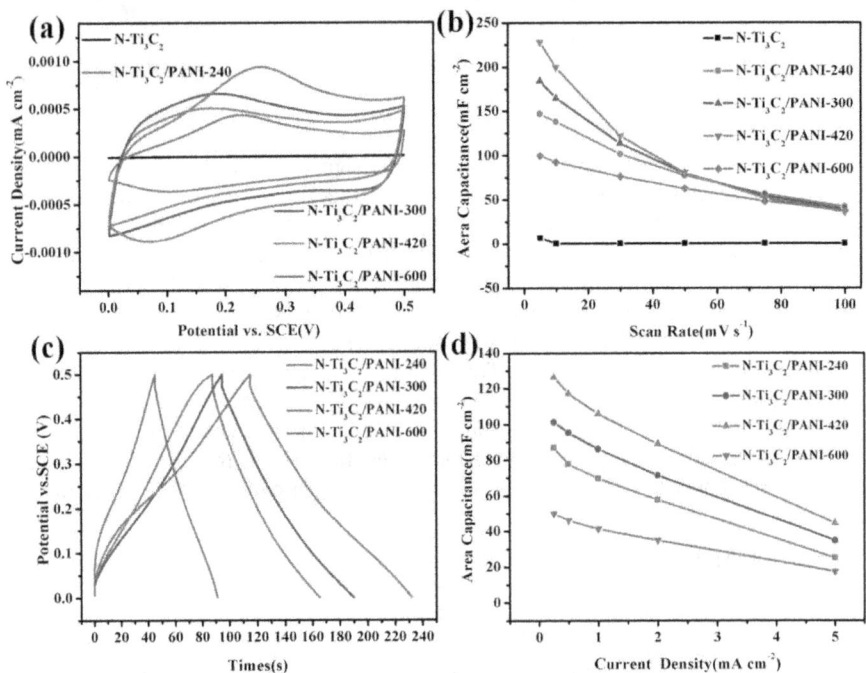

FIGURE 14.7   The N-Ti$_3$C$_2$ and N-Ti$_3$C$_2$/PANI electrodes of (a) CV curves at a scan rate of 5 mV s^{-1} and (b) specific capacitance at different scan rates from 5 to 100 mV s^{-1}. The N-Ti$_3$C$_2$/PANI electrodes of (c) typical GCD curves at a current density of 0.5 mA cm^{-2} and (d) specific capacitance at a different current density from 0.25 to 5 mA cm^{-2}.

decreases when the current density increases. The significant electrochemical performance of the electrodes rather than pristine Ti_3C_2 is attributed to the insertion of PANI chains into MXene layers, thereby avoiding restacking.

In another study, a Ti_3C_2/PANI nanotube (Ti_3C_2/PANI NTs) composite electrode fabricated by Wu et al. demonstrated a specific capacitance of 596 F g^{-1} and a retention rate of 94.7% after 5000 cycles [68]. The improved electrochemical performance of the Ti_3C_2/PANI NTs composite electrode was attributed to the introduction of PANI nanoparticles into the MXene layers, which not only prevents the restacking of MXene layers but also increases the space between the layers, providing sufficient surface area for ion transport. Moreover, the SSC assembled by Ti_3C_2/PANI NTs showed an energy density of 25.6 Wh kg^{-1} at a power density of 153.2 W kg^{-1} [68]. Zhou et al. adopted in situ non-oxidative polymerization and vacuum-assisted filtration method to synthesize i-PANI@$Ti_3C_2T_x$, Lig@$Ti_3C_2T_x$, and Lig@$Ti_3C_2T_x$/i-PANI@$Ti_3C_2T_x$ films [69]. The SSCs based on i-PANI@$Ti_3C_2T_x$, Lig@$Ti_3C_2T_x$, and Lig@$Ti_3C_2T_x$/i-PANI@ $Ti_3C_2T_x$ showed a specific capacitance of 310 F g^{-1} (~1001 F cm^{-3}), 271 F g^{-1} (~881 F cm^{-3}), and 295 F g^{-1} (~959 F cm^{-3}) at 1 A g^{-1}, respectively, in a PVA/H_2SO_4 electrolyte. In addition, the energy densities of these SSCs were 34.8 Wh L^{-1} (i-PANI@$Ti_3C_2T_x$//i-PANI@$Ti_3C_2T_x$), 30.6 Wh L^{-1} (Lig@$Ti_3C_2T_x$//Lig@ $Ti_3C_2T_x$), and 33.3 Wh L^{-1} (Lig@$Ti_3C_2T_x$/i-PANI@$Ti_3C_2T_x$//Lig@$Ti_3C_2T_x$/i-PANI@ $Ti_3C_2T_x$) at a power density of 1625 W L^{-1} [69].

14.4.2.3 MXene/PPy Composites

Incessant charging–discharging cycles lead to structural pulverization and capacitance degradation of MXene-based electrodes. However, this issue is alleviated by inserting polypyrrole (PPy) in the MXene [70, 71]. Moreover, the intercalation of PPy in MXene layers increases the interlayer distance, thereby enhancing the electrochemical performance.

Jian et al. employed a step co-electrodeposition method to fabricate MXene/PPy composite electrodes on ITO glass substrate for supercapacitor applications [72]. In this method, the MXene acts as the core of the polymerization due to various functional groups, and the PPy monomer gradually polymerizes in the MXene nanosheets, forming a carambola-like structure. The MXene/PPy electrode demonstrates a mass capacitance of 416 F g^{-1} at a current density of 0.5 A g^{-1} in 1 M H_2SO_4 electrolyte. The SSC device prepared by assembling two MXene/PPy-covered ITO glass electrodes displays significant capacitance performance in terms of low resistance and high reversibility (Figure 14.8a, b). As shown in Figure 14.8ci, cii, the device shows capacitances of 184 F g^{-1} and 160 F g^{-1} at a scan rate of 10 mV s^{-1} and a current density of 0.5 A g^{-1}, respectively. The ASC has capacitance retentions of 56.2% (Figure 14.8ci) and 76.3% (Figure 14.8cii). The device displays approximately 86.4% retention (Figure 14.8d), indicating an outstanding cycling performance [72].

Wei et al. demonstrate a simplistic one-step in situ strategy to synthesize a polypyrrole/Ti_3C_2 nanocomposite (PPy/Ti_3C_2-S2) electrode material for supercapacitor application. The PPy/Ti_3C_2 heterostructure electrode shows higher supercapacitor performance compared to pristine Ti_3C_2 in terms of high specific capacitance of 458 F g^{-1} (at a scan rate of 2 mV s^{-1}) and substantially lower intrinsic resistance (Rs) and

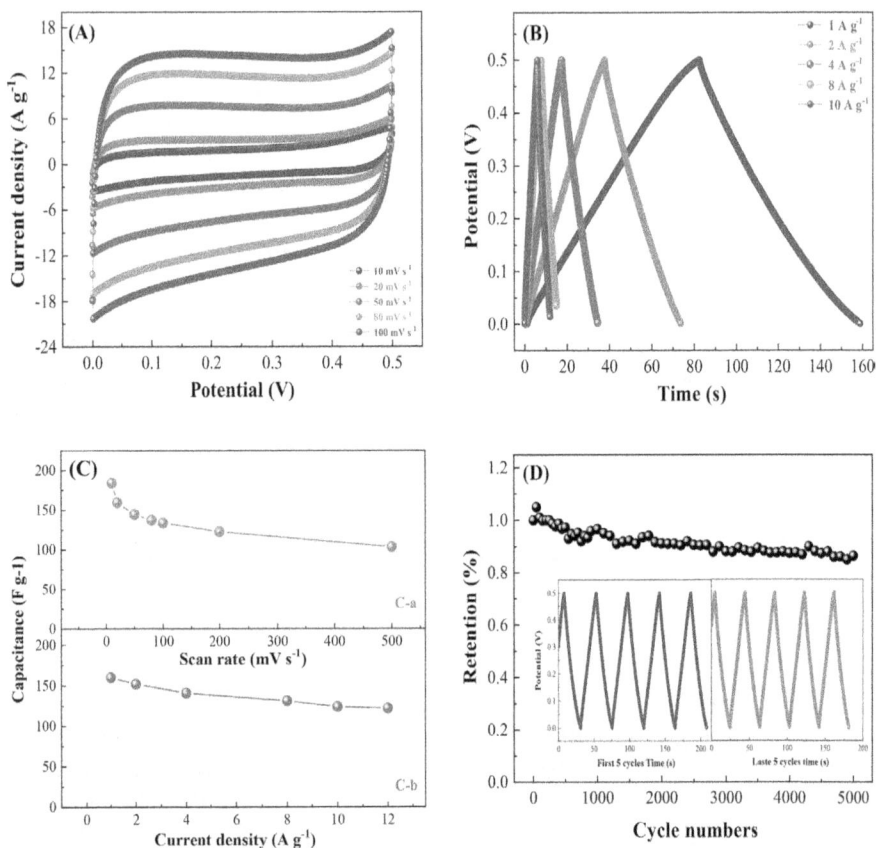

FIGURE 14.8 Electrochemical performance of the MXene/PPy symmetric supercapacitor device. (a) CV curves at various scan rates (10, 20, 50, 80, and 100 mV s^{-1}); (b) GCD curves at different current densities (1, 2, 4, 8, and 10 A g^{-1}); (c) capacitance of the MXene/PPy symmetric supercapacitor device at various scan rates (ci) and at different current densities (cii); (d) cycle stability of the device at a current density of 5 A g^{-1}, inset shows the GCD curves of the first and last five cycles.

charge transfer resistance (Rct). Moreover, the PPy/Ti$_3$C$_2$-S2 SSC attained a high energy density of 21.61 Wh kg^{-1} at the power density of 499.94 W kg^{-1} and capacitance retention of 73.68% at 1 A g^{-1} after 4000 charge and discharge cycles [73].

Liang et al. fabricated an ASC by using MXene/PPy (negative electrode) and PPy multiwalled carbon nanotubes (positive electrode) operating in a larger potential range in a 0.5 M Na$_2$SO$_4$ electrolyte [74]. The MXene/PPy electrode was prepared by in situ polymerization of PPy on the MXene surface. The pyrocatechol violet (PCV) used in the synthesis of the electrode acted as a dispersant for MXene particles and a dopant for PPy polymerization. The MXene/PPy electrode exhibited a capacitance of 2.11 F cm^{-2}, which was more than the reported results for the same.

The significant electrochemical performance of $Ti_3C_2T_x$/PPy as the negative electrode was attributed to the addition of PCV, which facilitates the charge transfer. Moreover, the ASC device displayed a total capacitance of 1.37 F cm^{-2} and 1.18 F cm^{-2} from CV and GCD data, respectively [74].

In a recent study, Fan et al. synthesized $Ti_3C_2T_x$ functionalized with PPy and ionic liquids (ILs) for high-performance SC applications [75]. The authors first used PPy and ILs as dual spacers, which prevented aggregation and restacking and provided a larger area for ion diffusion. The PPy-MXene-IL-mic ([EMIm][NTf2]) electrode displayed a high gravimetric capacitance of 51.85 F g^{-1} at a scan rate of 20 mV s^{-1}. The SSC device possessed a maximum gravimetric energy density of 31.2 Wh kg^{-1} (at 1030.4 W kg^{-1}) at room temperature and 91% retention of the initial specific capacitance after 2000 cycles, while Coulombic efficiency was 91%.

14.4.2.4 MXene/PEDOT Composites

The MXene/PEDOT composites display better charge storage capacity, electrochemical performance, and cyclic stability than their counterparts. Qin et al. prepared a high-performance solid-state supercapacitor electrode based on MXene poly(3,4-ethylene dioxythiophene):poly(styrenesulfonic acid) (PEDOT:PSS) by a hydrothermal and vacuum-assisted filtration process [76]. The electrode showed excellent electrochemical performance in terms of a high energy density of 33.2 mWh cm^{-3}, power density of 19470 mW cm^{-3}, and capacitance of 568 F cm^{-3}. The authors credited the layered structure to the confined alignment of PEDOT nanofiber between $M_{1.33}C$ layers. Inal et al. adopted electrochemical polymerization and co-doping techniques to fabricate PEDOT:PSS:MXene films [77]. It has been observed that the dual doping of PSS and MXene in PEDOT significantly improved its electrochemical performance and stability in comparison to the single doping of either PSS or MXene. The PEDOT:PSS:MXene exhibited high volumetric capacitance (607 ± 85.3 F cm^{-3}) and 78% capacity retention after 500 cycles, which was higher than the PEDOT:PSS (195.6 ± 1 F cm^{-3}, 37%) and PEDOT:MXene (358.9 ± 16.7 F cm^{-3}, 58%). The study demonstrated co-doping as an effective method to enhance electrochemical performance [77].

14.5 CONCLUSION AND OUTLOOK

Although much work has been done to improve the performance of MXene-based supercapacitors, research is still needed to fabricate transparent, flexible, small-sized, and low-cost devices with improved electrochemical performance and mechanical stability. A plethora of literature is available on Ti-based MXenes; however, other transition metal-based, double-M, and double-X MXenes have been less explored in supercapacitor applications. The exploration of these types of MXenes may deliver improved electrochemical performance towards supercapacitor applications.

The corrosive and hard acids such as HF and concentrated HCl are used in the etching process of the MAX phase for MXenes, and the morphology, size, and surface functional groups of the MAX phase are strongly dependent upon the etching technique, thus demanding a more convenient etching technique for the mass production of MXenes. Meanwhile, precise control over the preparation technique and polymerization time is highly essential to stop the agglomeration of conducting polymers.

The faster oxidation of MXenes in the presence of ambient air and water is another challenge to overcome. Although novel synthesis techniques, low temperature, antioxidants, and controlled storage conditions can alleviate the oxidation, they cannot completely prevent the degradation. Recently, Wu et al. employed an end-group protection technique to prevent the oxidation of MXene by functionalizing the edges of the MXene nanosheets with antioxidants [59]. Nevertheless, very few works have been reported in this regard.

MXenes have diversity in structure, morphology, composition, and properties. Nevertheless, some properties of MXenes, e.g., magnetic and mechanical properties, are obtained by only theoretical simulations and not corroborated by experimental evidence. In the near future, it will be possible to explore the unknown properties of MXene composites and their counterparts by combining computational simulation validated by experimental evidence. From a commercial perspective, it is highly essential to get an explicit understanding of the synthesis, properties, and energy storage mechanism if MXenes are ever to be used in the fabrication of flexible, stable, and robust supercapacitors, which is only possible by bringing both computational and experimental research together.

REFERENCES

1. Babak Anasori, and Yury Gogotsi. "2D metal carbides and nitrides (MXenes) for energy storage" Nat. Rev. Mater. 2, (2017): 16098.
2. Michel. W. Barsoum. "The MAX phases unique new carbide and nitride materials". Am. Sci. 89(4), (2001): 334–343.
3. Patrick Urbankowski, Babak Anasori, Taron Makaryan, Dequan Er, Sankalp Kota, Patrick L. Walsh, Mengqiang Zhao, Vivek B. Shenoy, Michel W. Barsoum, and Yury Gogotsi. "Synthesis of two-dimensional titanium nitride Ti_4N_3 (MXene)". Nanoscale. 22(8), (2016): 11385–11391.
4. Zheng Ming Sun, "Progress in research and development on MAX phases: A family of layered ternary compounds". Int. Mater. Rev. 56, (2013): 143–166.
5. Hans Högberg, Lars Hultman, Jens Emmerlich, Torbjörn Joelsson, Per Eklund, Jon M. Molina-Aldareguia, Jens P. Palmquistc, Ola Wilhelmsson, and Ulf Jansson. "Growth and characterization of MAX-phase thin films". Surf. Coat. Technol. 193, (2005): 6–10.
6. Martin Magnuson, and Maurizio Mattesini. "Chemical bonding and electronic-structure in MAX phases as viewed by X-ray spectroscopy and density functional theory". Thin Solid Films. 621, (2017): 108–130.
7. Zhimei Sun, Denis Music, Rajeev Ahuja, Sa Li, and Jochen M. Schneider. "Bonding and classification of nanolayered ternary carbides". Phys. Rev. B. 70, (2004): 092102–092103.
8. Joseph Halim, Kevin M. Cook, Michael Naguib, Per Eklund, Yury Gogotsi, Johanna Rosen, and Michel W. Barsoum. "X-ray photoelectron spectroscopy of select multi-layered transition metal carbides (MXenes)". Appl. Surf. Sci. 362, (2016): 406–417.
9. Jiahe Peng, Xingzhu Chen, Wee-Jun Ong, Xiujian Zhao, and Neng Li. "Surface and heterointerface engineering of 2D MXenes and their nanocomposites: Insights into electro and photocatalysis". Chem. 5 (2019): 18–50.

10. Ankita Sinha, Dhanjai, Huimin Zhao, Yujin Huang, Xianbou Lu, Jiping Chen, and Rajeev Jain. "MXene: An emerging material for sensing and biosensing". TrAC, Trends Anal. Chem. 105, (2018): 424–435.

11. Zhimou Liu, L. Zheng, L. Sun, Y. Qian, J. Wang, and M. Li. "$(Cr_{2/3}Ti_{1/3})_3$ AlC$_2$ and $(Cr_{5/8}Ti_{3/8})_4$ AlC$_3$: New MAX-phase compounds in Ti—Cr—Al—C system". J. Am. Ceram. Soc. 97, (2014): 67–69.

12. Nuala. M. Caffrey. "Effect of mixed surface terminations on the structural and electrochemical properties of two-dimensional $Ti_3C_2T_2$ and V_2CT_2 MXenes multilayers". Nanoscale. 10, (2018): 13520.

13. Mohammad Khazaei, Masao Arai, Taizo Sasaki, Mehdi Estili, and Yoshio Sakka. "Two-dimensional molybdenum carbides: Potential thermoelectric materials of the MXene family". Phys. Chem. Chem. Phys. 14: (2014), 7841–7849.

14. Ning Zhang, Yu Hong, Sanaz Yazdanparast, and Mohsen Asle Zaeem. "Superior structural, elastic and electronic properties of 2D titanium nitride MXenes over carbide MXenes: A comprehensive first principles study". 2D Mater. 5, (2018): 045004.

15. Subhasree Panda, Kalim Deshmukh, S.K. Khadheer Pasha, Jayaraman Theerthagiri, Sivakumar Manickam, and Myong Yong Choi. "MXene based emerging materials for supercapacitor applications: Recent advances, challenges, and future perspectives". Coord. Chem. Rev. 462, (2022): 214518–214568.

16. Jian Yang, Weizhai Bao, Pauline Jaumaux, Songtao Zhang, Chengyin Wang, and Guoxiu Wang. "MXene-based composites: Synthesis and applications in rechargeable batteries and supercapacitors". Adv. Mater. Interfaces. 6, (2019), 1802004.

17. Xu Zhang, Zihe Zhang, and Zhen Zhou. "MXene-based materials for electrochemical energy storage". J. Energy Chem. 27, (2018), 73.

18. Mustapha Nechiche, Thierry Cabioc'h, Elad N. Caspi, Oleg Rivin, Andreas Hoser, Veronique Gauthier-Brunet, Patrick Chartier, and Sylvain Dubois. "Evidence for symmetry reduction in $Ti_3(Al_{1-\delta}Cu_\delta)C_2$ MAX phase solid solutions". Inorg. Chem. 56, (2017): 14388–14395.

19. Chung-Chuan Lai, Quanzheng Tao, Hossein Fashandi, Ulf Wiedwald, Ruslan Salikhov, Michael Farle, Andrejs Petruhins, Jun Lu, Lars Hultman, Per Eklund, and Johanna Rosen. "Magnetic properties and structural characterization of layered $(Cr_{0.5}Mn_{0.5})_2AuC$ synthesized by thermally induced substitutional reaction in $(Cr_{0.5}Mn_{0.5})_2GaC$". APL Mater. 6(2), (2018): 026104.

20. Mian Li, Jun Lu, Kan Luo, Youbing Li, Keke Chang, Ke Chen, Johanna Rosen, Lars Hultman, Per Eklund, Per O. Å. Persson, Shiyu Du, Zhifang Chai, Zhengren Huang, and Qing Huang. "Element replacement approach by reaction with Lewis acidic molten salts to synthesize nanolaminated MAX phases and MXenes". J Am. Chem. Soc. 141(11), (2019): 4730–4737.

21. El'ad N. Caspi, Patrick Chartier, Florence Porcher, Françoise Damay, and Thierry Cabioc'h. "Ordering of (Cr, V) layers in nanolamellar $(Cr_{0.5}V_{0.5})_{n+1}AlC_n$ Compounds". Mater. Res. Lett. 3(2), (2015): 100–106.

22. Louisiane Verger, Chuan Xu, Varun Natu, Hui-Ming Cheng, Wencai Ren, and Michel W Barsoum. "Overview of the synthesis of MXenes and other ultrathin 2D transition metal carbides and nitrides". Curr. Opin. Solid State Mater. Sci. 23(3), (2019): 149–163.

23. Michael Naguib, Olha Mashtalir, Joshua Carle, Volker Presser, Jun Lu, Lars Hultman, Yury Gogotsi, and Michel W. Barsoum. "Two-dimensional transition metal carbides". ACS Nano. 6(2), (2012): 1322–1331.

24. Fanfan Liu, Jie Zhou, Shuwei Wang, Bingxin Wang, Cai Shen, Libo Wang, Qianku Hu, Qing Huang, and J. Aiguo Zhou. "Preparation of high-purity V$_2$C MXene and electrochemical properties as Li-ion batteries". Electrochem. Soc. 164, (2017): A709–A713.

25. Maria R. Lukatskaya, Olha Mashtalir, Chang E. Ren, Yohan Dall'agnese, Patrick Rozier, Pierre L. Taberna, Michael Naguib, Patrice Simon, Michel W. Barsoum, and Yury Gogotsi. "Cation intercalation and high volumetric capacitance of two-dimensional titanium carbide". Science. 341, (2013): 1502–1505.

26. Yuelei Bai, Kun Zhou, Narasimalu Srikanth, John H. Pang, Xiaodong He, and Rongguo Wang. "Dependence of elastic and optical properties on surface terminated groups in two-dimensional MXene monolayers: A first-principles study". RSC Adv. 6, (2016): 35731–35739.

27. Xian-Hu Zha, Kan Luo, Qiuwu Li, Qing Huang, Jian He, Xiaodong Wen, and Shiyu Du. " Role of the surface effect on the structural, electronic and mechanical properties of the carbide MXenes". Europhys. Lett. 111, (2015): 26007.

28. Michael Naguib, Vadym N. Mochalin, Michel W. Barsoum, and Yury Gogotsi. "25th anniversary article: MXenes: A new family of two-dimensional materials". Adv. Mater. 26, (2014): 992–1005.

29. Zheng Ling, Chang E. Ren, Meng-Qiang Zhao, Jian Yang, James M. Giammarco, Jieshan Qiu, Michel W. Barsoum, and Yury Gogotsi. "Flexible and conductive MXene films and nanocomposites with high capacitance". Proc. Natl. Acad. Sci. 111, (2014): 16676–16681.

30. Vadym N. Borysiuk, Vadym N. Mochalin, and Yury Gogotsi. "Molecular dynamic study of the mechanical properties of two-dimensional titanium carbides $Ti_{n+1}C_n$ (MXenes)". Nanotechnology. 26, (2015): 265705.

31. Michael Naguib, Tomonori Saito, Sophia Lai, Matthew S. Rager, Tolga Aytug, M. Parans Paranthaman, Meng Q. Zhao, and Yury Gogotsi. "$Ti_3C_2T_x$ (MXene)—Polyacrylamide nanocomposite films". RSC Adv. 6, (2016): 72069–72073.

32. Poulami Chakraborty, Tilak Das, Dhani Nafday, Lilia Boeri, and Tanusri Saha-Dasgupta "Manipulating the mechanical properties of Ti_2C MXene: Effect of substitutional doping". Phys. Rev. B. 95, (2017): 184106.

33. S. Milad Hatam-Lee, Ali Esfandiar, and Ali Rajabpour. "Mechanical behaviors of titanium nitride and carbide MXenes: A molecular dynamics study". Appl. Surf. Sci. 566, (2021): 150633.

34. Yuanyue Liu, Hai Xiao, and William A. Goddard. "Schottky-barrier-free contacts with two-dimensional semiconductors by surface-engineered MXenes". J. Am. Chem. Soc. 138, (2016): 15853–15856.

35. Michael Naguib, Murat Kurtoglu, Volker Presser, Jun Lu, Junjie Niu, Min Heon, Lars Hultman, Yury Gogotsi, and Michel W. Barsoum. "Two-dimensional nanocrystals produced by exfoliation of Ti_3AlC_2". Adv. Mater. 23, (2011): 4248–4253.

36. Babak Anasori, Chenyang Shi, Eun Ju Moon, Yu Xie, Cooper A. Voigt, Paul R. C. Kent, Steven J. May, Simon J. L. Billinge, Michel W. Barsoum, and Yury Gogotsi. "Control of electronic properties of 2D carbides (MXenes) by manipulating their transition metal layers". Nanoscale Horiz. 1, (2016): 227–234.

37. Youngbin Lee, Sung B. Cho, and Yong-Chae Chung. "Tunable indirect to direct band gap transition of monolayer Sc_2CO_2 by the strain effect". ACS Appl. Mater. Interfaces. 6, (2014): 14724–14728.

38. Alexey Lipatov, Mohamed Alhabeb, Maria R. Lukatskaya, Alex Boson, and Yury Gogotsi. "Effect of synthesis on quality, electronic properties and environmental stability of individual monolayer Ti_3C_2 MXene flakes". Adv. Electron. Mater. 2, (2016): 1600255.

39. Pooja Srivastava, Avanish Mishra, Hiroshi Mizuseki, Kwang-Ryeol Lee, and Abhishek K. Singh. "Mechanistic insight into the chemical exfoliation and functionalization of Ti_3C_2 MXene". ACS Appl. Mater. Interfaces. 8, (2016): 24256–24264.

40. Tyler S. Mathis, Kathleen Maleski, Adam Goad, Asia Sarycheva, Mark Anayee, Alexandre C. Foucher, Kanit Hantanasirisakul, Christopher E. Shuck, Eric A. Stach, and Yury Gogotsi. "Modified MAX phase synthesis for environmentally stable and highly conductive Ti_3C_2 MXene". ACS Nano. 15, (2021): 6420–6429.

41. Chuanfang John Zhang, Sergio Pinilla, Niall McEvoy, Conor P. Cullen, Babak Anasori, Edmund Long, Sang-Hoon Park, Andres Seral-Ascaso, Aleksey Shmeliov, Dileep Krishnan, Carmen Morant, Xinhua Liu, Georg S. Duesberg, Yury Gogotsi, and Valeria Nicolosi. "Oxidation stability of colloidal two-dimensional titanium carbides (MXenes)". Chem. Mater. 29, (2017): 4848–4856.

42. Estephania Lira, Stefan Wendt, Peipei Huo, Jonas Ø. Hansen, Regine Streber, Soren Porsgaard, Yinying Wei, Ralf Bechstein, Erik Lægsgaard, and Flemming Besenbacher. "The importance of bulk Ti^{3+} defects in the oxygen chemistry on titania surfaces". J. Am. Chem. Soc. 133, (2011): 6529–6532.

43. Xiaofei Zhao, Aniruddh Vashisth, Evan Prehn, Wanmei Sun, Smit A. Shah, Touseef Habib, Yexiao Chen, Zeyi Tan, Jodie L. Lutkenhaus, Miladin Radovic, and Micah J. Green, "Antioxidants unlock shelf-stable $Ti_3C_2T_x$ (MXene) nanosheet dispersions". Matter. 1, (2019): 513–526.

44. Roghayyeh Lotfi, Michael Naguib, Dundar E. Yilmaz, Jagjit Nanda, and Adri C.T. Van Duin. "A comparative study on the oxidation of two-dimensional Ti_3C_2 MXene structures in different environments". J. Mater. Chem. A. 6, (2018): 12733–12743.

45. Kathleen Maleski, Vadym N. Mochalin, and Yury Gogotsi. "Dispersions of two-dimensional titanium carbide MXene in organic solvents". Chem. Mater. 29, (2017): 1632–1640.

46. Guoying Gao, Guangqian Ding, Jie Li, K Kailun Yao, Menghao Wu, and Meichun Qian. "Monolayer MXenes: Promising half-metals and spin gapless semiconductors". Nanoscale. 8, (2016): 8986–8994.

47. Chen Si, Jian Zhou, and Zhimei Sun. "Half-metallic ferromagnetism and surface functionalization-induced metal—Insulator transition in graphene-like two-dimensional Cr_2C crystals". ACS Appl. Mater. Interfaces. 7, (2015): 17510–17515.

48. Guo Wang. "Theoretical prediction of the intrinsic half-metallicity in surface-oxygen-passivated Cr_2N MXene". J. Phys. Chem. C. 120, (2016): 18850–18857.

49. Lin Hu, Xiaojun Wu, and Jinlong Yang. "Mn_2C monolayer: A 2D antiferromagnetic metal with high Néel temperature and large spin—Orbit coupling". Nanoscale. 8, (2016): 12939–12945.

50. Junjie He, Pengbo Lyu, and Petr Nachtigall. "New two-dimensional Mn-based MXenes with room-temperature ferromagnetism and half-metallicity". J. Mater. Chem. C. 4, (2016): 11143–11149.

51. Jianhue Yang, Xuepiao Luo, Shaozheng Zhang, and Liang Chen. "Investigation of magnetic and electronic properties of transition metal doped Sc_2CT_2 (T=O, OH or F) using a first principles study". Phys. Chem. Chem. Phys. 18, (2016): 12914–12919.

52. Marina Mariano, Olha Mashtalir, Francisco Q. Antonio, Won-Hee Ryu, Bingchen Deng, Fengnian Xia, Yury Gogotsi, and André D. Taylor. "Solution-processed titanium carbide MXene films examined as highly transparent conductors". Nanoscale. 8, (2016): 16371–16378.

53. Joseph Halim, Maria R. Lukatskaya, Kevin M. Cook, Jun Lu, Cole R. Smith, Lars-Å. Näslund, Steven J. May, L. Hultman, Yury Gogotsi, Per Eklund, and Michel W. Barsoum. "Transparent conductive two-dimensionaln titanium carbide epitaxial thin films". Chem. Mater. 26, (2014): 2374–2381.

54. Hamed Lashgari, Mohammadreza Abolhassani, Arash Boochani, Seyedmohammad Elahi, and Jabbar Khodadadi. "Electronic and optical properties of 2D graphene-like

compounds titanium carbides and nitrides: DFT calculations'. Solid State Commun. 195, (2014): 61–69.

55. Golibjon R. Berdiyorov. "Optical properties of functionalized $Ti_3C_2T_2$(T= F, O, OH) MXene: First-principles calculations'. AIP Adv. 6, (2016) 55105.

56. Kanit Hantanasirisakul, and Yury Gogotsi. "Electronic and optical properties of 2D transition metal carbides and nitrides (MXenes)". Adv. Mater. 30, (2018): 1804779.

57. Qi Xia, Nanasaheb M. Shinde, Tengfei Zhang, Je M. Yun, Aiguo Zhou, Rajaram S. Mane, Sanjay Mathur, and Kwang H. Kim. "Seawater electrolyte-mediated high volumetric MXene-based electrochemical symmetric supercapacitors". Dalton Trans. 47, (2018): 8676–8682.

58. Kai Zhu, Yuming Jin, Fei Du, Shuang Gao, Zhongmin Gao, Xing Meng, Gang Chen, Yingjin Wei, and Yu Gao. "Synthesis of Ti_2CT_x MXene as electrode materials for symmetric supercapacitor with capable volumetric capacitance". J. Energy Chem. 31, (2019): 11–18.

59. Zhitan Wu, Xiaochen Liu, Tongxin Shang, Yaqian Deng, Ning Wang, Ximan Dong, Juan Zhao, Derong Chen, Ying Tao, and Quan-Hong Yang. "Reassembly of MXene hydrogels into flexible films towards compact and ultrafast supercapacitors". Adv. Funct. Mater. 31, (2021): 2102874–2102882.

60. Chao Zhu, and Fengxia Geng. "Macroscopic MXene ribbon with the oriented sheet stacking for high-performance flexible supercapacitors". Carbon Energy. 3, (2021): 142–152.

61. Sunil Kumar, Malik A. Rehman, Sungwon Lee, Minwook Kim, Hyeryeon Hong, Jun-Young Park, and Yongho Seo. "Supercapacitors based on $Ti_3C_2T_x$ MXene extracted from supernatant and current collectors passivated by CVD-graphene". Sci. Rep. 11, (2021): 1–9.

62. Anirban Sikdar, Pronoy Dutta, Sujit K. Deb, Abhisek Majumdar, Narayanan Padma, Subhradip Ghosh, Uday N. Maiti. "Spontaneous three-dimensional self-assembly of MXene and graphene for impressive energy and rate performance pseudocapacitors". Electrochim. Acta. 391, (2021): 138959–138970.

63. Ruochong Wang, Shaohong Luo, Chen Xiao, Zhenyu Chen, Houshen Li, Muhammad Asif, Vincent Chan, Kin Liao, and Yimin Sun. "MXene-carbon nanotubes layer-by-layer assembly based on-chip microsupercapacitor with improved capacitive performance". Electrochim. Acta. 386, (2021): 138420.

64. Huizhong Xu, Dehua Zheng, Faqian Liu, Wei Li, and Jianjian Lin. "Synthesis of an MXene/polyaniline composite with excellent electrochemical properties". J. Mater. Chem. A. 8, (2020): 5853–5858.

65. Armin VahidMohammadi, Jorge Moncada, Hengze Chen, Emre Kayali, Jafar Orangi, Carlos A. Carrero, and Majid Beidaghi. "Thick and freestanding MXene/PANI pseudocapacitive electrodes with ultrahigh specific capacitance". J. Mater. Chem. A. 6, (2018): 22123–22133.

66. Yuanming Wang, Xue Wang, Xiaolong Li, Yang Bai, Huanhao Xiao, Yang Liu, and Guohui Yuan. "Scalable fabrication of polyaniline nanodots decorated MXene film electrodes enabled by viscous functional inks for high-energy-density asymmetric supercapacitors". Chem. Eng. J. 405, (2021): 126664.

67. Wenling Wu, Dongjuan Niu, Jianfeng Zhu, Yan Gao, Dan Wei, Xiaohua Liu, Fen Wang, Lei Wang, and Liuqing Yang. "Organ-like Ti_3C_2 Mxenes/polyaniline composites by chemical grafting as high-performance supercapacitors". J. Electroanal. Chem. 847, (2019): 113203.

68. Wenling Wu, Chengwei Wang, Chunhui Zhao, Dan Wei, Jianfeng Zhu, and Youlong Xu. "Facile strategy of hollow polyaniline nanotubes supported on Ti_3C_2-MXene nanosheets for high performance symmetric supercapacitors'. J. Colloid Interface Sci. 580, (2020): 601–613.

69. Yang Zhou, Yubo Zou, Zhiyuan Peng, Chuying Yu, and Wenbin Zhong. "Arbitrary deformable and high strength electroactive polymer/MXene anti-exfoliative composite films assembled into high performance, flexible all-solid-state supercapacitors'. Nanoscale. 12, (2020): 20797–20810.

70. Muhammad Boota, Babak Anasori, Cooper Voigt, Meng-Qiang Zhao, Michel W. Barsoum, and Yury Gogotsi. "Pseudocapacitive electrodes produced by oxidant-free polymerization of pyrrole between the layers of 2D titanium carbide (MXene)". Adv. Mater. 28, (2016): 1517–1522.

71. Chao Zhang, Shuaikai Xu, Dong Cai, Junming Cao, Lili Wang, and Wei Han. "Planar supercapacitor with high areal capacitance based on Ti_3C_2/polypyrrole composite film". Electrochim. Acta. 330, (2020): 135277.

72. Xuan Jian, Min He, Lu Chen, Mi-mi Zhang, Rui Li, Lou-jun Gao, Feng Fu, and Zhen-hai Liang. . "Three-dimensional carambola-like MXene/polypyrrole composite produced by one-step co-electrodeposition method for electrochemical energy storage". Electrochim. Acta. 318, (2019): 820–827.

73. Dan Wei, Wenling Wu, Jianfeng Zhu, Chengwei Wang, Chunhui Zhao, and Lei Wang. "A facile strategy of polypyrrole nanospheres grown on Ti_3C_2-MXene nanosheets as advanced supercapacitor electrodes". J. Electroanal. Chem. 877, (2020): 114538.

74. Wenyu Liang, and Igor Zhitomirsky. "MXene-polypyrrole electrodes for asymmetric supercapacitors". Electrochim. Acta. 406, (2022): 139843.

75. Qi Fan, Ruizheng Zhao, Mengjiao Yi, Ping Qi, Chunxiao Chai, Hao Ying, and Jingcheng Hao. "Ti_3C_2-MXene composite films functionalized with polypyrrole and ionic liquid-based microemulsion particles for supercapacitor applications". Chem. Eng. J. 428, (2022): 131107.

76. Leiqiang Qin, Quanzheng Tao, Ahmed El Ghazaly, Julia Fernandez-Rodriguez, Per O.Å. Persson, Johanna Rosen, and Feling Zhang. "High-performance ultrathin flexible solid-state supercapacitors based on solution processable Mo1.33C MXene and PEDOT:PSS". Adv. Funct. Mater. 28, (2018): 1703808.

77. Shofarul Wustoni, Abdulelah Saleh, Jehad K. El-Demellawi, Anil Koklu, Adel Hama, Victor Druet, Nimer Wehbe, Yizhou Zhang, and Sahika Inal. "MXene improves the stability and electrochemical performance of electropolymerized PEDOT films". APL Mater. 8, (2020), 121105.

78. Chein-Wei Wu, Binesh Unnikrishnan, I-Wen P. Chen, Scott G. Harroun, Huan T. Chang, and Chai C. Huang, "Oxidation resistive MXene aqueous ink for micro-supercapacitor application". Energy Storage Mater. 25, (2020): 563–571.

15 MXene-Based Materials for Photothermal Therapy

Banendu Sunder Dash, Suprava Das, and Jyh-Ping Chen

15.1 INTRODUCTION

One of the most promising techniques for cancer treatment is laser photothermal ablation aided by nanoparticles. A substantial amount of research has been conducted in this sector, taking into account the different critical elements for the evaluation of a protocol that ensures the efficacy of therapy, safety, and appropriate tumor targeting all at the same time. The successful modification of the graphene family has driven the attention of various research groups towards thickness and lateral two-dimensional (2D) materials due to planar structure and various physicochemical properties, tunable surface functionalization ability, better nanocarrier ability, and a high surface area, making the research more progressive in recent past. Due to good biocompatibility and water dispersion, these materials show a more convincible impact in cancer nanomedicine, specifically photothermal therapy. At the near-infrared region (NIR), they absorbs heat and show a strong photothermal effect; such nanoparticles have shown promising efficacy in recent years with various nanomaterials like graphene-based materials, iron oxide-based materials, and some NIR photosensitizers [1–7].

The development of material chemistry in 2D materials has taken much attention and named transition elements nitrides or carbides (MXene) in 2011; these are carbon or nitrogen atom layers connected to a primary transition metal (M) such as Ti, Nb, Ta, Hf, V, Sc, Cr, or Mo. The chemical formula of MXenes is $M_{n+1}AX_n$ (n= 1–3), where A is an element from the right side of the periodic table, specifically belonging to group 13 (IIIA) or group 14 (IVA). X is a nitrogen or carbon component [8–11]. The M–A bond is weaker than the M–X bond because the A part is more active and hence easily removed by strong acid (hydrofluoric acid [HF] as an etching reagent during exfoliation), and the surface of MXene materials is usually terminated with carbonyl (=O), fluorine (-F), or hydroxide (-OH) due to surface energy. The final empirical formula is summarized as $M_{n+1}X_nT_x$ (n =1–3), where T_x is the surface functional group [12, 13]. For biomedical applications, specifically cancer treatment, Ti_3C_2 and Ti_2C are the most used MXene nanocarriers, but various combinations of M and A groups may be used for further research in MXene carriers [14–16].

Due to its incurability and mortality rate, cancer, which is unchecked cell development in the human body, has gravely jeopardized human health globally. Although there are innumerable causes for this fatal condition, the failure of a body's natural

 DOI: 10.1201/9781003366225-15

control system leads to the irregular expansion of new cells, which is the mechanism of cancer development. The advent of nanomaterials in biomedical research has enabled a ground-breaking use of biomaterials in cancer treatment owing to the severity of this disease [17–19].

Out of various cancer treatments, photothermal therapy (PTT) takes lots of attention due to its efficient strategy, especially with MXene nanoparticles in the near-infrared region at a wavelength of 650 nm to 1250 nm using laser light. There are two kind of NIR laser used in various research, NIR-I (650–900 nm) and NIR-II (950–1250 nm) wavelength infrared laser lights, which have the potential to penetrate and kill cancer cells. In the presence of laser light, the vibration relaxation of nanoparticles by absorbing heat generates a significant amount of heat energy, which causes cancer cell death [20, 21]. This so-called light-to-heat conversion process attracts more interest in NIR laser-guided PTT [22, 23]. Laser pyrotechnics have achieved extensive research and are considered a starting strategy in pyrotechnics because of their uncomplicated igniting sequence, large consistency, and effective laser induction. The elimination of heat energy, which calls for energetic materials or additives that contain both good photothermal and light spectral responses, is one of the most crucial prerequisites in order to use laser ignition practically [24, 25]. The larger surface area of nanoparticles makes them qualified as various cargo carriers for better combined therapies like chemotherapy, photodynamic therapy (PDT), immune therapy, and gene therapy for synergetic effects [26–29]. The hydrophilic nature of MXenes leads to biocompatibility in in vitro as well as in vivo applications [30–33]. As there are various applications for MXene nanocomposites, in this chapter, we focus on cancer therapies only, especially advanced applications of photothermal therapy.

15.2 MXENE CHEMISTRY

In chemistry, two-dimensional (2D) inorganic compounds known as MXenes are composed of a small number of atoms that organize into thick layers of transition metal carbides, nitrides, or carbonitrides. The hydrophilic nature of MXenes (oxygen- and hydroxyl-terminated surface) combined with metallic conductivity of transitional metal carbide was reported in 2011. There are various combinations of MXene arrangement possible to make a MAX phase with surface functionalization, as shown in Figure 15.1, which shows the possible chemistry formulation of MXenes.

15.2.1 STRUCTURE OF MXENES

The morphology of MXenes is classified as multi-layer MXenes or few-layer MXenes depending on the surface modified by various functional units, as discussed previously. The etching of HF makes a structure like an accordion. There are 100 types MXenes formulated theoretically, but only ~30 types of MXenes have been developed experimentally [34]. Out of these, some of them are useful for photothermal effect, which is discussed in the chapter. The possible MAX phases ($M_{n+1}AX_n$), where n = 1–3, have a layered structure and contain divacancy, mono-, and double-transition metal MXenes [35–37]. The structure of the MAX phase looks like M_2C, M_3C_2, and M_4C_3,

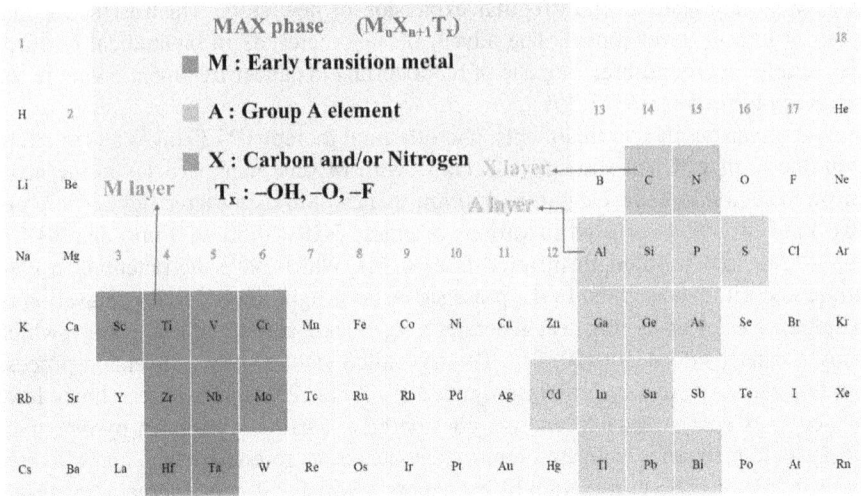

FIGURE 15.1 MAX phase elemental composition and possible arrangement. T_x: surface functional group.

for example, Sc_2C, Ti_3C_2, and V_4C_3. Instead of carbon, if nitrogen is used, the structure is M_2N, M_3N_2, and M_4N_3 and possible examples are Ti_2N, Ti_3N_2, and Ti_4N_3.

15.2.2 SYNTHESIS OF MXENES

MXenes are created from a MAX phase by selectively etching specific atomic layers from their layered antecedents. More than 70 ternary carbides and nitrides have been reported as MAX phases, in addition to a huge number of solid solutions and ordered double-transition metal structures [38]. There are basically three varieties of synthesis procedures such as top-down, bottom-up, and surface engineering modification. Here, we will just discuss the approaches of synthesis.

15.2.2.1 Top-Down Approach

MXene bulky layer structures are reduced to the nanoscale by chemical etching in top-down production techniques. Top-down synthesis techniques have been the main approach utilized to manufacture MXenes since the first successful selective extraction of aluminum (Al) from Ti_3AlC_2 to make Ti_3C_2. The process of reducing, making smaller, or removing a portion of the bulk structure from a material or powder to leave micro- to nano-sized particles is referred to as a top-down approach in this context. Among the top-down approaches, lithography, polishing, and wet chemical etching are the few examples.

15.2.2.2 Bottom-Up Approach

Bottom-up manufacturing techniques are more advantageous for the manufacture of two-dimensional materials than the top-down fabrication strategy. It is simpler method done by chemical vapor deposition to create MXene crystals with well-tailed structures by employing a single inorganic atom or molecule as a precursor. The inherent advantage of the bottom-up strategy is that it allows for exact control over MXene properties such composition, size, shape, and surface groups. This MXene synthesis approach helps in various cancer treatment.

15.2.2.3 Surface Modification

Surface terminations, which are attached to the metal layers, primarily consist of hydroxyl (-OH), fluorine (-F), and oxygen (-O), depending on the production method. Even though the surface terminations provide MXenes their hydrophilic feature, the delaminated ultrathin MXenes typically lack multi-functionalization and are unstable under complex physiological settings. In addition to improving circulation, targeting ability, biocompatibility, and loading efficiency, surface functionalization also prevents fast precipitation and aggregation in biological conditions. Thus, these nanosystems are given great environmental stability and dispersion in physiological settings by surface engineering, which then achieves numerous capabilities for successful biomedical application.

15.3 PHOTOTHERMAL THERAPY (PTT)

MXene-assisted PTT requires photothermal agents confined with NIR laser irradiation, and simultaneously, these agents also can be deployed as drug carriers. These PTT methods help to achieve the triggered release of the drug at the tumor site without damaging healthy tissue. Basically, the photothermal phenomenon confines with photothermal agents to absorb a certain wavelength upon laser irradiation and transform this light energy through a succession of photophysical reactions into heat, calculated as photothermal conversion efficiency (PCE).

15.3.1 PHOTOTHERMAL CONVERSION EFFICIENCY

Photothermal conversion efficiency can be varied depending on the different parameters such as the properties of photothermal agents and intensity of NIR irradiation. In the pursuit of ideal photothermal agents, the distinctive features that they must possess are good biocompatibility with low toxicity, high conversion efficiency, a high molar extinction coefficient, nano-range size for better uptake, and permeability and retention effects (EPR) [39, 40]. The heat conservation formulation shows a better approach in cancer cell necrosis and apoptosis. The PCE (η) is calculated by the heating and cooling curve of laser irradiation [41, 42], as follows.

$$\eta = \frac{hS\left(T_{max} - T_{sur}\right) - Q_0}{I\left(1 - 10^{-A\lambda}\right)} \qquad (15.1)$$

where S is the container surface area and h is the heat transfer coefficient. Maximum and surrounding temperatures are represented, respectively, by T_{max} and T_{sur}. Q_0 is a symbol for the heat released when light is absorbed by the sample cell. I stands for the laser's output power, and A for the absorbance at the laser's wavelength. The dimensionless parameter θ is calculated as follows:

$$\theta = \frac{T - T_{Sur}}{T_{Max} - T_{Sur}} \tag{15.2}$$

The time constant (τ_s) follows a linear cooling curve calculated from equations (15.2) and (15.3):

$$t = \tau_s ln(\theta) \tag{15.3}$$

$$hS = \frac{m_D C_D}{\tau_S} \tag{15.4}$$

where C_D is the heat capacity of solvent in j/g and m_D is the mass of solvent in g.

At some point after the laser exposure, the laser light will switch off. The cooling curve is used to compute the time constant (τ_s). The increase in laser temperature and the ambient temperature at time (t = 0$_s$) are used to compute the theta (θ) value.

15.4 FUNCTIONALIZED MXENE-BASED MATERIALS FOR PTT

There are various studies going on using MXene-based materials for photothermal therapy. Researchers in this field used various combinations of MXene-functionalized agents, using an NIR laser to study the photothermal effects. For the application part, they use various cancer cell lines to do in vitro and in vivo studies to optimize the suitable conditions for tumor ablation, as shown in Table 15.1.

In 2017, Lin et al. developed biodegradable niobium carbide (Nb₂C) modified polyvinyl-pyrrolidone (PVP) for tumor stamp out using NIR-1 and NIR-II laser

TABLE 15.1
The Photothermal Effects of MXene-Based Materials

Materials	Cancerous cells	Type of study	Wavelength (nm)	Intensity (W/cm²)	Ref.
Nb₂C-PVP	4T1, U87	In vitro, in vivo	808, 1064	1	[43]
Ti₃AlC₂-QD	HeLa	In vitro, in vivo	808	0.5	[44]
Ti₃C₂@Au-PEG	4T1	In vitro, in vivo	1064	0.75,1	[45]
Ta₄C₃-IONP-SP	4T1	In vitro, in vivo	808	1.5	[46]
Ti₂C_PEG	A375, MCF-7	In vitro	808	2	[47]
W₁.₃₃C -BSA	4T1	In vitro, in vivo	808, 1064	1.25	[48]
V₂C-TAT@Ex-RGD	MCF-7	In vitro, in vivo	1064	0.96	[49]
Mo₂C	HepG2	In vitro, in vivo	1064	2	[50]
Ti₃C₁.₁₅N₀.₈₅F₀.₈₈O₀.₅₆(OH)₀.₅₆	4T1	In vitro, in vivo	808,1064	1	[51]

light. The PCE was extremely high; NIR-I showed 36.4% efficacy, higher efficiency was shown by NIR-II (45.7%), and great photothermal stability were both displayed by the Nb_2C nanoparticles. The intrinsic properties of the Nb_2C-PVP NSs showed excellent biocompatibility and phototoxicity in 4T1 (breast cancer) and U87 (brain cancer) cell lines for in vitro cytotoxicity study. For an in vivo study on the 4T1 cell line used in both NIR bio windows, these surface-engineered Nb_2C NSs exhibited a high range of effects for in vivo photothermal ablation and tumor elimination. By logically planning their compositions with regard to phototherapy for cancer, this work greatly increases the application prospects for MXenes [43]. In another study done by Yu et al. for PTT applications, titanium aluminum carbide MXene (Ti_3AlC_2) quantum dots (QDs) were shown to be a biocompatible and highly effective nanoagent. The combination approach was discovered to modify numerous aluminum oxoanions on the surfaces of MXenes at an NIR absorbance at 808 nm with a PCE value of approximately 52.2%, leading to in vitro cell death and in vivo tumor regression using a HeLa (cervical cancer) cell line [44]. Another study successfully produced 2D core/shell Ti_3C_2-loaded gold (Au) nanocomposites using a simple seed growth approach. Due to the gold nanoparticles' growth on the Ti_3C_2 nanosheets' surface, the thiol chemistry considerably improved the stability of the MXenes and raised the nanocomposites' absorption in the NIR-II windows. PEG-SH conjugate (named Ti_3C_2@Au-PEG) nanocomposites were injected, and the tumor oxygenation was increased by utilizing the strong absorption. Using the NIR-II laser, the photothermal action of the nanocomposites showed toxic behavior in vivo as well as in vitro using 4T1 murine breast cancer cells [45]. In a different work, Liu and colleagues developed a novel theranostic nanoplatform based on superparamagnetic a MXene for effective breast cancer theranostics. Its foundation was a tantalum carbide (Ta_4C_3) MXene, and its development included functionalized iron oxide (Ta_4C_3-IONP-SPs). The soybean phospholipid (SP) used for good stability of Ta_4C_3-IONP's distinct theranostic functioning has been thoroughly examined at the intracellular level in vitro and in vivo using breast cancer tumor allografts on mice. To ensure their possible future clinical application, Ta_4C_3-IONP composite nanosheets' biocompatibility and biosafety have been carefully evaluated in vivo [46]. In a different study, Szuplewska and his teams prepared Ti_2C-coated PEG flakes that demonstrated acceptable PCE and good biocompatibility over a wide range of tested doses up to a concentration of 37.5 g mL^{-1}. In vitro, investigations using the PEG-modified Ti_2C showed a substantial NIR-induced ability to ablate tumor cells. When compared to other MXene-based photothermal agents, the applied dosages of Ti_2C-coated PEG in the study were 24 times lower. Through the creation of completely new agents for photothermal therapy, this work is anticipated to increase the usefulness of 2D MXenes in biological applications [47]. Zhou and co-researchers investigated to find unique $W_{1.33}C$-based i-MXenes with numerous theranostic functions, quick biodegradation, and good biocompatibility, and the nanosheets with ordered vacancies were successfully produced by constructing a parent bulk laminate with in-plane ordered $(W_{2/3}Y_{1/3})_2AlC$ ceramic and optionally etching aluminum (Al) and yttrium (Y) elements. Theoretical simulations, in particular, showed that $W_{1.33}C$ i-MXenes had a high majority of near-infrared absorbance. The created ultrathin $W_{1.33}C$ nanosheets have a good photothermal conversion efficiency at NIR-II (49.3%) as compared to

NIR-I (32.5%), good compatibility under biological conditions, and quick disintegration in healthy tissue rather than cancerous tissue. It's significant that the $W_{1.33}C$-BSA nanosheets' performance in photothermal ablation and both in vitro and in vivo multimodal imaging is methodically revealed and proven. Transcriptome and proteome sequencing further reveals the underlying mechanism and regulatory elements for the $W_{1.33}C$-BSA nanosheet-induced hyperthermia depletion. This research presents a framework for personalizing biological applications based on MXene for PTT [48]. Cao et al. designed a novel vanadium carbide quantum dot (V_2C QDs) photothermal agent in conjunction with an edited exosome (Ex) vector that effectively kills the tumor when used in the second NIR region's mild temperature nucleus-targeted PTT. The TAT peptides were used to customize the nanoscale NIR-II-compatible fluorescent V_2C-QDs with good photothermal properties and package them into exosomes with RGD modification (V_2C-TAT@Ex-RGD). The particles formed as a result (NPs) were highly biocompatible, had a prolonged circulation time, and could escape endosomes. Additionally, they possessed the capacity to target cells and enter their nuclei to perform low-temperature PTT with increased tumor removal efficiency and appropriate biocompatibility, suggesting the possibility for successful clinical application [49]. In another study, Zhang and his team developed a Mo_2C-based MXene used as a theranostic compound for imaging contrast and tumor phototherapy. The visible and near-infrared optical absorption spectra showed that Mo_2C produced a broad and strong absorbance, which was connected to its metalloid features, as predicted by theoretical models. Under laser stimulation, Mo_2C generated hyperthermia, which could cause blatant cell death. Cell experiments showed that the PTT had a superior cancer cell killing efficacy. An in vivo anticancer investigation, meanwhile, confirmed that the Mo_2C-mediated photothermal effect completely eradicated the mice's solid tumors without causing any systemic side effects or hemolysis [50]. Given that the metal element is primarily responsible for the metallic property, anionic modulation may provide a unique path for the advancement of effective PTAs with metallic properties and is anticipated to result in strong light-harvesting over the NIR-I and NIR-II wavelengths.

The unique metal carbonitride used for the anionic modulation of $Ti_3C_{1.15}N_{0.85}F_{0.88}O_{0.56}(OH)_{0.56}$ demonstrates a significant NIR absorption (43.5 L g/cm at 1064 nm, 36.6 L g/cm at 808 nm) by utilizing electron injection from nitrogen, producing effective photonic hyperthermia against malignant cells and in animal studies. An in-depth explanation of the relationship between atomic composition and physicochemical qualities is provided by this proof-of-concept when applied to a vast family of MXenes, further solidifying MXenes' promise for medicinal applications [51]. In this chapter, we examine several metal combinations used in the MAX phase of various research group, which are still the subject of ongoing studies on PTT effects.

15.5   CONCLUSION

In this chapter, we have discussed the unique properties of MXenes, which are used for photothermal synthesizing, surface modification, and other biomedicine. MXenes' overall photothermal performance appears to be improved when compared to conventional photothermal agents like graphene, MoS_2, and organic materials.

Specifically, in cancer therapy, these MXene-based carriers show better toxicity towards cancer cells. Even though the reported MXenes showed encouraging results in PTT, there are still issues that must be resolved before these materials can be used in healthcare situations. This chapter provides an overall idea for exploring the combination of different surface modification functional groups and termination groups that can enhance the photothermal effect from a future perspective.

REFERENCES

[1] B.S. Dash, G. Jose, Y.-J. Lu, J.-P. Chen, Functionalized Reduced Graphene Oxide as a Versatile Tool for Cancer Therapy, International Journal of Molecular Sciences 22(6) (2021) 2989.

[2] B.S. Dash, Y.-J. Lu, H.-A. Chen, C.-C. Chuang, J.-P. Chen, Magnetic and GRPR-Targeted Reduced Graphene Oxide/Doxorubicin Nanocomposite for Dual-Targeted Chemo-Photothermal Cancer Therapy, Materials Science and Engineering: C 128 (2021) 112311.

[3] B.S. Dash, S. Das, J.-P. Chen, Photosensitizer-Functionalized Nanocomposites for Light-Activated Cancer Theranostics Micromachines 22(13) (2021) 6658.

[4] A.T. Shivanna, B.S. Dash, J.-P. Chen, Functionalized Magnetic Nanoparticles for Alternating Magnetic Field- or Near Infrared Light-Induced Cancer Therapies, Micromachines 13(8) (2022) 1279.

[5] S. Soren, S. Chakroborty, L. Pradhan, P. Chandra, J. Sahu, P. Parhi, Hydrothermal Synthesis of Graphene Modified SnO Nanocomposite for Oxygen Reduction Reaction, Materials Today: Proceedings 57 (2022) 72–76.

[6] S. Soren, S. Chakroborty, K. Pal, Enhanced in Tunning of Photochemical and Electrochemical Responses of Inorganic Metal Oxide Nanoparticles Via rGO Frameworks (MO/rGO): A Comprehensive Review, Materials Science and Engineering: B 278 (2022) 115632.

[7] S. Chakroborty, P. Panda, Nanovaccinology Against Infectious Disease, Nanovaccinology as Targeted Therapeutics (2022) 95–113.

[8] G. Liu, J. Zou, Q. Tang, X. Yang, Y. Zhang, Q. Zhang, W. Huang, P. Chen, J. Shao, X. Dong, Surface Modified Ti3C2 MXene Nanosheets for Tumor Targeting Photothermal/Photodynamic/Chemo Synergistic Therapy, ACS Applied Materials & Interfaces 9(46) (2017) 40077–40086.

[9] D.-Y. Zhang, H. Liu, M.R. Younis, S. Lei, Y. Chen, P. Huang, J. Lin, In-Situ TiO2-x Decoration of Titanium Carbide MXene for Photo/Sono-Responsive Antitumor Theranostics, Journal of Nanobiotechnology 20(1) (2022) 53.

[10] J. Meng, Z. An, Y. Liu, X. Sun, J. Li, MXene-based Hydrogels towards the Photothermal Applications, Journal of Physics D: Applied Physics 55(37) (2022) 374003.

[11] J. Li, X. Cai, Y. Zhang, K. Li, L. Guan, Y. Li, T. Wang, T. Sun, MnO2 Nanozyme-Loaded MXene for Cancer Synergistic Photothermal-Chemodynamic Therapy, ChemistrySelect 7(23) (2022) e202201127.

[12] X. Li, H. Chang, L. Zeng, X. Huang, Y. Li, R. Li, Z. Xi, Numerical Analysis of Photothermal Conversion Performance of MXene Nanofluid in Direct Absorption Solar Collectors, Energy Conversion and Management 226 (2020) 113515.

[13] H. Zeng, L. Deng, L. Yang, H. Wu, H. Zhang, C. Zhou, B. Liu, Z. Shi, Novel Prussian Blue Analogues@MXene Nanocomposite as Heterogeneous Activator of Peroxymonosulfate for the Degradation of Coumarin: The Nonnegligible Role of Lewis-Acid Sites on MXene, Chemical Engineering Journal 416 (2021) 128071.

[14] A. Zamhuri, G.P. Lim, N.L. Ma, K.S. Tee, C.F. Soon, MXene in the Lens of Biomedical Engineering: Synthesis, Applications and Future Outlook, BioMedical Engineering OnLine 20(1) (2021) 33.

[15] A.M. Jastrzębska, A. Szuplewska, T. Wojciechowski, M. Chudy, W. Ziemkowska, L. Chlubny, A. Rozmysłowska, A. Olszyna, In Vitro Studies on Cytotoxicity of Delaminated Ti3C2 MXene, Journal of Hazardous Materials 339 (2017) 1–8.

[16] L. Wang, H. Zhang, B. Wang, C. Shen, C. Zhang, Q. Hu, A. Zhou, B. Liu, Synthesis and Electrochemical Performance of Ti3C2Tx with Hydrothermal Process, Electronic Materials Letters 12(5) (2016) 702–710.

[17] P. Anand, A.B. Kunnumakara, C. Sundaram, K.B. Harikumar, S.T. Tharakan, O.S. Lai, B. Sung, B.B. Aggarwal, Cancer is a Preventable Disease that Requires Major Lifestyle Changes, Pharmaceutical Research 25(9) (2008) 2097–2116.

[18] M. Goldberg, R. Langer, X. Jia, Nanostructured Materials for Applications in Drug Delivery and Tissue Engineering, Journal of Biomaterials Science, Polymer Edition 18(3) (2007) 241–268.

[19] X. Huang, I.H. El-Sayed, W. Qian, M.A. El-Sayed, Cancer Cell Imaging and Photothermal Therapy in the Near-Infrared Region by Using Gold Nanorods, Journal of the American Chemical Society 128(6) (2006) 2115–2120.

[20] J. Shao, J. Zhang, C. Jiang, J. Lin, P. Huang, Biodegradable Titanium Nitride MXene Quantum Dots for Cancer Phototheranostics in NIR-I/II Biowindows, Chemical Engineering Journal 400 (2020) 126009.

[21] B. Rashid, A. Anwar, S. Shahabuddin, G. Mohan, R. Saidur, N. Aslfattahi, N. Sridewi, A Comparative Study of Cytotoxicity of PPG and PEG Surface-Modified 2-D Ti(3)C(2) MXene Flakes on Human Cancer Cells and Their Photothermal Response, Materials (Basel, Switzerland) 14(16) (2021).

[22] Z. Huang, X. Cui, S. Li, J. Wei, P. Li, Y. Wang, C.-S. Lee, Two-Dimensional MXene-based Materials for Photothermal Therapy, Nanophotonics 9(8) (2020) 2233–2249.

[23] H. Lin, Y. Chen, J. Shi, Insights into 2D MXenes for Versatile Biomedical Applications: Current Advances and Challenges Ahead, Advanced Science 5(10) (2018) 1800518.

[24] H.-Y. Deng, L. Wang, D. Tang, Y. Zhang, L. Zhang, Review on the Laser-Induced Performance of Photothermal Materials for Ignition Application, Energetic Materials Frontiers 2(3) (2021) 201–217.

[25] Y. Zhang, S. Zhang, Z. Zhang, L. Ji, J. Zhang, Q. Wang, T. Guo, S. Ni, R. Cai, X. Mu, W. Long, H. Wang, Recent Progress on NIR-II Photothermal Therapy, Frontiers in Chemistry 9 (2021).

[26] W. Gao, W. Zhang, H. Yu, W. Xing, X. Yang, Y. Zhang, C. Liang, 3D CNT/MXene Microspheres for Combined Photothermal/Photodynamic/Chemo for Cancer Treatment, Frontiers in Bioengineering Biotechnology 10 (2022).

[27] A. Gazzi, L. Fusco, A. Khan, D. Bedognetti, B. Zavan, F. Vitale, A. Yilmazer, L.G. Delogu, Photodynamic Therapy Based on Graphene and MXene in Cancer Theranostics, Frontiers in Bioengineering Biotechnology 7 (2019).

[28] Y. Xu, Y. Wang, J. An, A.C. Sedgwick, M. Li, J. Xie, W. Hu, J. Kang, S. Sen, A. Steinbrueck, B. Zhang, L. Qiao, S. Wageh, J.F. Arambula, L. Liu, H. Zhang, J.L. Sessler, J.S. Kim, 2D-Ultrathin MXene/DOXjade Platform for Iron Chelation Chemo-Photothermal Therapy, Bioactive Materials 14 (2022) 76–85.

[29] Y. Guo, H. Wang, X. Feng, Y. Zhao, C. Liang, L. Yang, M. Li, Y. Zhang, W. Gao, 3D MXene Microspheres with Honeycomb Architecture for Tumor Photothermal/Photodynamic/Chemo Combination Therapy, Nanotechnology 32(19) (2021) 195701.

[30] G.P. Lim, C.F. Soon, N.L. Ma, M. Morsin, N. Nayan, M.K. Ahmad, K.S. Tee, Cytotoxicity of MXene-based Nanomaterials for Biomedical Applications: A Mini Review, Environmental Research 201 (2021) 111592.

[31] J. Huang, Z. Li, Y. Mao, Z. Li, Progress and Biomedical Applications of MXenes, Nano Select 2(8) (2021) 1480–1508.

[32] S. Iravani, R.S. Varma, MXenes in Cancer Nanotheranostics, Nanomaterials (Basel) 12(19) (2022).

[33] A. Sundaram, J.S. Ponraj, C. Wang, W.K. Peng, R.K. Manavalan, S.C. Dhanabalan, H. Zhang, J. Gaspar, Engineering of 2D Transition Metal Carbides and Nitrides MXenes for Cancer Therapeutics and Diagnostics, Journal of Materials Chemistry B 8(23) (2020) 4990–5013.

[34] B. Anasori, M.R. Lukatskaya, Y. Gogotsi, 2D Metal Carbides and Nitrides (MXenes) for Energy Storage, Nature Reviews Materials 2(2) (2017) 16098.

[35] M. Naguib, M.W. Barsoum, Y. Gogotsi, Ten Years of Progress in the Synthesis and Development of MXenes, Advanced Materials 33(39) (2021) 2103393.

[36] Y. Wei, P. Zhang, R.A. Soomro, Q. Zhu, B. Xu, Advances in the Synthesis of 2D MXenes, Advanced Materials 33(39) (2021) 2103148.

[37] W. Hong, B.C. Wyatt, S.K. Nemani, B. Anasori, Double Transition-Metal MXenes: Atomistic Design of Two-Dimensional Carbides and Nitrides, MRS Bulletin 45(10) (2020) 850–861.

[38] M. Naguib, V.N. Mochalin, M.W. Barsoum, Y. Gogotsi, 25th Anniversary Article: MXenes: A New Family of Two-Dimensional Materials, Advanced Materials 26(7) (2014) 992–1005.

[39] J.R. Melamed, R.S. Edelstein, E.S. Day, Elucidating the Fundamental Mechanisms of Cell Death Triggered by Photothermal Therapy, ACS Nano 9(1) (2015) 6–11.

[40] R.K. Jain, T. Stylianopoulos, Delivering Nanomedicine to Solid Tumors, Nature Reviews Clinical Oncology 7(11) (2010) 653–664.

[41] D.K. Roper, W. Ahn, M. Hoepfner, Microscale Heat Transfer Transduced by Surface Plasmon Resonant Gold Nanoparticles, The Journal of Physical Chemistry C 111(9) (2007) 3636–3641.

[42] B.S. Dash, Y.-J. Lu, P. Pejrprim, Y.-H. Lan, J.-P. Chen, Hyaluronic Acid-Modified, IR780-Conjugated and Doxorubicin-Loaded Reduced Graphene Oxide for Targeted Cancer Chemo/Photothermal/Photodynamic Therapy, Biomaterials Advances 136 (2022) 212764.

[43] H. Lin, S. Gao, C. Dai, Y. Chen, J. Shi, A Two-Dimensional Biodegradable Niobium Carbide (MXene) for Photothermal Tumor Eradication in NIR-I and NIR-II Biowindows, Journal of the American Chemical Society 139(45) (2017) 16235–16247.

[44] X. Yu, X. Cai, H. Cui, S.-W. Lee, X.-F. Yu, B. Liu, Fluorine-Free Preparation of Titanium Carbide MXene Quantum Dots with High Near-Infrared Photothermal Performances for Cancer Therapy, Nanoscale 9(45) (2017) 17859–17864.

[45] W. Tang, Z. Dong, R. Zhang, X. Yi, K. Yang, M. Jin, C. Yuan, Z. Xiao, Z. Liu, L. Cheng, Multifunctional Two-Dimensional Core—Shell MXene@Gold Nanocomposites for Enhanced Photo—Radio Combined Therapy in the Second Biological Window, ACS Nano 13(1) (2019) 284–294.

[46] Z. Liu, H. Lin, M. Zhao, C. Dai, S. Zhang, W. Peng, Y. Chen, 2D Superparamagnetic Tantalum Carbide Composite MXenes for Efficient Breast-Cancer Theranostics, Theranostics 8(6) (2018) 1648–1664.

[47] A. Szuplewska, D. Kulpińska, A. Dybko, A.M. Jastrzębska, T. Wojciechowski, A. Rozmysłowska, M. Chudy, I. Grabowska-Jadach, W. Ziemkowska, Z. Brzózka, A. Olszyna, 2D Ti2C (MXene) as a Novel Highly Efficient and Selective Agent for Photothermal Therapy, Materials Science and Engineering: C 98 (2019) 874–886.

[48] B. Zhou, H. Yin, C. Dong, L. Sun, W. Feng, Y. Pu, X. Han, X. Li, D. Du, H. Xu, Y. Chen, Biodegradable and Excretable 2D W1.33C i-MXene with Vacancy Ordering for Theory-Oriented Cancer Nanotheranostics in Near-Infrared Biowindow, Advanced Science 8(24) (2021) 2101043.

[49] Y. Cao, T. Wu, K. Zhang, X. Meng, W. Dai, D. Wang, H. Dong, X. Zhang, Engineered Exosome-Mediated Near-Infrared-II Region V2C Quantum Dot Delivery for Nucleus-Target Low-Temperature Photothermal Therapy, ACS Nano 13(2) (2019) 1499–1510.

[50] Q. Zhang, W. Huang, C. Yang, F. Wang, C. Song, Y. Gao, Y. Qiu, M. Yan, B. Yang, C. Guo, The Theranostic Nanoagent Mo2C for Multi-Modal Imaging-Guided Cancer Synergistic Phototherapy, Biomaterials Science 7(7) (2019) 2729–2739.

[51] Y. Zhu, X. Tang, Q. Liu, Y. Xia, X. Zhai, H. Zhang, D. Duan, H. Wang, W. Zhan, L. Wu, N. Zheng, W. Lv, Y. Wang, M. Zhou, Metallic Carbonitride MXene Based Photonic Hyperthermia for Tumor Therapy, Small 18(22) (2022) 2200646.

16 Exploring the Potential of MXene-Based Materials for Sodium-Ion Batteries

Manas Ranjan Panda, Sally El Meragawi, Jagabandhu Patra, and Mainak Majumder

16.1 BACKGROUND

The current battery market, which includes portable and mobile storage technologies such as those used in electric vehicles, is exclusively dominated by lithium-ion batteries (LIBs). [1–4] The reliance of this technology on the availability of lithium makes it a costly raw material further encumbered by geopolitics; thus, there is a substantial research commitment towards the development of alternate storage devices. [3] Sodium-ion batteries (SIBs) are at the vanguard of potential alternatives to LIBs due to the accessibility of unlimited abundant sodium resources worldwide. [3, 5–9] The prominent drawbacks in implementing SIBs are that the well-established anodes used for LIBs, such as graphite, are unsuitable due to poor sodium storage. [5] The conversion anodes are not suitable owing to their poor cycling life, and alloying anodes like Sn/P show enormous volume expansion, resulting in fast capacity decay. The other limitations are the lower specific charge capacity of existing anode materials such as hard carbon used for SIBs. We need to find new materials that can overcome current limitations. [5, 7, 8] There is a continuous effort placed on developing high-capacity anode materials, as this directly impacts the specific capacity of a battery. [5, 6] The intrinsically high surface area-to-volume ratio and the van der Waals interlayer bonding present in 2D materials such as transition metal dichalcogenides (TMDCs) and transition metal carbides or nitrites (MXenes) have driven interest in their use for energy storage. [10–15]

The general formula of TMDCs is MX_2, where M stands for the transition metal from group 4 to 10, while X is an element from the chalcogens family such as S, Se, Te, and so on. [6, 7, 16–18] TMDCs have three types of polystructures, 1T, 2H, and 3R, representing one, two, and three layers of tetragonal (T), hexagonal (H), or rhombohedral (R) unit cells. Interestingly, all types of electronic structures, like insulating (HfS_2), semi-conducting (MoS_2, WS_2), semi-metallic (VS_2, TiS_2), and superconducting ($TaSe_2$, $NbSe_2$) materials, are a part of the TMDC family of materials. [7, 8, 17, 18] Two-dimensional layered materials show superior conductivity and electrochemical storage performance when compared to bulk TMDCs, so exfoliation is a key step towards achieving high-performance devices. [18] Figure 16.1a shows

DOI: 10.1201/9781003366225-16

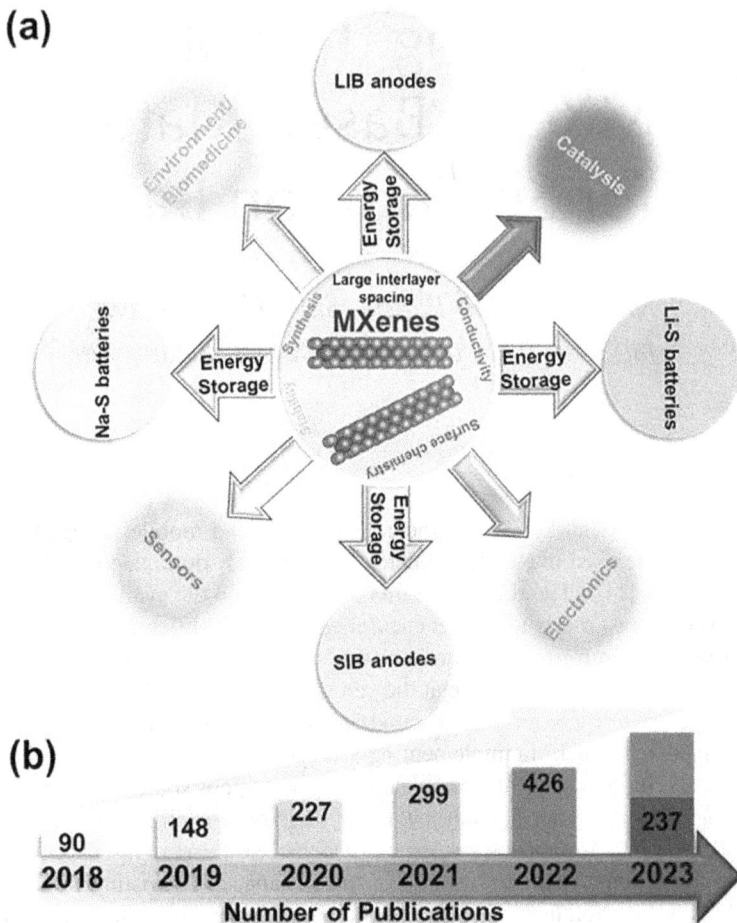

FIGURE 16.1 (a) MXenes are of growing interest in a number of applications such as elec-trodes in batteries due to their interlayer spacing and conductivity or as catalysts due to their rich surface chemistry. (b) There has been an increasing number of publications since 2018 on the use of MXenes as electrodes in batteries, with around 426 publications recorded in Scopus in 2022. (Keywords with MXene and battery).

the growing interest in MXenes for many applications, from their use in battery electrodes due to their high interlayer spacing and conductivity and their applications as catalysts due to their rich surface chemistry. [10, 12, 15] Figure 16.1b shows an increasing number of publications since 2018 on the use of MXenes as electrodes in different types of batteries.

MXenes have the general formula of $M_{n+1}X_nT_x$, wherein M corresponds to a transition metal such as Mo, Mn, Nb, Sc, Ti, Xr, or V; X represents either C and/or N; n lies in between 1 to 3 depending on the precursor MAX phase; and T_x is a surface-terminated functional group (oxygen (-O), fluorine (-F), and hydroxyl (-OH)).

Legend:
- M in MAX phase
- A in MAX phase
- X in MAX phase
- M in MXene
- X in MXene
- T in MXene

$M_{n+1}AX_n$ $M_{n+1}X_nT_x$

1 IA	2 IIA	3 IIIB	4 IVB	5 VB	6 VIB	7 VIIB	8	9 VIIB	10	11 IB	12 IIB	13 IIIA	14 IVA	15 VA	16 VIA	17 VIIA	18 VIIIA
1 H																	2 He
3 Li	4 Be											5 B	6 C	7 N	8 O	9 F	10 Ne
11 Na	12 Mg											13 Al	14 Si	15 P	16 S	17 Cl	18 Ar
19 K	20 Ca	21 Sc	22 Ti	23 V	24 Cr	25 Mn	26 Fe	27 Co	28 Ni	29 Cu	30 Zn	31 Ga	32 Ge	33 As	34 Se	35 Br	36 Kr
37 Rb	38 Sr	39 Y	40 Zr	41 Nb	42 Mo	43 Tc	44 Ru	45 Rh	46 Pd	47 Ag	48 Cd	49 In	50 Sn	51 Sb	52 Te	53 I	54 Xe
55 Cs	56 Ba	57-71	72 Hf	73 Ta	74 W	75 Re	76 Os	77 Ir	78 Pt	79 Au	80 Hg	81 Tl	82 Pb	83 Bi	84 Po	85 At	86 Rn
87 Fr	88 Ra	89-103	104 Rf	105 Db	106 Sg	107 Bh	108 Hs	109 Mt	110 Ds	111 Rg	112 Cn	113 Nh	114 Fl	115 Mc	116 Lv	117 Ts	118 Og

57 La	58 Ce	59 Pr	60 Nd	61 Pm	62 Sm	63 Eu	64 Gd	65 Tb	66 Dy	67 Ho	68 Er	69 Tm	70 Yb	71 Lu

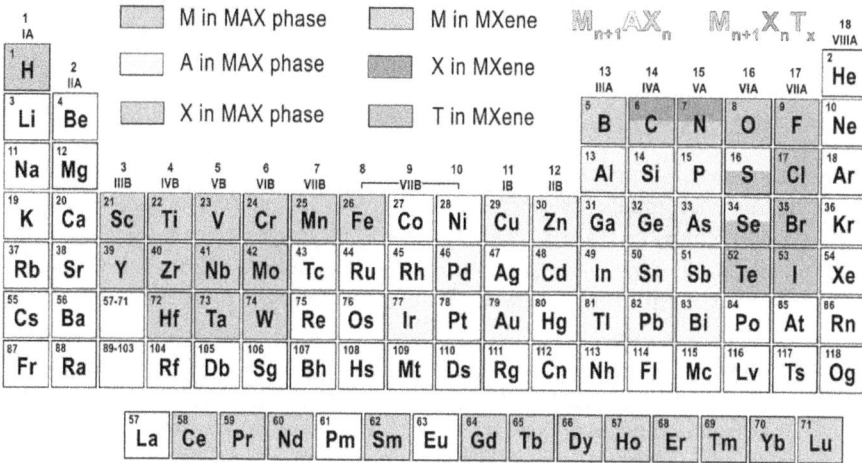

FIGURE 16.2 Elements of the known MAX phases and MXenes. Metals such as manganese, iron, and molybdenum (highlighted in green) can form MAX phases; however, it is currently unknown if MXenes can be synthesized with them. However, metals such as titanium, zirconium, and vanadium (highlighted in blue) have been effectively used for MXene synthesis. The formation of simple MXenes requires one transition metal (M), A, or X element. [19]

(No Permission is required to reuse the figure).

[10, 12, 14] MXenes are typically synthesized via selective etching of a corresponding MAX phase, which have the general formula $M_{n+1}AX_n$, where "A" is the group A element like Al, In, Sn, or Si and is generally the element substituted for a T_x functional group. [10, 12] Figure 16.2 illustrates the positions of elements in MAX and MXene structures, along with the associated surface functional groups in the MXene phase. The growth of a simple MXenes with the transition metal (M); A elements like Al, In, Sn, Si, Ge, Pb; or X elements (C or N) are all possible stable structures. [19]

16.2 MXENES FOR LI/NA ION STORAGE

MXenes have 2D layer structures, superior metallic conductivity, tuneable surface chemistry, and are known as the most suitable class of materials for application as field-effect transistors (FETs) [20], supercapacitors [21], oxygen evolution/reduction devices [22], sensors [20, 22], and, importantly, rechargeable batteries (for LIBs, SIBs, and beyond) [12, 13, 23]. MXenes are widely used as electrodes due to their large surface area, which enables the utilization of a large number of active sites and allows for highly reversible and fast redox reactions. [12–15, 23] However, this material is also disadvantaged in key areas such as its low stability during cycling and the irreversibility of the redox reactions at a high current rate. [12, 23] Various strategies like fabricating hybrid electrodes with a conductive additive like graphene or graphene oxide and doping with N_2 or Nb are being employed to overcome the challenges associated with using MXenes in an electrode. Still, much work is required before they can be used in practical devices. [12–15, 19, 23]

16.3   MXENES FOR SODIUM-ION BATTERIES (SIBS)

It has only been 10 years since MXenes have been researched for use as electrodes for SIBs; this is a significantly shorter period of research and development time when compared with more established electrode materials such as hard carbon, which has been under development for years. [10, 13, 24] Theoretically, it has been observed that the surface-terminated functional groups such as -O, -F, and -OH of MXene electrodes greatly influence the SIB storage properties. [13, 24] Single-layer pristine MXenes show a minimum diffusion barrier for the storage of Na$^+$ ions owing to the lower open circuit voltage (OCV). [24] The experimentally observed lower capacity of MXenes is due to the aggregation of layers hindering the transport of charges in the bulk electrodes. These difficulties hinder the required reversible capacity, limiting their further application as SIB anodes compared to the existing potential anodes for SIBs. [10, 13, 24] To overcome these limitations, strategies that involve modifying the 3D porous structure, expanding interlayer spacing, and/or synthesizing single/few-layer MXenes have been proposed to enhance the Na$^+$ storage capacity and the kinetics of the electrochemical reactions. [10, 13, 24] Many researchers have suggested that MXenes can offer a conductive network to improve the electronic conductivity of sulphur in lithium sulphur batteries. [25] The engineered MXene structures can benefit from electrochemical performance enhancement. Here, the conductive carbide core is sandwiched between the transition metal oxide/hydroxide-like surface, which makes a highly conductive, few-atoms-thick layered material highly suitable for use as an anode in SIBs. [10, 13, 24, 26–29]

16.4   SYNTHESIS STRATEGY OF MXENES

Figure 16.3a shows a schematic of the hydrofluoric (HF) acid-based exfoliation of a MAX phase into its respective carbides and carbonitrides. [30] The elimination of the "A" group from the MAX phases results in the respective MXenes, which show structural similarities with graphene and have sheet resistances on the order of multilayer graphene. [30] Naguib *et al.* investigated the MAX phase structure with the two layers of transition metal carbides and nitrides associated with an "A" element in a 2D configuration. [10] Huang *et al.* studied the freeze-and-thaw method to exfoliate multiple-layer MXene, which can be raised to an exfoliation yield of 81.4% through a combination of sonication and the water freezing-and-thawing (FAT) approach. [31] HCl and HF are effective etching chemicals able to remove the Al atom from Ti$_3$AlC$_2$. [31] Figure 16.3b shows the synthesis of Cl-terminated MXenes (Ti$_3$C$_2$Cl$_2$) using ZnCl$_2$ (Lewis acids in their molten state) to functionalize Ti$_3$C$_2$Cl$_2$. [32] Figure 16.3c depicts the synthesis and exfoliation of Ti$_3$CN. The process describes the synthesis of Ti$_3$AlCN by ball milling of Ti, Al, N, and graphite followed by thermal treatment under an inert argon atmosphere. [33] Ti$_3$CN was synthesized by treating the precursor MAX phases with HF etching to remove the Al, transforming the MAX phase to the final MXene phase. [33] The insertion of highly electronegative

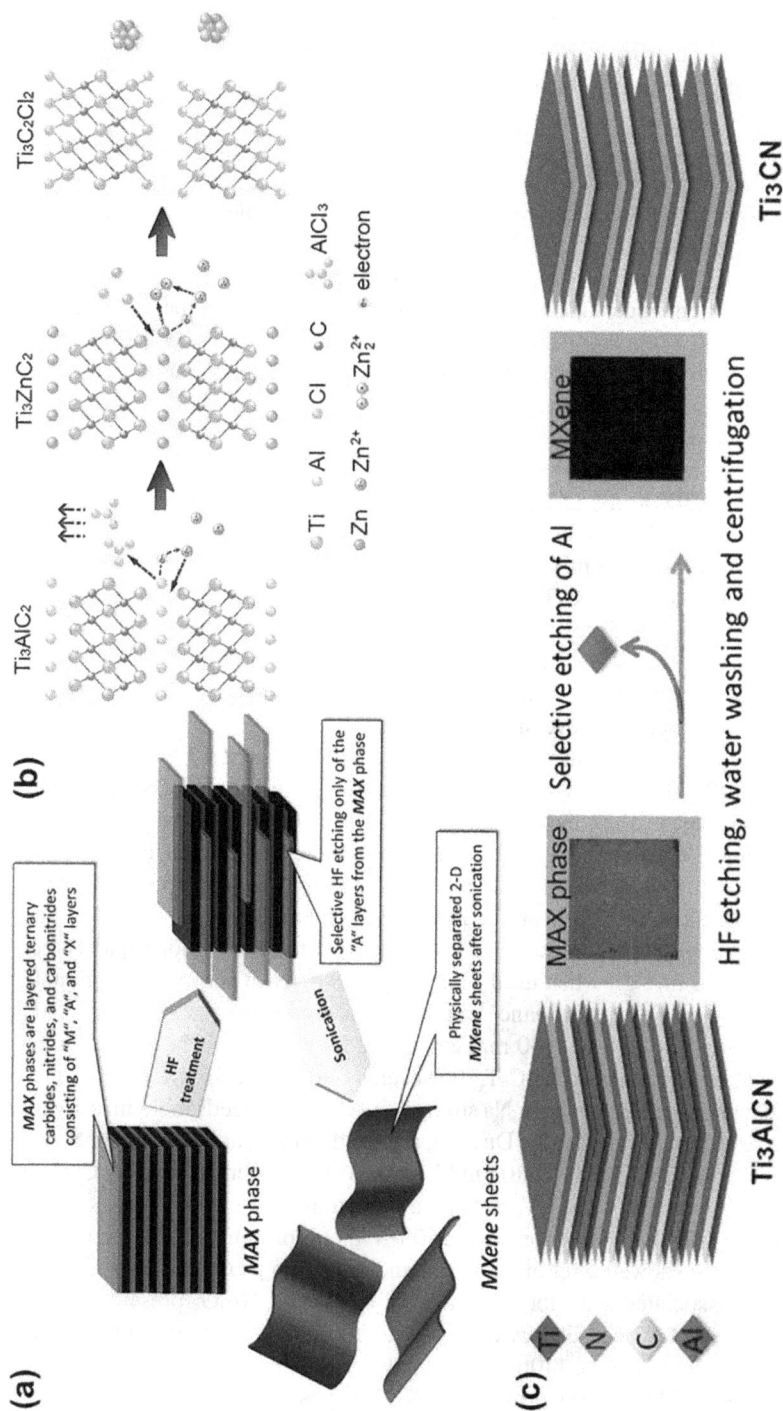

FIGURE 16.3 (a) The simple synthesis of MXenes from MAX phases in a two-stage process involving acid etching and sonication. (b) Synthesis of $Ti_3C_2Cl_2$ via reaction of MAX with Lewis acidic molten salts. (c) Selective etching of aluminium from the layered MAX phase (Ti_3AlCN) for preparation of Ti_3CN.

Reproduced with permission. Reprinted with permission from [30] (a), [32] (b), and [33] (c). Copyright 2023, American Chemical Society.

N_2 into the Ti_3C structure enhances the electron count of the MXene structure. [33] These are the most viable approaches for the synthesis of MXenes that also function to enhance MXene electrode conductivity and make them more appropriate anodes for SIBs having higher specific capacities and cycle lifespans.

16.5   LATEST PROGRESS ON MXENES IN SIBS

In 2013, Zhao et al. reported a 3D porous $Ti_3C_2T_x$ with a crumpled nanosheet structure produced by treatment with alkali ions to be an excellent anode for SIBs. [34] It was known that the restacking of MXene layers could cause structural instability, so this work aimed to enhance structural stability through converting the 2D MXene flakes into crumpled 3D porous structures using various alkali ions. [34] However, the exact reason driving the flocculation of sheets was not explained. The as-prepared 3D material maintained a capacity of 54 mAh g^{-1} at 1.5 A g^{-1} for 1000 cycles when employed as an anode against Na metal. [34] In the same year, Anayee et al. investigated the role of surface chemistry in Na^+ storage in $Ti_3C_2T_x$ using different etchants through solid-state NMR. In ^{13}C NMR, small differences were observed in the peak positions for carbides obtained using different etchants. The resonant peak for $Ti_3C_2T_x$ treated with HF or a mixture of HF/HCl was found to occur at 398 ppm, while the resonant peak for $Ti_3C_2T_x$ treated with a mixture of HF/H_2SO_4 was found at 389 ppm. [35] The differences can be explained by the presence of more electronegative atoms such as F, Cl, and O on the surface of $Ti_3C_2T_x$ (HF) and $Ti_3C_2T_x$ (HF/ HCl), which ensured overlap between Ti 3d and C 2p orbitals. The $Ti_3C_2T_x$ (HF/ H_2SO_4) sample had a reduced reversible capacity in the first cycle and an enhanced electrochemical performance attributed to the removal of etching products and structural water. [35] In 2017, Wu et al. synthesized a few-layered MXene through a solvent-assisted ball milling method that resulted in enhanced electron transport and Na^+ diffusion. [36] With the use of dimethyl sulfoxide, a multilayer $Ti_3C_2T_x$ was delaminated into a few-layer MXene without oxidation. [36] The as-prepared MXene showed a reversible cycle capacity of 267 mAh g^{-1} at 100 mA g^{-1}. [36] In 2018, Zhu et al. prepared titanium carbonitride (Ti_3CN) and reported a superior capacity of 211.5 mAh g^{-1} at 20 mA g^{-1} when used as an anode in SIBs. It showed exceptional rate capability and cycling performance, retaining 98.9 mAh g^{-1} at 500 mA g^{-1} and 85.6 mAh g^{-1} after 500 cycles at 100 mA g^{-1}. It was found that Ti_3CN followed the same intercalation mechanism as $Ti_3C_2T_x$; however, it outperformed $Ti_3C_2T_x$ in terms of its electrochemical performance in Na storage and therefore needs more investigation in future research. [37] Recently, Du et al. reported Nb_2CT_x as an anode for SIBs that delivered a specific capacity of 102 mAh g^{-1} at 1 A g^{-1} after 500 cycles. Cyclic voltammetry (CV) curves of Nb_2CT_x showed a reduction peak at 1.17 V associated with SEI formation and an oxidation peak at 0.1 V that represents the extraction of Na^+. CV curves also showed a set of oxidation and reduction peaks at 2.35 and 2.66 V, respectively, associated with the intercalation of Na^+ on Nb_2O_5 present on the exterior layer of the MXene. [38] Evidence of a short charge/discharge plateau at 2.66 and 2.35 V appeared at the 140th cycle in galvanostatic cycling, again matching with the CV results. Thus, it was proven that Nb_2CT_x possesses two different pseudocapacitance mechanisms in SIBs run at different potentials: intercalation-based pseudocapacitance at low potential and surface-controlled pseudocapacitance at high potential. [38] Kajiyama et al. demonstrated a Na^+ intercalation-type mechanism in $Ti_3C_2T_x$ through ^{23}Na magic angle spinning NMR and density functional theory (DFT) calculations. From this ex situ NMR, they concluded that there is a reversible intercalation and deintercalation of desolvated Na^+ between MXene layers. Few similar peaks were observed in the NMR spectra of every sample, as SEI will remain in the electrode once it is formed. So, it is difficult to differentiate between intercalation or adsorption. When the NMR spectra are normalized with respect to one peak, the areas of the other peaks increase or decrease upon sodiation or desodiation, respectively. The observation of these trends allows for confirmation that the primary sodiation mechanism involved with using $Ti_3C_2T_x$ as an electrode is intercalation and not adsorption. [39]

16.6   A THEORETICAL PERSPECTIVE ON MXENES IN SIBS

Along with the extensive experimental analysis, key properties such as sodium storage sites, diffusion coordinates, the effect of adsorption strength on the mobility of Na ions, and the sodium storage mechanisms involved in the use of MXenes as anodes have been explored theoretically to further justify their use in SIB application. Yang-Xin Yu methodically investigated the mobility, storage capacity, and volume change upon Na intake in a functionalized Ti_3C_2 MXene using theoretical analysis based on DFT. [40] The results reveal that the Ti_3C_2 MXene shows a comparatively low barrier for sodium diffusion, with a small variation in lattice constant during sodiation upon functionalization with -F, -O, and -OH. Feng Li et al. investigated carbides as an efficient storage material for SIBs based on first principle calculations. [41] According to this report, a 2D phosphorus carbide compound (β_0-PC) monolayer (PCM) can be considered an efficient anode material for Na and other alkali ions. The PCM shows good electric conductivity and a high charge/discharge rate owing to the low diffusion barrier Na. Notably, the PCM possesses high stiffness constants along the armchair and zigzag direction, which prevent any structural degradation during battery cycling. Another report by Haitao Huang et al. carried out a similar theoretical analysis on a V_3C_2 MXene. [42]

Figure 16.4a shows a DFT study of Na ion intercalated Ti_3CN nanosheets. [33] The relaxed crystal structures and the equivalent electron density plots are illustrated in Figure 16.4b. [33] The projected density of states (PDOS) and the split wave DOS of Ti 3d pass the Fermi level, confirming the metallic characteristic of Ti_3CN. [33] This observation confirms that the electronic states of the composite elements (Na, C, N, and Ti) overlap, allowing for the interaction of Na ions in the Ti_3CN structure. [33] The interaction of Na ions in Ti_3CN is preferred over other ions, as the MXene structure provides a larger number of active sites for the Na ions to occupy. [33] The Na storage performance in Ti_3CN further relates to the presence of nitrogen, which modifies the electronic structure of Ti_3C_2. Figure 16.4c depicts the change in charge of certain $Ti_3C_2O_2$ (C_B–C_D) elements, as determined by the Ti, C, and O atom distances to the nearest Na ions. The structural atoms, Ti, C, and O, in this instance, experience distinct oxidation states following sodiation as a result of alterations in their coordination environments caused by the presence of four different

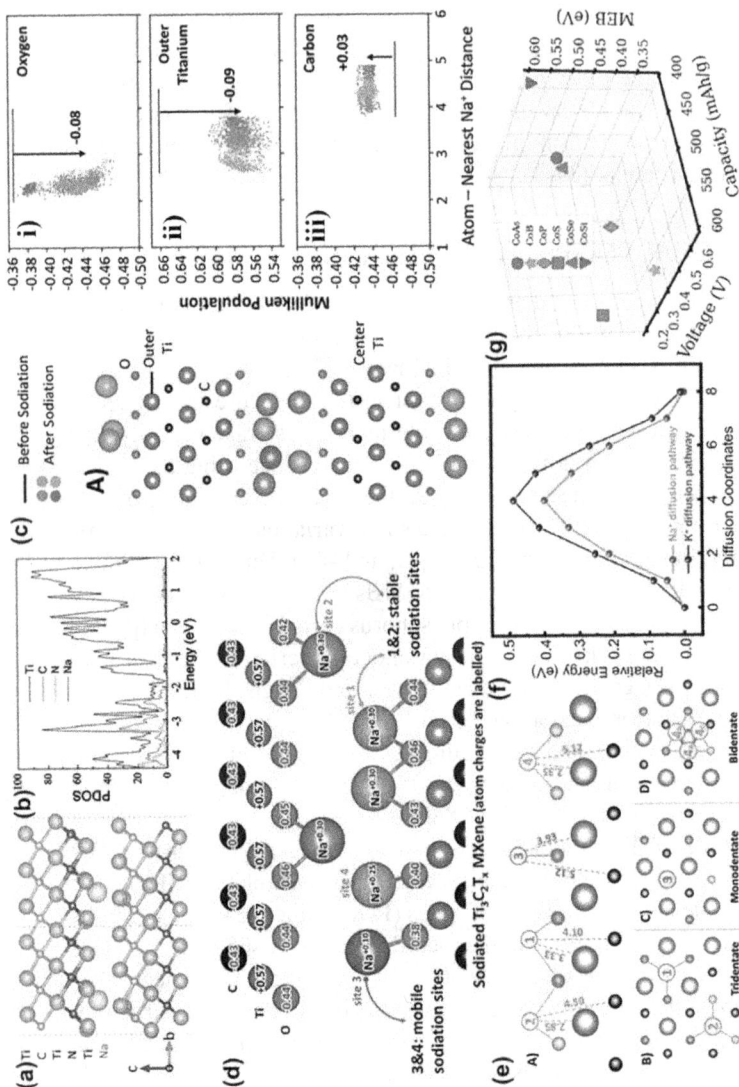

FIGURE 16.4 The sodium storage mechanism of MXenes and their effect on the structure and performance (a) Optimized geometry of Na intercalated Ti3CN. (b) The projected density of states of Ti3CN. (c) Sodiated MXene (Na1Ti3C2O2) structure with four sodium storage sites within the Ti3C2O2 layers. (ci–ciii) illustrate the change in Mulliken charge populations prior to (black line) and following (coloured dots) sodium intercalation. (d) An illustration of the structure of Ti₃C₂Tₓ and its behaviour as an anode in Na-ion capacitors. (e) Predicted tridentate, monodentate, and bidentate Na+ intercalation sites at the MXene interface. (f) Impact of the adsorption strength on the mobility of alkali ions as seen via the relative energy at different diffusion coordinates of Cu1.75Se and the MXene/CNRib. (g) Cobalt-based anti-MXene materials and their efficiency as anode materials for SIBs.

Reprinted with permission from [33] (b), [43] (e), and [45] (g). Copyright 2023, American Chemical Society. Panel (f) reprinted from [44]; no permission is required to reuse the figure.

Na ions occupying intercalation sites (C_A). According to the Mulliken population analysis, the Na ions are oxidized, with an average of 0.28 e$^-$ removed per Na ion occupying a site within the MXene structure. [43] The four sodium storage sites of $Ti_3C_2O_2$ represented in Figure 16.4c demonstrate that the redox kinetics occurring at the surface change drastically depending on the Na ion locations. [43] Figure 16.4d illustrates the titanium carbide ($Ti_3C_2T_x$) MXene structure and its behaviour as a promising electrode for Na-ion capacitors. [43] Understanding the correlation between surface redox reactions and cations inserted within the MXene structure is essential for better charge transfer and as well as determining sodium storage behaviour. Computational simulations have detected the presence of sodiation sites and the formation of sodium domains at interfaces through neutron pair distribution function analysis. [43] When used in a Na-ion capacitor, pre-sodiated MXenes exhibit highly reversible sodium storage performance, further confirming the practicality of chemical pre-intercalation as an appropriate priming strategy for MXene electrodes. [43] Figure 16.4e depicts the four potential sites for Na ions within the $Ti_3C_2T_x$ structure, along with the coordinates and distances to the nearby atoms. Among these sites, site 4 is the most stable intermediate path for the transport of Na ions within the MXene structure. [43] As a case study, Figure 16.4f illustrates the effect of the adsorption strength on alkali ion mobility and the relative energies of the most suitable diffusion paths for Na and K at the interface of $Cu_{1.75}Se$ and the MXene. [44] Because of its higher radius and atomic weight, K transport possesses a higher energy barrier. This theoretical calculation indicates that the synergetic effects of MSe and the MXene have an advantageous effect on the adsorption and diffusion kinetics of Na/K ions. [44] There is a competing class of materials known as anti-MXenes, wherein the transition metal "M" element is positioned between two nonmetal "X" element atomic layers in an MX_2-type configuration that is the inverse of MXenes. Figure 16.4g illustrates the performance of a series of six cobalt-based anti-MXene materials such as CoAs, CoB, CoP, CoS, CoSe, and CoSi and anode materials for SIBs. [45] These findings demonstrate that, in comparison to previously well-researched 2D materials like MoS_2 (146 mAh g^{-1}), Cr_2C (276 mAh g^{-1}), and expanded graphite (284 mAh g^{-1}), Co-anti-MXenes (a recently discovered 2D material) have superior specific charge capacities in the range of 390–590 mAh g^{-1}. They have shown a great affinity to sodium atoms (−0.55 to −1.16 eV), as well as, on average, low sodiation voltages (0.2–0.64 V) and small diffusion energy barriers (0.32–0.59 eV). This proves that Co-anti-MXenes can make ideal anode materials for SIB applications. [45]

16.7 STRATEGIES FOR LONG-CYCLE-LIFE AND HIGH-RATE-CAPABILITY SIBS

The porous structure of $Ti_3C_2T_x$ enables higher conductivity and enhanced Na ion storage kinetics, as shown in Figure 16.5a. Due to these improved properties enabled by the structure of MXenes, electrodes with an extended cycle life of over 1000 cycles at 1 A g^{-1} and a specific capacity of 166 and 124 mA h g^{-1} at 1 and 10 A g^{-1}, respectively, have been established (Figure 16.5b). Thus, MXenes with structures

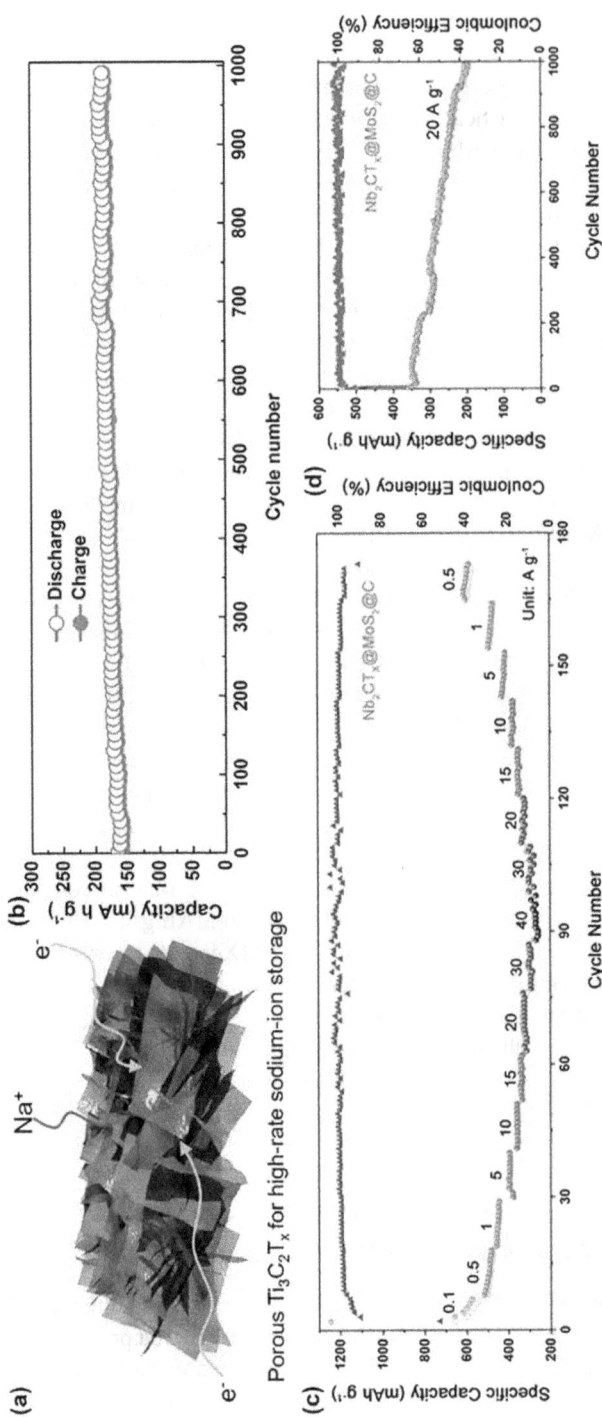

FIGURE 16.5 (a) Schematic of porous $Ti_3C_2T_x$ nanosheets. (b) $Ti_3C_2T_x$ nanosheets have exceptional cycling performance over 1000 cycles at 1 A g^{-1}. (c) The addition of a carbon coating to a Nb_2CT_x-MoS_2 framework demonstrated an enhanced rate performance at high current densities. (d) This Nb_2CT_x-MoS_2-C framework was able to maintain its performance even after 1000 cycles at 20 A g^{-1}.

Reprinted with permission from [49] (b) and [50] (d). Copyright 2023, American Chemical Society.

similar to $Ti_3C_2T_x$ have demonstrated a possibility for Na ion storage in large-scale applications. Figure 16.5c depicts carbon-coated Nb_2CT_x-MoS_2 nanosheets used to design a three-dimensional cross-linked structure which has high mechanical strength. In order to facilitate the Na ion kinetics and long-term stability, the hierarchical carbon covering also possesses a high degree of elasticity and electrical conductivity. The carbon-coated Nb_2CT_x-MoS_2 anode showed a specific capacity of 403 mA h g^{-1}, higher-order capacity retention (~88.4% from 0.1 to 1 A g^{-1}), and extremely long cycling stability over 1000 cycles at 1.0 A g^{-1}. Most crucially, at current densities between 20 and 40 A g^{-1}, the Nb_2CT_x@MoS_2@C anode can achieve fast charge and discharge, which is essential for the practical application of SIBs, as depicted in Figure 16.5d.

Yang et al. studied the in situ transformation of Ti_3C_2 to $NaTi_2(PO_4)_3$ through a two-step reaction consisting of oxidation with TiO_2 and a follow-up reaction with $(PO_4)_3$ and Na^+. [46] The synthesis of MXene@NTP-C utilizes an etching–exfoliation technique to obtain a Ti_3C_2 MXene which is then calcinated to produce the desired product. The MXene@NTP-C, as an anode material for SIBs, showed a superior rate capability and a specific capacity of 188 mAh g^{-1} while recovering to the initial current density (0.1 A g^{-1}). The outstanding rate capability at high current densities enabled by MXene@NTP-C anode demonstrates the significant role of the C coating, which enhances the overall conductivity by providing a conducting network that helps the maximum utilization of the active materials of the MXene@NTP-C anode. [46] The long cycle performance of the MXene@NTP-C anode over 2000 cycles shows a capacity retention of 74% at 1 A g^{-1} with a 100% Coulombic efficiency, further realizing the potential of the MXene@NTP-C anode for SIB applications. [46] Luo et al. investigated the intercalation of S-atoms into a Ti_3C_2 MXene (CT-S@Ti_3C_2) structure. [47] The process involves a simple procedure involving treating the MXene with cetyltrimethylammonium bromide (CTAB), followed by thermal diffusion with the S atom. [47] The S atom intercalation of Ti_3C_2 enhances the d spacing, which further facilitates sodium storage. The CT-S@Ti_3C_2_450 electrode showed an excellent rate capability at current densities between 0.1 and 15 A g^{-1}. [47] The cycling performance of an CT-S@Ti_3C_2_450//AC SIC electrode at 2 A g^{-1} has shown a capacity retention of 73.3% over 10,000 cycles with a 100% Coulombic efficiency, which illustrates the potential capability of these electrodes for large-scale energy storage applications. [47]

16.8   SODIUM STORAGE MECHANISM OF MXENES

Through the use of scanning transmission electron microscopy (STEM) with aberration-correction in combination with DFT computations, (Figure 16.6a) the intercalation of Na and Al into the Ti_3C_2X phase can be observed at the atomic level. From Figure 16.6a, the STEM results confirm the presence of OH-, F-, and O- functional groups, as well as Na ions' preference to occupy sites adjacent to Ti and C atoms of the Ti_3C_2 structure. [48] Extensive Na intercalation via two-phase transition and solid-solution interactions results in the formation of a dual layer of Na atoms within the Ti_3C_2X interlayer. [48] Furthermore, the horizontal sliding of the Ti_3C_2X monolayer that occurs upon aluminium (Al) ion intercalation facilitates

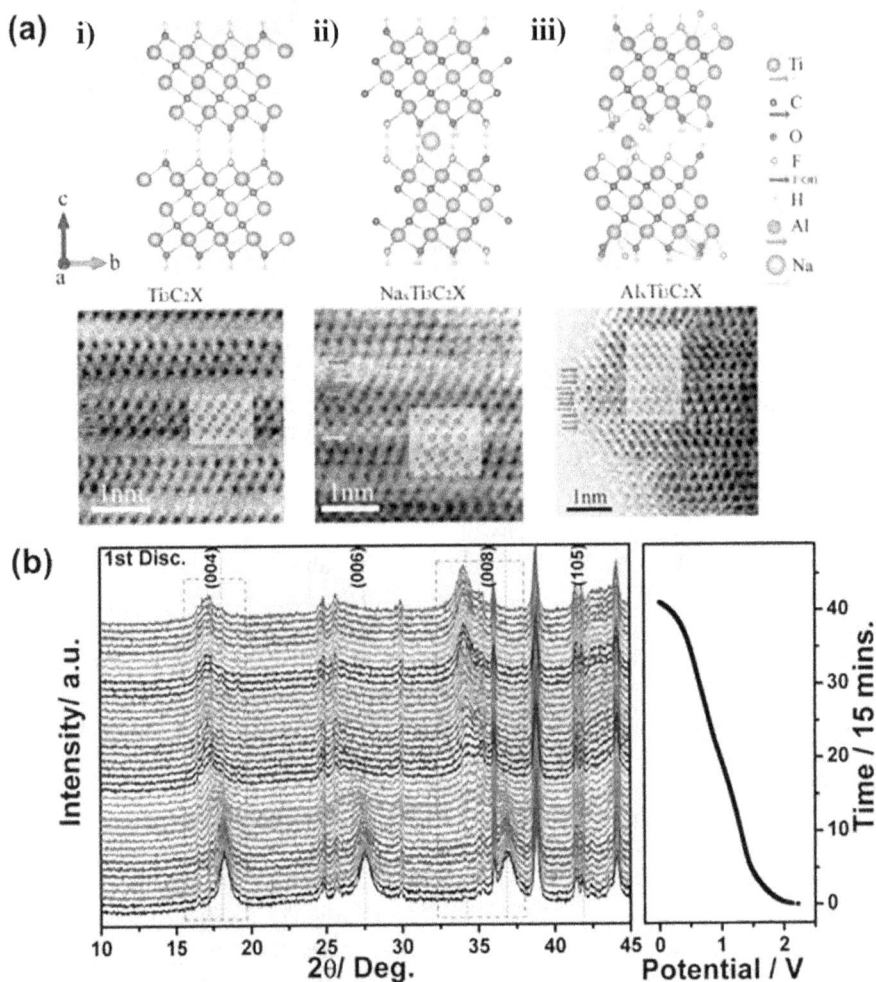

FIGURE 16.6 Intercalation of Na and Al ions into MXene structures result in varying the ion storage mechanisms. (a) DFT illustration and aberration-corrected STEM micrographs of the horizontal sliding of the MXene monolayer triggered by the intercalation of the Al ions (identified with green arrows). (b) In situ XRD demonstrating the reversible intercalation of Na ions through a combination of two-phase transitions and solid-solution reactions.

accelerated transport of the Na ions to and from the active sites. DFT calculations using the model described by Wang *et. al* help to develop a greater understanding of the physical and chemical properties of MXenes. [48] These results help to elucidate fundamental design strategies and their effects on applications of Ti_3C_2X as a potential anode material in terms of rate and long-term cycling SIBs. [48] Figure

16.6b uses in situ XRD to elucidate the Na storage mechanism of Ti_3C_2X at voltages between 0.01 and 2.50 V at 10 mA g^{-1}. [48] The phase transition from Ti_3C_2X to $NaTi_3C_2X$ was observed during Na ion intercalation at different Na ion intercalation potentials. The slow fading and shifting of the Ti_3C_2X peaks towards lower angles was demonstrated and, in combination with the emergence of new peaks, supports the occurrence of solid-solution reactions. These structural variations are highly reversible in the subsequent deintercalation in the interlayers of Ti_3C_2X, as reported by Wang *et al.* [48]

The requirements of such in situ investigations of MXene electrodes are to confirm the Na ion intercalation process through understanding the Na ion mobility near active sites of carbides regardless of elemental components of the MXene structure. Understanding the intercalation mechanism allows for the design of electrodes focused on diminishing volume change and minimizing the mechanical strain developed during repeated insertion and extraction of Na ions during the cycling of MXene-based anodes in SIBs. [10, 13, 14, 24, 48–50]

16.9 CONCLUSIONS

In summary, MXenes such as 2D transition metal carbides, nitrides, and carbonitrides have shown exceptional properties such as fast sodium storage kinetics, high electrical conductivity, and high specific surface area, revealing immense potential as alternatives to LIBs. The presence of surface functional groups facilitates enhanced interfacial charge transfer, while the 2D layers allow for easy access to active conducting sites, promoting intense redox activities. Further, theoretical DFT studies have provided insight into the sodium storage mechanism, and recent breakthroughs in MXene synthesis and exfoliation have spurred further development of these materials as electrodes. However, there are still challenges to address, including the irreversible capacity fading in the initial discharge cycle, which is much higher than the expected theoretical values, necessitating further development efforts. The low specific capacity after the first cycle and the intense capacity fading due to the irreversibility of the redox reactions in the following cycles require innovative solutions and a comprehensive understanding of sodium storage mechanisms through advanced experimental techniques. As discussed, it has been observed that the specific capacity, cycle life, and rate capabilities can be improved by enhancing the expanded interlayer spacing through surface functionalization, adjusting cationic composition between the 2D layers, and incorporating C composites to increase the conductivity of MXene-based electrodes. These modifications have enabled MXenes to compete with established materials such as carbon or other 2D materials such as MoS_2. Furthermore, recent investigations into cobalt-based anti-MXenes, heterolayer structures, and composites have demonstrated additional active sites, low diffusion energy barriers, and low average sodiation voltages for the intercalation of Na ions, resulting in excellent sodium storage properties. Among the alternatives to LIBs, the use of MXenes as anodes in SIBs is a field in its infancy which still has tremendous potential that should be expanded on in the design of high-performance storage devices.

REFERENCES

1. H. Li, "Practical evaluation of Li-ion batteries," Joule, 3, 4, 911–914, 2019, doi: 10.1016/j. joule.2019.03.028.
2. G. Crabtree, "The coming electric vehicle transformation," Science, 366, 6464, 422–424, 2019, doi: 10.1126/science. aax0704.
3. C. Vaalma, D. Buchholz, M. Weil, and S. Passerini, "A cost and resource analysis of sodium-ion batteries," Nat. Rev. Mater., 3, 18013, 2018, doi: 10.1038/natrevmats.2018.13.
4. J. Sun, H. W. Lee, M. Pasta, H. T. Yuan, G. Y. Zheng, Y. M. Sun, Y. Z. Li, and Y. Cui, "A phosphorene-graphene hybrid material as a high-capacity anode for sodium-ion batteries," Nat. Nanotechnol., 10, 980, 2015, doi: 10.1038/nnano.2015.194.
5. M. D. Slater, D. Kim, E. Lee, and C. S. Johnson, "Sodium-ion batteries," Adv. Funct. Mater., 23, 947–958, 2013, doi: 10.1002/adfm.201200691.
6. D. Wu, X. Li, B. Xu, N. Twu, L. Liu, and G. Ceder, "$NaTiO_2$: A layered anode material for sodium-ion batteries," Energy Environ. Sci., 8, 195–202, 2015, doi: 10.1039/C4EE03045A.
7. M. R. Panda, A. R. Kathribail, A. Ghosh, A. Kumar, D. Muthuraj, S. Sau, W. Yu, Y. Zhang, A. K. Sinha, M. Weyland, Q. Bao, and S. Mitra, "Blocks of molybdenum ditelluride: A high-rate anode for sodium-ion battery and full cell prototype study," Nano Energy, 64, 103951, 2019, doi: 10.1016/j.nanoen.2019.103951.
8. M. R. Panda, A. R. Kathribail, A. Sarkar, Q. Bao, and S. Mitra, "Electrochemical investigation of $MoTe_2$/rGO composite materials for sodium-ion battery application," AIP Conf. Proc., 1961, 030033, 2018, doi: 10.1063/1.5035235.
9. S. Guo, J. Yi, Y. Sun, and H. Zhou, "Recent advances in titanium based electrode materials for stationary sodium-ion batteries," Energy Environ. Sci., 9, 2978–3006, 2016, doi: 10.1039/C6EE01807F.
10. M. Naguib, V. N. Mochalin, M. W. Barsoum, and Y. Gogotsi, "MXenes: A new family of two-dimensional materials," Adv. Mater., 26, 992–1005, 2014, doi: 10.1002/adma.201304138.
11. O. Mashtalir, M. Naguib, V. N. Mochalin, Y. Dall'Agnese, M. Heon, M. W. Barsoum, and Y. Gogotsi, "Intercalation and delamination of layered carbides and carbonitrides," Nat. Commun., 4, 1716, 2013, doi: 10.1038/ncomms2664.
12. X. Zhang, Z. Zhang, and Z. Zhou, "MXene-based materials for electrochemical energy storage," J. Energy Chem., 27, 73–85, 2018, doi: 10.1016/j.jechem.2017.08.004.
13. D. Er, J. Li, M. Naguib, Y. Gogotsi, and V. B. Shenoy, "Ti_3C_2 MXene as a high-capacity electrode material for metal (Li, Na, K, Ca) ion batteries," ACS Appl. Mater. Interfaces, 6, 11173–11179, 2014. doi: 10.1021/am501144q.
14. R. Garg, A. Agarwal, and M. Agarwal, "A review on MXene for energy storage application: Effect of interlayer distance" Mater. Res. Express, 7, 022001, 2020, doi: 10.1088/2053-1591/ab750d.
15. Y. Gogotsi, and B. Anasori, "The rise of MXenes," ACS Nano, 13, 8491–8494, 2019, doi: 10.1021/acsnano.9b06394.
16. Y. Shi, M. Li, Y. Yu, and B. Zhang, "Recent advances in nanostructured transition metal phosphides: Synthesis and energy-related applications," Energy Environ. Sci., 13, 4564–4582, 2020, doi: 10.1039/D0EE02577A.
17. M. R. Panda, S. Sau, R. Gangwar, D. Pandey, D. L. Muthuraj, W. Chen, A. Chakrabarti, A. Banerjee, A. Sagdeo, Q. Bao, M. Majumder, and S. Mitra, "An excellent and fast anodes for lithium-ion batteries based on the 1T'-$MoTe_2$ phase material," ACS Appl. Energy Mater., 5, 8, 9625–9640, 2022, doi: 10.1021/acsaem.2c01280.

18. X. Zhang, L. Hou, A. Ciesielski, and P. Samorì, "2D materials beyond graphene for high-performance energy storage applications," Adv. Energy Mater., 6, 1600671, 2016, doi: 10.1002/aenm.201600671.

19. M. Pogorielov, K. Smyrnova, S. Kyrylenko, O. Gogotsi, V. Zahorodna, and A. Pogrebnjak, "MXenes—A new class of two-dimensional materials: Structure, properties and potential applications," Nanomaterials, 11, 3412, 2021, doi: 10.3390/nano11123412.

20. Y. Li, Z. Peng, N. J. Holl, M. R. Hassan, J. M. Pappas, C. Wei, O. H. Izadi, Y. Wang, X. Dong, C. Wang, Y. W. Huang, D. H. Kim, and C. Wu, "MXene—Graphene field-effect transistor sensing of influenza virus and SARS-CoV-2," ACS Omega, 6, 10, 6643–6653, 2021, doi: 10.1021/acsomega.0c05421.

21. M. R. Lukatskaya, O. Mashtalir, C. E. Ren, Y. Dall'Agnese, P. Rozier, P. L. Taberna, M. Naguib, P. Simon, M. W. Barsoum, and Y. Gogotsi, "Cation intercalation and high volumetric capacitance of two-dimensional titanium carbide," Science, 341, 1502–1505, 2013, doi: 10.1126/science.1241488.

22. P. B. Michelle, D. Tyndall, and V. Nicolosi, "The potential of MXene materials as a component in the catalyst layer for the oxygen evolution reaction," Curr. Opin. Electrochem., 34, 101021, 2022, doi: 10.1016/j.coelec.2022.101021.

23. K. A. Muhammad, and X. Maowen, "A mini-review: MXene composites for sodium/potassium-ion batteries," Nanoscale, 2020, doi: 10.1039/D0NR04111D.

24. M. K. Aslam, Y. Niu, and M. Xu, "MXenes for non-lithium-ion (Na, K, Ca, Mg, and Al) batteries and supercapacitors," Adv. Energy Mater., 2000681, 2020, doi: 10.1002/aenm.202000681.

25. W. Bao, L. Liu, C. Wang, S. Choi, D. Wang, and G. Wang, "Facile synthesis of crumpled nitrogen-doped MXene nanosheets as a new sulfur host for lithium–sulfur batteries," Adv. Energy Mater., 8, 1702485, 2018, doi: 10.1002/aenm.201702485.

26. M. Naguib, M. Kurtoglu, V. Presser, J. Lu, J. Niu, M. Heon, L. Hultman, Y. Gogotsi, and M. W. Barsoum, "Two-dimensional nanocrystals produced by exfoliation of Ti_3AlC_2," Adv. Mater., 23, 4248–4253, 2011, doi: 10.1002/adma.201102306.

27. X. Sang, Y. Xie, M. W. Lin, M. Alhabeb, K. L. Van Aken, Y. Gogotsi, P. R. C. Kent, K. Xiao, and R. R. Unocic, "Atomic defects in monolayer titanium carbide ($Ti_3C_2T_x$) MXene," ACS Nano, 10, 9193–9200, 2016, doi: 10.1021/acsnano.6b05240.

28. M. Ghidiu, M. R. Lukatskaya, M. Q. Zhao, Y. Gogotsi, and M. W. Barsoum, "Conductive two-dimensional titanium carbide 'Clay' with high volumetric capacitance," Nature, 516, 78–81, 2014, doi: 10.1038/nature13970.

29. Y. Dall'Agnese, P. L. Taberna, Y. Gogotsi, and P. Simon, "Two-dimensional vanadium carbide (MXene) as positive electrode for sodium-ion capacitors," J. Phys. Chem. Lett., 6, 2305–2309, 2015, doi:10.1021/acs.jpclett.5b00868.

30. N. Michael, M. Olha, C. Joshua, P. Volker, L. Jun, H. Lars, Y. Gogotsi, and W. B. Michel, "Two-dimensional transition metal carbides," ACS Nano, 6, 2, 1322–1331, 2012, doi:10.1021/nn204153h.

31. X. Huang, and P. Wu, "A facile high-yield, and freeze-and-thaw-assisted approach to fabricate MXene with plentiful wrinkles and its application in on-chip micro-supercapacitors," Adv. Funct. Mater., 30, 1910048, 2020, doi: 10.1002/adfm.201910048.

32. M. Li, J. Lu, K. Luo, Y. Li, K. Chang, K. Chen, J. Zhou, J. Rosen, L. Hultman, P. Eklund, P. O. A. Persson, S. Du, Z. Chai, Z. Huang, and Q. Huang, "Element replacement approach by reaction with lewis acidic molten salts to synthesize nanolaminated MAX phases and MXenes," J. Am. Chem. Soc., 141, 4730–4737, 2019, doi: 10.1021/jacs.9b00574.

33. J. Zhu, M. Wang, M. Lyu, Y. Jiao, A. Du, B. Luo, I. Gentle, and L. Wang, "Two-dimensional titanium carbonitride Mxene for high-performance sodium ion batteries," ACS Appl. Nano Mater., 1, 6854–6863, 2018, doi: 10.1021/acsanm.8b01330.

34. D. Zhao, M. Clites, G. Ying, S. Kota, J. Wang, V. Natu, X. Wang, E. Pomerantseva, M. Cao, and W. B. Michel, "Alkali-induced crumpling of $Ti_3C_2T_x$ (MXene) to form 3D porous networks for sodium ion storage" Chem. Chomm., 2013, doi: 10.1039/c8cc00649k.

35. M. Anayee, N. Kurra, M. Alhabeb, M. Seredych, M. N. Hedhili, A. H. Emwas, H. N. Alshareef, B. Anasori, and Y. Gogotsi, "Role of acid mixtures etching on the surface chemistry and sodium ion storage in $Ti_3C_2T_x$ Mxene," Chem. Chomm., 2013, doi: 10.1039/D0CC01042A.

36. Y. Wu, P. Nie, J. Wang, H. Dou, and X. Zhang, "Few-layer MXenes delaminated via high-energy mechanical milling for enhanced sodium-ion batteries performance," ACS Appl. Mater. Interfaces, 9, 39610–39617, 2017, doi: 10.1021/acsami.7b12155.

37. J. Zhu, M. Wang, M. Lyu, Y. Jiao, A. Du, B. Luo, I. R. Gentle, and L. Wang, "Two-dimensional titanium carbonitride Mxene for high-performance sodium-ion batteries," ACS Appl. Nano Mater., 2018, doi: 10.1021/acsanm.8b01330.

38. L. Du, H. Duan, Q. Xia, C. Jiang, Y. Yan, and S. Wu, "Hybrid charge-storage route to Nb_2CT_x MXene as anode for sodium-ion batteries," Chemistry Select, 5, 1186–1192, 2020, doi: 10.1002/slct.201903888.

39. S. Kajiyama, L. Szabova, K. Sodeyama, H. Iinuma, R. Morita, K. Gotoh, Y. Tateyama, M. Okubo, and A. Yamada, "Sodium-ion intercalation mechanism in Mxene nanosheets," ACS Nano, 10, 3, 3334–3341, 2016, doi: 10.1021/acsnano.5b06958.

40. Y.-X. Yu, "Prediction of mobility, enhanced storage capacity, and volume change during sodiation on interlayer-expanded functionalized Ti_3C_2 MXene anode materials for sodium-ion batteries," J. Phys. Chem. C, 120, 5288–5296, 2016, doi: 10.1021/acs.jpcc.5b10366.

41. F. Li, X. Liu, J. Wang, X. Zhang, B. Yang, Y. Qu, and M. Zhao "A promising alkali-metal ion battery anode material: 2D metallic phosphorus carbide ($ß_0$-PC)," Electrochim. Acta, 258, 582–590, 2017, doi: 10.1016/j.electacta.2017.11.101.

42. K. Fan, Y. Ying, X. Li, X. Luo, and H. Huang "Theoretical investigation of V_3C_2 MXene as prospective high-capacity anode material for metal-ion (Li, Na, K, and Ca) batteries" J. Phys. Chem. C, 123, 18207–18214, 2019, doi: 10.1021/acs.jpcc.9b03963.

43. A. Brady, K. Liang, V. Q. Vuong, R. Sacci, K. Prenger, M. Thompson, R. Matsumoto, P. Cummings, S. Irle, H. W. Wang, and M. Naguib, "Pre-sodiated $Ti_3C_2T_x$ MXene structure and behavior as electrode for sodium-ion capacitors," ACS Nano, 15, 2, 2994–3003, 2021, doi: 10.1021/acsnano.0c09301.

44. J. Cao, J. Li, and D. Li, et al., "Strongly coupled 2D transition metal chalcogenide-MXene-carbonaceous nanoribbon heterostructures with ultrafast ion transport for boosting sodium/potassium ions storage," Nano-Micro Lett., 13, 113, 2021, doi: 10.1007/s40820-021-00623-5.

45. S. Banerjee, K. Ghosh, S. K. Reddy, S. R. Sharma, and K. C. Yamijala, "Cobalt anti-MXenes as promising anode materials for sodium-ion batteries," J. Phys. Chem. C, 126, 25, 10298–10308, 2022, doi: 10.1021/acs.jpcc.2c02459.

46. Q. Yang, T. Jiao, M. Li, Y. Li, L. Ma, F. Mo, G. Liang, D. Wang, Z. Wang, Z. Ruan, W. Zhang, Q. Huang, and C. Zhi, "In situ formation of $NaTi_2(PO_4)_3$ cubes on Ti_3C_2 MXene for dual-mode sodium storage," J. Mater. Chem. A, 6, 18525, 2018, doi: 10.1039/C8TA06995F.

47. J. Luo, J. Zheng, J. Nai, C. Jin, H. Yuan, O. Sheng, Y. Liu, R. Fang, W. Zhang, H. Huang, Y. Gan, Y. Xia, C. Liang, J. Zhang, W. Li, and X. Tao, "Atomic sulfur covalently engineered interlayers of Ti_3C_2 MXene for ultra-fast sodium-ion storage by enhanced pseudocapacitance," Adv. Funct. Mater., 29, 1808107, 2019, doi: 10.1002/adfm.201808107.

48. X. Wang, X. Shen, Y. Gao, Z. Wang, R. Yu, and L. Chen, "Atomic-scale recognition of surface structure and intercalation mechanism of Ti_3C_2X," J. Am. Chem. Soc., 137, 2715–2721, 2015, doi: 10.1021/ja512820k.

49. X. Xie, K. Kretschmer, B. Anasori, B. Sun, G. Wang, and Y. Gogotsi, "Porous $Ti_3C_2T_x$ MXene for ultrahigh-rate sodium-ion storage with long cycle life," ACS Appl. Nano Mater., 1, 505–511, 2018, doi: 10.1021/acsanm.8b00045.

50. Z. Yuan, L. Wang, D. Li, J. Cao, and W. Han, "Carbon-reinforced Nb_2CT_x MXene/ MoS_2 nanosheets as a superior rate and high-capacity anode for sodium-ion batteries," ACS Nano, 15, 4, 7439–7450, 2021, doi: 10.1021/acsnano.1c00849.

17 MXene-Based Flexible and Wearable Piezoresistive Physical Sensors

Dipak Dutta

17.1 INTRODUCTION TO PIEZORESISTIVITY

Among various types of pressure sensors like piezoresistive, [1–3] capacitive, [4–7] piezoelectric, [8,9] and triboelectric sensors, [10,11] piezoresistive pressure sensors have broad market prospects because of their uncomplicated structure, cost-effectiveness, comparatively low energy consumption, wide measurement range, very-high sensitivity, quick response time, fantastic thermal stability, etc. [12–16] Consequently, piezoresistive sensors have become vital components of numerous wearable and flexible electronic devices including smart glasses, smart watches, smart bracelets, etc., and received significant attention for industrial applications, including biomedical monitoring, Internet of Things, and human–computer interaction. [12,17,18]

Piezoresistivity is the term used to describe the change in electrical resistance exhibited by a material when it is subjected to mechanical strain or force. The roots of piezoresistivity can be traced back to the late 19th century, when Lord Kelvin made the groundbreaking observation of variation in the resistance of copper and iron wires when they were subjected to stress. [19] The term "piezo" finds its origins in the Greek word "piezen," which conveys the meaning of "to press." Conversely, it was Cookson who first introduced the term "piezoresistance" in 1935, characterizing it as the change in conductivity resulting from the application of stress. [20] Since then, tremendous progress has been made in developing piezoresistive materials for sensing applications; the first of this kind was based on semiconductor silicon.

Conductivity, or, better to say, a change in conductive pathways as a function of stress, is an essential criterion for a material to exhibit the piezoresistive effect. In materials that exhibit the piezoresistive effect, two main components are crucial: the geometric component and the resistive component. The geometric aspect arises from the observation that when a material is subjected to strain, changes in length and cross-sectional area directly impact its resistance.

The gauge factor (GF) of a material symbolizes the correlation between the applied strain (ϵ) and the resulting fractional change in resistance (R). It can be

DOI: 10.1201/9781003366225-17

mathematically expressed by using resistivity (ρ) and Poisson's ratio (v) in the following manner:

$$GF = \frac{\Delta R / R}{\Delta L / L} = \frac{\Delta R / R}{\epsilon} = 1 + 2v + \frac{\Delta \rho / \rho}{\epsilon} \tag{17.1}$$

Here, ϵ represents the ratio of absolute change in length and the original length ($\epsilon = \Delta L/L$), and ΔR is the change in resistance.

In a strain gauge, which is a device used for measuring strain, the GF is a crucial element. A higher value of the gauge factor results in increased sensitivity of strain detection. For different materials, the GF depends on material properties and the conduction mechanism. In general, metals have a gauge factor between 2 and 4, and for an incompressible liquid, e.g., mercury, it is 2.0. This means that if a (incompressible) liquid is stretched by 1%, its resistance raises by 2%. [21]

In metals, the geometric effect accounts for the majority (1.4–2.0) of the GF, while the change in resistivity is considered insignificant. On the contrary, in single-crystal semiconductors like Si and Ge, there is a remarkable change in resistivity that is 50- to 100-fold greater than the geometric term. This significant disparity prompted the introduction of a new concept called the "piezoresistive coefficient" (π) for strain gauges based on semiconductors. Thus, the piezoresistive coefficient now can be defined as follows:

$$\frac{\Delta \rho}{\rho} = \pi_t \sigma_t + \pi_l \sigma_l \tag{17.2}$$

where σ stands for stress and the subscripts t and l correspond to transverse and longitudinal, respectively. [22] Remarkably, n- and p-type piezoresistors display contrasting behaviors regarding changes in resistance and direction-dependent strain. The piezoresistive coefficient (π), in terms of magnitude and sign, is subject to the influence of several factors, including temperature, doping level, semiconductor type (n or p), crystallographic direction, and the alignment between current and stress directions. [22,23]

17.2 MATERIALS AND SENSING MECHANISMS OF PIEZORESISTIVE SENSORS

17.2.1 MATERIALS

Piezoresistive materials typically encompass brittle metals or semiconductors and are frequently manufactured on sturdy supports or combined with suitable insulating matrices. Among these materials, silicon (Si) holds a prominent position as the favored choice for room-temperature piezoresistive sensors. This preference is attributed to its abundant availability (Si is the second most abundant element, constituting roughly 27.7% of the Earth's crust), remarkable GF, and the presence of well-established, sophisticated fabrication technologies. [24] However, the performance of piezoresistors based on silicon (Si) experiences a notable degradation beyond 125°C.

This decline can be attributed, in part, to challenges such as current leakage. The low band gap of Si (1.1 eV) plays a role in facilitating the effortless thermal excitation of electrons, further contributing to this deterioration.

In high-temperature applications like aircraft and automobile engines, there is a consistent need for materials with wider band gaps. As a substitute for silicon (Si), silicon carbide (SiC) has emerged as a promising option. SiC offers advantages such as a higher sublimation temperature (1800°C) and chemically super inertness as compared to Si. Depending on the polytype used, SiC exhibits varying band gaps ranging from 2.3 to 3.4 eV, enabling it to withstand demanding conditions effectively. [25] Although the fabrication of SiC devices is more complex as compared to Si, and challenges arise in growing oxide films due to the presence of carbon, SiC-based devices have achieved noteworthy advancements. Notably, these devices have attained a GF of 35 [26] and exhibited excellent resilience in challenging environments, including high working temperatures (200–500°C) [27] and high-impact conditions (40,000 × g). [28]

Diamond, with its superior strength, Young's modulus, and thermal conductivity as compared to Si, presents an intriguing alternative. Remarkably, single-crystal diamond has the potential to achieve an exceptional GF of up to 3836. However, the main challenge lies in the intricate process of growing single-crystal diamond. On the other hand, polycrystalline diamond can exhibit similar sensitivity to SiC while being more feasible for production. [29]

Carbon nanotubes (CNTs), because of their unique structure and strong carbon–carbon bonds, can exhibit the combined best properties of polymers, carbon fibers, and metals. [30] CNTs are also the most flexible and strongest materials known, and can be metallic, semiconducting, or insulating in nature depending on the way the graphene sheets are rolled with respect to the axis of the nanotube. Interestingly, a GF of 400–850 could be achieved for a (small-gap) semiconducting CNT. [31] However, the performance was limited by the difficult ohmic connection to the CNT. Later on, improved growth and integration approaches [32] alleviated this deficiency, leading to the development of more sensitive CNT and CNT-composite piezoresistive sensors.

Other carbonaceous materials that showed huge potential for piezoresistive sensors are graphene, carbon black, carbon fibers, etc. Composites of two or more of these materials have also been reported to show good piezoresistive responses. [33,34] While Rahimi et al. reported a GF of 20,000 for a composite of graphene and CNT in polydimethylsiloxane (PDMS), [35] Li et al., by embedding graphene woven fabrics (GWFs) in PDMS reported an extremely high GF of 10^6. [36]

Conductive polymer composites (CPCs) have gained significant recognition as exceptionally promising piezoresistive sensors across a wide spectrum of applications. These composites usually consist of conductive components integrated into an insulating polymer matrix. Their versatility makes them suitable for many applications, including health monitoring, human motion detection, wearable electronic devices, and human–machine interfaces. Interestingly, conductive polymers, because of being organic in nature, have inexhaustible sources and offer excellent flexibility in terms of both ease of processability and tunability of chemical compositions and surface functionality modification, which eventually opens better opportunity for improving the performance of CPCs. Conductive polymers like polypyrrole (PPy), polyaniline (PANI), and poly(3,4-ethylenedioxythiophene)-polystyrenesulfonate (PEDOT:PSS), along with their derivatives, are widely used in various applications. However, it is noteworthy to mention that in terms of overall effectiveness and electronic properties, they are unable to outperform their metallic counterparts, graphene, carbon nanotubes (CNTs), or doped semiconductors.

17.2.2 SENSING MECHANISMS

17.2.2.1 Electronic Circuits for Piezoresistive Devices: Wheatstone Bridge Circuits

A piezoresistive sensor's active layer material not only needs to have certain conductivity properties; it also needs to be sensitive to external force stimuli.

The electrical output signal produced by a piezoelectric strain gauge is generally insufficient for practical applications. Therefore, it necessitates amplification using a low-noise circuit before undergoing any analog or digital conditioning processes. A general form of conditioning circuit comprises a Wheatstone bridge circuit that

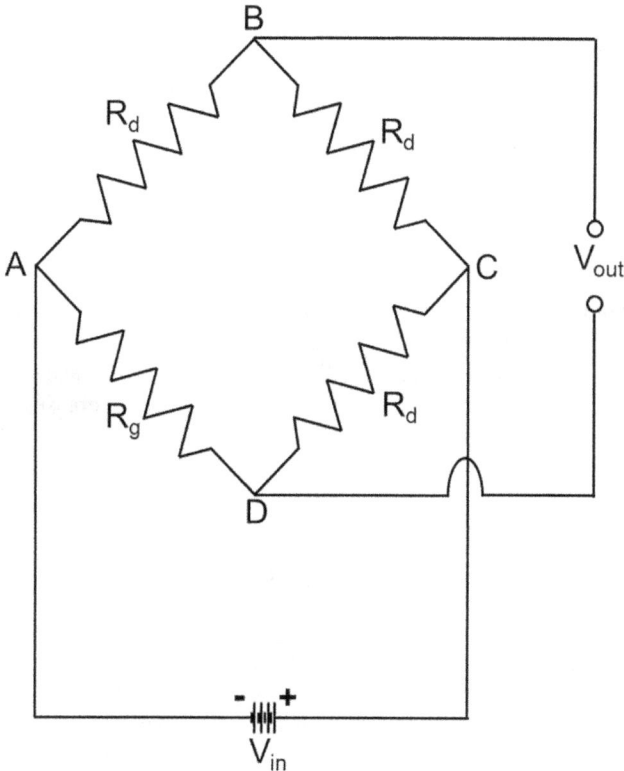

SCHEME 17.1 Wheatstone bridge circuit.

is usually used to monitor the change in resistance of the piezoresistive sensor is shown in Scheme 17.1. The arms of the Wheatstone bridge configuration comprise of dummy resistances (R_d) for compensation, which are positioned on the unstrained section of the test specimen and the strain gauges $(R_g = R_d + \Delta R)$. The input supply (V_{in}) is connected to two nodes of the bridge in a differential DC configuration, while the electronic preamplifier via the output is connected to the remaining two nodes. The output of the circuit (V_{out}) is correlated to the change in resistance of the strain gauge by Equation 17.3. [37]

$$V_{out} = \left(\frac{R_F}{R_D}\right)\left(\frac{?}{1+?}\right)\left(\frac{V}{R_D / R_F \left(2+?\right)/\left(1+?\right)}\right) \tag{17.3}$$

where, $? = \dfrac{\Delta R}{R_d}$.

17.2.2.1.1 Semiconductors

In semiconductors, the GF is mainly influenced by the changes in conductivity (resistivity) resulting from alterations in the number of free electrons and their mobility due to lattice deformation under strain. This effect contributes significantly to the GF in semiconductors, and its geometric contribution is comparable to that observed in metal strain gauges. The effect of piezoresistivity in semiconductors is explainable on the basis of the band theory, which provides quite different explanations for n- and p-type semiconductors. By taking p-type semiconductors (p-silicon, p-Si) as an example, the conduction phenomenon is dependent on the movement of holes in the valence band. The valence band in this case consists of two different energy functions corresponding to two different energy bands. When a traction is applied along the [111] direction of p-Si, the energy bands and the hole distribution change. The valence band splits apart into two bands. Those holes losing energy relocate to the top band, where they possess diminished mobility, thus producing "heavy holes;" on the other hand, the "light holes" move to the lower bands (Figure 17.1a). As the number of holes with reduced mobility increases (the average hole mobility, μ decreases), the resistivity (ρ) increases and the conductivity decreases [37,38]; these are correlated by Equation (17.4):

$$\rho = \frac{1}{ne\mu} \tag{17.4}$$

where, n and e are the concentration and charge of the carrier, respectively.

For n-type Si, the mobility of an electron is influenced by the direction of its motion in the lattice. Thus, when the lattice spacing is modified by employing external stress, the energy levels are modified as well. Upon application of a tensile stress in one direction, the minimum energy level (of the constant energy surfaces) rises along that direction and diminishes along the two other directions (longitudinal effect). The opposite, however, happens for compressive stress (transverse effect). The transverse effect, in fact, breaks the equivalence of the energy minima, resulting in upward and downward shifts in energy. In order to minimize the free energy,

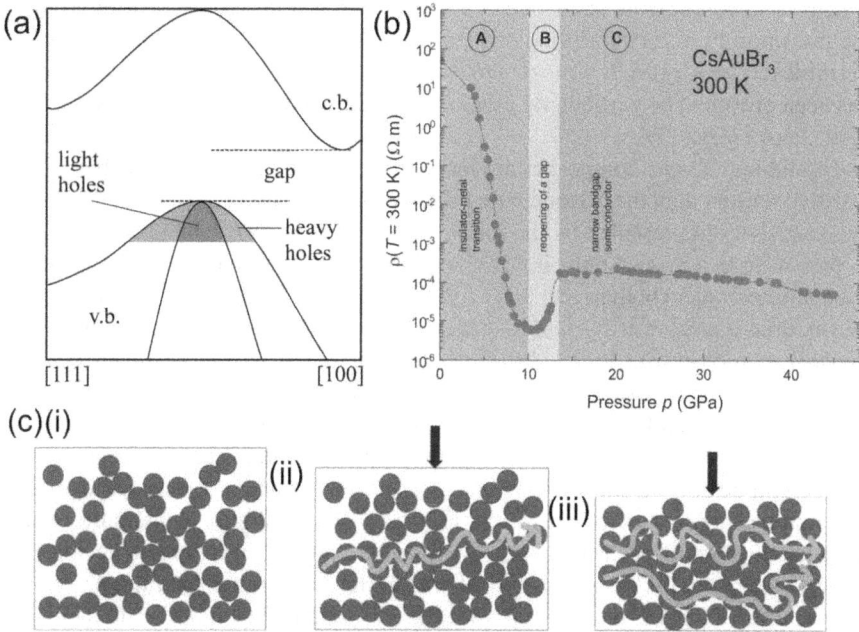

FIGURE 17.1   (a) Energy band structure of silicon along the [111] and [100] k directions [37]. (b) Evolution of the room temperature (300 K) resistivity of the compound $CsAuBr_3$ in a range of pressure up to 45 GP [39]. (c) Schematic illustration of the reduction in resistivity (in the perpendicular direction) when a polymer–metal composite is pressed; (i) randomly distributed metallic particles (filled circles) loaded insulating polymer matrix resulting in a high-resistance material. (ii) Upon application of external pressure, the metallic particles are forced together, generating conducting percolation pathways and (iii) supplementary conducting pathways generated due to the add-on compressive force quickly decreasing the resistance of the material [40].

electrons are transported to the lower positioned minima, resulting in decreased mobility and hence a corresponding increase in resistivity. [37]

This change in band structure as a function of pressure can be better understood by considering a mixed-valence compound, $CsAuBr_3$, which transforms from a mixed-valence state to a single-valence state when subjected to pressure. Its pressure dependent (room temperature) resistivity can be categorized into three clearly distinguishable regions. A change in resistivity higher than six orders of magnitude and a corresponding transition from an insulator to metal phase is observed at pressures of less than 10 GPa. For intermediate pressures in the range (10–14) GPa, the resistivity moves through a minimum of less than 10^{-5} Ωm and then starts to increase again. Above a pressure of 14 GPa up to 45 GPa, a semiconducting phase is realized with nearly no change in resistivity (Figure 17.1b). The investigation of the band structure and density of states has revealed that the significant alteration in resistivity under pressure is not due to any band crossings at the Fermi level. Rather, the transition from mixed to single valence causes two bands to be anchored at the Fermi level, which results in their movement towards each other. This mechanism has been proven to be particularly efficient in rapidly increasing the density of states at the Fermi level. [39]

Insulating polymer composites are more attractive candidates for (flexible) piezoresistive sensors than the pristine bulk materials because of their excellent flexibility and sensitivity (represented by the ratio of the change in resistivity due to variation in pressure). In a recent study, it was shown that a polymer matrix filled with semiconductor particles (both n- and p-type Bi_2Te_3 particles, for a volume fraction ~ 35 ± 1%) exhibits a piezoresistive response that is 10^5 times greater than that exhibited by the bulk semiconductor itself. [40] To explain the excellent piezoresistive response of the polymer composite, a percolation model where the formation of percolation pathways is induced by the pressure has been suggested (Figure 17.1c). When pressure is applied, the volume fraction of the filler particles in the composite increases, resulting in an enhancement in percolation connectivity and a corresponding order-of-magnitude decrease in resistivity (Figure 17.1cii). Additional pressure results in creating more of such percolation pathways, resulting in a further decrease in resistivity (Figure 17.1ciii). This model was found to be better than the previously reported tunneling mechanism both in terms of its simplicity and the data being fitted. [40,41]

17.3   MXENES AS PIEZORESISTIVE SENSOR MATERIALS

The main limitations of conventional materials used in (flexible) wearable piezoresistive sensors stem from the difficulties encountered during their processing. Traditional approaches, such as blending and coating methods, often yield fragile sensors that are unsuitable for durable and robust wearable devices. Consequently, there is a pressing need for advanced processing techniques and meticulous designs to fulfill the requirements of wearable applications. Notably, the extended use of a sensing device under external force, which is crucial for real-time applications, inevitably causes a permanent shift in the original positions of the active sensing material. This alteration severely undermines the sensitivity of the device. High-modulus graphene and CNTs, having excellent mechanical strength, electronic properties, and strain sensitivity, are excellent candidates for piezoresistive sensors; however, their structural restrictions hinders atomic mobility within them. Moreover, the traditional rigid piezoresistive sensors available on the market face the limitations of being large, not easily bendable, uncomfortable for wear, and fairly unportable. On the other hand, wearable sensors necessitate exceptional flexibility, good elasticity, high sensitivity, and excellent stability. [42,43] The active sensing materials being the heart of the sensor device, it is very essential to reasonably design their geometric configuration in order to achieve the best performance. [44] Sensors with outstanding performance demand high sensitivity, a broad range of detection, high integration, and also multifunctionality. Achieving this level of performance surpasses the capabilities of preexisting devices and demands innovative design and technology. [45,46] These limitations of preexisting piezoresistive sensors have triggered scientists to search for alternative materials, which have much simplified formulation and fabrication steps and can, in fact, fulfill the requirements of real-world applications.

MXenes, two-dimensional (2D) materials known for outstanding characteristics such as high mechanical strength, superior hydrophobicity, metallic conductivity, and large surface area, have gained considerable attention as highly promising materials for pressure-sensitive sensors. Their remarkable properties facilitate the development of sensors that demonstrate exceptional sensitivity, performance, and potential for future applications. The large specific surface area of MXenes imparts an excellent binding force to the substrate, resulting in exceptional mechanical properties. Additionally, the material's metallic conductivity and tunable layer spacing enable a wide range of adjustable resistance, rendering it extremely sensitive to pressure.

In the year 2017, Ma *et al.* first realized that the interlayer spacings in MXenes could be simply altered by the application of external compressive forces, and that gave MXenes a new platform to act as piezoresistive sensors. [47] Figure 17.2a shows the working micromechanism of a MXene-based sensor where the changed interlayer distance leads to a change in internal resistance (R_c) and, hence, conductivity as a consequence of external force. Total resistance is defined as $R_{Total} = R_I + R_c$, where R_I is the MXene's initial resistance. The corresponding circuit diagram is shown in Figure 17.2b.

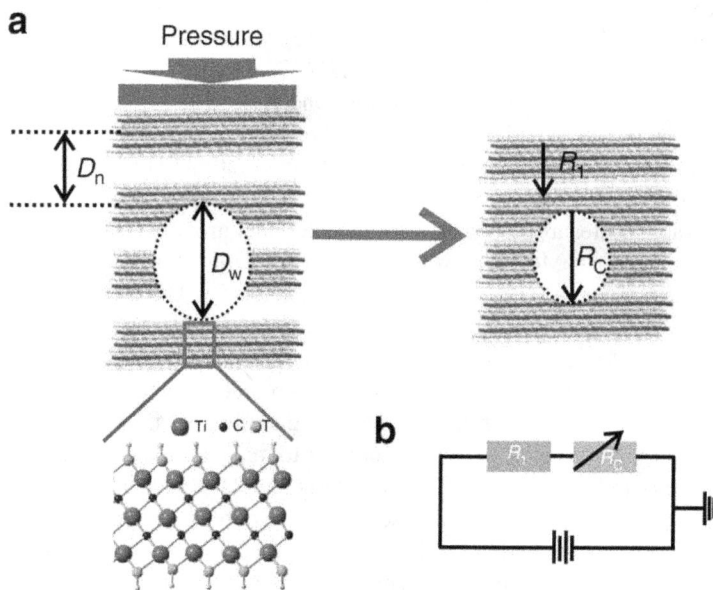

FIGURE 17.2 (a) Working micromechanism of a MXene when used in a sensor, whereby the interlayer spacings decrease under compressive pressure, and (b) corresponding equivalent circuit diagram used to detect minute changes in resistance of the sensor.

To investigate microstructural changes in a MXene material under pressure, in situ transmission electron microscopy (TEM) was used, where pressure is exerted through a tapered needle-point nanoindenter, which reveals changes in lamellar arrangements. The TEM results clearly demonstrate that the interlayer distance is greatly changed under the application of external force. The resultant device also has shown great sensitivity ($GF \sim 180.1$), quick response time (< 30 ms), and extremely high mechanical reversibility ($> 4,000$-fold increase in magnitude). The MXene-based sensors were utilized to perceive and differentiate minuscule human activities, viz., coughing, swallowing, and joint bending. [47]

Further proof of the alteration in resistance on account of changed interlayer spacing of a MXene driven by the applied pressure was obtained from the experiments of Yue *et al.*, where an LED-RGB light was connected to form a series circuit with a MXene sponge, which in turn was fabricated with a polyvinyl alcohol (PVA) nanowire spacer. [48] During the experiment, it was realized that enhancing the applied force on the sensor resulted in a proportional increase in the brightness of the LED light and a change in its color. These changes indicate a decrease in the sensor's resistance.

With their exceptional electronic properties, distinct morphological characteristics, customizable surface functionalities, and controllable synthesis methodologies, MXenes exhibit desirable qualities as physical sensors. These attributes position MXenes as compelling options for the advancement of physical sensor technology. Additionally, they have demonstrated the ability to detect delicate human body motions, viz., blinking, pulse vocal movements, coughing, and bending, and could potentially be used as electronic skin. [12,49–53] MXene-based composite systems are capable of achieving high sensitivity ($GF \sim 772.6$) over a wide sensing range of 30–130% strain. [54] Nonetheless, a major concern is also the stretchability of the sensing platform. The broad sensing potential of MXenes is hindered by several significant limitations, including the tendency of the ultrathin 2D layers to restack and the absence of a constricted porous structure. These limitations reduce the number of surface-active sites available for surface functional groups, which can limit their potential applications. Moreover, when additional functional materials are integrated with MXenes, it can lead to unfavorable properties. [50] The limited compression space between MXene layers is also a concern for achieving high sensitivity under small amounts of pressure. To address these issues, creating porous 3D interconnected networks is a potential solution that can provide sensors with improved flexibility, lighter weight, and significantly increased surface area.

Four strategies are usually adopted in order to make a MXene-based piezoresistive framework to be porous. [50] By employing templates such as ice, water, sulfur, or polymers, it is possible to create a self-standing porous structure. Through commonly used fabrication methods like freeze drying, hard templating, or chemical reduction and subsequent template removal, a highly resilient and interconnected porous network of MXenes can be formed. For instance, in the freeze-drying process, the nanosheet of the MXene precursor is expelled into the boundary of the ice crystal such that after subliming the ice, a free-standing, porous, interconnected framework can be obtained. Strong interactions between the MXene sheets, even strong H-bonding or van der Waals interactions, are reported to produce porous

MXene frameworks with large surface areas. [55] The second approach to a porous MXene framework involves depositing or inserting MXene nanosheets into a porous substrate that can hold the former, where the porous substrate is pre-synthesized or simultaneously forms with the deposition and insertion of MXene. The third process involves loading or coating a porous framework on the surface of MXene, and the fourth involves drilling in-plane pores on the MXene sheets. Porous MXene frameworks exhibit sensitivity as piezoresistive sensors due to the phenomenon in which the application of pressure causes the pore size to decrease, consequently increasing the number of conductive pathways. [56]

17.4 MXENE-BASED COMPOSITES AS PIEZORESISTIVE SENSOR MATERIALS

Even though MXenes show great potential for piezoresistive sensors, their actual prospects to bear great pressure and exhibit bifunctionality are still limited much beyond expectation due to challenges regarding the flexibility (fragile nature), functionality, and stability of MXenes. Ongoing efforts are focused on improving the functionalization/modification and optimizing the synthesis/reaction conditions to achieve further advancements. One of the possibilities to alleviate this issue is to blend MXenes with high-strength/elastomeric materials into a composite. Over the last few years, several composites of MXenes have been reported to perform as excellent piezoresistive sensors, including MXene–polymer composites, MXene–oxide composites, MXene–carbon nanotube (CNT) composites, MXene–graphene composites, etc.

17.4.1 MXene–Carbonaceous Material Composites

17.4.1.1 MXene–Carbon Nanotube (CNT) Composites

The integration of MXenes and carbon nanotubes (CNTs) in a composite structure presents a highly promising and innovative platform for advancing the development of strain sensors. This composite exhibits tremendous potential for various applications, including interactive processing, prosthetic feedback, and wearable sensing. Cai et al. revolutionized strain sensing with their pioneering work on a sandwich-like strain sensor that utilizes a MXene–CNT composite. Their innovative approach involved the application of a layer-by-layer spray-coating process to deposit exfoliated $Ti_3C_2T_x$ MXene flakes and hydrophilic single-walled carbon nanotubes (SWCNTs) onto a flexible latex rubber film, resulting in a remarkable advancement in the field. [57] This strain sensor demonstrates remarkable versatility, offering a diverse range of features. These include an impressively high GF ranging from 4.4 to 772.6, an exceptionally low limit of detection (≤0.1%), a customizable sensing range that spans from 30% to 130%, and outstanding durability with a lifespan exceeding 5000 cycles. The sandwich-like structure is clearly visible in the SEM images (Figure 17.3a and 17.3b). When affixed to human skin, the MXene–CNT–latex composite device shows great potential in real-time monitoring of subtle physiological signals and capturing extensive body motions. In Figure 17.3, it is shown how two-syllable words (carbon, Figure 17.3d, and sensor, Figure 17.3e) and three-syllable

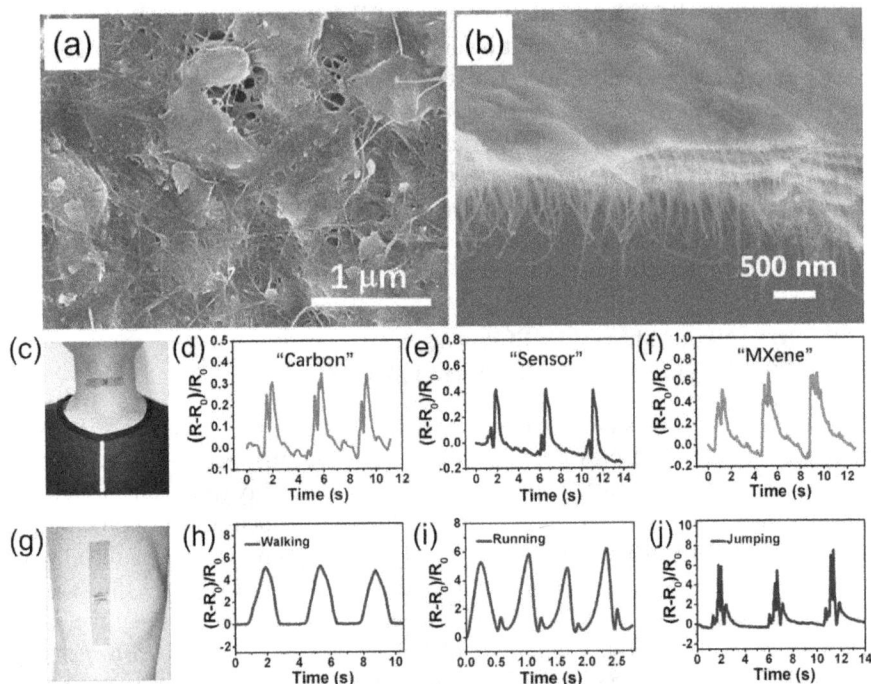

FIGURE 17.3 SEM images depicting the (a) top view and (b) cross-sectional view of a sandwich-like MXene/CNT composite. (c) Photograph of MXene/CNT/latex strain sensor when fixed on the throat of a person and (d–f) corresponding responses obtained during speaking "carbon," "sensor," and "MXene," respectively, by the person. Photograph of the sensor fixed at the knee of a person (g) and (h–j) corresponding resistive changes from detecting leg movement of the person while walking, running, and jumping respectively.

words (MXene, Figure 17.3f) can reproducibly produce two and three peaks with different intensities, suggesting a great potential for this device in phonation rehabilitation activities and human–machine interactions. Additionally, the MXene–CNT–latex strain sensor demonstrates a dependable capacity for monitoring physiological signals and tracking body movements, as illustrated in Figure 17.3g–j. This characteristic renders it exceptionally well-suited for integration into robotic systems and prosthetic devices.

In their study, Chen *et al.* utilized a solvent evaporation method to produce a flexible composite material consisting of a MXene, SWCNT, and PVP. Within the film, the resulting composite showcased a hierarchical dendritic-lamellar structure, with the MXene nanosheets intercalated among the dendritic branches formed by the SWCNT bundles. [58] The piezoresistive electrode sensor comprising MXene/SWCNT/PVP was printed onto natural rubber substrates. The dendritic-lamellar structure was clearly seen in the cross-sectional SEM images of the composite when compared with the pristine MXene. The developed tactile composite sensor was utilized for voice recognition and pulse sensing, exhibiting excellent sensitivity (165.35 kPa^{-1}), an extremely low detection limit (0.69 Pa), and outstanding stability for > 10,000 cycles. Moreover, it has been shown that this sensor can be used as an artificial skin that is able to locate the touch point, with a sensitivity of n^2 by using 2n connection ports.

Fan *et al.* developed a SWCNT–Ti$_3$C$_2$T$_x$ MXene composite by using a vacuum-assisted filtration technique followed by a thermal shrinkage method in which SWCNT prevents the MXene sheets from restacking and improves electrical performance. [59] The films displayed remarkable sensitivity over a broad pressure range, spanning from 33–130 kPa. Sensitivity values of 116.15 kPa^{-1} were observed below 40 kPa, while between 40 kPa and 130 kPa, the sensitivity was measured at 12.7 kPa^{-1}. Furthermore, the strain sensors exhibited a rapid response time (13 ms) and demonstrated long-term stability (> 6000 cycles). By conducting real-time monitoring of human physiological signals, such as finger movements, voice detection, and wrist pulse, the sensor's performance was validated. This verification underscores its potential for diverse applications, including electronic skin, medical devices, and various wearable devices.

Zheng *et al.* were pioneers in the development of a versatile sensor that utilized a bark-shaped composite film of CNT–MXene composite on a fiber surface. The film was fabricated by employing a roll-to-roll, layer-by-layer assembly method. The sensor demonstrated exceptional capabilities in detecting small objects, including plastic toys, coins, and ceramic toys, while also monitoring human movements. Notably, it exhibited remarkable sensitivity, with values of 0.245 kPa^{-1} for wide range of pressures (0.128–1.9 kPa) and 0.060 kPa^{-1} for a pressure range of 1.9–12.9 kPa. [60]

Similarly, a few more reports also recently appeared in the literature where composites of CNT and MXene were able to act as efficient pressure sensors; one such reports shows a GF as high as 9022 with wide sensing range of ~210%. [61]

Tracking, monitoring, and reconstructing full-body motions are becoming increasingly vital across a diverse array of applications. These include high-precision movement detection, motor sign recognition, athlete performance analysis, rehabilitation assessment, human–machine interaction, and personalized avatar reconstruction in augmented and virtual reality. Particularly during the COVID-19 pandemic, there has been a growing necessity for technologies capable of capturing subtle body motions like trembling and shivering. Wearable strain sensors have emerged as a promising alternative to conventional imaging systems, which face limitations such as immobility, high cost, bulky equipment, and inadequate suitability for distant and dynamic objects. However, current wearable devices and commercial sensors offer limited options for customization, resulting in challenges in aligning the sensor's working range with the strain changes occurring in joints and muscles. This disparity leads to inaccurate sensing outcomes and a low signal-to-noise ratio. Furthermore, the accurate transmission, storage, and processing of collected raw data present additional complexities due to the involvement of multiple signal acquisition channels. Addressing these challenges will require a multidisciplinary effort, encompassing advancements in hardware and software development, as well as optimization of the sensor–circuit interface. Yang *et al.* developed a wearable piezoresistive sensor that was composed of a nanolayer composite of $Ti_3C_2T_x$ MXene, SWCNT, and PVA, which overcomes the limitations of existing wearable devices by incorporating in-sensor machine learning (ML) models. By leveraging wireless streaming or edge computing, these models possess the ability to perform comprehensive classification of full-body motion and facilitate the reconstruction of avatars. [62] Basically, they monitored the strain window of the wearable sensors by a combination of topographies that adjusts the crack propagation phenomena in the film and an adjustment of the nanolayer thickness and composition (relative fraction of MXene/SWCNT/PVA) in the composite. For this purpose, they fabricated four self-suspended films designated as M_p, M_w, $M_{p\text{-}w\text{-}p}$, and $M_{w\text{-}p\text{-}w}$ (Figure 17.4a–d), where M represents the polystyrene–MXene nanolayer, "p" the planar topography, and "w" the wrinkle-like feature in the nanolayer. The M_p, $M_{p\text{-}w\text{-}p}$, M_w, and $M_{w\text{-}p\text{-}w}$ sensors exhibit linear working windows of 3–6% (ideal for back waist bending, with an average strain change of ~5%), 8–24% (suitable for shoulders, with an average strain change of ~10%), 25–39% (ideal for elbows, with an average strain change of ~30%), and 35–50% (suitable for knees, with an average strain change of ~50%), respectively, when subjected to parallel (to the wrinkle axis) stretching. These sensors possess high *GF* values, with M_p, M_w, $M_{p\text{-}w\text{-}p}$, and $M_{w\text{-}p\text{-}w}$ exhibiting values of 3400, 1160, 1230, and 1470, respectively. By integrating wearable sensors from different body regions through a multi-channel connection and utilizing a machine learning (ML) chip, a cutting-edge edge sensor

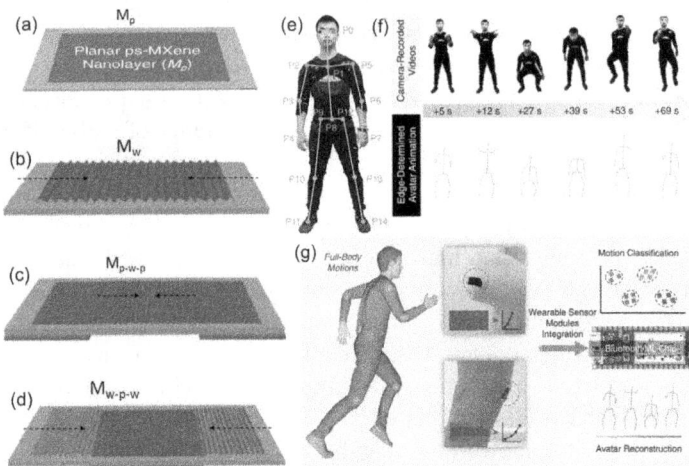

FIGURE 17.4 Four types of sensor films: (a) M_p, (b) M_w, (c) $M_{p\text{-}w\text{-}p}$, and (d) $M_{w\text{-}p\text{-}w}$. (e) A stickman avatar, consisting of 15 joints, implemented using a camera-recorded video as a reference. (f) A comparison made between the full-body motions of a volunteer and the avatar created using corresponding data from an edge sensor module. (g) The classification of full-body motion and the subsequent reconstruction of an avatar achieved by utilizing the data from the integrated Bluetooth unit and the ML chip.

module was created. This module empowers the in-sensor reconstruction of remarkably precise avatar animations, faithfully capturing continuous full-body movements and eliminating the need for supplementary computing devices (Figure 17.4f–g). In addition, a wireless sensor unit was developed by combining wearable sensors with Bluetooth chips, enabling seamless streaming of multi-channel strain sensing data. This data is utilized to train an artificial neural network (ANN) model, enabling accurate detection of various full-body motions.

17.4.1.2 MXene–Graphene Composite

Graphene, a 2D sp^2-hybridized allotrope of carbon that is nearly transparent and has outstanding mechanical properties and specific surface area, is also an excellent conductor of both heat and electricity. Piezoresistive sensors have previously been employed in combination with graphene and its derivatives, viz., graphene oxide (GO) and reduced graphene oxide (rGO), to enable various applications. A hybrid of graphene and MXene is considered an ideal piezoresistive platform due to their synergistic effect, provided by the highly active surface area of graphene and the strong electron conduction pathways provided by the MXene sheets. Several composites of graphene and MXenes were reported over last few years in the literature.

Ma *et al.* produced a MXene/rGO aerogel composite with 3D architecture having micro- and mesoporous networks by using an ice-templated freeze-drying technique followed by annealing in an inert atmosphere. [56] In a 3D aerogel, MXenes were wrapped by the large rGO sheets, thus avoiding the oxidation of the former. The aerogel could withstand a strain of 60% and recovered when the pressure was released, suggesting its excellent mechanical stability, which is much better than the single counterparts. The sensor operated based on the principle that compression alters the contact within the aerogel structure, resulting in an augmented number of conductive pathways (Figure 17.1c). The sensor fabricated from a MXene/rGO composite, showcased remarkable piezoresistive properties. These included an exceptional sensitivity of up to 22.56 kPa^{-1}, a quick response time (<200 ms), and remarkable cycle stability (>10,000 cycles). The sensor also exhibited voice recognition and identified jugular venous and wrist pulse beating.

Yang *et al.* reported a layered composite structure consisting of $Ti_3C_2T_x$ MXenes, graphene, and PDMS, which, during stretching, can be categorized into two distinct layers: a brittle top layer composed of tightly stacked irregular MXenes (together with few sheets of multilayer graphene) and a flexible bottom layer consisting of sheets of multilayer graphene inserted in a polymer (PDMS) substrate. [63] The operation of the sensor primarily is based on the principle of crack generation and propagation in the upper layer during the stressing process. The dissipation of tensile stress through this mechanism leads to a significant increase in the *GF*. Additionally, the flexible bottom layer ensures the preservation of conductive pathways throughout the entire working range (Figure 17.5a). The sensor employs the cooperative action of the upper and lower layers to exhibit high sensitivity (detection limit ~0.025%) and a linear strain response across a wide detection range (*GF* ~ 190.8 in the 0–52.6% strain range, and 1148.2 in the 52.6–74.1% range), as well as excellent cycling stability (>5000 cycles). The sensor exhibited effective detection of diverse human motions when attached to various body parts. Furthermore, the sensor showed high sensitivity when placed on the chest and abdomen, capable of distinguishing between three different breathing patterns during yoga exercises (see Figure 17.5b–c).

Later, more reports also appeared in the literature where graphene–MXene composites were shown to exhibit excellent piezoelectric sensor performance. [64–67] Among these reports, Xu *et al.* showed an interesting prospect of rGO/MXene-based sandwich-structured piezoresistive sensors with hierarchical microspines. The lower and upper films each consisted of 280 mesh sandpaper as a template on which PDMS, rGO, MXene particles, and MXene nanosheets were sequentially coated. The middle film consisted of an 800-mesh sandpaper template on which PDMS microspines followed by MXene nanosheets were coated. [64] The sensor's multi-level and multi-layered architecture allowed for a phased response and a wide detection window up to 70 kPa. The device exhibited remarkable performance, boasting response and recovery times of 40 ms and 80 ms, respectively, all while maintaining exceptional stability throughout 1000 fatigue cycles. Moreover, it effectively detected various

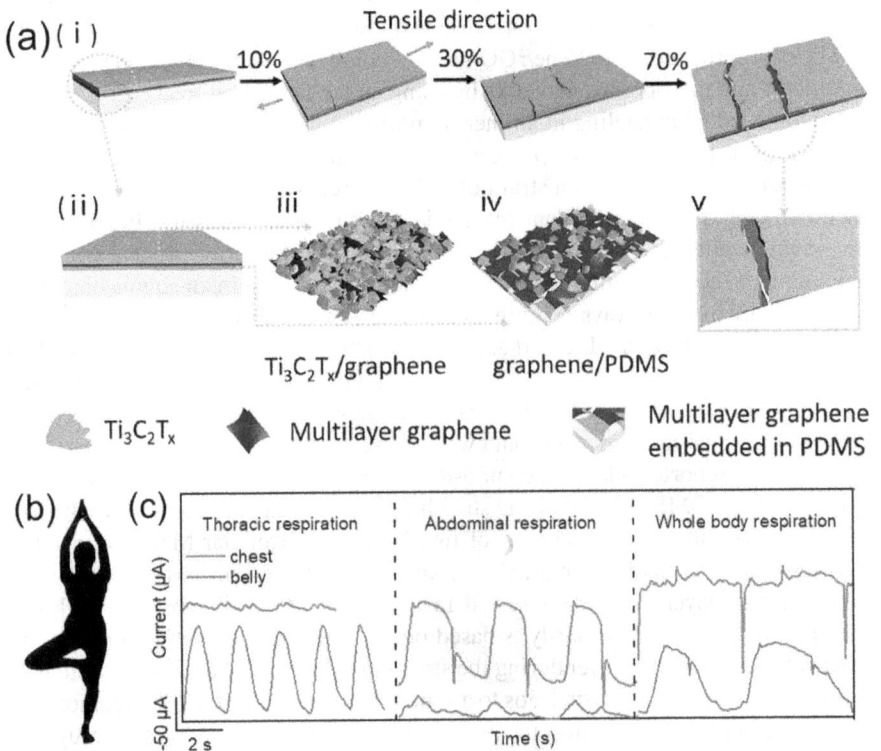

FIGURE 17.5   (a) Schematic illustration of the $Ti_3C_2T_x/G_{0.5}$/PDMS-based ($Ti_3C_2T_x/G_{0.5}$ top layer is brittle while the PDMS bottom layer is flexible) strain sensor at various stages of stretching. (b) Schematic representation of a mannequin and (c) corresponding recorded current signals illustrating breathing modes during yoga practice.

human motions such as pulse beats, cheek bulging, nodding, finger bending, and even speech recognition.

17.4.1.3 MXene–Carbon Nanoparticle Composites

In carbonaceous materials, in addition to CNTs and graphene, composites of MXenes with other forms like carbon nanofibers (CNFs), [68] carbon black, [69] carbon nanospheres, [70] etc., have also been reported to perform as excellent piezoresistive pressure sensors. Zhuo *et al.* reported a 3D carbon aerogel derived from the composite of a MXene and cellulose nanocrystals (CNCs) bound together through strong hydrogen bonds using a simple directional freeze-drying technique followed by annealing; the carbonized 3D architecture of c-MXene/c-CNC consists of continuous and oriented wave-shaped lamellar architectures. [71] The primary objectives of this study were threefold. Firstly, the investigation involved the incorporation of CNCs to mitigate the restacking of MXene sheets during the freeze-drying procedure. Secondly, the mechanical strength of the 3D aerogel was bolstered by fostering robust interactions between the MXene and CNCs. Lastly, wave-shaped layers were meticulously designed to facilitate substantial elastic deformation and ensure sufficient interspace for the reliable detection of response signals, even under minimal pressures or strains. The resulting aerogel showcased remarkable performance attributes, including an exceptionally high compression strain of 95%, outstanding long-term compression stability over 10,000 cycles at a 50% strain, unprecedented sensitivity (114.6 kPa^{-1}), and a broad linear pressure range spanning from 50 Pa to 10 kPa. Additionally, it was capable of detecting pressure changes as small as 1.0 Pa.

17.4.2 MXENE COMPOSITES WITH NON-CARBONACEOUS MATERIALS

17.4.2.1 MXene–Polymer Composites

Because they possess several desirable qualities, viz., flexibility, biocompatibility, transparency, and excellent stability at normal atmospheric conditions, which govern the essential characteristics of a flexible sensor, polymers remain vital components of many MXene-based composites. The common polymers used in MXene composites include polyurethane (PU), polyvinyl alcohol (PVA), chitosan (CS), polyvinylpyrrolidone (PVP), polydimethylsiloxane (PDMS), polyacrylate, poly(vinylidene-fluoride-co-trifluoroethylene) (P(VDF-co-TrFE)), polylactic acid (PLA), etc.

Song *et al.* reported a flexible piezoresistive pressure sensor consisting of a hollow-structured composite of MXene and polydimethylsiloxane (PDMS) that has excellent bending capabilities. [72] To commence the process, a skeleton-shaped Ni foam was immersed in an aqueous MXene dispersion. Subsequently, the MXene-coated Ni foam was dried, and PDMS was then infiltrated into the foam. The Ni foam was finally etched with acid, resulting in the formation of the 3D hollow-structured MXene–PDMS composite. The composite showed satisfactory structural stability, excellent sensitivity, long-term reliability (1000 cycles), and large-angle deformability (0–180°). Additionally, the composite showcased remarkable potential for wearable sensor devices, as evidenced by its exceptional performance in monitoring a wide range of activities. It effectively detected subtle actions like swallowing, facial muscle movements, stereo sound, and ultrasonic vibrations while reliably tracking more vigorous human movements, viz., finger twisting, bending, compression, and wrist and neck movements.

Li *et al.* successfully designed a versatile piezoresistive sensor by creating a composite material composed of polyurethane (PU), chitosan (CS), and a MXene. The researchers accomplished this by depositing CS onto the backbone of a PU sponge, which resulted in the formation of a positively charged CS/PU sponge. They then dip-coated the sponge with a negatively charged $Ti_3C_2T_x$ MXene, leading to the development of a MXene@CS@PU sponge-based sensor. This sensor demonstrates the ability to detect pressure signals across a wide range of magnitudes, encompassing both small and large variations. [73]

Acquiring flexible wearable sensors that possess both multifunctional features and deliver satisfactory performance presents a substantial challenge, mainly attributed to their inherent vulnerability to water. In this respect, Xu *et al.* reported a super hydrophobic 3D bifunctional (physical and chemical stimuli sensitive) flexible sensor based on a composite of a MXene with melamine sponge. [74] Typically, cleaned melamine sponge was impregnated with MXene flakes from an aqueous suspension with ultrasonication followed by drying and annealing at 280°C under N_2. The rich porous structure of the composite maintained superior flexibility, making it both an excellent pressure sensor for long-term testing of human respiration and a humidity sensor that quickly responses to change in humidity, thus making it an excellent bifunctional sensor.

To enable the seamless integration of pressure sensors into wearable electronics, point-of-care testing, and soft robotics applications, it is vital for these sensors to possess specific essential characteristics. These desirable attributes encompass high sensitivity, stretchability, the capacity to adhere comfortably to diverse and complex surfaces, and the ability to self-heal when damaged. To meet these requirements, Zhang *et al.* developed a composite hydrogel of MXene–PVA, which includes PVA, water, and anti-dehydration additives. The hydrogel was produced through a process of repeated mixing, rolling into a ball, and flattening by hand. [75] Under compressive strains, the MXene–hydrogel developed in this way demonstrates significantly higher sensitivity than under tensile strains. By combining this anisotropic property of the hydrogel, typically considered to be disadvantageous, with its viscoelastic nature, an intriguing material is created. This unique combination enables the convenient detection of motion direction and speed on the surface of the hydrogel. As a result, this material is well-suited for tasks that require precise and sensitive detection of subtle motion and traces, including handwriting, facial expressions, and vocal signals. In a separate study investigating the PVA composite of a MXene, researchers successfully showcased adjustable sensitivity and sensing range, achieving an outstanding sensitivity of 2320.9 kPa^{-1} and long-term durability spanning over 10,000 cycles. [76]

Even though MXenes have controllable structure and excellent electrical conductivity, which are beneficial for piezoresistive sensors, it is quite easy to lose the connection between the sheets during the stretching process, thus resulting in limited performance and limited span of detection. Conjugated conductive polymers such as PANI, PPy, and polyacetylene have exceptional electrical and flexible mechanical properties, making them potential hosts for MXene sheets, thus leading to outstanding device performance. Yang *et al.* developed a strain sensor that possesses stretchable, flexible, and bendable properties, resulting in a remarkable achievement. To construct the sensor, a transparent PDMS substrate was utilized along with a composite of MXene, PPy, and hydroxyethyl cellulose serving as conductive filler. Notably, this sensor demonstrated outstanding sensing capabilities, including exceptional long-term stability, an impressive detection range, and a rapid and efficient response. [77] Furthermore, by employing machine learning methods such as wavelet scattering and long short-term memory artificial neural network (WTSN-LSTM), this device achieved an impressive accuracy rate of over 96% in accurately recognizing a wide variety of English words, Arabic numerals, and Chinese characters. This significant accomplishment highlights its potential for various applications, including wireless human motion detection, medical health monitoring, and handwriting recognition.

17.4.2.2 MXene–Plasmonic Particle Nanocomposites

One of the vital requirements of wearable and transparent sensor devices is their self-healing after experiencing unexpected ruptures or undesired scratches without losing any sort of original performance. The incorporation of active material in the self-healing polymer, which is an alternative, however, is quite challenging because of their internal contradiction and tricky manufacturing procedure; the dispersibility and compatibility of the active component in the polymer matrix is also an inevitable factor for achieving the best performance. [78] Plasmonic particles possess several desirable properties, viz., broadband absorption, high light-to-heat conversion, high thermal conductivity, large aspect ratio, and ease of processing, making them an ideal option for developing transparent, healable, and wearable sensor devices. Silver nanowires (Ag NWs) and different shaped nanoparticles (Ag NPs) are highly conductive, mechanically flexible, and exhibit plasmonic characteristics, making them popular choices for developing strain and pressure sensors that can form percolation networks. Blending MXenes with Ag NWs/Ag NPs in a healable polymer matrix could be a great option for self-healable pressure sensor devices. Achieving a strain sensor that encompasses properties like self-healing, high sensitivity, and extensive stretchability presents considerable challenges. Zhang *et al.* developed a robust and healable strain sensor using a facile spraying process. This was achieved by using 1D semi-embedded Ag NWs, a 2D MXene, and a self-healing elastomer. [79] The self-healing dynamic cross-link elastomer, known as PUA-Bx, is a composite of the branch prepolymer PUA (PUA: poly(urethane acrylate)) that has been functionalized with furfurylamine and bismaleimide. This distinctive combination undergoes a Dials–Alder reaction, resulting in remarkable self-healing efficiency of over 88%. Regarding this matter, the Ag NWs offer dependable conductivity and durable interfacial bonding with the substrate under strain, and the MXene nanosheets tightly envelop the Ag NW scaffold, with the inherent brittleness of the latter providing considerable resistance variation during stretching. The sensor exhibits remarkable pressure detection capabilities, covering a wide range from 183 to 2260 kPa. It showcases high sensitivity, ranging from 0.5% to 96%. With a rapid response time of approximately 71±4.9 ms, the sensor provides timely feedback. Furthermore, it demonstrates exceptional durability, ensuring reliable performance over extended periods.

When constructing a strain gauge sensor using smart textiles, opting for soft, skin-friendly, biocompatible, and tensile-stable yarns and fabrics is highly advantageous. However, the self-reduction process of the conductive yarn usually prepared by using conductive polymers like PAni, PPy, etc., is difficult, tedious, and time-consuming. To mitigate this issue, Li *et al.* developed a 0D–1D–2D multidimensional composite of Ag and MXene (0D: Ag NPs, 1D: Ag NWs, and 2D: MXenes) on an elastic textile (yarn) as a strain sensor with excellent stretchability and sensitivity. [80] The composite comprises flexible 0D Ag nanoparticles that act as connectors between the 1D Ag nanowires and 2D MXene. By incorporating 1D Ag nanowires, the material's conductivity is significantly enhanced, ensuring its continuity and maintaining a high *GF* even when the yarn is subjected to substantial strains (200%). When fabricating a smart glove, by individually placing the device on the thumb, index, middle, ring, and little finger joints, it has been shown that the sensor can effectively recognize sign language, thus promoting a way to care for deaf and mute people for barrier free communication.

A few more strain sensors were developed based on the silver nanoarchitectures that involve Ag NWs/$Ti_3C_2T_x$ aerogel, [81] multilayer structured Ag NWs/waterborne PU/MXene fiber, [82] layer-by-layer assembled MXene nanosheets/Ag nanoflowers, [83] etc., that also exhibited excellent performance for piezoresistive pressure and strain sensors.

17.4.2.3 MXene–Biodegradable Material Composites

The utilization of flexible, breathable, and biodegradable pressure sensors has gained significant traction in a broad window of applications, encompassing wearable artificial skins, healthcare monitoring, and artificial intelligence. These sensors exhibit outstanding sensing performance while being lightweight, environmentally friendly, and contributing to a reduction in electronic waste. However, the traditionally used plastic or elastomers, with characteristics of impermeability, uncomfortableness, mechanical mismatches, and non-degradability, lack such feasibility for practical realization, thus always urging, despite being challenging, for the development of a flexible, breathable, and degradable pressure sensor. Chao *et al.* have made significant progress in this regard by creating a sensor using a composite material that was composed of MXene and silk fibroin (SF), which is a fibrous protein manufactured by the silkworm *Bombyx mori*. [84] Typically, two parts were constructed separately: First, by using dip-coating method, few-layer MXene nanosheets were uniformly deposited on porous and biodegradable silk fibroin nanofiber (MXene–SF) membrane through supramolecular interaction. The second part consists of a fabrication process that involved creating an interdigitated electrode, referred to as MXene ink–SF, by screen printing MXene ink onto an SF nanofiber membrane. The resulting two components were then put together face-to-face, forming the sensor. In this configuration, the MXene–SF membrane serves as the sensing layer, while the MXene ink–SF interdigitated setup functions as the electrode layer. The sensing principle is based on the modulation of contact resistance between the two layers, which can be adjusted during the application and release of external pressure. To assess the sensor's potential in detecting human motion, particularly in tactile signal detection and mapping pressure distributions, the researchers integrated the sensors into an artificial E-skin setup. When different fingers touched the E-skin, it exhibited remarkable capability for accurately distinguishing the pressure location and weight. Additionally, by integrating the sensor with a wireless transmitter, real-time pressure detection, such as wireless monitoring of human motion, became possible. It is notable to mention that the MXene/protein-based sensor demonstrated complete degradation in a 0.1 mol/L NaOH solution within 28 days, highlighting its degradability and environmental friendliness.

In their pioneering work, Zhang *et al.* accomplished a remarkable milestone in the field of highly sensitive pressure sensors. They successfully developed a composite aerogel by combining degradable cross-linked collagen fiber (CCF) with a MXene, leading to significant advancements in this area. This sensor showcased outstandingly rapid response times (0.30 s) and recovery times (0.15 s). The composite's 3D architecture, characterized by its porous nature, along with a precisely controlled interface between the CCF and MXene, imparts it with robust mechanical strength capable of withstanding applied pressure. Notably, the sensor exhibits a wide range of pressure detection capabilities (0–2.8 kPa), a low detection limit (0.4 kPa), and exceptional sensitivity (61.99 kPa^{-1}). These impressive characteristics position the sensor as a promising candidate for various applications, including monitoring of human motions and protecting the environment. [85]

In a different method, MXene nanosheets were incorporated into tissue paper, capitalizing on its porous structure, exceptional recyclability, affordability, biodegradability, and reliable elasticity. The objective of this integration was to create a flexible, wearable transient pressure sensor that offers outstanding sensitivity, reproducibility, wireless functionality, degradability, and a wide range of human–machine interfacing capabilities. The sensor could remarkably detect strains up to a range of 30 kPa. To achieve this, MXene-infused tissue paper was positioned between two biodegradable polylactic acid (PLA) sheets, with one of them featuring a patterned interdigitated conductive electrode. [86] The sensor displayed exceptional characteristics, including remarkable sensitivity, an impressive detection limit as low as 10.2 Pa, an exceptionally fast response time of 11 ms, minimal power consumption (10^{-8} W), outstanding reproducibility over 10,000 cycles, reliable biocompatibility, and environmentally friendly degradability. Additionally, the sensor proved its versatility by successfully inputting Morse code through touch on the device's surface and assembling an E-skin to detect diverse tactile signals. Additionally, the integration of the sensor with a wireless transmitter was demonstrated, enabling seamless wireless sensing for human–machine interfaces.

17.5 FUTURE PERSPECTIVES OF PIEZORESISTIVE SENSORS

Currently, the predominant method for preparing MXenes involves etching using highly corrosive hydrofluoric acid (HF), which poses significant safety hazards in both laboratory- and industrial-scale production. Thus, by considering the huge application prospects of MXenes, an alternative, much safer, and gentle technique is urgently required for their production in large quantity. Even though MXenes prepared by the CVD method are realized to be of the best quality in terms of purity, fewer surface defects, and no surface ligands, until now, only Mo$_2$C has been prepared by this method. Thus, the successful preparation of other MXenes by CVD and their facile transfer for the fabrication of devices is an urgent need.

The stability of MXenes is a genuine concern, as they undergo complete oxidation within a matter of days when exposed to open air. Even when stored in argon at low temperatures, the slow oxidation of MXenes cannot be effectively halted. Thus, the stability of MXenes needs to be improved for better and long-term performance of the device without any sort of degradation of performance.

To meet the requirements of real-life applications like robotic systems and prosthetics, it is crucial to optimize and appropriately engineer the sensing interfaces. This necessitates a thorough comprehension of the surface chemistry of MXenes, an area that requires further exploration.

For the exploration of MXene-based piezoresistive sensors where a direct contact with biomolecules and the organelles will happen, exploring the biosafety of MXenes and understanding their impacts and interactions with the former is immensely important.

17.6 SUMMARY

MXenes offer excellent conductivity, rich surface chemistry, and highly controllable interlayer spacing, thanks to their interconnected, layered, accordion-like structural aspects. These characteristics position MXenes as highly promising candidates for futuristic ultrasensitive pressure and strain sensor applications. However, MXenes encounter notable drawbacks: the tendency of ultrathin 2D layers to restack and the absence of a confined hollow structure. These limitations reduce the availability of surface-active sites and hinder the effective loading of other functional materials onto MXenes, impeding the attainment of desired properties. To overcome this challenge, researchers have developed porous 3D interconnected networks, which enhance flexibility, reduce weight, and provide a vast surface area, thereby improving sensor performance. Another significant challenge lies in the ability of MXenes to withstand high pressure and exhibit desired bifunctionality. Currently, MXenes face limitations in these areas due to issues related to flexibility (fragility), functionality, and stability. However, blending MXenes with other functional materials such as CNTs, graphene, conductive polymers, and plasmonic nanoparticles presents a viable solution. These composite materials enable the development of advanced piezoresistive strain and pressure sensors. This chapter provides a comprehensive explanation of piezoresistive materials, including their mechanisms and suitable materials for achieving piezoresistivity. It also explores how MXenes, in combination with other functional materials, can serve as piezoresistive sensors for cutting-edge applications, viz., wearable artificial skins, healthcare monitoring, artificial intelligence, and more.

REFERENCES

1. Zhang W., Xiao Y., Duan Y., Li N., Wu L., Lou Y., Wang H., Peng Z. A High-Performance Flexible Pressure Sensor Realized by Overhanging Cobweb-Like Structure on a Micropost Array. ACS Appl. Mater. Interfaces. 2020;12(43):48938–48947.

2. Guan H., Meng J., Cheng Z., Wang X. Processing Natural Wood into a High-Performance Flexible Pressure Sensor. ACS Appl. Mater. Interfaces. 2020;12(41):46357–46365.
3. Cao M., Su J., Fan S., Qiu H., Su D., Li L. Wearable Piezoresistive Pressure Sensors Based on 3d Graphene. Chem. Eng. J. 2021;406:126777.
4. Zhao S., Ran W., Wang D., Yin R., Yan Y., Jiang K., Lou Z., Shen G. 3d Dielectric Layer Enabled Highly Sensitive Capacitive Pressure Sensors for Wearable Electronics. ACS Appl. Mater. Interfaces. 2020;12(28):32023–32030.
5. Chen L., Lu M., Yang H., Salas Avila J. R., Shi B., Ren L., Wei G., Liu X., Yin W. Textile-Based Capacitive Sensor for Physical Rehabilitation Via Surface Topological Modification. ACS Nano. 2020;14(7):8191–8201.
6. Li X., Ma Y., Yue Y., Li G., Zhang C., Cao M., Xiong Y., Zou J., Zhou Y., Gao Y. A Flexible Zn-Ion Hybrid Micro-Supercapacitor Based on Mxene Anode and V_2O_5 Cathode with High Capacitance. Chem. Eng. J. 2022;428:130965.
7. Lee J., Kwon H., Seo J., Shin S., Koo J. H., Pang C., Son S., Kim J. H., Jang Y. H., Kim D. E., Lee T. Conductive Fiber-Based Ultrasensitive Textile Pressure Sensor for Wearable Electronics. Adv. Mater. 2015;27(15):2433–2439.
8. Kim N.-I., Chen J., Wang W., Moradnia M., Pouladi S., Kwon M.-K., Kim J.-Y., Li X., Ryou J.-H. Highly-Sensitive Skin-Attachable Eye-Movement Sensor Using Flexible Nonhazardous Piezoelectric Thin Film. Adv. Funct. Mater. 2021;31(8):2008242.
9. Li T., Qu M., Carlos C., Gu L., Jin F., Yuan T., Wu X., Xiao J., Wang T., Dong W., Wang X., Feng Z.-Q. High-Performance Poly(Vinylidene Difluoride)/Dopamine Core/Shell Piezoelectric Nanofiber and Its Application for Biomedical Sensors. Adv. Mater. 2021;33(3):2006093.
10. Wang X., Zhang H., Dong L., Han X., Du W., Zhai J., Pan C., Wang Z. L. Self-Powered High-Resolution and Pressure-Sensitive Triboelectric Sensor Matrix for Real-Time Tactile Mapping. Adv. Mater. 2016;28(15):2896–2903.
11. Tan X., Zhou Z., Zhang L., Wang X., Lin Z., Yang R., Yang J. A Passive Wireless Triboelectric Sensor Via a Surface Acoustic Wave Resonator (Sawr). Nano Energy. 2020;78:105307.
12. Wang Y., Yue Y., Cheng F., Cheng Y., Ge B., Liu N., Gao Y. $Ti_3c_2t_x$ Mxene-Based Flexible Piezoresistive Physical Sensors. ACS Nano. 2022;16(2):1734–1758.
13. Li X., Li X., Ting L., Lu Y., Shang C., Ding X., Zhang J., Feng Y., Xu F. J. Wearable, Washable, and Highly Sensitive Piezoresistive Pressure Sensor Based on a 3d Sponge Network for Real-Time Monitoring Human Body Activities. ACS Appl. Mater. Interfaces. 2021;13(39):46848–46857.
14. Zheng Y., Yin R., Zhao Y., Liu H., Zhang D., Shi X., Zhang B., Liu C., Shen C. Conductive Mxene/Cotton Fabric Based Pressure Sensor with Both High Sensitivity and Wide Sensing Range for Human Motion Detection and E-Skin. Chem. Eng. J. 2021;420:127720.
15. Yang Z., Li H., Zhang S., Lai X., Zeng X. Superhydrophobic Mxene@Carboxylated Carbon Nanotubes/Carboxymethyl Chitosan Aerogel for Piezoresistive Pressure Sensor. Chem. Eng. J. 2021;425:130462.
16. Yan J., Ma Y., Jia G., Zhao S., Yue Y., Cheng F., Zhang C., Cao M., Xiong Y., Shen P., Gao Y. Bionic Mxene Based Hybrid Film Design for an Ultrasensitive Piezoresistive Pressure Sensor. Chem. Eng. J. 2022;431:133458.
17. Wang C., Xia K., Wang H., Liang X., Yin Z., Zhang Y. Advanced Carbon for Flexible and Wearable Electronics. Adv. Mater. 2019;31(9):1801072.
18. Cheng Y., Wang K., Xu H., Li T., Jin Q., Cui D. Recent Developments in Sensors for Wearable Device Applications. Anal. Bioanal. Chem. 2021;413(24):6037–6057.

19. Thomson W. On the Electro-Dynamic Qualities of Metals: Effects of Magnetization on the Electric Conductivity of Nickel and of Iron. Proc. R. Soc. Lond. 1856;8:546–550.
20. Cookson J. W. Theory of the Piezo-Resistive Effect. Phys. Rev. 1935;47(2):194–195.
21. Kenny T. Sensor Technology Handbook. Wilson, J. S. (Ed) Oxford: Newnes; 2004.
22. Park W.-T. Piezoresistivity In: Bhushan, B. (Ed) Encyclopedia of Nanotechnology. Dordrecht: Springer; 2016.
23. Mason W. P. Crystal Physics of Interaction Processes Vol. New York: Academic Press; 1966.
24. Barlian A. A., Park W. T., Mallon J. R., Jr., Rastegar A. J., Pruitt B. L. Semiconductor Piezoresistance for Microsystems. Proc. IEEE Inst. Electr. Electron Eng. 2009;97(3):513–552.
25. Phan H.-P., Dao D. V., Nakamura K., Dimitrijev S., Nguyen N.-T. The Piezoresistive Effect of Sic for Mems Sensors at High Temperatures: A Review. J. Microelectromechanical Syst. 2015;24(6):1663–1677.
26. Okojie R. S., Ned A. A., Kurtz A. D., Carr W. N. Characterization of Highly Doped N- and P-Type 6h-Sic Piezoresistors. Electron. Dev. IEEE Trans. 1998;45:785–790.
27. Eickhoff M., Möller H., Kroetz G., v. Berg J., Ziermann R. A High Temperature Pressure Sensor Prepared by Selective Deposition of Cubic Silicon Carbide on Soi Substrates. Sens. Actuator A Phys. 1999;74(1):56–59.
28. Atwell A. R., Okojie R. S., Kornegay K. T., Roberson S. L., Beliveau A. Simulation, Fabrication and Testing of Bulk Micromachined 6h-Sic High-G Piezoresistive Accelerometers. Sens. Actuator A Phys. 2003;104(1):11–18.
29. Taher I., Aslam M., Tamor M. A., Potter T. J., Elder R. C. Piezoresistive Microsensors Using P-Type Cvd Diamond Films. Sens. Actuator A Phys. 1994;45(1):35–43.
30. Wang J. N., Luo X. G., Wu T., Chen Y. High-Strength Carbon Nanotube Fibre-Like Ribbon with High Ductility and High Electrical Conductivity. Nat. Commun. 2014;5:3848.
31. Grow R. J., Wang Q., Cao J., Wang D., Dai H. Piezoresistance of Carbon Nanotubes on Deformable Thin-Film Membranes. Appl. Phys. Lett. 2005;86(9):093104.
32. Dai H. Carbon Nanotubes: Opportunities and Challenges. Surf. Sci. 2002;500:218–241.
33. Liu M., Shenga Y., Huang C., Zhou Y., Jiang L., Tian M., Chen S., Jerrams S., Zhou F., Yu J. Highly Stretchable and Sensitive Sbs/Gr/Cnts Fibers with Hierarchical Structure for Strain Sensors. Compos.—A Appl. Sci. Manuf. 2023;164:107296.
34. Tang Z.-H., Wang D.-Y., Li Y.-Q., Huang P., Fu S.-Y. Modeling the Synergistic Electrical Percolation Effect of Carbon Nanotube/Graphene/Polymer Composites. Compos. Sci. Technol. 2022;225:109496.
35. Rahimi R., Ochoa M., Yu W., Ziaie B. Highly Stretchable and Sensitive Unidirectional Strain Sensor Via Laser Carbonization. ACS Appl. Mater. Interfaces. 2015;7(8):4463–4470.
36. Li X., Zhang R., Yu W., Wang K., Wei J., Wu D., Cao A., Li Z., Cheng Y., Zheng Q., Ruoff R. S., Zhu H. Stretchable and Highly Sensitive Graphene-on-Polymer Strain Sensors. Sci. Rep. 2012;2:870.
37. Fiorillo A. S., Critello C. D., Pullano S. A. Theory, Technology and Applications of Piezoresistive Sensors: A Review. Sens. Actuator A Phys. 2018;281:156–175.
38. Sze S. M. Physics of Semiconductor Devices. First Edition. New York: A John Wiley & Sons, Inc.; 1969.
39. Naumov P., Huangfu S., Wu X., Schilling A., Thomale R., Felser C., Medvedev S., Jeschke H. O., von Rohr F. O. Large Resistivity Reduction in Mixed-Valent Csaubr$_3$ under Pressure. Phys. Rev. B. 2019;100(15):155113.

40. Wang M., Gurunathan R., Imasato K., Geisendorfer N. R., Jakus A. E., Peng J., Shah R. N., Grayson M., Snyde G. J. A Percolation Model for Piezoresistivity in Conductor—Polymer Composites. Adv. Theory Simul. 2019;2:1800125.
41. Zhang X.-W., Pan Y., Zheng Q., Yi X.-S. Time Dependence of Piezoresistance for the Conductor Filled Polymer Composites. J. Polym. Sci. Part B: Polym. Phys. 2000;38(21):2739–2749.
42. Tan C., Dong Z., Li Y., Zhao H., Huang X., Zhou Z., Jiang J. W., Long Y. Z., Jiang P., Zhang T. Y., Sun B. A High Performance Wearable Strain Sensor with Advanced Thermal Management for Motion Monitoring. Nat. Commun. 2020;11(1):3530.
43. Li X., Fan Y. J., Li H. Y., Cao J. W., Xiao Y. C., Wang Y., Liang F., Wang H. L., Jiang Y., Wang Z. L., Zhu G. Ultracomfortable Hierarchical Nanonetwork for Highly Sensitive Pressure Sensor. ACS Nano. 2020;14(8):9605–9612.
44. Chen W., Liu L. X., Zhang H. B., Yu Z. Z. Kirigami-Inspired Highly Stretchable, Conductive, and Hierarchical $Ti_3C_2t_x$ Mxene Films for Efficient Electromagnetic Interference Shielding and Pressure Sensing. ACS Nano. 2021;15(4):7668–7681.
45. An B., Ma Y., Li W., Su M., Li F., Song Y. Three-Dimensional Multi-Recognition Flexible Wearable Sensor Via Graphene Aerogel Printing. Chem. Commun. 2016;52(73):10948–10951.
46. Joo Y., Byun J., Seong N., Ha J., Kim H., Kim S., Kim T., Im H., Kim D., Hong Y. Silver Nanowire-Embedded Pdms with a Multiscale Structure for a Highly Sensitive and Robust Flexible Pressure Sensor. Nanoscale. 2015;7(14):6208–6215.
47. Ma Y., Liu N., Li L., Hu X., Zou Z., Wang J., Luo S., Gao Y. A Highly Flexible and Sensitive Piezoresistive Sensor Based on Mxene with Greatly Changed Interlayer Distances. Nat. Commun. 2017;8(1):1207.
48. Yue Y., Liu N., Liu W., Li M., Ma Y., Luo C., Wang S., Rao J., Hu X., Su J., Zhang Z., Huang Q., Gao Y. 3d Hybrid Porous Mxene-Sponge Network and Its Application in Piezoresistive Sensor. Nano Energy. 2018;50:79–87.
49. Szuplewska A., Kulpinska D., Dybko A., Chudy M., Jastrzebska A. M., Olszyna A., Brzozka Z. Future Applications of Mxenes in Biotechnology, Nanomedicine, and Sensors. Trends Biotechnol. 2020;38(3):264–279.
50. Bu F., Zagho M. M., Ibrahim Y., Ma B., Elzatahry A., Zhao D. Porous Mxenes: Synthesis, Structures, and Applications. Nano Today. 2020;30:100803.
51. Xin M., Li J., Ma Z., Pan L., Shi Y. Mxenes and Their Applications in Wearable Sensors. Front. Chem. 2020;8:297.
52. Yang J., Zhang Z., Zhou P., Zhang Y., Liu Y., Xu Y., Gu Y., Qin S., Haick H., Wang Y. Toward a New Generation of Permeable Skin Electronics. Nanoscale. 2023;15(7):3051–3078.
53. Chen S., Huang W. A Review Related to Mxene Preparation and Its Sensor Arrays of Electronic Skins. Analyst. 2023;148(3):435–453.
54. Sinha A., Dhanjai, Zhao H., Huang Y., Lu X., Chen J., Jain R. Mxene: An Emerging Material for Sensing and Biosensing. TrAC Trends Anal. Chem. 2018;105:424–435.
55. Bao W., Tang X., Guo X., Choi S., Wang C., Gogotsi Y., Wang G. Porous Cryo-Dried Mxene for Efficient Capacitive Deionization. Joule. 2018;2(4):778–787.
56. Ma Y., Yue Y., Zhang H., Cheng F., Zhao W., Rao J., Luo S., Wang J., Jiang X., Liu Z., Liu N., Gao Y. 3d Synergistical Mxene/Reduced Graphene Oxide Aerogel for a Piezoresistive Sensor. ACS Nano. 2018;12(4):3209–3216.
57. Cai Y., Shen J., Ge G., Zhang Y., Jin W., Huang W., Shao J., Yang J., Dong X. Stretchable Ti(3)C(2)T(X) Mxene/Carbon Nanotube Composite Based Strain Sensor with Ultrahigh Sensitivity and Tunable Sensing Range. ACS Nano. 2018;12(1):56–62.

58. Chen M., Hu X., Li K., Sun J., Liu Z., An B., Zhou X., Liu Z. Self-Assembly of Dendritic-Lamellar Mxene/Carbon Nanotube Conductive Films for Wearable Tactile Sensors and Artificial Skin. Carbon. 2020;164:111–120.

59. Fan Z., Zhang L., Tan Q., Yao X., Lin B., Wang Y., Xiong J. Wearable Pressure Sensor Based on Mxene/Single-Wall Carbon Nanotube Film with Crumpled Structure for Broad-Range Measurements. Smart Mater. Struct. 2021;30(3).

60. Zheng X., Hu Q., Wang Z., Nie W., Wang P., Li C. Roll-to-Roll Layer-by-Layer Assembly Bark-Shaped Carbon Nanotube/$Ti_3c_2t_x$ Mxene Textiles for Wearable Electronics. J. Colloid Interface Sci. 2021;602:680–688.

61. Zhang D., Yin R., Zheng Y., Li Q., Liu H., Liu C., Shen C. Multifunctional Mxene/Cnts Based Flexible Electronic Textile with Excellent Strain Sensing, Electromagnetic Interference Shielding and Joule Heating Performances. Chem. Eng. J. 2022;438:135587.

62. Yang H., Li J., Xiao X., Wang J., Li Y., Li K., Li Z., Yang H., Wang Q., Yang J., Ho J. S., Yeh P. L., Mouthaan K., Wang X., Shah S., Chen P. Y. Topographic Design in Wearable Mxene Sensors with in-Sensor Machine Learning for Full-Body Avatar Reconstruction. Nat. Commun. 2022;13(1):5311.

63. Yang Y., Cao Z., He P., Shi L., Ding G., Wang R., Sun J. $Ti_3c_2t_x$ Mxene-Graphene Composite Films for Wearable Strain Sensors Featured with High Sensitivity and Large Range of Linear Response. Nano Energy. 2019;66.

64. Xu J., Zhang L., Lai X., Zeng X., Li H. Wearable Rgo/Mxene Piezoresistive Pressure Sensors with Hierarchical Microspines for Detecting Human Motion. ACS Appl. Mater. Interfaces. 2022;14:27262–27273.

65. Zhu M., Yue Y., Cheng Y., Zhang Y., Su J., Long F., Jiang X., Ma Y., Gao Y. Hollow Mxene Sphere/Reduced Graphene Aerogel Composites for Piezoresistive Sensor with Ultra-High Sensitivity. Adv. Electron. Mater. 2019;6(2).

66. Yang N., Liu H., Yin X., Wang F., Yan X., Zhang X., Cheng T. Flexible Pressure Sensor Decorated with Mxene and Reduced Graphene Oxide Composites for Motion Detection, Information Transmission, and Pressure Sensing Performance. ACS Appl. Mater. Interfaces. 2022;14(40):45978–45987.

67. Li L., Cheng Y., Cao H., Liang Z., Liu Z., Yan S., Li L., Jia S., Wang J., Gao Y. Mxene/Rgo/Ps Spheres Multiple Physical Networks as High-Performance Pressure Sensor. Nano Energy. 2022;95:106986.

68. Qin L., Yang D., Zhang M., Zhao T., Luo Z., Yu Z.-Z. Superelastic and Ultralight Electrospun Carbon Nanofiber/Mxene Hybrid Aerogels with Anisotropic Microchannels for Pressure Sensing and Energy Storage. J. Colloid Interface Sci. 2021;589:264–274.

69. Xia H., Zhang D., Wang D., Tang M., Zhang H., Chen X., Mao R., Ma Y., Cai H. High Sensitivity, Wide Range Pressure Sensor Based on Layer-by-Layer Self-Assembled Mxene/Carbon Black@Polyurethane Sponge for Human Motion Monitoring and Intelligent Vehicle Control. IEEE Sens. J. 2022;22(22):21561–21568.

70. Chen A., Wang C., Ali O. A. A., Mahmoud S. F., Shi Y., Ji Y., Algadi H., El-Bahy S. M., Huang M., Guo Z., Cui D., Wei H. Mxene@Nitrogen-Doped Carbon Films for Supercapacitor and Piezoresistive Sensing Applications. Composite A-Appl Sci Manufac. 2022;163:107174.

71. Zhuo H., Hu Y., Chen Z., Peng X., Liu L., Luo Q., Yi J., Liu C., Zhong L. A Carbon Aerogel with Super Mechanical and Sensing Performances for Wearable Piezoresistive Sensors. J. Mater. Chem. A. 2019;7(14):8092–8100.

72. Song D., Li X., Li X. P., Jia X., Min P., Yu Z. Z. Hollow-Structured Mxene-Pdms Composites as Flexible, Wearable and Highly Bendable Sensors with Wide Working Range. J. Colloid Interface Sci. 2019;555:751–758.

73. Li X. P., Li Y., Li X., Song D., Min P., Hu C., Zhang H. B., Koratkar N., Yu Z. Z. Highly Sensitive, Reliable and Flexible Piezoresistive Pressure Sensors Featuring Polyurethane Sponge Coated with Mxene Sheets. J. Colloid Interface Sci. 2019;542:54–62.

74. Xu Y., Qiang Q., Zhao Y., Li H., Xu L., Liu C., Wang Y., Xu Y., Tao C., Lang T., Zhao L., Liu B. A Super Water-Resistant Mxene Sponge Flexible Sensor for Bifunctional Sensing of Physical and Chemical Stimuli. Lab Chip. 2023;23(3):485–494.

75. Zhang Y.-Z., Lee K. H., Anjum D. H., Sougrat R., Jiang Q., Kim H., Alshareef H. N. Mxenes Stretch Hydrogel Sensor Performance to New Limits. Sci. Adv. 2018;4:eaat0098.

76. Qin R., Li X., Hu M., Shan G., Seeram R., Yin M. Preparation of High-Performance Mxene/Pva-Based Flexible Pressure Sensors with Adjustable Sensitivity and Sensing Range. Sens. Actuator A Phys. 2022;338:113458.

77. Yang C., Zhang D., Wang D., Luan H., Chen X., Yan W. In Situ Polymerized Mxene/Polypyrrole/Hydroxyethyl Cellulose-Based Flexible Strain Sensor Enabled by Machine Learning for Handwriting Recognition. ACS Appl. Mater. Interfaces. 2023;15(4):5811–5821.

78. Dutta D., Ganda A. N. F., Chih J. K., Huang C. C., Tseng C. J., Su C. Y. Revisiting Graphene-Polymer Nanocomposite for Enhancing Anticorrosion Performance: A New Insight into Interface Chemistry and Diffusion Model. Nanoscale. 2018;10(26):12612–12624.

79. Zhang L., Zhang X., zhang H., Xu L., Wang D., Lu X., Zhang A. Semi-Embedded Robust Mxene/Agnw Sensor with Self-Healing, High Sensitivity and a Wide Range for Motion Detection. Chem. Eng. J. 2022;434,:134751.

80. Li H., Du Z. Preparation of a Highly Sensitive and Stretchable Strain Sensor of Mxene/ Silver Nanocomposite-Based Yarn and Wearable Applications. ACS Appl. Mater. Interfaces. 2019;11(49):45930–45938.

81. Bi L., Yang Z., Chen L., Wu Z., Ye C. Compressible Agnws/$Ti_3c_2t_x$ Mxene Aerogel-Based Highly Sensitive Piezoresistive Pressure Sensor as Versatile Electronic Skins. J. Mater. Chem. A. 2020;8(38):20030–20036.

82. Pu J.-H., Zhao X., Zha X.-J., Bai L., Ke K., Bao R.-Y., Liu Z.-Y., Yang M.-B., Yang W. Multilayer Structured Agnw/Wpu-Mxene Fiber Strain Sensors with Ultrahigh Sensitivity and a Wide Operating Range for Wearable Monitoring and Healthcare. J. Mater. Chem. A. 2019;7(26):15913–15923.

83. Zhang H., Zhang D., Zhang B., Wang D., Tang M. Wearable Pressure Sensor Array with Layer-by-Layer Assembled Mxene Nanosheets/Ag Nanoflowers for Motion Monitoring and Human-Machine Interfaces. ACS Appl. Mater. Interfaces. 2022;14(43):48907–48916.

84. Chao M., He L., Gong M., Li N., Li X., Peng L., Shi F., Zhang L., Wan P. Breathable $Ti_3c_2t_x$ Mxene/Protein Nanocomposites for Ultrasensitive Medical Pressure Sensor with Degradability in Solvents. ACS Nano. 2021;15(6):9746–9758.

85. Zhang W., Pan Z., Ma J., Wei L., Chen Z., Wang J. Degradable Cross-Linked Collagen Fiber/Mxene Composite Aerogels as a High-Performing Sensitive Pressure Sensor. ACS Sustainable Chem. Eng. 2022;10:1408–1418.

86. Guo Y., Zhong M., Fang Z., Wan P., Yu G. A Wearable Transient Pressure Sensor Made with Mxene Nanosheets for Sensitive Broad-Range Human-Machine Interfacing. Nano Lett. 2019;19(2):1143–1150.

18 MXene-Based Materials for Fuel Cell Applications

Debajani Tripathy and Srikanta Moharana

18.1 INTRODUCTION

Recent advances in science and nanotechnology are altering our understanding of nanoparticles and the composites made from them. Due to their size and composition-dependent features, nanomaterials have the potential to shed light on and perhaps address a number of pressing scientific problems. Depending on their structure, size, form, dimensionality, aggregation state, and chemical composition, nanomaterials offer a broad range of possible uses [1–2]. When it comes to energy storage, electronics, sensors, catalysis, and biological applications, two-dimensional (2D) nanomaterials stand out due to their atomic thickness, numerous active surface sites, massive surface area-to-volume ratio, and exceptional mechanical qualities. Graphene-based materials, effectively isolated from graphite in 2004, are two-dimensional (2D) nanomaterials that have attracted a lot of interest due to their unusual physical and chemical properties [2–3]. Two-dimensional (2D) materials like graphene and transition metal dichalcogenides (TMDs) show great potential for broad applications in electrocatalysis and energy storage because of their remarkable electrical, mechanical, and surface features [3–4]. There are several advantages for practical applications of 2D materials, including self-supporting, flexible, thin films with controllable thickness, high surface area for easy coupling with other functional materials, and strong in-plane covalent bonds that provide outstanding mechanical strength and flexibility. Two-dimensional materials are promising for broad usage in electrocatalysis and energy storage due to their advantageous features [5]. The ability of an energy storage device to store charge from energy harvesting devices when microsensors are in sleep mode and give continuous power to them while they are active makes it a critical short-term accumulator in the design of such self-powered power units. In portable electronics, lithium-ion intercalation microbatteries and thin-film batteries are often employed to store energy [6]. However, they need regular maintenance and replacement due to their limited lifetime and poor power density. Additionally, because of their poor power density, they are ineffective in applications requiring abrupt or sustained spikes in current [7]. Electrolyte ions may be adsorbed onto the surface of electrode materials to store charge in energy storage devices known as electrochemical capacitors (ECs), and pseudo-capacitive/Faradaic processes may be used to store charge between the electrode material's surface and the ions in the electrolyte. Although ECs have a lower energy density than batteries, they may nevertheless be able to provide enough power to operate

                    DOI: 10.1201/9781003366225-18

a range of devices for extended periods of time. Electrochemical capacitors are a potential replacement for batteries as a source of energy storage in microelectronics because of their better power density and potentially endless lifespan (>10^5 cycles) [7–8]. Standard electrolytic capacitors (ECs) are too big for use in microdevices, making them incompatible with the strict tolerances necessary for use in micro-electronics. On the other hand, traditional supercapacitor assembly processes are not suited for mass production and are thus not an option. That is why scientists are putting forth so much effort to develop supercapacitors that can be packaged in very small sizes. In the field of microelectronics, supercapacitor devices are often referred to as "micro-supercapacitor" (MSC) devices. The most prevalent forms of these devices are thin-film electrodes that have a sandwich structure (10 micrometers in thickness) and arrays of planar microelectrodes that have microscale sizes in at least two dimensions. In more recent times, the concept of MSCs has been expanded to incorporate core–shell fiber-based electrodes [7–9]. Compared to competing tech-nologies, the in-plane interdigital design has many benefits, such as a shorter ion diffusion distance, enhanced exposure of the electrode materials to the electrolyte, and simple integration of the microelectronic components. Micro-supercapacitors have been created using carbon-based materials with very large surface areas [8–9]. For instance, carbons generated from carbides [10], onion-structured carbons [11], activated carbons [12], carbons formed from photo resists [13–16], carbons derived from carbon nanotubes [17–19], carbons derived from graphene [20–21], and laser-scribed carbons [22] have been used. Although carbon-based materials are advanta-geous for these devices because of their high conductivity and the enormous surface areas with low energy densities, these problems caused by poor energy density at high power densities have inspired the development of pseudo-capacitive materials. Among them are conducting polymers [23–26] and transition metal oxides ($RuO2$, $MnO2$, $MoO3$, $Nb2O5$) [27–28]. Some of the materials that have been considered for use in the creation of micro supercapacitor devices [29–35] include conductive 2D metal-organic frameworks (MOFs) [30–35], black phosphorous (BP) [34], and transition metal dichalcogenides [36]. Metal oxides or hydroxides and metal-organic frameworks (MOFs) have a low electrical conductivity, which may result in poor cycle and power performance [30–34].

Since 2011, there has been a proliferation of two-dimensional transition metal carbides, nitrides, and carbonitrides (MXenes). The "ene" suffix is intended to facili-tate associations with other 2D materials such as graphene, phosphorene, silicene, etc. [37]. The general formula is $M_{n+1}AX_n T_x$, where M refers to an early transition metal (Ti, Mo, Cr, Nb, V, Sc, Zr, Hf, or Ta), and T_x refers to the surface terminal, which might be hydroxyl, oxygen, or fluorine. X = carbon or nitrogen, and the value of n is an integral number between 1 and 3. The electrical characteristics and hydro-philicity of MXenes are strongly influenced by the density of states at the Fermi level, which is strongly affected by these surface terminations [38–39]. MXenes are created from MAX phases by selectively removing the "A" layer. Moreover, MXenes have garnered a lot of interest as potential electrode materials for use in energy storage applications [40]. This is primarily due to the distinctive structural charac-teristics that they possess, such as (1) an inner conductive transition metal carbide layer that allows for effective electron transportation and (2) a transitional metal oxide-like surface that provides active sites for rapid redox reactions. Because of its ultrahigh conductivity (2.4×10^4 S cm^{-1}), volumetric capacitance (1500 F cm^{-3}), and ultrahigh rate capability (10 V/s) in acidic conditions, titanium carbide (Ti$_3$C$_2$) has been the focus of intensive research for use in MSCs. This is in spite of the fact that over 20 distinct MXenes have been artificially created [41–44]. MXenes are excellent candidates for use as MSCs because of the high volumetric capacitance and two-dimensional structure of these materials. These features reduce the number of pathways for ion transport between the positive and negative electrodes, which in turn increases mechanical stability. The most recent developments in MXene-based micro-supercapacitors include improvements in electrode material design, deposition and patterning processes, and device architecture [45].

18.2 AN OVERVIEW OF MXENE-BASED 2D MATERIALS

In 2011, scientists made the exciting discovery of MXenes, a unique kind of 2D nanomaterial that has desired features such as hydrophilicity, conductivity, and exceptionally microscopic size. In the case of transition metals, examples of 2D sheets include carbides, nitrides, and carbonitrides. The MAX phases, which are stacked ternary $M_{n+1}AX_n$ phases, are where this phenomenon first manifests itself. The elements that make up the periodic table as transition metals are denoted by the letter "M," an element in the group IV–V range is denoted by the letter "A," and either carbon or nitrogen is denoted by the letter "X." The "A" layer of the MAX phase has to be selectively etched away in order to produce MXenes. It is important to highlight the 2D nature of the MAX stages, as well as their connection to one another; therefore, they are each given a name. It is common practice to use hydrofluoric acid (HF) or a mixture of HF and lithium fluoride (LiF), ammonium bifluoride (HF+AmF), or another chemical when attempting to etch "A" layers from MAX powders [46–47]. At this time, there is a significant amount of focus placed on investigations into the production of MXenes without the usage of potentially harmful acids. For further information on the synthetic processes that are employed to produce MXenes, we strongly suggest reading the recent paper that was written by Verger and colleagues [48]. The $Mn_1X_nT_z$ form of MXenes, where T_z denotes the surface functional groups formed after etching, is the most desired end product. Oxygen (O), hydroxyl (OH), and fluorine (F) groups make up the terminations, while water and exchangeable cations (including protons) populate the interlayer space [49].

The first name for surface terminations was "T_x," but this was changed to "T_z" so that there would not be any misunderstanding with the "X" in the equations for $M_{n+1}AlX_n$ and $M_{n+1}X_n$. The hydrophilic properties of MXenes are due to the presence of these functional groups. A stable suspension may be produced by sonicating a solution of a MXene that has a pH close to neutral in a probe or a bath [50]. This suspension can then be employed in water or polar solvents. These groups have the potential to interact with and react with many biopolymers, including polyvinyl alcohol (PVA), polydiallyldimethylammonium chloride (PDDA), polyacrylic acid (PAA), and others. Polymers combined with various fillers have been used extensively in the production of high-performance composite materials with outstanding properties ever since the 1960s [51]. In the late 1980s and early 1990s, nanoparticles were first utilized as additives to form a new class of materials known today as nanocomposites (NCs). This was primarily owing to Toyota's work on nylon or clay hybrids, which led to the development of this new class of materials. These nanocomposites have an exceptional property due to the fact that at least one of their filler dimensions is on the nanoscale and often falls below 100 nm [52]. Clays and other 2D materials have advantages over 3D materials due to their larger surface area and, in some situations, their smaller particle size, which makes them appropriate for usage as nanomaterials. These advantages allow clays and other 2D materials to outperform 3D materials. It is possible to produce considerable improvements in the mechanical, thermal, rheological, and gas barrier properties of a material while only using low nanofiller loadings of 2 vol% [52]. This is assuming that the nanosheets can be exfoliated and distributed uniformly across a polymer matrix. Conductivity, enhanced catalytic activity, and usage in sensing systems are just some of the possible benefits that might be shown by high-performance nanoparticles. High-performance nanoparticles also have the ability to display a broad range of other advantageous qualities. Since their discovery, there has been consistent growth in the amount of research conducted into the usage of MXenes as nanofillers owing to the unique qualities and compositional variation that they possess. This is particularly true if there is an improvement made to the chemical, physical, or processing interactions of MXenes [53]. MXene-based polymer NCs were first described and published in the year 2014 [54] (Figure 18.1); they were manufactured by combining the hydrophilic polymers PDDA and PVA with $Ti_3C_2T_z$, which is the MXene that has received the

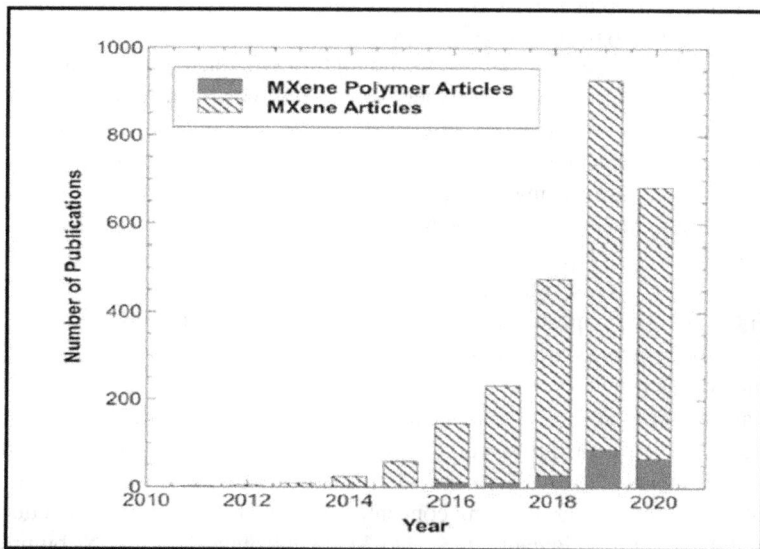

FIGURE 18.1 Published research articles on MXenes and MXene-based polymer nanocomposites [54].

most research up to this point. $Ti_3C_2T_z$ was produced by etching Ti_3AlC_2 in a mixture of 50% HF at room temperature for 18 hours while stirring constantly. Etching was performed on the MXene, and then it were rinsed in water until a pH of 6 was reached. After being exposed to room temperature air for 24 hours, the powder was then immersed in DMSO for 18 hours before being separated using centrifugation and water. The ML MXene powder that was produced after being treated with dimethyl sulfoxide (DMSO) was then diluted 1:300 with water before being sonicated for a total of five hours in the presence of flowing argon (Ar). The procedure of centrifugation was used in order to separate the colloidal supernatant, which, as a result, did not call for any further purification prior to its application. Colloidal suspensions that were treated with PDDA or PVA solutions of varying percentages were filtered using vacuum-assisted filtration to produce films with MXene concentrations of 0%, 40%, 60%, 80%, 90%, and 100 wt%. These films were made from colloidal suspensions. In spite of the broad use of this technique for the production of MXene colloidal suspensions, the amount of acid has been brought down to 10–20%, and the amount of DMSO that is used has been cut down. In order to facilitate delamination, while sonicating, the etching process now includes the addition of lithium as well as other cations. In order to accomplish this goal, HCl and fluoride salts are combined and applied in such a way as to inhibit the direct application of HF by introducing lithium or other cations into the interlayer area. There are various polymeric material-based MXene nanocomposites have been reported in recent times (as shown in Figure 18.1), which include polyethylene (PE), polyethylene oxide (PEO), polypropylene (PP), polystyrene (PS), polyamide (PA), polyimide (PI), acrylamide (AA), acrylates (AA), urethanes (UH), and silicones (Si) with epoxies, among others [54]. The majority of the current literature focuses on conductive polymers like polyaniline, poly(vinylpyrrolidone) (PVP), polypyrrole (PPy), poly(3,4-ethylene dioxythiophene) (PEDOT), and a variation including the component polystyrene sulfonate (PSS), as well as fluoride-based polymers such as polyvinylidene difluoride (PVDF), poly(vinylidene fluoride-trifluoro ethylene)(PVDF-TrFE) and poly(vinylidene fluoride-trifluoroethylenechlorofluoroethylene) (PVDF-TrFE-CFE) [54–55].

Since MXenes are 2D materials, a photocatalytic system that likewise uses 2D materials might maximize their potential. Due to this, Su and co-workers [56] have developed a g-C_3N_4/Ti_3C_2 2D/2D composite for photocatalytic H_2 production. In this experiment, g-C_3N_4 was synthesized by a calcination sequence. The g-C_3N_4 and Ti_3C_2 2D/2D composite was then fabricated using electrostatic self-assembly. However, the transmission electron microscopy (TEM) images show that a 2D/2D structure was formed when 2D Ti_3C_2 was put into 2D g-C_3N_4. Their novel 2D/2D shape allows for a large contact area, which might speed up the transport of charge carriers in response to illumination. Therefore, the photocatalytic activity of the produced samples steadily increased along with the loading percentage of Ti_3C_2 from 1% to 3% by weight. Loading concentrations of Ti_3C_2 over 3% may reduce the photocatalytic activity. Because Ti_3C_2 blocks the advantages of g-C_3N_4 by preventing it from absorbing incident light, g-C_3N_4 is useless. In addition to optimizing the MXene's intrinsic advantages, the innovative 2D/2D structure of the resulting g-C_3N_4/Ti_3C_2 may open the way for the development of complex MXene-based photocatalysts with sizable contact surfaces. Ultra-fast photogenerated charge carrier separation was shown to be possible in g-C_3N_4 and Ti_2C due to their 2D/2D structures, as revealed by Shao and co-workers [57]. However, Ti_2C may be preferable to g-C_3N_4 because its Fermi level is greater, resulting in a Schottky barrier. Ti_3C_2 was also used by Yuan et al. [58] as a precursor in the development of 2D layered carbon–TiO_2 hybrids. Ti_3C_2 oxidation is typical in a CO_2 flow. After that, the carbon layer and the TiO_2 complex was formed by the bonding of C and Ti atoms. In particular, the 2D carbon–TiO_2 structure maintained the layered structure of Ti_3C_2. The resulting 2D layered carbon–TiO_2 was much thinner than the Ti_3C_2 predecessor. These 2D layered carbon-based TiO_2 hybrids may enable light to flow through readily, facilitating the photocatalytic process. Moreover, Ti_3C_2 was evaluated for its transformation into C/TiO_2 by measuring its thermogravimetric and differential scanning calorimetric curves in a CO_2 flow. CO_2 started oxidizing at about 700°C. When the calcination temperature reached 800°C, the following Equation (18.1) predicted a loss in sample mass due to carbon reduction:

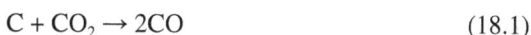

$$C + CO_2 \rightarrow 2CO \qquad (18.1)$$

Optimized C–TiO_2 has a significantly better photocatalytic activity for H_2 generation (24.0 mol h^{-1}) than both carbon QDs–TiO_2 (14.1 mol h^{-1}) and graphene–TiO_2 (21.7 mol h^{-1}). Using Ti_3C_2 as a precursor allowed for the creation of C/TiO_2 with a high specific surface area and a novel porous structure, both of which contribute to the material's outstanding photocatalytic activity. This study might provide insight into MXene's potential and lead to the creation of innovative photocatalysts [58].

18.3 SYNTHESIS TECHNIQUES FOR MXENE-BASED 2D MATERIALS

Since the discovery of graphene and its extraordinary properties, two-dimensional (2D) materials have drawn a lot of interest in the field of materials research [59–65]. A brand-new family of 2D materials known as transition metal carbides, nitrides, and carbonitrides was very recently identified [66]. To produce these novel materials, layers of sp elements are carefully etched from the appropriate 3D MAX phases. In ternary metal carbides, nitrides, and carbonitrides, layered structures called MAX phases may form. Although there are many different types of these compounds, they may all be described by the formula $M_{n+1}AX_n$ (n = 1, 2, or 3), where M is an early d-block transition metal, A is a main-group sp element (often IIIA or IVA), and X is either C or N. The MXene family comprises Ti_3C_2, Ti_2C, $(Ti_{0.5}, Nb_{0.5})2C$, $(V_{0.5}, Cr_{0.5})3C_2$, Ti_3CN, Ta_4C_3 [67], Nb_2C, V_2C [68], and Nb_4C_3 [69], even though more than 70 MAX phases have thus far been discovered [70]. More MXene compounds are anticipated to be produced by the enormous MAX phase family. After the etching procedure, the exfoliated layers are always left with groups of F, OH, and/or O on the surface. $M_{n+1}X_nT_x$, where x is the number of terminations and T is the sum of all surface groups (F, OH, or O), is the abbreviation used to refer to species of terminated MXenes. Since then, MXenes' additional features have been announced. MXenes, like layered graphene, have excellent conductivity [68]. Density functional theory (DFT) calculations demonstrate that the in-plane elastic constants of MXene are more than 500 GPa [71]. For instance, Khazaei and co-workers [72] have revealed that very high Seebeck coefficients in semiconducting MXenes may be produced at low temperatures. Researchers from many fields are intrigued by the possible far-reaching ramifications of these peculiar traits. MXenes have been discovered to offer potential as electrode materials in energy storage [73–74]. Xie et al. [75] have developed a very stable catalyst for fuel cell applications by using MXenes as a support material for platinum nanoparticles. The strong enzyme immobilization efficiency and compatibility with redox proteins reported in MXenes led by Wang et al. [76] to conclude that they are potential substrates for electrochemical biosensors. Furthermore, reports on MXenes derived from both theoretical and experimental studies have been published, focusing on the prospective uses of this variety of material [77]. The study suggests that pre-treatment agents like nitrides, carbides, and carbo-nitrides can create up to 20 MXenes through selective chemical etching of a few atomic layers, thereby promoting further research in this innovative area of materials science. Etchants may be broken down into two classes, one of which include fluoride-based aqueous salts [78]. In order for MAX systems to begin identifying MXenes, the M–A connections must be broken by prolonged exposure to certain acids. How long the corrosion continues and how much of a mixture there is will determine the outcome. Table 18.1 provides the details of the production techniques MXene-based materials [79–88].

The two most frequent approaches to making 2D MXenes are from the top down and the bottom up. Top-down methods include, for instance, the exfoliation of large crystal volumes into single-layered MXene sheets, whereas bottom-up methods include the creation of MXenes from atoms or molecules. A considerable amount of the created MXenes to date are the result of phase exfoliation that follows the wet application of an etching technique. Using small particles of Ti_3AlC_2 and 50%

TABLE 18.1
Synthesis Techniques of MXene-Based Materials

MXene-based materials	Synthesis technique	References
MXene–NiFe$_2$O$_4$ nanocomposites	Hydrothermal	79
NiCo–LDH–MXene hybrids	Heterojunction surface	80
2D Ti$_2$CT$_x$ MXene	3D Spheroid-type cultures	81
Ti$_3$C$_2$T$_x$ MXene	Wet chemical etching	82
MXene-derived nanoflower shaped TiO$_2$-Ti$_3$C$_2$ Mo$_2$CT$_x$ MXene	In situ transformation	83
Ti$_3$C$_2$	Hot-press method	84
Ti$_3$C$_2$T$_x$ MXene–graphene nanocomposites	Hydrothermal technique	85
MXene hybrids	Heterojunction	86
MXene-RGO	Hydrothermal technique	87
MO$_2$CT$_x$	Vacuum-assisted filtration technique	88

hydrofluoric acid for two hours at room temperature, Naguib et al. [89] synthesized the first MXene. The resulting Equations (18.2) and (18.3) reveal the response:

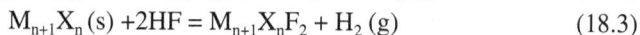

$$M_{n+1}AX_{n+3}HF = M_{n+1}X_n (s) + AF_3 + 1.5H_2 (g) \qquad (18.2)$$

$$M_{n+1}X_n (s) + 2HF = M_{n+1}X_nF_2 + H_2 (g) \qquad (18.3)$$

The exfoliation of MXene-adjoined layers may occur when the accumulated A layers in MAX phases combine with fluoride ions to produce AF and release hydrogen gas, as predicted by the aforementioned equation. Substituting atoms with -O, -F, and -OH functional groups in the preceding scheme results in graphite-like structures called MXenes [90–91], which are created by weakening the effective connections between $M_{n+1}Xn$ sheets. These reactions lead to the formation of functional groups on the surface of multi-layer MXenes, which negatively charge the whole surface and greatly promote the development of stable colloidal dispersions [89, 92]. When compared to their unmodified associated counterparts, terminated MXene sheets have the best possible degree of thermodynamic stability. It has been shown that raw Ti_3AlC_2 may be etched selectively in concentrated HF acid at room temperature [93]. The most flexible MXene is $Ti_3C_2T_x$. It was made by removing layers from a large number of MAX systems. Novel MXene materials such as MO_2C, Ti_2C, Ti_3SiC_2, and Cr_2Ti_2 have all benefited from extensive usage of HF etching in their manufacturing. It is difficult to generate stable nitride MXenes from similar MAX systems by HF etching, as detailed in [94]. This is due to the low energy of cohesion in the geometric structure of stable nitride MXenes. Therefore, a lot of power is needed to get them airborne. Developing less powerful and harmful etchants for delamination and exfoliation [95] is necessary since the hydrofluoric acid used to make MXenes is poisonous. For etching aluminum, the alternative approach known as Ti_3AlC_2 was speculated to feature innovative methodology and a simple, single-stage, scalable operation. Less dangerous acid salts like lithium hydroxide and HCl are added to the mixture. MXenes with a bigger interlayer gap, higher yield, and fewer crystal defects may be synthesized by allowing water molecules to intercalate with cations during the slow etching process [96–97]. $Ti_3C_2T_x$ MXene crystals with well-defined smooth edges and minimum crystal defects may be synthesized by adjusting the molar ratio of Ti_3AlC_2 and LiF in a manner similar to the synthesis of MXenes formed by pure HF acid etching. In a related work [98], a light, less harmful etchant like ammonium bifluoride was recommended for etching epitaxial Ti_3AlC_2. Ammonium ions (-NH$_3$/-NH$_4^+$) were intercalated into the proper $Ti_3C_2T_x$ interlayers, which led to an increase in lattice properties [99]. Combining HCl or sulfuric acid with fluoride species like calcium fluoride, potassium fluoride, sodium fluoride, cesium fluoride, and tetra butyl ammonium fluoride yields aqueous etchants that can be used to create multi-patterned transition state metallic carbides with obscured compositions and distinguishing properties. It has also been hypothesized that the characteristics of MXenes may be altered by varying the surface process and the molecules or ions that are pre-intercalated into the material [100]. The hydrothermal technique is suitable for mass producing multi-patterned, two-dimensional MXene components because it eliminates the need for workers to come into contact with very harmful HF vapor. However, Ti_3AlC_2 is immersed in aqueous NaOH at around 85°C for 100 h, followed by hydrothermal treatment with 1 M sulfuric acid at 85°C for 1.5 h, resulting in a fluorine-free technique to leach aluminum layers created by the MAX system [101]. In the presence of a binary electrolytic solution in aqueous form, a fluoride-free alternative strategy sensitive to the anodic corrosion process of Ti_3AlC_2 exists. $Ti_3C_2T_x$ may also be synthesized by combining finely powdered Ti_3AlC_2 powder with aqueous ammonium fluoride [102]. A previous finding suggested that the interaction between sodium borofluoride and hydrochloric acid might be etched only by a hydrothermal process, rather than directly by HF. This resulted in the laboratory synthesis of Nb_2C and Ti_3C_2 [103]. The Ti_3C_2 exhibited improvements in aluminum layer elimination, increased lattice parameter, maximum interlayer width between 2D MXenes with better BET surface distribution, and a facile exfoliation pattern rendered by sonication, all of which are highly favorable for excellent adsorptive efficacy thanks to the hydrothermal reaction's step-by-step release mechanism. Combining Baeyer's method with an alkali-induced hydrothermal technique allowed for the first time the purification of multi-layered $Ti_3C_2T_x$ to 90% [104]. This was accomplished by disabling the fluoride ion termination. Studies demonstrate that there are viable alternatives to fluoric acid in MXene production, such as chemical vapor deposition, salt template, and the molten salt process [105]. Because of the importance of synthesizing MXene-based adsorbent materials with proper control towards surface functional groups, maximum specific surface distribution, and well-defined chemical stability assured to a wide range of environmental issues, a practical and effective modern protocol is outlined with essential mechanisms to achieve these goals [106]. Furthermore, wet etching using HF, HCl-LiF, and HCl-NaF has attracted more attention in recent times due to its potential to create multi-layered MXene flakes and nanomaterials [107–108]. The schematic illustration for the production of MXenes from 2011 to 2019 is shown in Figure 18.2.

In order to create a lamellar structure, MXenes may be employed as (1) a skeleton material, (2) as a coating material to modify a membrane support layer, and (3) as a membrane support layer in and of themselves, in the form of mixed-matrix membranes. The fabrication of MXene-based membranes on various substrates, such as anodized aluminum oxide, polyvinylidene fluoride, and polycarbonate is shown in Figure 18.2. A new lamellar membrane was developed by stacking $Ti_3C_2T_x$ MXene nanosheets. Since the removal of Al as AlF_3 weakens the interlayer connection, MXene powder was created by coating Ti_3AlC_2 particles with an Al layer and then etching them with HF. By filtering a solution of MXene nanosheets and $Fe(OH)_3$ colloidal solutions over a 0.2 MPa pore size anodic aluminum oxide membrane, an MXene membrane was produced. To expand the pore space accessible for water molecules, $Fe(OH)_3$ colloids were dissolved in HCl. It has been reported that atomic force microscopy (AFM) can be used to verify the thickness (2 nm) and size distribution (100–400 nm) of MXene nanosheets, and Fourier-transform infrared spectroscopy (FTIR) and X-ray photoelectron spectroscopy (XPS) can be employed to verify the presence of -F, -OH, and -O functional groups on the MXene surfaces [109]. $Ti_3C_2T_x$ was filtered on a polyethersulfone ultrafiltration membrane with a molecular weight cutoff of 10,000 Da, which is lower than previous methods for two-dimensional $Ti_3C_2T_x$ MXene membranes [110]. Rasool and co-workers [111] studied a polyvinylidene difluoride membrane with a 450 nm pore size to create

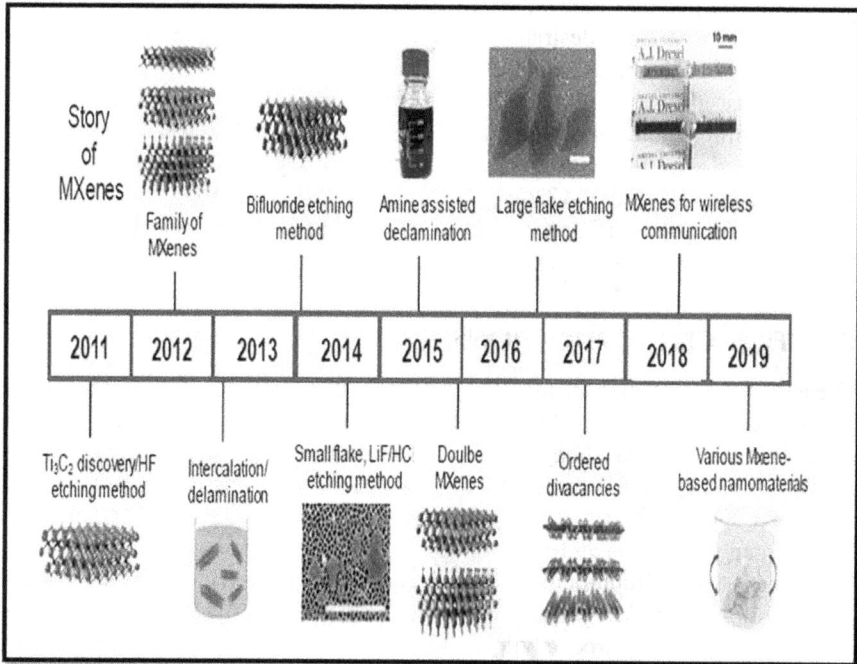

FIGURE 18.2 Manufacturing schedule for MXenes from 2011 to 2019 [108].

Adapted with permission from Ref. [108]. Copyright (2017) (American Chemical Society).

MXene membranes that were efficient against *Escherichia coli* and *Bacillus subtilis*. The membrane thicknesses ranged from 0.6 mm to 1.8 mm after 2–6 mg $Ti_3C_2T_x$ nanosheets were deposited on a polyvinylidene difluoride substrate. The water contact angle with the polyvinylidene difluoride substrate was 81°, whereas it was only 37° with the $Ti_3C_2T_x$ nanoparticle-coated membrane. Standard manufacturing techniques for non-cross-linked polyamide membranes do not include solvent-induced phase inversion; therefore, these membranes are not completely stable when exposed to oxidants like chlorine and organic solvents like methanol, acetone, isopropanol, and n-heptane [111]. An MXene mixed-matrix membrane with high hydrophilicity, selectivity, and solvent resistance was developed by Han et al. [112] utilizing $Ti_3C_2T_x$. $Ti_3C_2T_x$ nanoparticles were used to boost the selectivity and solvent resistance of polyimide (P84) polymer matrices cross-linked with triethylenetetramine. Based on scanning electron microscopy findings, a polyimide–MXene mixed-matrix membrane with channels of around 200 nm in size may be beneficial for mass transfer. As the MXene loading increases, the active top layers get denser, causing the membrane to shrink and smooth out. Due to the hydrophilicity of MXenes, water molecules may move more quickly between the hydrophobic barrier and the solvent during the phase-inversion process [113]. Thin-film nanofiltration membranes may be useful as a solvent filter if they are functionalized at the film filler stage rather than the polymer matrix, which is expected to create the necessary functionality and minimize undesirable swelling [114]. Isopropanol flow was increased by 30%, whereas butanone, ethyl acetate, and n-heptane flux were unaffected by an MXene-based membrane with 2% unfunctionalized $Ti_3C_2T_x$ nanosheets over a robust thin-film nanofiltration substrate. Polyacrylonitrile ultrafiltration membranes were employed with a support material made of $Ti_3C_2T_x$ nanosheets that had been functionalized with $-NH_2$, $-COOR$, $-C_6H_6$, and $-C_{12}H_{26}$ groups. Notable shifts in solvent flow occurred during the removal of n-heptane, toluene, isopropanol, ethyl acetate, and polyethylene glycol [115].

18.4 FUEL CELL APPLICATIONS

MXene-based two-dimensional (2D) transition metal nitrides and carbides with desirable features have gained popularity in recent times due to their wide range of possible applications. Multi-layer MXenes feature P63/MMC-symmetrical hexagonal forms with $M_{n+1}X_nT_x$ structures [116], where M may be any transition metal (Ti, V, Nb, etc.), X can be either carbon or nitrogen, T_x denotes surface terminations (like -O, -OH, and -F), and n can be any integer from 1 to 4. These materials have superior conductivity, hydrophilicity, and mechanical properties that may be achieved in large quantities using low-cost ingredients to create MXenes [117]. Different methods have shown the electrocatalytic activity of MXenes materials. Supercapacitors, Li-ion batteries, Na–S batteries, fuel cells, oxygen evolution processes, and other renewable energy systems have all utilized MXenes for energy storage and conversion [118]. Electrocatalysis, energy storage and conversion technologies (supercapacitors/fuel cells), bio or chemical sensors, antimicrobial research, and so on have all seen an uptick in interest in conducting polymers in recent years [119]. Polyaniline (PANI)-based MXene composite materials have many desirable characteristics for use in scientific research including water solubility, low-cost synthesis, chemical and environmental stability, reversible redox performance and conductivity, protonation-induced changes in optical and electrical properties, mechanical stability, and the possibility of bonding to electrode surfaces. [120]. Because Pd has excellent electrocatalytic performance, including electrode stability, homogeneous dispersion, and rapid, efficient electron transfer kinetics [121], PANI improves the electrocatalytic behavior of nanohybrid composites. MXene-based PANI nanocomposites have been used in a wide variety of applications, including microwave absorption, asymmetric supercapacitors, flexible sensors, and free-standing pseudo-capacitive electrodes with very high specific capacitance. When it comes to dispersing and stabilizing Pd-based electrocatalysis applications, MXene nanocomposites are an excellent support system. Heterojunction electrocatalysts may be created when functional groups on the surface of MXenes engage with certain arrangements of transition metal atoms, therefore enhancing lattice mismatch and interfacial electron transport [121–122]. The study investigates the potential of MXenes and Pd electrocatalysts to create a novel composite for DMFCs, considering their malleability, conductivity, and efficiency. This study presents a straightforward one-pot electrochemical co-deposition method for the rapid creation of a PANI-Pd-MXene nanocomposite. Electrochemically synthesized Pd-MXene electrocatalysts and PANI-Pd-MXene nanocomposite electrocatalysts were tested for their MEOR activity. The electro-deposited materials' functional elemental composition and electrochemical finger-prints were verified using X-ray photoelectron spectroscopy and electrochemical techniques [121]. Chronoamperometry was used to check the dependability through-out both cyclic and mechanical conditions. Using a novel combinational approach, the results suggest that MEOR may be a useful electrocatalytic material in DMFC applications. Elancheziyan et al. provide evidence that the current electrocatalysts' relatively sluggish methanol oxidation reaction kinetics pave the way for the creation of indirect methanol fuel cells (DMFCs). This research introduces a nanocompos-ite electrocatalyst for the methanol electro-oxidation reaction (MEOR) made from polyaniline, palladium, and titanium dioxide (PANI-Pd-MXene). An aniline and palladium chloride ($PdCl_2$) acidic electrolyte solution was applied to a pre-anodized screen-printed electrode (SPE) to form the PANI–$PdCl_2$–MXene nanocomposite [123], as illustrated in Figure 18.3a, b. Cyclic voltammetry, Fourier-transform infra-red spectroscopy, X-ray photoelectron spectroscopy, and field-emission scanning electron microscopy were all employed to learn more about the PANI-Pd-MXene nanocomposite. The electrochemical response of the PANI-Pd-MXene nanocom-posite displays enhanced electrocatalytic reactivity towards the oxidation of metha-nol, with a peak current density of 291 mA cm², which is about three times higher than that of Pd-MXene (106 mA cm²). Furthermore, it remained unchanged after one hundred cycles. Since MXene nanosheets include PANI-Pd sites that are elec-trochemically active, their MEOR is significantly enhanced. This promising result

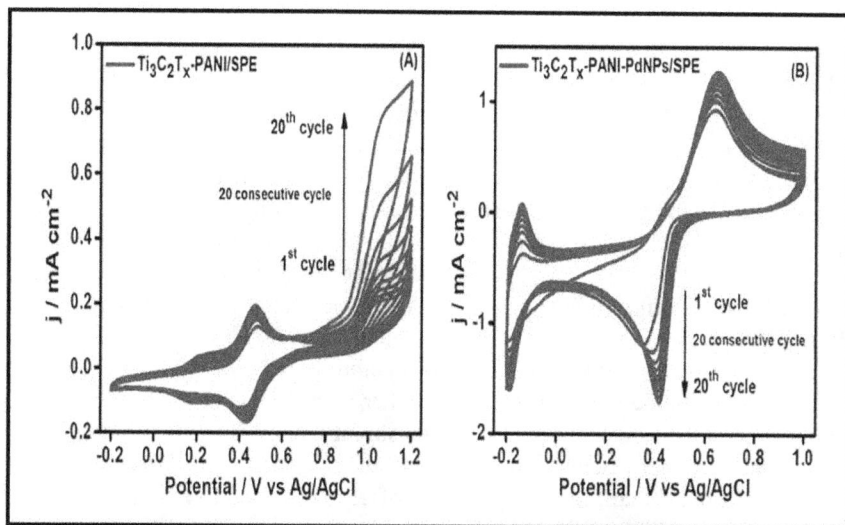

FIGURE 18.3 The electrodeposition of (a) $Ti_3C_2T_x$-PANI-SPE and (b) $Ti_3C_2T_x$-PANI-PdNP-SPE in 0.5 M H_2SO_4 at a sweep rate of 20 mV s⁻¹ [123].

is the result of strong metal-support interactions between PANI-Pd and MXene nanosheets, which optimize methanol adsorption on the electrode surface for efficient electrocatalytic oxidation. Improved electrocatalytic performance is a result of the $Ti_3C_2T_x$ support altering the interface and surface properties of the metal electrocatalyst. The findings of this study simplify the design of direct methanol fuel cell MXene-supported noble metal electrocatalysts for methane oxidation–reduction (MEOR) [123].

The oxygen reduction reaction (ORR) is the mechanism through which fuel cells transfer energy during the first half of their operational lifetime. However, the ORR process at the cathode is lengthy and laborious. The most significant challenge facing PEMFC technology is the development of catalyst materials that are both stable in the highly corrosive ORR and chemically active in the fuel cell cathode reaction. Reaching the ORR in water may either (1) use the four-electron path from O_2 to H_2O or (2) take the two-electron path from O_2 to H_2O_2. The ORR pathway is affected by a number of factors, including the catalyst's particle shape and geometry, particle size, electronic state, catalyst support, and electrolyte pH [124]. Commercial PEMFCs often use platinum (Pt)-based electrocatalysts because they have high ORR activity and are very stable in both acidic and alkaline conditions [125]. However, this expensive metal accounts for more than half of the total cost associated with manufacturing PEMFCs on an industrial scale. This is because PEMFCs haven't become widely used in the power industry. Although it is widely recognized as the most catalytically active substance and is often linked to Pt catalysts, it undergoes degradation and agglomeration when exposed to the cyclic conditions often seen in fuel cells. Catalytically active catalysts require materials that facilitate their uniform and efficient dispersion. For ORR catalysts, carbon-based support materials have been the subject of several recent studies. Corrosion occurs rapidly on carbon supports, and they don't work well with catalysts. Since carbon is thermodynamically unstable at high potentials, corrosion may occur [126], especially during starting and shutdown. Graphene, the weakest substance because of its vast surface area, high conductivity, and intensive interaction with catalysts, is the thinnest two-dimensional (2D) carbon material with a thickness of 0.335 nm. While graphene nanosheets, carbon nanotubes, and carbon nanofibers are all more durable carbon materials than carbon supports, they nonetheless present their own unique set of obstacles [127]. Guo and co-workers have reported Ti_3C_2 MXene-based anode materials for microbial fuel cells [128]. As shown in Figure 18.4a, it is observed that after 25 hours of inoculation, the Ti_3C_2 carbon cloth (CC) anode's output voltage rapidly closed and achieved a steady state. However, after 500 hours of operating time, a steady cell voltage of 640 8 mV was attained, which is 16.4% higher than that recorded for CC (550±8 mV), showing a quick active bacterial adhesion. Further evidence that Ti_3C_2-CC has a substantially longer discharge lifespan than the CC anode is provided by Figure 18.4b, which shows that the current density of Ti_3C_2/CC is 3.13 A m^2, 26% greater than that of CC (2.49 A m^2). By altering the external loading resistance, the authors were able to analyze power output performance by measuring polarization curves and power density. Compared to a CC anode, the Ti_3C_2-CC MFC's output voltage dropped more slowly when external loading was reduced (Figure 18.4c), suggesting a lower internal resistance made possible by the Ti_3C_2 MXene's high conductivity

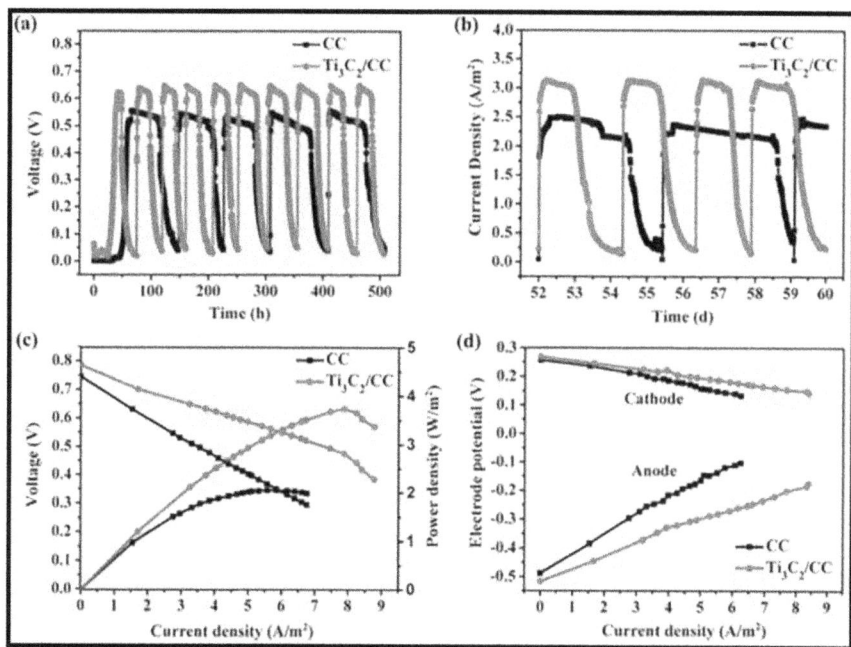

FIGURE 18.4 (a) Ti_3C_2-CC- (red) and CC- (black) based MFC output voltage vs. external loading resistance of 1000 Ω. (b) Long-term operation with an external loading resistance of 1000 Ω and the current discharging performance of various Ti_3C_2/CC- (red) and CC. (black) based MFCs. (c) Power density and polarization curves for anodes made of Ti_3C_2-CC (red) and CC (black) and (d) polarization profiles of the anode and cathode [128].

Adapted with permission from Ref. [128]. Copyright (2018) (Royal Society of Chemistry).

and rapid charge transfer. As can be seen in Figure 18.4d, whereas the potentials of the anodes varied substantially, those of the cathodes were altered just minimally. A power density of 3.74 W m^{-2} was achieved by the MFC with a Ti_3C_2-CC anode (at a higher current density of 7.88 A m^{-2} and an open circuit voltage of 0.785 V), which is an increase of 82.4% over the power density of 2.05 W m^{-2} achieved with a CC anode (at a lower current density of 5.85 A m^{-2} and an open circuit voltage of 0.745 V) [128].

Lee and co-workers reported that a nickel ferrite-based MXene deposited carbon felt anodes for improved microbial fuel cell properties. It was observed that in the scan range of −200 to +600 mV, the CV anodic current was greatly enhanced by changing the CF electrode with $NiFe_2O_4$ (Figure 18.5a), MXene (Figure 18.5b), and $NiFe_2O_4$-MXene (Figure 18.5c). Figure 18.5d displays the current output for each electrode type, with $NiFe_2O_4$-MXene@CF (60 mA), MXene@CF (30 mA), $NiFe_2O_4$@CF (13 mA), and CF (7 mA) producing the maximum current. Compared to the current densities produced by the CF (153 mA/m^2), $NiFe_2O_4$@CF (285 mA/m^2), and MXene@CF (463 mA/m^2) anodes, the $MXeneNiFe_2O_4$@CF was much more powerful [129].

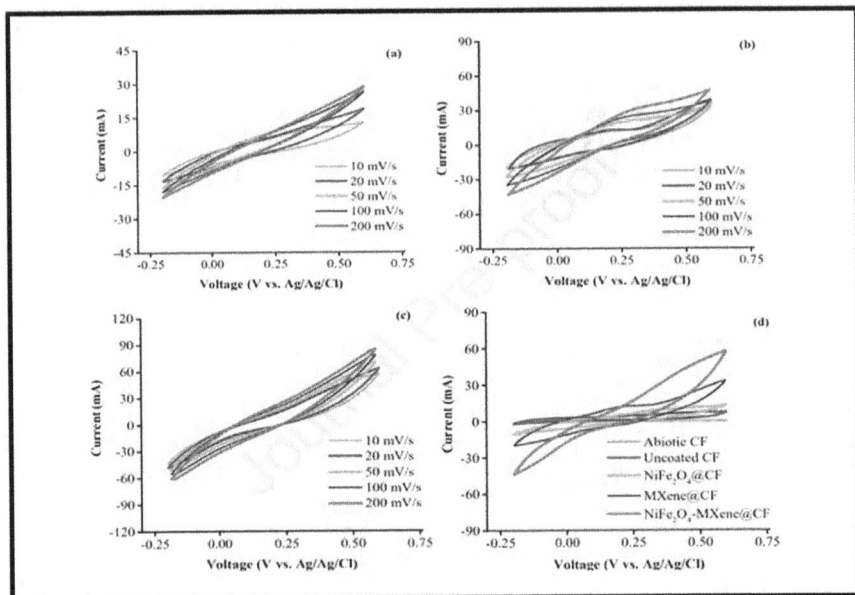

FIGURE 18.5 CV profiles of (a) NiFe$_2$O$_4$@CF, (b) MXene@CF, (c) NiFe$_2$O$_4$-MXene@CF, and (d) biotic and abiotic electrode systems were recorded at various scan speeds (from 10 to 200 mV/s) [129].

Adapted with permission from Ref. [129]. Copyright (2021) (Elsevier).

18.5 FUTURE PERSPECTIVE

MXenes have shown significant applications in various fields, including biological, mechanical, electrical, and electromagnetic fields [37]. They are used in lubricants to reduce wear and friction and are crucial for the fabrication of 2D magnets. Recent computational quantum mechanical research is expected to improve their electrical and magnetic properties. This has led to the emergence of several novel compounds in the MXene subfamily of high-entropy 2D metal carbides. The discovery of oxy-carbide MXenes in late 2022 established a new subfamily of MXenes, and research-ers are exploring its practical applications. The scientific community is eager to learn how this "game-changing compound in the world of materials" will impact the cre-ation of cutting-edge tools [130].

18.6 CONCLUSIONS

New approaches to material design for fuel cells and energy storage are crucial for continuing the steady expansion of modern civilization. Among all the many materi-als out there, 2D MXenes are particularly important due to their peculiar properties. MXenes' compositional variability, which arises from their many possible combina-tions and surface terminations, is very useful for obtaining excellent performance in the aforementioned applications. However, MXene materials are affected by their MAX phase, etching approach, functional groups (T_x), intercalants, and delamination process. The most common synthesis procedure involves the top-down separation of stacked MXene sheets from the MAX phase. The large specific surface area of carbon-based materials and the electroactive sites of MXene in the composite greatly enhance electronic/ion transport capabilities and supercapacitance with the synergistic contributions of both double layers and Faradaic capacitances, making MXene composites/hybrids superior to pure MXenes in terms of capacity, cyclability, and cyclic stability. In this chapter, we give an exhaustive analysis of the current cutting-edge advancements in the synthesis of diverse MXene-based materials, with a focus on their characteristics and, in particular, their usefulness in fuel cells. We do this by concentrating on their properties and, in particular, their suitability in fuel cells. In light of their synthesis and applications in fuel cells, both the promise and the limitations posed by materials based on MXenes are investigated further. Since the number of studies devoted to MXenes has grown over the past year, we hope that this overview will be useful in encouraging their further development and drawing the attention of researchers in related fields.

18.7   ACKNOWLEDGMENT

The authors gratefully acknowledge the support provided by Centurion University of Technology and Management, Odisha, India for carrying out the present research work.
Conflicts of Interest: The authors declare no conflict of interest.

REFERENCES

1. Mohajer, F., Ziarani, G. M., Badiei, A., Iravani, S., & Varma, R. S. (2023). MXene-carbon nanotube composites: Properties and applications. *Nanomaterials*, *13*(2), 345.
2. Long, M. Q., Tang, K. K., Xiao, J., Li, J. Y., Chen, J., Gao, H., & Liu, H. (2022). Recent advances on MXene based materials for energy storage applications. *Materials Today Sustainability*, *19*, 100163.
3. Zhang, Y., Zhan, R., Xu, Q., Liu, H., Tao, M., Luo, Y., & Xu, M. (2019). Circuit board-like CoS/MXene composite with superior performance for sodium storage. *Chemical Engineering Journal*, *357*, 220–225.
4. Feng, A., Yu, Y., Wang, Y., Jiang, F., Yu, Y., Mi, L., & Song, L. (2017). Two-dimensional MXene Ti3C2 produced by exfoliation of Ti3AlC2. *Materials & Design*, *114*, 161–166.
5. Verger, L., Natu, V., Carey, M., & Barsoum, M. W. (2019). MXenes: An introduction of their synthesis, select properties, and applications. *Trends in Chemistry*, *1*(7), 656–669.
6. Long, J. W., Dunn, B., Rolison, D. R., & White, H. S. (2004). Three-dimensional battery architectures. *Chemical Reviews*, *104*(10), 4463–4492.
7. Beidaghi, M., & Gogotsi, Y. (2014). Capacitive energy storage in micro-scale devices: Recent advances in design and fabrication of micro-supercapacitors. *Energy & Environmental Science*, *7*(3), 867–884.
8. Zheng, S., Shi, X., Das, P., Wu, Z. S., & Bao, X. (2019). The road towards planar micro-batteries and micro-supercapacitors: From 2D to 3D device geometries. *Advanced Materials*, *31*(50), 1900583.

9. Yu, D., Goh, K., Zhang, Q., Wei, L., Wang, H., Jiang, W., & Chen, Y. (2014). Controlled functionalization of carbonaceous fibers for asymmetric solid-state micro-supercapacitors with high volumetric energy density. *Advanced Materials*, *26*(39), 6790–6797.

10. Chmiola, J., Largeot, C., Taberna, P. L., Simon, P., & Gogotsi, Y. (2010). Monolithic carbide-derived carbon films for micro-supercapacitors. *Science*, *328*(5977), 480–483.

11. Pech, D., Brunet, M., Durou, H., Huang, P., Mochalin, V., Gogotsi, Y., & Simon, P. (2010). Ultrahigh-power micrometre-sized supercapacitors based on onion-like carbon. *Nature Nanotechnology*, *5*(9), 651–654.

12. Pech, D., Brunet, M., Taberna, P. L., Simon, P., Fabre, N., Mesnilgrente, F., Conédéra, V., & Durou, H. (2010). Elaboration of a microstructured inkjet-printed carbon electrochemical capacitor. *Journal of Power Sources*, *195*, 1266–1269.

13. Hsia, B., Kim, M. S., Vincent, M., Carraro, C., & Maboudian, R. (2013). Photoresist-derived porous carbon for on-chip micro-supercapacitors. *Carbon*, *57*, 395–400.

14. Beidaghi, M., Chen, W., & Wang, C. (2011). Electrochemically activated carbon micro-electrode arrays for electrochemical micro-capacitors. *Journal of Power Sources*, *196*(4), 2403–2409.

15. Wang, S., Hsia, B., Carraro, C., & Maboudian, R. (2014). High-performance all solid-state micro-supercapacitor based on patterned photoresist-derived porous carbon electrodes and an ionogel electrolyte. *Journal of Materials Chemistry A*, *2*(21), 7997–8002.

16. Kim, M. S., Hsia, B., Carraro, C., & Maboudian, R. (2014). Flexible micro-supercapacitors with high energy density from simple transfer of photoresist-derived porous carbon electrodes. *Carbon*, *74*, 163–169.

17. Beidaghi, M., & Wang, C. (2012). Micro-supercapacitors based on interdigital electrodes of reduced graphene oxide and carbon nanotube composites with ultrahigh power handling performance. *Advanced Functional Materials*, *22*(21), 4501–4510.

18. Lin, J., Zhang, C., Yan, Z., Zhu, Y., Peng, Z., Hauge, R. H., & Tour, J. M. (2013). 3-Dimensional graphene carbon nanotube carpet-based microsupercapacitors with high electrochemical performance. *Nano Letters*, *13*(1), 72–78.

19. Xiong, G., Meng, C., Reifenberger, R. G., Irazoqui, P. P., & Fisher, T. S. (2014). Graphitic petal micro-supercapacitor electrodes for ultra-high power density. *Energy Technology*, *2*(11), 897–905.

20. Liang, J., Mondal, A. K., Wang, D. W., & Iacopi, F. (2019). Graphene-based planar microsupercapacitors: Recent advances and future challenges. *Advanced Materials Technologies*, *4*(1), 1800200.

21. Zhang, G., Han, Y., Shao, C., Chen, N., Sun, G., Jin, X., & Qu, L. (2018). Processing and manufacturing of graphene-based microsupercapacitors. *Materials Chemistry Frontiers*, *2*(10), 1750–1764.

22. Kurra, N., Jiang, Q., Nayak, P., & Alshareef, H. N. (2019). Laser-derived graphene: A three-dimensional printed graphene electrode and its emerging applications. *Nano Today*, *24*, 81–102.

23. Kurra, N., Hota, M. K., & Alshareef, H. N. (2015). Conducting polymer micro-supercapacitors for flexible energy storage and Ac line-filtering. *Nano Energy*, *13*, 500–508.

24. Du, J., & Cheng, H. M. (2012). The fabrication, properties, and uses of graphene/polymer composites. *Macromolecular Chemistry and Physics*, *213*(10–11), 1060–1077.

25. Meng, C., Maeng, J., John, S. W., & Irazoqui, P. P. (2014). Ultrasmall integrated 3D micro-supercapacitors solve energy storage for miniature devices. *Advanced Energy Materials*, *4*(7), 1301269.

26. Wang, K., Zou, W., Quan, B., Yu, A., Wu, H., Jiang, P., & Wei, Z. (2011). An all-solid-state flexible micro-supercapacitor on a chip. *Advanced Energy Materials*, *1*(6), 1068–1072.

27. Dinh, T. M., Armstrong, K., Guay, D., & Pech, D. (2014). High-resolution on-chip supercapacitors with ultra-high scan rate ability. *Journal of Materials Chemistry A, 2*(20), 7170–7174.

28. Si, W., Yan, C., Chen, Y., Oswald, S., Han, L., & Schmidt, O. G. (2013). On chip, all solid-state and flexible micro-supercapacitors with high performance based on MnO x/ Au multilayers. *Energy & Environmental Science, 6*(11), 3218–3223.

29. Brezesinski, T., Wang, J., Tolbert, S. H., & Dunn, B. (2010). Ordered mesoporous α-MoO3 with iso-oriented nanocrystalline walls for thin-film pseudocapacitors. *Nature Materials, 9*(2), 146–151.

30. Augustyn, V., Come, J., Lowe, M. A., Kim, J. W., & Taberna, P. L. (2013). SH Tolbert, HD Abrua, P. Simon, B. Dunn. *Nature Materials, 12*, 518.

31. Choi, D., Blomgren, G. E., & Kumta, P. N. (2006). Fast and reversible surface redox reaction in nanocrystalline vanadium nitride supercapacitors. *Advanced Materials, 18*(9), 1178–1182.

32. Lu, S., Zhu, T., Wu, H., Wang, Y., Li, J., Abdelkader, A., . . . & Kumar, R. V. (2019). Construction of ultrafine ZnSe nanoparticles on/in amorphous carbon hollow nanospheres with high-power-density sodium storage. *Nano Energy, 59*, 762–772.

33. Liu, L., Wang, S., Huang, H., Zhang, Y., & Ma, T. (2020). Surface sites engineering on semiconductors to boost photocatalytic CO2 reduction. *Nano Energy, 75*, 104959.

34. Yang, J., Pan, Z., Yu, Q., Zhang, Q., Ding, X., Shi, X., & Zhang, Y. (2019). Free-standing black phosphorus thin films for flexible quasi-solid-state micro-supercapacitors with high volumetric power and energy density. *ACS Applied Materials & Interfaces, 11*(6), 5938–5946.

35. Wu, H., Zhang, W., Kandambeth, S., Shekhah, O., Eddaoudi, M., & Alshareef, H. N. (2019). Conductive metal—organic frameworks selectively grown on laser-scribed graphene for electrochemical microsupercapacitors. *Advanced Energy Materials, 9*(21), 1900482.

36. Kurra, N., Xia, C., Hedhili, M. N., & Alshareef, H. N. (2015). Ternary chalcogenide micro-pseudocapacitors for on-chip energy storage. *Chemical Communications, 51*(52), 10494–10497.

37. Yury, G., & Anasori, B. (2019). The rise of MXenes. *ACS Nano, 13*, 8491–8494.

38. Hart, J. L., Hantanasirisakul, K., Lang, A. C., Anasori, B., Pinto, D., Pivak, Y., & Taheri, M. L. (2019). Control of MXenes' electronic properties through termination and intercalation. *Nature Communications, 10*(1), 522.

39. Zhang, C. J., & Nicolosi, V. (2019). Graphene and MXene-based transparent conductive electrodes and supercapacitors. *Energy Storage Materials, 16*, 102–125.

40. Naguib, M., Kurtoglu, M., Presser, V., Lu, J., Niu, J., Heon, M., & Barsoum, M. W. (2011). Two-dimensional nanocrystals produced by exfoliation of Ti3AlC2. *Advanced Materials, 23*(37), 4248–4253.

41. Kim, H., Wang, Z., & Alshareef, H. N. (2019). MXetronics: Electronic and photonic applications of MXenes. *Nano Energy, 60*, 179–197.

42. Zhong, Y., Xia, X., Shi, F., Zhan, J., Tu, J., & Fan, H. J. (2016). Transition metal carbides and nitrides in energy storage and conversion. *Advanced Science, 3*(5), 1500286.

43. Anasori, B., & Gogotsi, Û. G. (2019). *2D metal carbides and nitrides (MXenes)* (Vol. 2549). Springer.

44. Michael, N., Mochalin, V. N., Barsoum, M. W., & Yury, G. (2014). Two-dimensional materials: 25th anniversary article: MXenes: A new family of two-dimensional materials. *Advanced Materials, 26*(7), 992–1005.

45. Wu, Z. S., Feng, X., & Cheng, H. M. (2014). Recent advances in graphene-based planar micro-supercapacitors for on-chip energy storage. *National Science Review, 1*(2), 277–292.

46. Naguib, M., Mashtalir, O., Carle, J., Presser, V., Lu, J., Hultman, L., . . . & Barsoum, M. W. (2011). ACS Nano 2012, 6, 1322–1331; b) M. *Naguib, M. Kurtoglu, V. Presser, J. Lu, J. Niu, M. Heon, L. Hultman, Y. Gogotsi, MW Barsoum. Advanced Materials, 23*, 4248–4253.

47. Halim, J., Lukatskaya, M. R., Cook, K. M., Lu, J., & Smith, C. R. (2014). L.-. Näslund, SJ May, L. Hultman, Y. Gogotsi, P. Eklund, MW Barsoum. *Chemistry of Materials, 26*, 2374–2381.

48. Verger, L., Natu, V., Carey, M., & Barsoum, M. W. (2019). MXenes: An introduction of their synthesis, select properties, and applications. *Trends in Chemistry, 1*(7), 656–669.

49. Hope, M. A., Forse, A. C., Griffith, K. J., Lukatskaya, M. R., Ghidiu, M., Gogotsi, Y., & Grey, C. P. (2016). NMR reveals the surface functionalisation of Ti 3 C 2 MXene. *Physical Chemistry Chemical Physics, 18*(7), 5099–5102.

50. Mashtalir, O., Naguib, M., Mochalin, V. N., Dall'Agnese, Y., Heon, M., Barsoum, M. W., & Gogotsi, Y. (2013). Intercalation and delamination of layered carbides and carbonitrides. *Nature Communications, 4*(1), 1716.

51. Scala, E. P. (1996). A brief history of composites in the US-The dream and the success. *JOM, 48*(2), 45–48.

52. Okada, A., & Usuki, A. (2006). Twenty years of polymer-clay nanocomposites. *Macromolecular Materials and Engineering, 291*(12), 1449–1476.

53. Iqbal, N., Ghani, U., Liao, W., He, X., Lu, Y., Wang, Z., & Li, T. (2022). Materials Today Advances.

54. Carey, M., & Barsoum, M. W. (2021). MXene polymer nanocomposites: A review. *Materials Today Advances, 9*, 100120.

55. Tu, S., Jiang, Q., Zhang, J., He, X., Hedhili, M. N., Zhang, X., & Alshareef, H. N. (2019). Enhancement of dielectric permittivity of Ti3C2T x MXene/polymer composites by controlling flake size and surface termination. *ACS Applied Materials & Interfaces, 11*(30), 27358–27362.

56. Su, T., Hood, Z. D., Naguib, M., Bai, L., Luo, S., Rouleau, C. M., & Wu, Z. (2019). 2D/2D heterojunction of Ti_3C_2/gC_3N_4 nanosheets for enhanced photocatalytic hydrogen evolution. *Nanoscale, 11*(17), 8138–8149.

57. Shao, M., Shao, Y., Chai, J., Qu, Y., Yang, M., Wang, Z., & Pan, H. (2017). Synergistic effect of 2D Ti 2 C and gC_3N_4 for efficient photocatalytic hydrogen production. *Journal of Materials Chemistry A, 5*(32), 16748–16756.

58. Yuan, W., Cheng, L., Zhang, Y., Wu, H., Lv, S., Chai, L. & Zheng, L. (2017). 2D-Layered Carbon/TiO_2 hybrids derived from Ti_3C_2 MXenes for photocatalytic hydrogen evolution under visible light irradiation. *Advanced Materials Interfaces, 4*(20), 1700577.

59. Guo, S., & Dong, S. (2011). Graphene nanosheet: Synthesis, molecular engineering, thin film, hybrids, and energy and analytical applications. *Chemical Society Reviews, 40*(5), 2644–2672.

60. Singh, V., Joung, D., Zhai, L., Das, S., Khondaker, S. I., & Seal, S. (2011). Graphene based materials: Past, present and future. *Progress in Materials Science, 56*(8), 1178–1271.

61. Kuila, T., Bose, S., Mishra, A. K., Khanra, P., Kim, N. H., & Lee, J. H. (2012). Chemical functionalization of graphene and its applications. *Progress in Materials Science, 57*(7), 1061–1105.

62. Tang, Q., Zhou, Z., & Chen, Z. (2013). Graphene-related nanomaterials: Tuning properties by functionalization. *Nanoscale, 5*(11), 4541–4583.

63. Tang, Q., & Zhou, Z. (2013). Graphene-analogous low-dimensional materials. *Progress in Materials Science, 58*(8), 1244–1315.

64. Naguib, M., & Gogotsi, Y. (2015). Synthesis of two-dimensional materials by selective extraction. *Accounts of Chemical Research, 48*(1), 128–135.

65. Jing, Y., Zhou, Z., Cabrera, C. R., & Chen, Z. (2014). Graphene, inorganic graphene analogs and their composites for lithium ion batteries. *Journal of Materials Chemistry A*, 2(31), 12104–12122.

66. Naguib, M., Kurtoglu, M., Presser, V., Lu, J., Niu, J., Heon, M., & Barsoum, M. W. (2011). Two-dimensional nanocrystals produced by exfoliation of Ti_3AlC_2. *Advanced Materials*, 23(37), 4248–4253.

67. Naguib, M., Mashtalir, O., Carle, J., Presser, V., Lu, J., Hultman, L., & Barsoum, M. W. (2012). Two-dimensional transition metal carbides. *ACS Nano*, 6(2), 1322–1331.

68. Naguib, M., Halim, J., Lu, J., Cook, K. M., Hultman, L., Gogotsi, Y., & Barsoum, M. W. (2013). New two-dimensional niobium and vanadium carbides as promising materials for Li-ion batteries. *Journal of the American Chemical Society*, 135(43), 15966–15969.

69. Ghidiu, M., Naguib, M., Shi, C., Mashtalir, O., Pan, L. M., Zhang, B., & Barsoum, M. W. (2014). Synthesis and characterization of two-dimensional Nb_4C_3 (MXene). *Chemical Communications*, 50(67), 9517–9520.

70. Barsoum, M. W. (2013). *MAX phases: Properties of machinable ternary carbides and nitrides*. John Wiley & Sons.

71. Kurtoglu, M., Naguib, M., Gogotsi, Y., & Barsoum, M. W. (2012). First principles study of two-dimensional early transition metal carbides. *Mrs Communications*, 2, 133–137.

72. Khazaei, M., Arai, M., Sasaki, T., Chung, C. Y., Venkataramanan, N. S., Estili, M., & Kawazoe, Y. (2013). Novel electronic and magnetic properties of two-dimensional transition metal carbides and nitrides. *Advanced Functional Materials*, 23(17), 2185–2192.

73. Come, J., Naguib, M., Rozier, P., Barsoum, M. W., Gogotsi, Y., Taberna, P. L., & Simon, P. (2012). A non-aqueous asymmetric cell with a Ti_2C-based two-dimensional negative electrode. *Journal of the Electrochemical Society*, 159(8), A1368.

74. Hu, J., Xu, B., Ouyang, C., Yang, S. A., & Yao, Y. (2014). Investigations on V_2C and V_2CX_2 (X= F, OH) monolayer as a promising anode material for Li ion batteries from first-principles calculations. *The Journal of Physical Chemistry C*, 118(42), 24274–24281.

75. Xie, X., Chen, S., Ding, W., Nie, Y., & Wei, Z. (2013). An extraordinarily stable catalyst: Pt NPs supported on two-dimensional $Ti_3C_2X_2$(X= OH, F) nanosheets for oxygen reduction reaction. *Chemical Communications*, 49(86), 10112–10114.

76. Wang, F., Yang, C., Duan, C., Xiao, D., Tang, Y., & Zhu, J. (2014). An organ-like titanium carbide material (MXene) with multilayer structure encapsulating hemoglobin for a mediator-free biosensor. *Journal of the Electrochemical Society*, 162(1), B16.

77. Naguib, M., Mochalin, V. N., Barsoum, M. W., & Gogotsi, Y. (2014). 25th anniversary article: MXenes: A new family of two-dimensional materials. *Advanced Materials*, 26(7), 992–1005.

78. Soundiraraju, B., & George, B. K. (2017). Two-dimensional titanium nitride (Ti_2N) MXene: Synthesis, characterization, and potential application as surface-enhanced Raman scattering substrate. *ACS Nano*, 11(9), 8892–8900.

79. Qiu, F., Wang, Z., Liu, M., Wang, Z., & Ding, S. (2021). Synthesis, characterization and microwave absorption of MXene/$NiFe_2O_4$ composites. *Ceramics International*, 47(17), 24713–24720.

80. Gao, X., Jia, Z., Wang, B., Wu, X., Sun, T., Liu, X., & Wu, G. (2021). Synthesis of NiCo-LDH/MXene hybrids with abundant heterojunction surfaces as a lightweight electromagnetic wave absorber. *Chemical Engineering Journal*, 419, 130019.

81. Lim, G. P., Soon, C. F., Jastrzębska, A. M., Ma, N. L., Wojciechowska, A. R., Szuplewska, A., & Tee, K. S. (2021). Synthesis, characterization and biophysical evaluation of the 2D Ti_2CTx MXene using 3D spheroid-type cultures. *Ceramics International*, 47(16), 22567–22577.

82. Qiao, C., Wu, H., Xu, X., Guan, Z., & Ou-Yang, W. (2021). Electrical conductivity enhancement and electronic applications of 2D $Ti_3C_2T_x$ MXene materials. *Advanced Materials Interfaces, 8*(24), 2100903.

83. Noor, U., Mughal, M. F., Ahmed, T., Farid, M. F., Ammar, M., Kulsum, U., & Waqar, K. (2023). Synthesis and applications of MXene-based composites: A review. *Nanotechnology, 34*, 262001.

84. Peng, C., Wei, P., Chen, X., Zhang, Y., Zhu, F., Cao, Y., & Peng, F. (2018). A hydrothermal etching route to synthesis of 2D MXene (Ti_3C_2, Nb_2C): Enhanced exfoliation and improved adsorption performance. *Ceramics International, 44*(15), 18886–18893.

85. Liu, Y., Yu, J., Guo, D., Li, Z., & Su, Y. (2020). $Ti_3C_2T_x$ MXene/graphene nanocomposites: Synthesis and application in electrochemical energy storage. *Journal of Alloys and Compounds, 815*, 152403.

86. Prasad, C., Yang, X., Liu, Q., Tang, H., Rammohan, A., Zulfiqar, S., & Shah, S. (2020). Recent advances in MXenes supported semiconductors based photocatalysts: Properties, synthesis and photocatalytic applications. *Journal of Industrial and Engineering Chemistry, 85*, 1–33.

87. Saha, S., Arole, K., Radovic, M., Lutkenhaus, J. L., & Green, M. J. (2021). One-step hydrothermal synthesis of porous $Ti_3C_2T_z$ MXene/rGO gels for supercapacitor applications. *Nanoscale, 13*(39), 16543–16553.

88. Bayhan, Z., El-Demellawi, J. K., Yin, J., Khan, Y., Lei, Y., Alhajji, E., & Alshareef, H. N. (2023). A laser-induced Mo_2CT_x MXene hybrid anode for high-performance Li-Ion batteries. *Small*, 2208253.

89. Naguib, M., Saito, T., Lai, S., Rager, M. S., Aytug, T., Paranthaman, M. P., & Gogotsi, Y. (2016). $Ti_3C_2T_x$ (MXene)—polyacrylamide nanocomposite films. *RSC Advances, 6*(76), 72069–72073.

90. Wu, Y., Nie, P., Wang, J., Dou, H., & Zhang, X. (2017). Few-layer MXenes delaminated via high-energy mechanical milling for enhanced sodium-ion batteries performance. *ACS Applied Materials & Interfaces, 9*(45), 39610–39617.

91. Zhang, T., Pan, L., Tang, H., Du, F., Guo, Y., Qiu, T., & Yang, J. (2017). Synthesis of two-dimensional Ti3C2Tx MXene using HCl+ LiF etchant: Enhanced exfoliation and delamination. *Journal of Alloys and Compounds, 695*, 818–826.

92. Liang, C., Meng, Y., Zhang, Y., Zhang, H., Wang, W., Lu, M., & Wang, G. (2023). Insights into the impact of interlayer spacing on MXene-based electrodes for supercapacitors: A review. *Journal of Energy Storage, 65*, 107341.

93. Ren, C. E., Hatzell, K. B., Alhabeb, M., Ling, Z., Mahmoud, K. A., & Gogotsi, Y. (2015). Charge-and size-selective ion sieving through $Ti_3C_2T_x$ MXene membranes. *The Journal of Physical Chemistry Letters, 6*(20), 4026–4031.

94. Yorulmaz, U., Özden, A., Perkgöz, N. K., Ay, F., & Sevik, C. (2016). Vibrational and mechanical properties of single layer MXene structures: A first-principles investigation. *Nanotechnology, 27*(33), 335702.

95. Khazaei, M., Arai, M., Sasaki, T., Ranjbar, A., Liang, Y., & Yunoki, S. (2015). OH-terminated two-dimensional transition metal carbides and nitrides as ultralow work function materials. *Physical Review B, 92*(7), 075411.

96. Peng, Q., Si, C., Zhou, J., & Sun, Z. (2019). Modulating the Schottky barriers in MoS_2/MXenes heterostructures via surface functionalization and electric field. *Applied Surface Science, 480*, 199–204.

97. Zhang, P., Wang, L., Du, K., Wang, S., Huang, Z., Yuan, L., & Shi, W. (2020). Effective removal of U (VI) and Eu (III) by carboxyl functionalized MXene nanosheets. *Journal of Hazardous Materials, 396*, 122731.

98. Ahmed, B., Anjum, D. H., Gogotsi, Y., & Alshareef, H. N. (2017). Atomic layer deposition of SnO_2 on MXene for Li-ion battery anodes. *Nano Energy, 34,* 249–256.

99. Jiang, J., Li, Y., Liu, J., Huang, X., Yuan, C., & Lou, X. W. (2012). Recent advances in metal oxide-based electrode architecture design for electrochemical energy storage. *Advanced Materials, 24*(38), 5166–5180.

100. Xiong, D., Li, X., Bai, Z., & Lu, S. (2018). Recent advances in layered $Ti_3C_2T_x$ MXene for electrochemical energy storage. *Small, 14*(17), 1703419.

101. Cai, Y., Shen, J., Ge, G., Zhang, Y., Jin, W., Huang, W., & Dong, X. (2018). Stretchable $Ti_3C_2T_x$ MXene/carbon nanotube composite based strain sensor with ultrahigh sensitivity and tunable sensing range. *ACS Nano, 12*(1), 56–62.

102. Wu, H. B., Chen, J. S., Hng, H. H., & Lou, X. W. D. (2012). Nanostructured metal oxide-based materials as advanced anodes for lithium-ion batteries. *Nanoscale, 4*(8), 2526–2542.

103. Guo, X., Xie, X., Choi, S., Zhao, Y., Liu, H., Wang, C., & Wang, G. (2017). $Sb_2O_3/$ MXene ($Ti_3C_2T_x$) hybrid anode materials with enhanced performance for sodium-ion batteries. *Journal of Materials Chemistry A, 5*(24), 12445–12452.

104. Cheng, Y., Zhang, Y., Li, Y., Dai, J., & Song, Y. (2019). Hierarchical $Ni_2P/Cr_2 CT_x$ (MXene) composites with oxidized surface groups as efficient bifunctional electrocatalysts for overall water splitting. *Journal of Materials Chemistry A, 7*(15), 9324–9334.

105. Kumar, S., & Schwingenschlögl, U. (2016). Thermoelectric performance of functionalized Sc 2 C MXenes. *Physical Review B, 94*(3), 035405.

106. He, Y., Zhang, M., Shi, J. J., Cen, Y. L., & Wu, M. (2019). Improvement of visible-light photocatalytic efficiency in a novel $InSe/Zr_2CO_2$ heterostructure for overall water splitting. *The Journal of Physical Chemistry C, 123*(20), 12781–12790.

107. Anasori, B., Lukatskaya, M. R., & Gogotsi, Y. (2017). 2D metal carbides and nitrides (MXenes) for energy storage. *Nature Reviews Materials, 2*(2), 1–17.

108. Alhabeb, M., Maleski, K., Anasori, B., Lelyukh, P., Clark, L., Sin, S., & Gogotsi, Y. (2017). Guidelines for synthesis and processing of two-dimensional titanium carbide ($Ti_3C_2T_x$ MXene). *Chemistry of Materials, 29*(18), 7633–7644.

109. Ding, L., Wei, Y., Wang, Y., Chen, H., Caro, J., & Wang, H. (2017). A two-dimensional lamellar membrane: MXene nanosheet stacks. *Angewandte Chemie International Edition, 56*(7), 1825–1829.

110. Han, R., Ma, X., Xie, Y., Teng, D., & Zhang, S. (2017). Preparation of a new 2D MXene/ PES composite membrane with excellent hydrophilicity and high flux. *RSC Advances, 7*(89), 56204–56210.

111. Rasool, K., Mahmoud, K. A., Johnson, D. J., Helal, M., Berdiyorov, G. R., & Gogotsi, Y. (2017). Efficient antibacterial membrane based on two-dimensional Ti3C2Tx (MXene) nanosheets. *Scientific Reports, 7*(1), 1–11.

112. Han, R., Xie, Y., & Ma, X. (2019). Crosslinked P84 copolyimide/MXene mixed matrix membrane with excellent solvent resistance and permselectivity. *Chinese Journal of Chemical Engineering, 27*(4), 877–883.

113. Kusumawati, N., Koestiari, T., & Muslim, S. (2016). The development of a new polymer membrane: PSf/PVDF blended membrane. *Research Journal of Pharmaceutical Biological and Chemical Sciences, 7*(4), 69–77.

114. Xue, S. M., Xu, Z. L., Tang, Y. J., & Ji, C. H. (2016). Polypiperazine-amide nanofiltration membrane modified by different functionalized multiwalled carbon nanotubes (MWCNTs). *ACS Applied Materials & Interfaces, 8*(29), 19135–19144.

115. Wu, X., Hao, L., Zhang, J., Zhang, X., Wang, J., & Liu, J. (2016). Polymer-Ti3C2Tx composite membranes to overcome the trade-off in solvent resistant nanofiltration for alcohol-based system. *Journal of Membrane Science, 515,* 175–188.

116. Deysher, G., Shuck, C. E., Hantanasirisakul, K., Frey, N. C., Foucher, A. C., Maleski, K., & Gogotsi, Y. (2019). Synthesis of $Mo_4V_{Al}C_4$ MAX phase and two-dimensional Mo_4VC_4 MXene with five atomic layers of transition metals. *ACS Nano*, *14*(1), 204–217.

117. Lipatov, A., Alhabeb, M., Lu, H., Zhao, S., Loes, M. J., Vorobeva, N. S., & Sinitskii, A. (2020). Electrical and elastic properties of individual single-layer $Nb_4C_3T_x$ MXene flakes. *Advanced Electronic Materials*, *6*(4), 1901382.

118. Li, Z., Yu, L., Milligan, C., Ma, T., Zhou, L., Cui, Y., & Wu, Y. (2018). Two-dimensional transition metal carbides as supports for tuning the chemistry of catalytic nanoparticles. *Nature Communications*, *9*(1), 5258.

119. Muthusankar, E., Lee, S. C., & Ragupathy, D. (2018). Enhanced electron transfer characteristics of surfactant wrapped SnO_2 nanorods impregnated poly (diphenylamine) matrix. *Sensor Letters*, *16*(12), 911–917.

120. Yang, L., Tang, Y., Yan, D., Liu, T., Liu, C., & Luo, S. (2016). Polyaniline-reduced graphene oxide hybrid nanosheets with nearly vertical orientation anchoring palladium nanoparticles for highly active and stable electrocatalysis. *ACS Applied Materials & Interfaces*, *8*(1), 169–176.

121. Soleimani-Lashkenari, M., Rezaei, S., Fallah, J., & Rostami, H. (2018). Electrocatalytic performance of $Pd/PANI/TiO_2$ nanocomposites for methanol electrooxidation in alkaline media. *Synthetic Metals*, *235*, 71–79.

122. VahidMohammadi, A., Moncada, J., Chen, H., Kayali, E., Orangi, J., Carrero, C. A., & Beidaghi, M. (2018). Thick and freestanding MXene/PANI pseudocapacitive electrodes with ultrahigh specific capacitance. *Journal of Materials Chemistry A*, *6*(44), 22123–22133.

123. Elancheziyan, M., Eswaran, M., Shuck, C. E., Senthilkumar, S., Elumalai, S., Dhanusuraman, R., & Ponnusamy, V. K. (2021). Facile synthesis of polyaniline/titanium carbide (MXene) nanosheets/palladium nanocomposite for efficient electrocatalytic oxidation of methanol for fuel cell application. *Fuel*, *303*, 121329.

124. Sui, S., Wang, X., Zhou, X., Su, Y., Riffat, S., & Liu, C. J. (2017). A comprehensive review of Pt electrocatalysts for the oxygen reduction reaction: Nanostructure, activity, mechanism and carbon support in PEM fuel cells. *Journal of Materials Chemistry A*, *5*(5), 1808–1825.

125. Lin, R., Cai, X., Zeng, H., & Yu, Z. (2018). Stability of High-Performance Pt-Based Catalysts for Oxygen Reduction Reactions. *Advanced Materials*, *30*(17), 1705332.

126. Nørskov, J. K., Rossmeisl, J., Logadottir, A., Lindqvist, L. R. K. J., Kitchin, J. R., Bligaard, T., & Jonsson, H. (2004). Origin of the overpotential for oxygen reduction at a fuel-cell cathode. *The Journal of Physical Chemistry B*, *108*(46), 17886–17892.

127. Farmani, A., & Mir, A. (2019). Graphene sensor based on surface plasmon resonance for optical scanning. *IEEE Photonics Technology Letters*, *31*(8), 643–646.

128. Liu, D., Wang, R., Chang, W., Zhang, L., Peng, B., Li, H., & Guo, C. (2018). Ti 3 C 2 MXene as an excellent anode material for high-performance microbial fuel cells. *Journal of Materials Chemistry A*, *6*(42), 20887–20895.

129. Tahir, K., Miran, W., Jang, J., Maile, N., Shahzad, A., Moztahida, M., & Lee, D. S. (2021). Nickel ferrite/MXene-coated carbon felt anodes for enhanced microbial fuel cell performance. *Chemosphere*, *268*, 128784.

130. Gogotsi, Y., & Anasori, B. (2019). The rise of MXenes. *ACS Nano*, *13*(8), 8491–8494.

19 MXene-Based Nanostructured Materials for Gas Sensing Applications

Prakash Chandra, Dojalisa Sahu, and Subhendu Chakroborty

19.1 INTRODUCTION

Recently, gas sensors have emerged for the identification and quantification of gases. Therefore, gas sensors have become indispensable for contemporaneous human society in defending against obnoxious gases (combustible, flammable and toxic) existing in the atmosphere. [1,2] A diverse array of sensor techniques has been developed like ultrasound, infrared, electrochemical and chemiresistive gas sensing. [3,4,5,6] Among a diversity of well-known sensing techniques, chemiresistive sensors have received considerable attention because of their affordability, facile configuration and high accuracy. [7,8,9] Chemiresistive sensors function via the interaction of the gas molecules with the material used for gas sensing, which leads to a change in electrical signal due to the change in resistance depending on the concentration of gaseous material used for gas sensing. [10,11] Due to the aforementioned important features, chemiresistive sensors are becoming integral parts of human life due to their utility in human safety and healthcare. [12] Moreover, additional functionalities of these chemiresistive sensors like flexibility, reusability, high sensitivity and selectivity can make them more suitable for practical applications. [13,14]

The important physicochemical properties for chemiresistive sensor materials include a high surface area that assists efficient interaction between the gas molecules and a solid surface for superior suitability for the selective and highly sensitive technique of signal transduction. [15,16] Among the diversity of materials investigated to date, metal oxides are the most commonly investigated semiconductor materials for signal transduction for gas sensing applications. [17,18,19] These metal oxide-based sensors furnish to these devices' excellent sensitivity, long-term consistency and fast response time. [20,21] However, these metal oxide-based semiconductors exhibit some weaknesses like inadequate sensitivity, inferior electrical conductivity and high working temperature. To tackle such problems, two-dimensional (2D) materials have recently emerged as the appropriate potential successors as materials for gas sensing applications via sorting out the aforementioned problems with exclusive material

DOI: 10.1201/9781003366225-19

properties that are obviously different from orthodox metal oxide-based materials. [22,23] These 2D materials are frequently defined as a single layer or few layers (1–5 nm thick) in atomic thickness. Beginning with graphene, a diversity of 2D materials have been recently explored, with a boom in the family of 2D materials that include M chalcogenides, hexagonal boron nitrides, phosphorene and so on. [24,25] For discrete sensing properties, individual 2D materials possess different morphological characteristics and surface chemistries. [26,27] These two-dimensional (2D) materials have appeared as capable contenders for gas detection due to their exclusive characteristics such as high surface-to-volume ratio, large surface area and excellent mechanical flexibility. Here are some examples of 2D materials used for gas sensing:

- Graphene demonstrates excellent physicochemical and sensitivity to gas adsorption, making it a good candidate for gas detecting capabilities. [28,29]
- TMDs are the category of 2D materials with the general formula MX_2, where M is a TM and X is a chalcogen (sulfur, selenium or tellurium). TMDs have been shown to have high sensitivity and selectivity towards various gases, including nitrogen dioxide, ammonia and hydrogen sulfide. [27,30]
- Black phosphorus is a 2D material with a layered structure that has high sensitivity towards gases due to its high surface-to-volume ratio. It has been shown to have high sensitivity towards nitrogen dioxide and ammonia. [31,32]
- MOFs are a class of porous materials that have been used for gas sensing due to their high surface area and tunable properties. MOFs have been shown to have high sensitivity and selectivity towards various gases, including carbon dioxide, methane and ammonia. [33,34]

Overall, these 2D materials demonstrate outstanding gas sensing abilities due to their unique properties, and ongoing research is focused on exploring their capability for gas sensing.

In the endless pursuit for 2D materials, MXenes, a recent family of 2D TM carbides and nitrides, is one of the most contemporary as well as distinguished 2D materials. MXenes were the first 2D materials revealed in 2011, shaped by careful abstraction of robustly attached layered crystal morphology. [35,36] The term "MXene" initiated from the selective exclusion of "A" from parent MAX phases, and the "ene" was introduced to highlight the character of 2D materials analogous to graphene. [37] The MAX phases incorporate an extended collection of ternary carbides and nitrides possessing hexagonal layered assemblies following the formula $M_{n+1}AX_n$, where n = 1, 2 or 3 (like M_2AX, M_3AX_2 or M_3AX_3); "M" represents an early TM; "A" represents an A-group element (frequently groups 13 and 14); and "X" represents C, N or both of them. [38,39] To date, roughly 70 distinct MXenes have been synthesized by experimentation. Moreover, for these MXenes, numerous structures and properties have been projected hypothetically. Unexpectedly, recent research has discovered that the MXene surface is haphazardly terminated by OH, O and/or F, with the ratios regulated principally by miscellaneous synthesis techniques. [40] Due to MXenes' adaptable chemistry, their properties can be tailor-made for specific applications like energy integration, electromagnetic interference shielding, water decontamination, gas and biosensors, material for friction reduction and catalysis. [41,42]

According to the present scenario, there is a surge in attention to MXenes for miscellaneous applications as discussed earlier; however, limited attention of the researchers has been paid to utilize MXenes as advanced materials for gas sensing applications. [43,44] Synthesis and structural properties of the MXenes are discussed somewhere else. MXene-based nanomaterials possess electrical conductivity and high-density surface termination groups, with excellent gas detection capabilities under moderate conditions. The unique electronic and surface properties of MXenes respond extremely sensitively to the presence of gas molecules, allowing for their expansion in applications to highly sensitive detectors for a particular gas molecule. Two of the key advantages of MXenes for gas sensing are their high surface area and ability to form thin films, which provide a large contact area with the gas molecules and facilitate their detection. MXenes can be synthesized with various surface terminations, which can be tailored to selectively interact with specific gas molecules. This tunability facilitates their selective gas sensing capability. In addition to their sensing properties, MXenes also exhibit good mechanical and thermal stability, which make them suitable for use in harsh environments. These properties, combined with their low cost and ease of synthesis, make MXenes promising materials for use in the development of next-generation gas sensors. Due to the negative surface charge of the MXene sheet, cationic species can be effortlessly intercalated into MXene interlayers to principally augment the film properties. Moreover, the transparency and sheet-like 2D morphology of MXenes make them interesting materials for gas sensing applications. Surface adsorption and interlayer diffusion administer the sensing properties of MXenes. [45] The present chapter of this book provides an overview of the recent progress in the computational and experimental potential of MXenes for gas sensing applications. Moreover, the present chapter also discusses the future prospects of MXene-based materials for gas sensing applications.

19.2 APPLICATIONS OF MXENES IN GAS SENSING: THEORETICAL CONSIDERATIONS

Gas detection devices are essential to identify and examine biomarkers and noxious and dangerous gases, which comprise volatile organic compounds like ethanol, toluene, acetone, fluorinated organic compounds, etc., and inorganic gases like carbon dioxide, ammonia, carbon monoxide, sulfur dioxide, hydrogen sulfide, etc. Therefore, successfully distinguishing these gases at a fixed safe level for a definite application is crucial. Numerous investigations into the adsorption of gases on the surface of several MXenes have discovered that functionalities on the MXenes' surface demonstrate high selectivity to definite organic or inorganic gas molecules. However, a limited number of these MXenes demonstrate experimentally the potential to become candidates for gas sensing.

Yu et al. examined the adsorption of ammonia on monolayer Ti_2C by means of O termination (Ti_2CO_2) with the help of first-principles simulations and performed a comparison of adsorption of other inorganic gases like carbon dioxide, nitrogen, oxygen, methane, hydrogen and nitrogen dioxide. The exclusive ammonia gas absorption capability of Ti_2CO_2 in the presence of a mixture of other gases makes it a capable and selective ammonia gas sensor. [46–48,49] The MXenes possessing O-terminals (M_2CO_2,

M = titanium, zirconium, hafnium and scandium) can powerfully confine and liberate ammonia gas molecules by means of electron injection as compared to nitrogen, methane, carbon dioxide, carbon monoxide and oxygen. V_2CO_2 and Mo_2CO_2 confirmed greater selectivity to NO gas, although Ti_2CO_2 and Nb_2CO_2 established greater selectivity to carbon monoxide, water, nitrogen, hydrogen, nitrogen dioxide, hydrogen sulfide, sulfur dioxide and carbon dioxide. Ammonia gas sensing has been carried out with the help of alternative semiconducting Hf_2CO_2-based MXene materials. The Hf_2CO_2 due to the existence of 0.834 eV binding energy displays extraordinary selectivity to ammonia, outdoing the reported adsorption of ammonia on Ti_2CO_2 (0.37 eV).[50] Additionally, external strains and vertical electric fields can directly modify the ammonia adsorption strength and alternate the electronic structures, permitting the precise trapping and liberation of ammonia molecules from the Hf_2CO_2 monolayer. The vacancy (including Hf, O, and C vacancies) and strain coexistence directly impact the electronic properties of ammonia adsorbed on the Hf_2CO_2 MXene surface. Under strain circumstances, N atoms of the ammonia molecule situated on top of the Hf atom in a C-vacant Hf_2CO_2 were revealed to be valuable as recyclable gas sensors.[51] Ti-deficient $Ti_3C_2O_2$ MXenes have a comparatively robust physical contact with ammonia, making them more appropriate for extremely sensitive ammonia gas sensors. These aforementioned MXenes possess one surface termination functional group.

The surface termination groups of MXene materials include mixtures of F, O and OH, with the variation in the surface functional group based on etching and synthesis conditions. The contribution of the Ti- and C-based atomic defects influence the physicochemical and optoelectronic properties of Ti_2CT_2 (T = F, O and OH) MXenes and measure their phosgene sensing capabilities.[52,53,54] These studies clearly demonstrate that MXenes with low -F content demonstrate superior ammonia sensing capability as compared to the MXenes with high -F content. In the same way, the surface termination group affects the adsorption/absorption of a few polar and nonpolar analytes over 2D V_2CT_x MXenes using three dissimilar surface group content configurations corresponding to variable synthesis methods. The V_2CT_x MXene with fluoride content offered healthier sensitivity for hydrogen and methane molecules devoid of polarity and exposed the maximum reaction to methylamine.[55]

In addition to ammonia, poisonous gaseous molecules like sulfur dioxide adsorbed on O-terminated M_2CO_2 monolayers (M = scandium, titanium, hafnium and zirconium) were considered. Among various MXenes investigated, Sc_2CO_2 has a greater sensing capability for toxic sulfur dioxide gas compared to other investigated M_2CO_2 monolayers. The interactions were further improved by using tensile strains or negative electric fields. Yang and coworkers used the same set of MXenes to investigate the detection of other toxic gases such as nitrogen monoxide and carbon monoxide. Sc_2CO_2 is, as expected, tremendously selective for nitrogen monoxide sensing. This is principally due to the greater adsorption energy of nitrogen monoxide gas on Sc_2CO_2 (0.47 eV) as compared to ammonia on Ti_2CO_2 (0.37 eV). Moreover, an additional reason can be better charge transferred to nitrogen monoxide molecules from Sc_2CO_2 (0.303 e) as compared with ammonia absorption on Ti_2CO_2 (0.174 e). Mn doping meaningfully improves CO adsorption on Sc_2CO_2. Titanium, zirconium, hafnium, tungsten and vanadium (MXene)-based carbides have also been investigated for the carbon dioxide adsorption, and could demonstrate adsorption energy (-3.96 eV). These M_2C MXenes can theoretically adsorb carbon dioxide at temperatures and partial pressures ranging from 2.34 to 8.25 mol carbon dioxide/kg of substrate. Oxygen, nitrogen monoxide, nitrogen dioxide and sulfur dioxide gases were chemisorbed on the Sc_2CO_2 monolayer, where the adsorption selectivity for oxygen molecules limits its applicability in gas sensing as well as capture. This increases the likelihood of using vacancies to alter the gas adsorption capabilities of the Sc_2CO_2 monolayer. Moreover, the adsorption properties of different inorganic gases like carbon monoxide, nitrogen, nitrogen monoxide, hydrogen, nitrogen dioxide, hydrogen sulfide, water, sulfur dioxide and ammonia on the Sc_2CO_2 monolayer by means of O vacancy was also investigated. These investigations clearly demonstrate that inorganic gases like ammonia, water, nitrogen, hydrogen and hydrogen sulfide where physiosorbed, whereas gases like nitrogen monoxide, sulfur dioxide, carbon monoxide, carbon dioxide and oxygen were chemisorbed of the surface of the Sc_2CO_2 sensor. In comparison to the pristine Sc_2CO_2 monolayer, the oxygen vacancies on the Sc_2CO_2 monolayer significantly augmented the strength of carbon monoxide and carbon dioxide adsorption. The Sc_2CO_2 monolayer in the O vacancy opened the possibility to make nitrogen monoxide gas resistive sensors. [56,57,58,59] Cheng and coworkers also looked into the selectivity of nitrogen dioxide molecules towards Sc_2CF_2 MXenes. The selectivity for nitrogen dioxide molecules has also been investigated using a Sc_2CF_2 MXene sensor. Moreover, the simulations established that nitrogen dioxide can possibly be soundly adsorbed on the Sc_2CF_2 surface under ambient conditions. The biaxial strain can assist in improving the adsorption energy and charge transfer quantity for nitrogen dioxide, augmenting the sensitivity of Sc_2CF_2 in the detection of nitrogen dioxide molecules. These discoveries suggest that Sc_2CF_2 can possibly be a capable contender for nitrogen dioxide sensors with excellent selectivity and selectivity. The Sc_2CF_2 monolayer has been identified as an efficient sensor for nitrogen monoxide gas. [60] CaC_2 has also been identified as a novel sensor for hazardous gases like PH_3, NH_3, SbH_3 and AsH_3 molecules. [61] The pristine Sn_3C_2 monolayer in corroboration of with TMs like copper, cobalt, iron, nickel, ruthenium, rhodium, palladium and silver have been studied for gas sensing ability. Theoretical calculations clearly demonstrate that Rh/Sn_3C_2, Ru/Sn_3C_2 and Pd/Sn_3C_2 monolayer-based sensors functioned well for carbon dioxide, carbon monoxide, nitrogen dioxide, nitrogen mono oxide and sulfur dioxide molecules. [62] Additionally, Kong and coworkers inspected the adsorption performance of $Ti_3C_2T_x$ with diverse termination groups like -O, -F and -OH on the foremost decomposition characteristic components of sulfur hexafluoride, together with hydrogen sulfide, sulfur dioxide, hydrogen sulfide, SOF_2 and SO_2F_2. [63] According to the findings, $Ti_3C_2F_2$ and $Ti_3C_2O_2$ with point vacancy are potential novel sensing materials for detecting sulfur hexafluoride disintegrated species with excellent sensitivity and less electronic noise. [64] The hydroxyl group-terminated $Ti_3C_2T_x$ MXene ($Ti_3C_2(OH)_2$) displayed improved sensing performance for sulfur hexafluoride disintegration characteristic components as compared to $Ti_3C_2O_2$ and $Ti_3C_2F_x$. As a consequence, the type and quantity of $Ti_3C_2T_x$ terminal functionality can be controlled by an appropriate synthesis tactic, and gas detection for numerous gases can be attained. Additionally, metal carbide- and metal nitride-based MXene siblings were investigated. Naqvi and coworkers have examined sulfur-terminated M_2N (M = Ti, V) for sensing gaseous contaminants like sulfur dioxide, carbon monoxide, methane, carbon monoxide, hydrogen sulfide, ammonia and NO_x. Ti_2NS_2 and V_2NS_2 MXene sheets were observed to be the most sensitive to nitrogen monoxide and nitrogen dioxide. [65] Furthermore, sulfur dioxide adsorption on Ti_2NS_2 is stronger than on a Ti_2CO_2 MXene, whereas Sc_2CO_2 monolayers have been described to be extremely sensitive to sulfur dioxide molecules. Gases like nitrogen monoxide, nitrogen dioxide and ammonia have been examined for sensing applications using pristine Ti_2N monolayers as well as functionalized Ti_2NT_2 (T composed of -O, -OH and -F). A great volume of charge collects due to the functionalization of the pristine Ti_2N monolayer, significantly lowering the absolute value of adsorption energy for nitrogen-containing gases. [66] Ti_2NO_2 may be a better candidate for nitrogen monoxide and ammonia sensing based on adsorption structure and energy, whereas Ti_2NF_2 may be used for nitrogen dioxide.

The progression in the field of computational material disciplines has permitted computational approaches applicable to offer accurate and complete descriptions for experimental information. Nevertheless, the theoretical investigations also provide assistance in the plan of original material structures.

19.3 APPLICATIONS OF MXENES IN GAS SENSING: PRACTICAL CONSIDERATIONS

In the recent past, the family of MXenes has revolutionized the field of gas sensing via the advancement of novel materials. To make the most of the selectivity of gas sensors, numerous n- or p-type semiconducting or conducting behaviors must be created via inducing the dopants into the MXene matrix and improvising the sensing capabilities. In the n–p-conductor three-phase transition of gas sensing activity by means of a Mo_2CT_x MXene, the occurrence of organic intercalants like tetramethylammonium hydroxide (TMAOH) intercalant was capable of exhibiting a p-type gas sensing reaction; however, n-type semiconductor properties were exhibited without the any such intercalation. Mo_2CT_x films possessing width more than 700 nm demonstrated conducting properties. These MXenes demonstrate excellent conductivity as well as the existence of high concentrations of the surface functionalities. The Mo_2CT_x MXene composed of TMAOH demonstrated superior sensitivity for volatile organic compounds, nitriles and ammonia. The experimental investigations were further corroborated via DFT calculations. [67]

Triethylamine (TMA) is an indicator that determines how freshly a fish is caught. Therefore, in order to determine the freshness of the fish during transportation, Au@MXene-based sensors with elastic properties have been developed. These Au@MXene hydrogels possess outstanding sensing, stretching, tensile strength, stretchability, durability and antifreeze properties. The resulting Au@MXene, in corroboration with the polymerization of N,N'-methylenebisacrylamide, carboxyl methylcellulose, acrylamide and ethylene glycol, was capable of sensing TMA from 0 °C to 25 °C. [68]

PEDOT:PSS/MXene synthesized via the in situ oxidative polymerization of 3,4-ethylenedioxythiophene (EDOT) and poly(4-styrenesulfonate) (PSS) was considered for its gas sensing capabilities. The gas sensing capabilities reveal a 36.6%

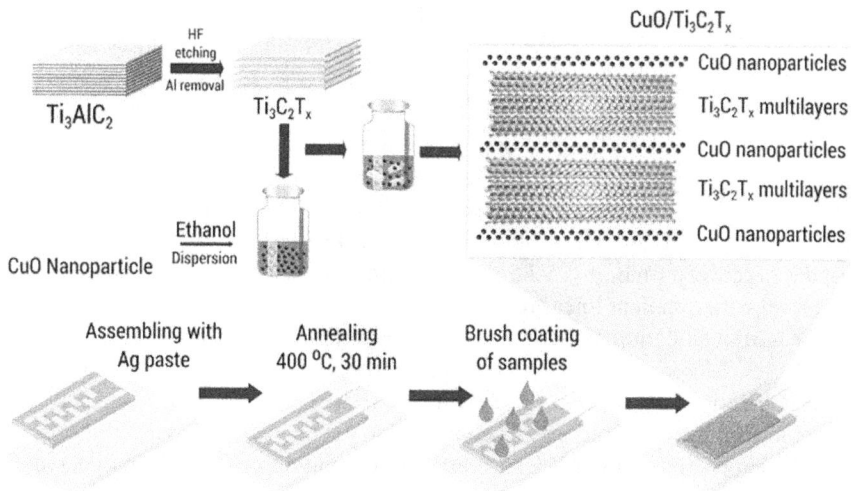

FIGURE 19.1 The scheme describes a simple method for preparing CuO nanoparticles/ $Ti_3C_2T_x$ hybrid heterostructures and the fabrication of gas sensor devices.

Adapted with permission from Ref. [70].

response in favor of 100 ppm NH_3 (the response and retrieval time for ammonia was recorded to be 116 and 40 s, correspondingly). The cooperative effect among PEDOT:PSS polymers and the aforementioned MXene was responsible for the superior gas sensing capability of the PEDOT:PSS/MXene. [69] CuO/$Ti_3C_2T_x$ MXene hybrid materials demonstrate ameliorated gas sensing capabilities as compared to bare pristine CuO nanoparticles along with improved response rate, selectivity and response/recovery times. The prime reason responsible for the enhanced sensing capabilities was due to the amelioration in the Schottky junction generated at the interface of CuO and a $Ti_3C_2T_x$ MXene, as shown in Figure 19.1.[70]

Doping with ions improvises the gas sensing performance of MXenes. $Ti_3C_2T_x$ MXene sensors with alkaline additives demonstrated ameliorated gas sensing capabilities. With potassium hydroxide, the sensor was 20× more sensitive to ethanol vapor and 700× more sensitive to ammonia gas, signifying high sensitivity. [71] Intercalating monovalent or divalent metal cations into a MXene enhances its gas sensing properties, transparency and conductivity. Ion-intercalated MXene films have an optical transmission of 90% at 550 nm and superior ammonia gas sensing capabilities. [72]

ZnO nanoparticles were amalgamated on $Ti_3C_2T_x$ MXene-derived titania nanosheets to form ZnO@MTiO$_2$ (0D–2D heterostructure) for extremely sensitive nitrite recognition. ZnO nanoparticles can not only act as spacers to prevent the restacking of MTiO$_2$ nanosheets and ensure effective gas molecule transfer, but they can also improve sensor sensitivity by trapping electrons. In the meantime, MTiO$_2$ nanosheets encourage gas diffusion, permitting quicker sensor responses. The ZnO@MTiO$_2$ 0D–2D heterostructure-based sensors established extraordinary sensitivity and outstanding selectivity to low-concentration nitrogen dioxide under ambient conditions due to the cooperative outcome of distinct components. [73]

In_2O_3 nanocubes and a layered $Ti_3C_2T_x$ MXene were amalgamated to form a $In_2O_3/Ti_3C_2T_x$ MXene nanocomposite using a hydrothermal self-assembly technique. The physicochemical characterization of the $In_2O_3/Ti_3C_2T_x$ MXene nanocomposite revealed the formation of In_2O_3 nanocubes of 20 to 130 nm dimensions uniformly dispersed over the $Ti_3C_2T_x$ possessing a laminar structure. The synergistic impact of the electrical characteristics and gas adsorption capabilities of $In_2O_3/Ti_3C_2T_x$ MXene demonstrated an amplification in response (29.6%, 5 ppm) and high selectivity to methanol under moderate conditions. Moreover, the material exhibited a very low response/recovery time of 6.5/3.5 s and lowered the detection concentration to the ppm level with excellent linear response. [74]

The partial oxidation of the $Ti_3C_2T_x$ MXene at 350 °C greatly enhances its sensing capabilities. Moreover, annealing these thin films assists in retaining their electrical conductivity as well as their chemiresistive sensitivity in the sensing of alcohols at low ppm concentrations. As compared to the pure $Ti_3C_2T_x$, partially oxidized $Ti_3C_2T_x$ demonstrates a much faster response time and is qualitatively different. In addition to all this, the sensor is capable of identifying analytes of similar chemical characteristics, like low-molecular-weight alcohols. [75]

The chemiresistive response ($S = \Delta G/Gair$) is an average of all MXene sensing elements on the chip with respect to the concentration of organic vapors. In the main panel, the vector signals for three organic vapors (10 ppm) and dry air projected by LDA into a two-component discrimination illustration are depicted. The gas-related circles are shaped with 0.7 confidence around them, matching with the gravity centers. These experiments were conducted at 350 °C.

$Ti_3C_2T_x/ZnO$ MXene nanocomposites exhibit excellent optoelectronic nitrate sensing property at the sub-ppb level with excellent responsiveness, sensing capability and detection limits. Moreover, these meso-structures demonstrate high surface area when produced via facile anchoring growth technique. The hybrids were capable of sensing NO_2 gas at the ppb level under UV irradiation as compared to pure ZnO nanorods. Moreover, the sensor was capable of demonstrating a reaction time of 17 seconds and a recovery time of 24 seconds for the nitrogen dioxide concentration of 5 to 200 ppb. The hybrid material meso-structure as well as the photoconductive property of the $Ti_3C_2T_x$ MXene were primarily responsible for the outstanding reversibility and sub-ppm-level selective detection limit for nitrous oxide gas. [76]

MXene/SnO_2 heterojunction nanocomposites were synthesized via the facile hydrothermal technique and were used as chemiresistive ammonia gas sensors at 0.5 to 100 ppm concentrations. Their excellent capability to adsorb ammonia and high electrical conductivity played a critical role in high sensing performance. Mechanistic studies revealed that the MXene/SnO_2 heterojunction at the Fermi level permitted charge transfer to tin oxide and enhanced ammonia gas sensing capabilities under ambient conditions. The aforementioned semiconductor heterojunction facilitated a quick response (30 s), and the sensor demonstrated steady performance under a range of ammonia gas concentrations. A low cofired ceramic technology was used to create a wireless sensor. [77]

A facile hydrothermal method was used to fabricate 2H-MoS_2 nanosheets over a few-layer $Ti_3C_2T_x$ MXene and were investigated for NO_2 gas sensing under ambient conditions. When the performance of the 2H-$MoS_2/Ti_3C_2T_x$ MXene was compared

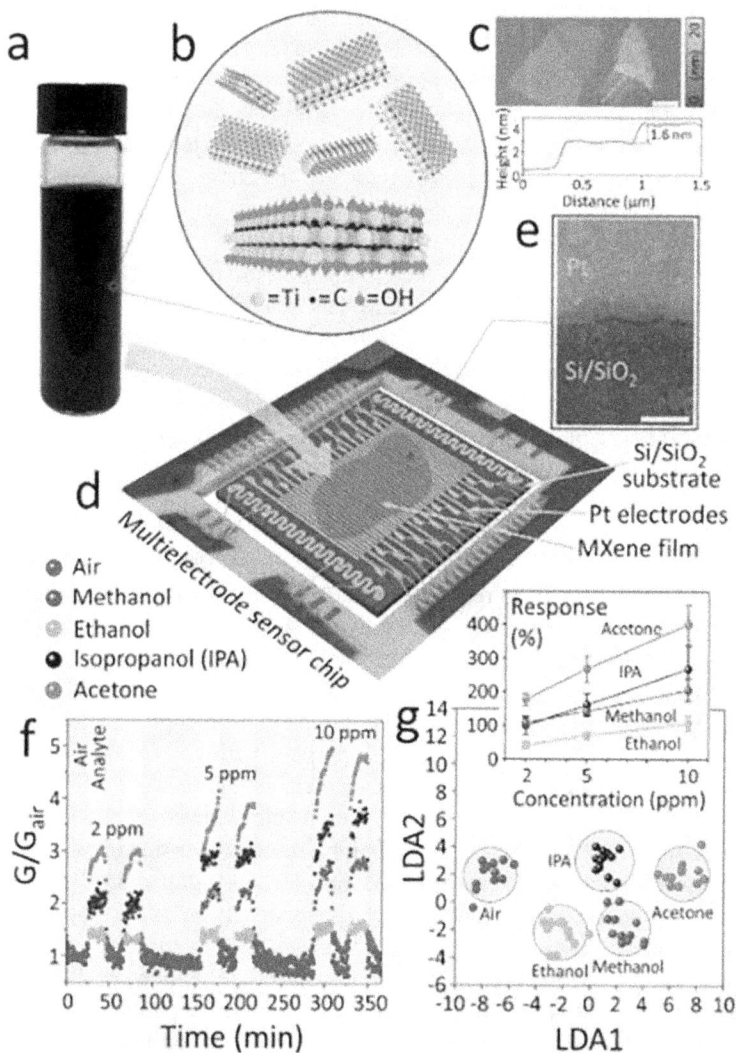

FIGURE 19.2 (a–b) Stable delaminated aqueous MXene, including the assembly of mono-layer $Ti_3C_2T_x$ MXene flakes. (c) AFM image of $Ti_3C_2T_x$ MXene flakes on Si/SiO$_2$ including the height profile evaluated in between the two MXene monolayers. (d–e) A multielectrode chip encompassing MXene flake synthesis assisted via drop-casting, together with an SEM image of a MXene film covering the connection area between the Pt electrode and Si/SiO$_2$ substrate. (f–g) The graph represents the local conductance variance (G(t)) of a chemiresistive element on a multisensor chip made up of two adjacent Pt electrodes when subjected to a consecutive dose up of acetone, IPA, EtOH and MeOH at concentrations varying between 2 and 10 ppm comparative to the conductance in air (Gair) in a dry air atmosphere.

Adapted with permission from Ref. [75].

with that of the bare MoS_2, it was observed that this composite showed a superior response by 65.6% to 100 ppm NO_2 gas in ambient surroundings. Mechanistic studies revealed that the outstanding activity of the aforementioned composite was due to the presence of multiple active sites and rapid charge transportation between the 2H-MoS_2 and the thin layers of $Ti_3C_2T_x$ MXene. In the composite system, the few layers of the $Ti_3C_2T_x$ MXene enabled speedy transportation of the charge carriers, and the distinguishing morphology of 2H-MoS_2 nanosheets fashioned on the $Ti_3C_2T_x$ MXene surface delivers a variety of active sites capable of sensing the gas molecules. [78]

Two-dimensional MXene/SnO_2 heterojunctions were generated via a facile hydrothermal technique having variable MXene loading [MXene = 10–40 wt%] to recognize NO_2 under moderate conditions. The incorporation of the MXene into the SnO_2 nanoparticles effectively enhanced the surface area of the 2D-MXene/SnO_2 heterostructures and revealed a diversity of sites available for adsorption. The heterostructures composed of 20 wt% SnO_2/MXene exhibited a five-fold superior response (231%) for a NO_2 concentration of 30 ppb NO_2 under ambient conditions, and a rapid response time (146 s) as well as recovery time (102 s) were observed as compared to the pristine SnO_2. Moreover, under humid conditions, the sensor exhibited superior selectivity, sensitivity, reproducibility, repeatability and durable sensing response. [79]

Ti_2CT_x MXenes have been recently investigated as novel highly active and selective sensing materials for methane estimation in the presence of visible light. When the $Ti_3C_2T_x$ MXene was illuminated with visible light, the material exhibited a sevenfold enhanced response towards the methane gas and remarkable reduction in the detection and recovery time. The extraordinary methane recognition performance at room temperature was due to the visible-light photocatalytic oxidation of methane using the $Ti_3C_2T_x$ MXene. The kinetics of methane oxidation were investigated with the help of photocatalytic assays, O_2-TPD and in situ IR spectroscopy. [80]

An S-doped $Ti_3C_2T_x$ MXene exhibited highly selective sensing of toluene as compared to the undoped $Ti_3C_2T_x$ MXene. Moreover, this S-doped Ti_3C_2 MXene was also used for detection under ambient conditions under the dynamic impedimetric detection of VOCs (for example, hexane, toluene, ethanol and hexyl-acetate). The S-doped $Ti_3C_2T_x$ confirmed a response improvement in the range of 214% at 1 ppm to 312% at 50 ppm (3–4-fold increase), although both undoped and doped $Ti_3C_2T_x$ MXenes had exclusive selectivity to toluene. An exclusive distinguished response to 500 ppb toluene was also reached with the S-doped $Ti_3C_2T_x$ MXene sensors, along with excellent longstanding steadiness. [81]

In order to fabricate self-powered transducing devices, a polyvinylalcohol/silver (PVA/Ag) nanofiber-based triboelectric nanogenerator (TENG) exhibited ameliorated peak-to-peak open-circuit voltage and power density. A $Ti_3C_2T_x$ MXene/tungsten oxide transducer coupled with a TENG power generator demonstrated outstanding sensitivity to NO_2 gas at room temperature, with 15-fold improved detection sensitivity as compared to the chemiresistive MXene/WO_3 transducer. Moreover, a versatile automated energy as well as the sensing arrangement detection of windborne NO_2 was achieved via the coupling of TENGs with a gas sensing device. Peak-to-peak open-circuit voltage and power-driven density for the airstream-based PVA/Ag-based TENG were 530 V and 359 mW/m^2, correspondingly. The outstanding response (Us/Usa = 510% @ 50 ppm) of the self-powered $Ti_3C_2T_x$ MXene/tungsten

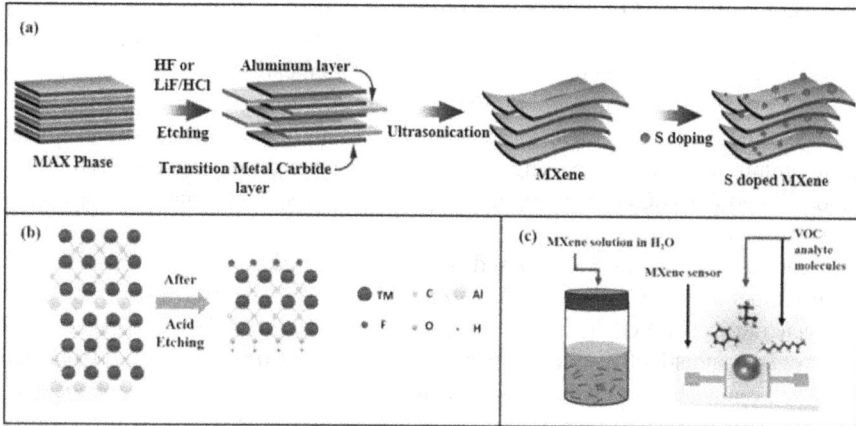

FIGURE 19.3 (a) The process of synthesizing the $Ti_3C_2T_x$ MXene and sulfur-doped $Ti_3C_2T_x$ MXene, as well as preparing the MXene. (b) The illustration shows the exfoliation-based method used for synthesizing the MXene and the sulfur doping process applied to the MXene flakes. (c) The MXene solution is then stored, and undoped and doped MXene sensors are fabricated for detecting VOC molecules.

Adapted with permission from Ref. [81].

oxide transducer driven by TENG for nitrogen dioxide gas at ambient temperature was 15 times stronger than that of the resistive MXene/tungsten oxide sensor. The TENGs coupled with the sensor demonstrate the potential of creating self-powered, workable transducers that can be employed in challenging circumstances or remote locations where traditional power sources would not be feasible or accessible. Even more promise for real-world applications is provided by the capacity to combine diverse functions into a single self-powered system. [82]

TiO_2-C/g-C_3N_4 (TC-CN) composites synthesized via mixing $Ti_3C_2T_x$ along with melamine were used to sense VOCs under ambient conditions and UV irradiation. The composites showed a high response to ethanol at room temperature under UV light. The high sensitivity is attributed to the UV light, energy level dislocations and lamellar structure of the composites. [83]

$W_{18}O_{49}/Ti_3C_2T_x$ composites have been used for the detection of acetone. The 1D $W_{18}O_{49}$ nanorods were grown in situ on the surfaces of 2D $Ti_3C_2T_x$ MXene sheets, resulting in a homogeneous dispersion of the nanorods on the $Ti_3C_2T_x$ surface. This increases the surface area available for gas molecules to interact with the sensing material, leading to a high sensitivity. The removal of fluorine-containing groups from the $Ti_3C_2T_x$ after the solvothermal process contributes to enhanced sensing ability. The removal of these groups creates active sites on the $Ti_3C_2T_x$ surface, allowing for more effective gas molecule adsorption and chemical reactions. The synergistic interfacial interactions between the $W_{18}O_{49}$ nanorods and the $Ti_3C_2T_x$ sheets enhance the sensing ability of the composite material. These interactions result in improved electron transfer and increased conductivity, leading to a speedy response time of 5.6 seconds and a recovery time of 6.0 seconds for detecting 170 ppb acetone. The $W_{18}O_{49}/Ti_3C_2T_x$ composites demonstrated outstanding selectivity, longstanding constancy and a very low limit of detection for 170 ppb acetone. These features make them highly promising for practical applications in detecting low-concentration acetone in various environments. Overall, the $W_{18}O_{49}/Ti_3C_2T_x$ composites show significant enhancement in acetone sensing ability, which can be attributed to the unique properties resulting from the synergistic combination of 1D $W_{18}O_{49}$ nanorods and 2D $Ti_3C_2T_x$ MXene sheets, as well as the solvothermal process used to prepare the composites. [84]

A formaldehyde sensor composed of a MXene/Co_3O_4 composite that works at room temperature and is powered by ZnO/MXene nanowire arrays piezoelectric nanogenerators (PENG) has been revealed. The MXene/Co_3O_4 composite sensor displayed an indistinct response when HCHO concentration augmented from 0.01 ppm to 10 ppm. The ZnO/MXene NW array-based piezoelectric output voltage demonstrated excellent self-powered capability to operate the MXene/Co_3O_4 composite sensor. The synergistic interactions between the MXene and Co_3O_4 at the interface were believed to establish the foundation for the potential gas detection mechanism. [85] The high acetone sensing performance of a 3D MXene/rGO/CuO aerogel is due to several key factors. First, the aerogel is constructed from highly interconnected porous networks of MXene and reduced graphene oxide (rGO) nanosheets, which provides a large specific surface area for gas molecules to interact with the sensing material. The MXene and CuO-based aerogel composite system has CuO nanoparticles evenly dispersed throughout the aerogel, which enhances the sensitivity of the sensor. The homogeneity of the CuO nanoparticle distribution over the MXene and rGO nanosheets allows for efficient and uniform nucleation and growth of the nanoparticles. Furthermore, the decent electron conductivity of the aerogel facilitates the rapid transfer of charge carriers, resulting in a quick response as well as a recovery time of around 6.5 seconds and 7.5 seconds, correspondingly. The sensor response towards 100 ppm acetone is high at 52.09%, even at room temperature under ambient conditions, which is advantageous for practical applications. The excellent reproducibility and selectivity of the sensor can be attributed to the 3D architecture and the unique composition of the MXene/rGO/CuO aerogel. [86]

An α-Fe_2O_3 composite with a $Ti_3C_2T_x$ MXene was successfully produced using the simple hydrothermal method, and the shape and microstructure were studied using various characterizations. According to the findings, the $Ti_3C_2T_x$ MXene nanosheet's surface was created using α-Fe_2O_3 nanocubes that were consistently disseminated and had widths of approximately 250 nm. The results of the gas sensing tests revealed that the aforementioned composites had transducer properties and demonstrated remarkable selectivity to acetone when related to other common gases. They also displayed a high response (16.6% to 5 ppm acetone), rapid response and restoration time (5/5 s), superb linearity and prominent repeatability at room temperature. Additionally, calculations using density functional theory (DFT) were used to analyze the gas sensing mechanism of the composite transducer towards acetone. [87]

The capability of a $Ti_3C_2T_x$-decorated, WO_3 nanorod-based nanocomposite to produce $WO_3/Ti_3C_2T_x$ via hydrothermal synthesis was examined. The $WO_3/Ti_3C_2T_x$ sensors have been revealed to be noticeably more sensitive than pristine WO_3 nanorods and to be highly selective to nitrogen dioxide at ambient temperature.

Furthermore, $Ti_3C_2T_x$ sheets that had been treated with sodium l-ascorbate in WO_3/$Ti_3C_2T_x$ enhanced the stability and reversibility of the sensor towards nitrogen dioxide even in settings of varied humidity (0–99% relative humidity). [88] A 3D $Ti_3C_2T_x$ MXene/rGO/SnO_2 was used as formaldehyde sensor that exhibited a 10 ppm formaldehyde response at room temperature with a 54.97% response rate. It also showed excellent selectivity, repeatability and stability. [89]

A Cu_2O/$Ti_3C_2T_x$-based semiconductor sensor material demonstrated a 3.5x better response to 10 ppm under ambient conditions as compared with bare Cu_2O, with a rapid response recovery rate along with excellent selectivity, stability and repeatability. [90] A $Ti_3C_2T_x$ MXene in corroboration with a NiO nanodisk served as an ethanol sensor with an improved response (14.68) as compared to a bare NiO sensor (1.96) at 100 ppm ethanol and 200 °C, along with excellent selectivity, low detection limit and moisture resistance, making it suitable for IoT applications. [91] A $Ti_3C_2T_x$–ZnO-based semiconductor heterojunction demonstrated superior response and recovery times (10 s and 22 s), high sensitivity (367.63%) and selectivity to nitrogen dioxide. It was also anti-humidity and reproducible. The active sites, Schottky barriers and photogenerated charge carriers contributed to the gas sensing performance. [92]

A CO_2 sensor was fabricated with the help of porous silicon and Mo_2CT_x. The sensor was efficient from ambient temperature to 250 °C and had twice the response

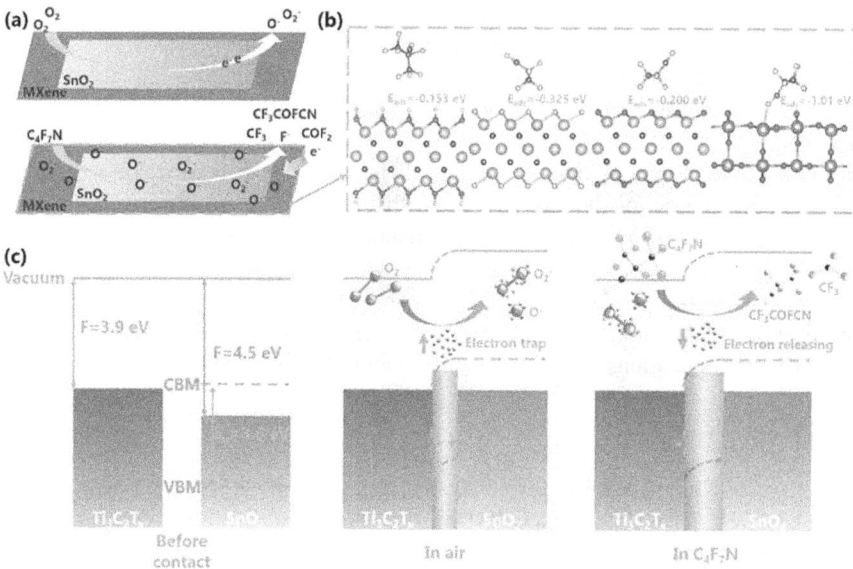

FIGURE 19.4 (a) $Ti_3C_2T_x$–tin oxide heterostructure gas sensing explained by a plane model depicting the adsorption and reaction process of the material. (b) DFT calculations; the most energy advantageous structure for each adsorption system can be observed from the side views. (c) Additionally, the energy band illustration of the $Ti_3C_2T_x$–tin oxide heterostructure can be seen before contact, in air and in C_4F_7N.

Adapted with permission from Ref. [95].

of a crystalline silicon sensor. Humidity improved the sensor response and it could detect 50 ppm CO_2 at room temperature with a response and recovery time of 32 and 45 seconds. [93] A $WO_3/Ti_3C_2T_x$ nanocomposite sensor was developed using a hydrothermal method. The sensor showed a high response and selectivity to nitrogen dioxide at room temperature. It had a response of 78, which is eight times higher than the pure WO_3-based sensor due to the cooperativity between WO_3 nanoparticles and the 2D $Ti_3C_2T_x$. [94]

$Ti_3C_2T_x$ and tin oxide were investigated as the practical materials for hybridization and room-temperature recognition of fluorinated nitrile (C_4F_7N), a gas insulating medium with microtoxicity. The characteristics of a $Ti_3C_2T_x$–tin oxide nanocomposite sensor include great long-term stability, high selectivity and superior sensitivity. The synergistic influence of tin oxide and $Ti_3C_2T_x$ and the significant adsorption capacity of tin oxide to C_4F_7N, which is analogous to fish bait, are attributed to the augmented sensing mechanism.[95]

19.4   CONCLUSIONS

In the present book chapter, we have thoughtfully addressed the ground-breaking application of MXenes and their hybrids as state-of-the-art gas sensors from both an experimental and a theoretical perspective. Owing to their attractive properties, which include excellent conductivity (close to metals), graphene-like morphology, high aspect ratios, hydrophilic properties, flexible mechanical properties, miscellaneous elemental arrangement and plentiful surface termination, MXenes have been acknowledged as fascinating materials for gas sensing applications. Of late, research has demonstrated that MXenes' morphology, surface functional groups, precursors and interlayer structures may all be adjusted to improve the gas sensing capabilities of MXene-based devices. Around 20 different MXenes have been experimentally synthesized out of the many MXenes that have been theoretically introduced. Titanium carbide, the original MXene, is mentioned in the majority of experimental studies on MXene-based gas sensors to date. Unfortunately, there aren't many publications that establish the gas sensing capabilities of V- or Mo-based carbides in the literature. Therefore, its useful to scrutinize the potential of other synthesized MXenes for applications such as gas sensing and VOCs.

REFERENCES

(1) Lee, E.; Kim, D.-J. Recent exploration of two-dimensional MXenes for gas sensing: From a theoretical to an experimental view. *Journal of The Electrochemical Society* **2020**, *167* (3), 037515.

(2) Janata, J. Chemical sensors. *Analytical Chemistry* **1992**, *64* (12), 196.

(3) Buszewski, B.; Kęsy, M.; Ligor, T.; Amann, A. Human exhaled air analytics: Biomarkers of diseases. *Biomedical Chromatography* **2007**, *21* (6), 553.

(4) Singh, E.; Meyyappan, M.; Nalwa, H. S. Flexible graphene-based wearable gas and chemical sensors. *ACS Applied Materials & Interfaces* **2017**, *9* (40), 34544.

(5) Arasaradnam, R. P.; Covington, J. A.; Harmston, C.; Nwokolo, C. U. Review article: Next generation diagnostic modalities in gastroenterology—gas phase volatile compound biomarker detection. *Alimentary Pharmacology & Therapeutics* **2014**, *39* (8), 780.

(6) Pirondini, L.; Dalcanale, E. Molecular recognition at the gas—solid interface: A powerful tool for chemical sensing. *Chemical Society Reviews* **2007**, *36* (5), 695.

(7) Ghosh, R.; Aslam, M.; Kalita, H. Graphene derivatives for chemiresistive gas sensors: A review. *Materials Today Communications* **2022**, *30*, 103182.

(8) Srinivasan, P.; Ezhilan, M.; Kulandaisamy, A. J.; Babu, K. J.; Rayappan, J. B. B. Room temperature chemiresistive gas sensors: Challenges and strategies—a mini review. *Journal of Materials Science: Materials in Electronics* **2019**, *30* (17), 15825.

(9) Majhi, S. M.; Mirzaei, A.; Kim, H. W.; Kim, S. S.; Kim, T. W. Recent advances in energy-saving chemiresistive gas sensors: A review. *Nano Energy* **2021**, *79*, 105369.

(10) Reddy, B. K. S.; Borse, P. H. Review—recent material advances and their mechanistic approaches for room temperature chemiresistive gas sensors. *Journal of the Electrochemical Society* **2021**, *168* (5), 057521.

(11) Aswal, D. K.; Gupta, S. K. *Science and technology of chemiresistor gas sensors*; Nova Publishers, 2007.

(12) Folke, M.; Cernerud, L.; Ekström, M.; Hök, B. Critical review of non-invasive respiratory monitoring in medical care. *Medical and Biological Engineering and Computing* **2003**, *41* (4), 377.

(13) Das, M.; Roy, S. Polypyrrole and associated hybrid nanocomposites as chemiresistive gas sensors: A comprehensive review. *Materials Science in Semiconductor Processing* **2021**, *121*, 105332.

(14) Pirsa, S. *Handbook of research on nanoelectronic sensor modeling and applications*; IGI Global, 2017.

(15) Ramgir, N.; Datta, N.; Kaur, M.; Kailasaganapathi, S.; Debnath, A. K.; Aswal, D. K.; Gupta, S. K. Metal oxide nanowires for chemiresistive gas sensors: Issues, challenges and prospects. *Colloids and Surfaces A: Physicochemical and Engineering Aspects* **2013**, *439*, 101.

(16) Yang, T.; Liu, Y.; Wang, H.; Duo, Y.; Zhang, B.; Ge, Y.; Zhang, H.; Chen, W. Recent advances in 0D nanostructure-functionalized low-dimensional nanomaterials for chemiresistive gas sensors. *Journal of Materials Chemistry C* **2020**, *8* (22), 7272.

(17) Mondal, B.; Gogoi, P. K. Nanoscale heterostructured materials based on metal oxides for a chemiresistive gas sensor. *ACS Applied Electronic Materials* **2022**, *4* (1), 59.

(18) Yang, B.; Myung, N. V.; Tran, T.-T. 1D metal oxide semiconductor materials for chemiresistive gas sensors: A review. *Advanced Electronic Materials* **2021**, *7* (9), 2100271.

(19) Das, S.; Mojumder, S.; Saha, D.; Pal, M. Influence of major parameters on the sensing mechanism of semiconductor metal oxide based chemiresistive gas sensors: A review focused on personalized healthcare. *Sensors and Actuators B: Chemical* **2022**, *352*, 131066.

(20) Liu, L.; Wang, Y.; Liu, Y.; Wang, S.; Li, T.; Feng, S.; Qin, S.; Zhang, T. Heteronanostructural metal oxide-based gas microsensors. *Microsystems & Nanoengineering* **2022**, *8* (1), 85.

(21) Franke, M. E.; Koplin, T. J.; Simon, U. Metal and metal oxide nanoparticles in chemiresistors: Does the nanoscale matter? *Small* **2006**, *2* (1), 36.

(22) Lee, E.; Yoon, Y. S.; Kim, D.-J. Two-dimensional transition metal dichalcogenides and metal oxide hybrids for gas sensing. *ACS Sensors* **2018**, *3* (10), 2045.

(23) Choi, S.-J.; Kim, I.-D. Recent developments in 2D nanomaterials for chemiresistive-type gas sensors. *Electronic Materials Letters* **2018**, *14* (3), 221.

(24) Shin, H.; Ahn, J.; Kim, D.-H.; Ko, J.; Choi, S.-J.; Penner, R. M.; Kim, I.-D. Rational design approaches of two-dimensional metal oxides for chemiresistive gas sensors: A comprehensive review. *MRS Bulletin* **2021**, *46* (11), 1080.

(25) Joshi, N.; Hayasaka, T.; Liu, Y.; Liu, H.; Oliveira, O. N.; Lin, L. A review on chemiresistive room temperature gas sensors based on metal oxide nanostructures, graphene and 2D transition metal dichalcogenides. *Microchimica Acta* **2018**, *185* (4), 213.

(26) Zhang, J.; Liu, L.; Yang, Y.; Huang, Q.; Li, D.; Zeng, D. A review on two-dimensional materials for chemiresistive-and FET-type gas sensors. *Physical Chemistry Chemical Physics* **2021**, *23* (29), 15420.

(27) Liu, X.; Ma, T.; Pinna, N.; Zhang, J. Two-dimensional nanostructured materials for gas sensing. *Advanced Functional Materials* **2017**, *27* (37), 1702168.

(28) Park, H.; Kim, W.; Lee, S. W.; Park, J.; Lee, G.; Yoon, D. S.; Lee, W.; Park, J. Flexible and disposable paper-based gas sensor using reduced graphene oxide/chitosan composite. *Journal of Materials Science & Technology* **2022**, *101*, 165.

(29) Schedin, F.; Geim, A. K.; Morozov, S. V.; Hill, E. W.; Blake, P.; Katsnelson, M. I.; Novoselov, K. S. Detection of individual gas molecules adsorbed on graphene. *Nature Materials* **2007**, *6* (9), 652.

(30) Cao, P.-J.; Li, M.; Rao, C. N.; Han, S.; Xu, W.-Y.; Fang, M.; Liu, X.-K.; Zeng, Y.-X.; Liu, W.-J.; Zhu, D.-L. High sensitivity NO2 gas sensor based on 3D WO3 microflowers assembled by numerous nanoplates. *Journal of Nanoscience and Nanotechnology* **2020**, *20* (3), 1790.

(31) Abbas, A. N.; Liu, B.; Chen, L.; Ma, Y.; Cong, S.; Aroonyadet, N.; Kopf, M.; Nilges, T.; Zhou, C. Black phosphorus gas sensors. *ACS Nano* **2015**, *9* (5), 5618.

(32) Shinde, P. V.; Kumar, A.; Late, D. J.; Rout, C. S. Recent advances in 2D black phosphorus based materials for gas sensing applications. *Journal of Materials Chemistry C* **2021**, *9* (11), 3773.

(33) Li, Y.; Xiao, A.-S.; Zou, B.; Zhang, H.-X.; Yan, K.-L.; Lin, Y. Advances of metal—organic frameworks for gas sensing. *Polyhedron* **2018**, *154*, 83.

(34) Wang, X.-F.; Song, X.-Z.; Sun, K.-M.; Cheng, L.; Ma, W. MOFs-derived porous nano-materials for gas sensing. *Polyhedron* **2018**, *152*, 155.

(35) Gao, L.; Li, C.; Huang, W.; Mei, S.; Lin, H.; Ou, Q.; Zhang, Y.; Guo, J.; Zhang, F.; Xu, S. MXene/polymer membranes: Synthesis, properties, and emerging applications. *Chemistry of Materials* **2020**, *32* (5), 1703.

(36) Zhan, X.; Si, C.; Zhou, J.; Sun, Z. MXene and MXene-based composites: Synthesis, properties and environment-related applications. *Nanoscale Horizons* **2020**, *5* (2), 235.

(37) Shao, B.; Liu, Z.; Zeng, G.; Wang, H.; Liang, Q.; He, Q.; Cheng, M.; Zhou, C.; Jiang, L.; Song, B. Two-dimensional transition metal carbide and nitride (MXene) derived quantum dots (QDs): Synthesis, properties, applications and prospects. *Journal of Materials Chemistry A* **2020**, *8* (16), 7508.

(38) Rasheed, P. A.; Pandey, R. P.; Banat, F.; Hasan, S. W. Recent advances in niobium MXenes: Synthesis, properties, and emerging applications. *Matter* **2022**, *5* (2), 546.

(39) Zhang, C.; Ma, Y.; Zhang, X.; Abdolhosseinzadeh, S.; Sheng, H.; Lan, W.; Pakdel, A.; Heier, J.; Nüesch, F. Two-dimensional transition metal carbides and nitrides (MXenes): Synthesis, properties, and electrochemical energy storage applications. *Energy & Environmental Materials* **2020**, *3* (1), 29.

(40) Ayodhya, D. A review of recent progress in 2D MXenes: Synthesis, properties, and applications. *Diamond and Related Materials* **2022**, 109634.

(41) Bhargava Reddy, M. S.; Kailasa, S.; Marupalli, B. C.; Sadasivuni, K. K.; Aich, S. A family of 2D-MXenes: Synthesis, properties, and gas sensing applications. *ACS Sensors* **2022**, *7* (8), 2132.

(42) Babu, A. M.; Kumar, S.; Rajeev, R.; Thadathil, D. A.; Varghese, A. New horizons in the synthesis, properties, and applications of MXene quantum dots. *Advanced Materials Interfaces* **2023**, *10* (5), 2202139.

(43) Azadmanjiri, J.; Roy, P. K.; Děkanovský, L.; Regner, J.; Sofer, Z. Ti3C2Tx MXene anchoring semi-metallic selenium atoms: Self-powered photoelectrochemical-type photodetector, hydrogen evolution, and gas-sensing applications. *2D Materials* **2022**, *9* (4), 045019.

(44) Chaudhary, V.; Awan, H. T. A.; Khalid, M.; Bhadola, P.; Tandon, R.; Khosla, A. Progress in engineering interlayer space modulated MXenes to architect next-generation airborne pollutant sensors. *Sensors and Actuators B: Chemical* **2023**, *379*, 133225.

(45) Solangi, N. H.; Mubarak, N. M.; Karri, R. R.; Mazari, S. A.; Jatoi, A. S. Advanced growth of 2D MXene for electrochemical sensors. *Environmental Research* **2023**, *222*, 115279.

(46) Xiao, B.; Li, Y.-C.; Yu, X.-F.; Cheng, J.-B. MXenes: Reusable materials for NH3 sensor or capturer by controlling the charge injection. *Sensors and Actuators B: Chemical* **2016**, *235*, 103.

(47) Hajian, S.; Khakbaz, P.; Moshayedi, M.; Maddipatla, D.; Narakathu, B. B.; Turkani, V. S.; Bazuin, B. J.; Pourfath, M.; Atashbar, M. Z. 2018 IEEE Sensors, 2018; p. 1.

(48) Junkaew, A.; Arróyave, R. Enhancement of the selectivity of MXenes (M2C, M = Ti, V, Nb, Mo) via oxygen-functionalization: Promising materials for gas-sensing and -separation. *Physical Chemistry Chemical Physics* **2018**, *20* (9), 6073.

(49) Yu, X.-F.; Li, Y.-C.; Cheng, J.-B.; Liu, Z.-B.; Li, Q.-Z.; Li, W.-Z.; Yang, X.; Xiao, B. Monolayer Ti2CO2: A promising candidate for NH3 sensor or capturer with high sensitivity and selectivity. *ACS Applied Materials & Interfaces* **2015**, *7* (24), 13707.

(50) Wang, Y.; Ma, S.; Wang, L.; Jiao, Z. A novel highly selective and sensitive NH3 gas sensor based on monolayer Hf2CO2. *Applied Surface Science* **2019**, *492*, 116.

(51) Li, S.-S.; Cui, X.-H.; Li, X.-H.; Yan, H.-T.; Zhang, R.-Z.; Cui, H.-L. Effect of coexistence of vacancy and strain on the electronic properties of NH3 adsorption on the Hf2CO2 MXene from first-principles calculations. *Vacuum* **2022**, *196*, 110774.

(52) Li, L.; Cao, H.; Liang, Z.; Cheng, Y.; Yin, T.; Liu, Z.; Yan, S.; Jia, S.; Li, L.; Wang, J. et al. First-principles study of Ti-deficient Ti3C2 MXene nanosheets as NH3 gas sensors. *ACS Applied Nano Materials* **2022**, *5* (2), 2470.

(53) Khakbaz, P.; Moshayedi, M.; Hajian, S.; Soleimani, M.; Narakathu, B. B.; Bazuin, B. J.; Pourfath, M.; Atashbar, M. Z. Titanium Carbide MXene as NH3 sensor: Realistic first-principles study. *The Journal of Physical Chemistry C* **2019**, *123* (49), 29794.

(54) Hajian, S.; Tabatabaei, S. M.; Narakathu, B. B.; Maddipatla, D.; Masihi, S.; Panahi, M.; Fleming, P. D.; Bazuin, B. J.; Atashbar, M. Z. 2021 IEEE International Conference on Flexible and Printable Sensors and Systems (FLEPS), 2021; p. 1.

(55) Salami, N. First-principles realistic prediction of gas adsorption on two-dimensional Vanadium Carbide (MXene). *Applied Surface Science* **2022**, *581*, 152105.

(56) Ma, S.; Yuan, D.; Jiao, Z.; Wang, T.; Dai, X. Monolayer Sc2CO2: A promising candidate as a SO2 gas sensor or capturer. *The Journal of Physical Chemistry C* **2017**, *121* (43), 24077.

(57) Yang, D.; Fan, X.; Zhao, D.; An, Y.; Hu, Y.; Luo, Z. Sc2CO2 and Mn-doped Sc2CO2 as gas sensor materials to NO and CO: A first-principles study. *Physica E: Low-Dimensional Systems and Nanostructures* **2019**, *111*, 84.

(58) Pham, K. D.; Ly, T. H.; Vu, T. V.; Hai, L. L.; Nguyen, H. T. T.; Le, P. T. T.; Khyzhun, O. Y. Gas adsorption properties (N2, H2, O2, NO, NO2, CO, CO2, and SO2) on a Sc2CO2 monolayer: A first-principles study. *New Journal of Chemistry* **2020**, *44* (43), 18763.

(59) Khang, P. D.; Hai, L. L.; Hong, N. T. T.; Tuan, V. V. Study on gas adsorption properties (N2, H2, O2, NO, NO2, CO, CO2, SO2, H2S, H2O and NH3) on the O-vacancy-containing Sc2CO2 Monolayer. *VNU Journal of Science: Mathematics—Physics* **2022**, *38* (1).

(60) Cheng, K.; Wang, M.; Wang, S.; Liu, N.; Xu, J.; Wang, H.; Su, Y. Monolayer Sc2CF2 as a potential selective and sensitive NO2 sensor: Insight from first-principles calculations. *ACS Omega* **2022**, *7* (11), 9267.

(61) Yan, Y.; Wei, Z. Ca2C MXene monolayer as a superior material for detection of toxic pnictogen hydrides. *Materials Chemistry and Physics* **2022,** *281,* 125869.

(62) Obodo, K. O.; Ouma, C. N. M.; Obodo, J. T.; Gebreyesus, G.; Rai, D. P.; Ukpong, A. M.; Bouhafs, B. Sn3C2 monolayer with transition metal adatom for gas sensing: A density functional theory studies. *Nanotechnology* **2021,** *32* (35), 355502.

(63) Zeng, F.; Feng, X.; Chen, X.; Yao, Q.; Miao, Y.; Dai, L.; Li, Y.; Tang, J. First-principles analysis of Ti3C2Tx MXene as a promising candidate for SF6 decomposition characteristic components sensor. *Applied Surface Science* **2022,** *578,* 152020.

(64) Kong, L.; Liang, X.; Deng, X.; Guo, C.; Wu, C.-M. L. Adsorption of SF6 Decomposed Species on Ti3C2O2 and Ti3C2F2 with Point Defects by DFT Study. *Advanced Theory and Simulations* **2021,** *4* (7), 2100074.

(65) Naqvi, S. R.; Shukla, V.; Jena, N. K.; Luo, W.; Ahuja, R. Exploring two-dimensional M2NS2 (M = Ti, V) MXenes based gas sensors for air pollutants. *Applied Materials Today* **2020,** *19,* 100574.

(66) Zhang, H.; Du, W.; Zhang, J.; Ahuja, R.; Qian, Z. Nitrogen-containing gas sensing properties of 2-D Ti2N and its derivative nanosheets: Electronic structures insight. *Nanomaterials* **2021,** *11* (9), 2459.

(67) Choi, J.; Chacon, B.; Park, H.; Hantanasirisakul, K.; Kim, T.; Shevchuk, K.; Lee, J.; Kang, H.; Cho, S.-Y.; Kim, J.et al. N—p-conductor transition in gas sensing behaviors in Mo2CTx MXene. *ACS Sensors* **2022,** *7* (8), 2225.

(68) Li, X.; Jin, L.; Ni, A.; Zhang, L.; He, L.; Gao, H.; Lin, P.; Zhang, K.; Chu, X.; Wang, S. Tough and antifreezing MXene@Au hydrogel for low-temperature trimethylamine gas sensing. *ACS Applied Materials & Interfaces* **2022,** *14* (26), 30182.

(69) Jin, L.; Wu, C.; Wei, K.; He, L.; Gao, H.; Zhang, H.; Zhang, K.; Asiri, A. M.; Alamry, K. A.; Yang, L.et al. Polymeric Ti3C2Tx MXene composites for room temperature ammonia sensing. *ACS Applied Nano Materials* **2020,** *3* (12), 12071.

(70) Hermawan, A.; Zhang, B.; Taufik, A.; Asakura, Y.; Hasegawa, T.; Zhu, J.; Shi, P.; Yin, S. CuO Nanoparticles/Ti3C2Tx MXene hybrid nanocomposites for detection of toluene gas. *ACS Applied Nano Materials* **2020,** *3* (5), 4755.

(71) Lee, J.; Kang, Y. C.; Koo, C. M.; Kim, S. J. Ti3C2Tx MXene nanolaminates with ionic additives for enhanced gas-sensing performance. *ACS Applied Nano Materials* **2022,** *5* (8), 11997.

(72) Kim, S.; Lee, J.; Doo, S.; Kang, Y. C.; Koo, C. M.; Kim, S. J. Metal-Ion-Intercalated MXene nanosheet films for NH3 gas detection. *ACS Applied Nano Materials* **2021,** *4* (12), 14249.

(73) Li, H.-P.; Wen, J.; Ding, S.-M.; Ding, J.-B.; Song, Z.-H.; Zhang, C.; Ge, Z.; Liu, X.; Zhao, R.-Z.; Li, F.-C. Synergistic coupling of 0D–2D heterostructure from ZnO and Ti3C2Tx MXene-derived TiO2 for boosted NO2 detection at room temperature. *Nano Materials Science* **2023,** https://doi.org/10.1016/j.nanoms.2023.02.001.

(74) Liu, M.; Wang, Z.; Song, P.; Yang, Z.; Wang, Q. In2O3 nanocubes/Ti3C2Tx MXene composites for enhanced methanol gas sensing properties at room temperature. *Ceramics International* **2021,** *47* (16), 23028.

(75) Pazniak, H.; Plugin, I. A.; Loes, M. J.; Inerbaev, T. M.; Burmistrov, I. N.; Gorshenkov, M.; Polcak, J.; Varezhnikov, A. S.; Sommer, M.; Kuznetsov, D. V.et al. Partially oxidized Ti3C2Tx MXenes for fast and selective detection of organic vapors at part-per-million concentrations. *ACS Applied Nano Materials* **2020,** *3* (4), 3195.

(76) Wang, J.; Yang, Y.; Xia, Y. Mesoporous MXene/ZnO nanorod hybrids of high surface area for UV-activated NO2 gas sensing in ppb-level. *Sensors and Actuators B: Chemical* **2022,** *353,* 131087.

(77) He, T.; Liu, W.; Lv, T.; Ma, M.; Liu, Z.; Vasiliev, A.; Li, X. MXene/SnO2 heterojunction based chemical gas sensors. *Sensors and Actuators B: Chemical* **2021,** *329,* 129275.

(78) Yan, H.; Chu, L.; Li, Z.; Sun, C.; Shi, Y.; Ma, J. 2H-MoS2/Ti3C2Tx MXene composites for enhanced NO2 gas sensing properties at room temperature. *Sensors and Actuators Reports* **2022,** *4,* 100103.

(79) Gasso, S.; Sohal, M. K.; Mahajan, A. MXene modulated SnO2 gas sensor for ultra-responsive room-temperature detection of NO2. *Sensors and Actuators B: Chemical* **2022,** *357,* 131427.

(80) Wang, J.; Xu, R.; Xia, Y.; Komarneni, S. Ti2CTx MXene: A novel p-type sensing material for visible light-enhanced room temperature methane detection. *Ceramics International* **2021,** *47* (24), 34437.

(81) Shuvo, S. N.; Ulloa Gomez, A. M.; Mishra, A.; Chen, W. Y.; Dongare, A. M.; Stanciu, L. A. Sulfur-doped titanium carbide MXenes for room-temperature gas sensing. *ACS Sensors* **2020,** *5* (9), 2915.

(82) Wang, D.; Zhang, D.; Guo, J.; Hu, Y.; Yang, Y.; Sun, T.; Zhang, H.; Liu, X. Multifunctional poly(vinyl alcohol)/Ag nanofibers-based triboelectric nanogenerator for self-powered MXene/tungsten oxide nanohybrid NO2 gas sensor. *Nano Energy* **2021,** *89,* 106410.

(83) Hou, M.; Gao, J.; Yang, L.; Guo, S.; Hu, T.; Li, Y. Room temperature gas sensing under UV light irradiation for Ti3C2Tx MXene derived lamellar TiO2-C/g-C3N4 composites. *Applied Surface Science* **2021,** *535,* 147666.

(84) Sun, S.; Wang, M.; Chang, X.; Jiang, Y.; Zhang, D.; Wang, D.; Zhang, Y.; Lei, Y. W18O49/Ti3C2Tx Mxene nanocomposites for highly sensitive acetone gas sensor with low detection limit. *Sensors and Actuators B: Chemical* **2020,** *304,* 127274.

(85) Zhang, D.; Mi, Q.; Wang, D.; Li, T. MXene/Co3O4 composite based formaldehyde sensor driven by ZnO/MXene nanowire arrays piezoelectric nanogenerator. *Sensors and Actuators B: Chemical* **2021,** *339,* 129923.

(86) Liu, M.; Wang, Z.; Song, P.; Yang, Z.; Wang, Q. Flexible MXene/rGO/CuO hybrid aerogels for high performance acetone sensing at room temperature. *Sensors and Actuators B: Chemical* **2021,** *340,* 129946.

(87) Liu, M.; Ji, J.; Song, P.; Liu, M.; Wang, Q. α-Fe2O3 nanocubes/Ti3C2Tx MXene composites for improvement of acetone sensing performance at room temperature. *Sensors and Actuators B: Chemical* **2021,** *349,* 130782.

(88) Gasso, S.; Mahajan, A. Development of highly sensitive and humidity independent room temprature NO2 gas sensor using two dimensional Ti3C2Tx nanosheets and one dimensional WO3 nanorods nanocomposite. *ACS Sensors* **2022,** *7* (8), 2454.

(89) Liu, M.; Song, P.; Liang, D.; Ding, Y.; Wang, Q. 3D porous Ti3C2Tx MXene/rGO/SnO2 aerogel for formaldehyde detection at room temperature. *Journal of Alloys and Compounds* **2022,** *925,* 166664.

(90) Zhou, M.; Yao, Y.; Han, Y.; Xie, L.; Zhu, Z. Cu2O/Ti3C2Tx nanocomposites for detection of triethylamine gas at room temperature. *Nanotechnology* **2022,** *33* (41), 415501.

(91) Shao, Z.; Zhao, Z.; Chen, P.; Chen, J.; Liu, W.; Shen, X.; Liu, X. Enhanced ethanol response of Ti3C2TX MXene derivative coupled with NiO nanodisk. *Inorganic and Nano-Metal Chemistry* **2022,** https://doi.org/10.1080/24701556.2022.2078363, 1.

(92) Fan, C.; Shi, J.; Zhang, Y.; Quan, W.; Chen, X.; Yang, J.; Zeng, M.; Zhou, Z.; Su, Y.; Wei, H. et al. Fast and recoverable NO2 detection achieved by assembling ZnO on Ti3C2Tx MXene nanosheets under UV illumination at room temperature. *Nanoscale* **2022,** *14* (9), 3441.

(93) Thomas, T.; Ramos Ramón, J. A.; Agarwal, V.; Méndez, A. Á.; Martinez, J. A. A.; Kumar, Y.; Sanal, K. C. Highly stable, fast responsive Mo2CTx MXene sensors for room temperature carbon dioxide detection. *Microporous and Mesoporous Materials* **2022,** *336*, 111872.

(94) Gasso, S.; Mahajan, A. MXene decorated tungsten trioxide nanocomposite-based sensor capable of detecting NO2 gas down to ppb-level at room temperature. *Materials Science in Semiconductor Processing* **2022,** *152*, 107048.

(95) Wu, P.; Li, Y.; Xiao, S.; Chen, D.; Chen, J.; Tang, J.; Zhang, X. Room-temprature detection of perfluoroisobutyronitrile with SnO2/Ti3C2Tx gas sensors. *ACS Applied Materials & Interfaces* **2022,** *14* (42), 48200.

20 MXenes for Textile Industry Applications

Debajani Tripathy, Tarun Yadav,
and Srikanta Moharana

20.1 INTRODUCTION

In the dynamic industry of textiles, innovation is one of the most important factors in fostering forward movement and satisfying the requirements of contemporary consumers. MXenes are a promising innovation that has the potential to completely transform the textile industry as it moves towards more eco-friendly and sustainable practices. MXenes, the result of extensive development over many years, are a state-of-the-art textile technology with several applications across the textile industry's value chain. The absence of durable fibres with adequate electrical, electrochemical, and sensing qualities slows the progress of functional textile-based electronics. MXenes, a recently created and investigated 2D material, have proven to be very significant for the construction of purposeful fibres owing to their extraordinary electrical and electrochemical properties and their capacity to be treated in solutions. Wet spinning can be used to make additive-free fibres from MXenes because of their peculiar feature of generating liquid crystal phases spontaneously in a wide range of solvents. The advancements in the production of pure MXene fibres, as well as the processing difficulties in MXenes, affect their macroscopic properties and the efficacy of devices created from them. The future production of MXene fibres in the textile industry presents opportunities and challenges, allowing for their wider application in advanced wearable technologies [1]. In general, MXenes (2D inorganic compounds) have the formula $M_{n+1}X_nT_x$, where M is an early transition metal, X is carbon or nitrogen, and T is a functional group on the surface of a MXene, most commonly oxygen, hydroxide, and fluorine. These MXene-based materials are hydrophilic in nature due to their surfaces being terminated in hydroxyl and oxygen groups, and because of the strong metallic conductivity of transition metal carbides [2].

MXenes are a class of 2D transition metal carbides and nitrides with unusual features that have attracted a lot of attention in the fields of materials science and engineering. Drexel University scientists made the initial discovery of MXenes in 2011, and ever since then, the field has been a hotbed of activity [3]. In order to improve their surface area and accessibility, multi-layered MXenes are typically further treated into solutions of delaminated MXenes [4]. The hydrophilic MXenes can be treated in polar organic and aqueous solvents to generate stable colloidal

DOI: 10.1201/9781003366225-20

solutions, which can then be used to create self-supporting films or transparent conductive coatings. During the last few years, MXenes have been wonderful materials and have been used for a wide variety of electrochemical applications such as batteries and supercapacitors [5]. MXenes' many advantages provide an extensive solution to some of the most pressing problems facing the textile industry today, from increased durability and breathability to greater colour retention and reduced water use. Manufacturers may adapt to the ever-changing market and satisfy the growing demand for eco-friendly products by incorporating this breakthrough technology.

20.2 MXENE PROPERTIES

MXenes have exceptional visual properties, exquisite electric and magnetic characteristics, outstanding wettability, and intercalation capacities that can be attributed to their surface features [6–8]. Several applications such as wastewater treatment [9–11], biosensing [12–15], photocatalysis [16–17], modern electronics [18–19], and power conversion technologies [20–21] can all benefit from the use of MXenes because of their unique properties and ease of production. MXenes' prospective uses in energy storage devices like supercapacitors (SCs) and flexible batteries have piqued the curiosity of the scientific and technical communities. Due to their outstanding mechanical, physicochemical, optical, electrical, and electrochemical characteristics, MXenes and their composites are to be used in a variety of SCs. A number of MXene-based composites with carbon, metals, and conducting polymers have been reported [22–23]. In addition, MXenes can be used as effective filtration materials owing to their unusual molecular structure. Textile manufacturers now have more options for developing functional textiles that provide protection from dangerous pollutants, allergens, and even viruses thanks to their incorporation into the fabric. The creation of textiles with built-in air and water cleansing capabilities could result from this application, meeting the rising demand for such goods amongst customers [24].

20.3 STRUCTURE AND OVERVIEW OF MXENES

MXenes are one-of-a-kind materials because of their atomic structure, which has a layered arrangement. Ternary compounds of early transition metals (M), group A elements (A), and carbon or nitrogen (X) are where this term originates. By selectively etching away the A layer, the MXene family is formed; this 2D structure has a general formula of $M_{n+1}X_nT_x$, where "M" stands for the transition metal, "X" is carbon or nitrogen, "T" denotes surface terminations (such as -OH or -O), and "n" can vary. Figure 20.1 is a simplified diagram showing the elements that make up both MAX phases and MXenes. Due to the combined covalent/metallic/ionic character of the M–X bond against the metallic nature of the M–A bond, M–X bonds are revealed to be stronger in the MAX phase [25].

(a)

M$_2$XT$_x$ M$_3$X$_2$T$_x$ M$_4$X$_3$T$_x$

(b)

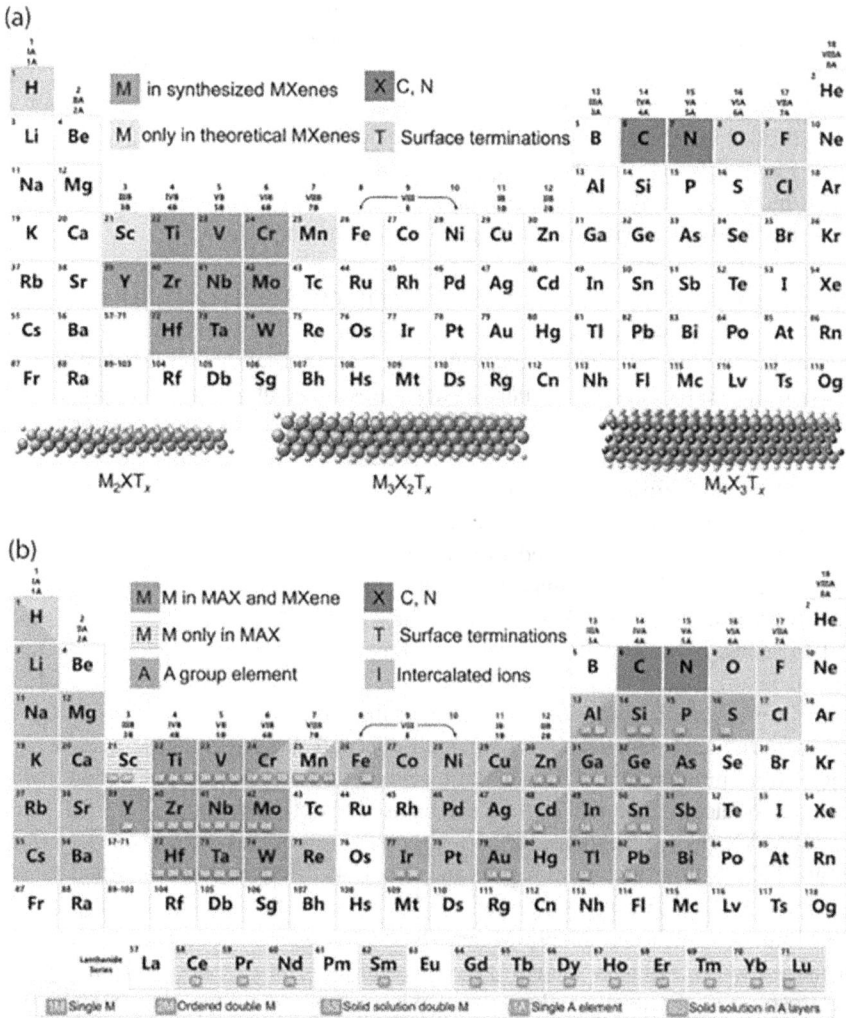

FIGURE 20.1 MXene and MAX phase elemental composition tables [25].

MXenes are well-known for their exceptional characteristics, which make them useful in numerous contexts. MXenes are extremely desirable materials due to a number of factors, including:

- **Conductivity:** MXenes are well suited for usage in electronics and energy storage due to their excellent electrical and thermal conductivity.
- **Mechanical Strength:** They can sustain stress and strain thanks to their high mechanical strength, making them useful for structural applications.

- **Surface Functionality:** MXenes' surface terminations are amenable to engineering, allowing for a wide range of control over features including hydrophobicity and chemical reactivity.
- **High Surface Area:** MXenes' performance in catalysis and sensing applications is boosted by their 2D structure's enormous surface area.
- **Flexibility:** Due to their adaptability, MXenes can be easily incorporated into a wide range of matrices, including polymers, ceramics, and composites, without sacrificing the original materials' performance.

MXenes have a wide range of potential applications, including communication, entertainment, fitness tracking, and health monitoring. The next generation of wearable technology is expected to make use of electronic fabrics (E-textiles) that are as comfortable to wear as regular clothes. Functional fibres have the potential to enhance social interaction, enjoyment, and health by facilitating greater human involvement with technology. The market for smart textiles is projected to increase from 2015's \$0.5 billion to \$9.3 billion by 2024 [26]. This suggests that in the near future, innovative functional fibres can be used to make garments. A wide variety of functional materials are used to make functional fibres for textiles [27]. There are several examples of 1D materials including carbon nanotubes and gold nanowires, while examples of 2D materials include graphene, molybdenum disulphide, and MXenes. MXene-based 2D materials have high electrical conductivity (up to 15,000 S/cm), electrochemical characteristics (volumetric capacitance up to 1,500 F/cm^3), and solution processability, and their implementation in fibre-based systems is expanding [28]. Chemical etching with concentrated hydrofluoric acid, a solution of lithium fluoride and hydrochloric acid (the least intensive layer delamination method), or electrochemical etching can be used to obtain MXenes ($M_{n+1}X_nT_x$) from parent MAX ($M_{n+1}AX_n$) phases ("M", early transition metal; "A", group 13–16 elements; "X", carbon and or nitrogen; and n = 1–4). MXenes are simplified for use in solutions when termination groups "T_x" (where x is the number of T groups) are added during synthesis [25,29]. MXene-based fibres can be made using a variety of fabrication techniques, including coating, inkjet printing, solution spinning, and electrospinning. By coating silver-plated nylon yarns and carbon fibres with $Ti_3C_2T_x$ MXene, Hu and co-workers created fibre-based electrodes with an aerial capacitance of 328 mF/cm^2 and a length capacitance of 253 mF/cm [30–31]. Since it is simple to make functional fibres by coating commercial fibres or yarns, this approach was used to produce $Ti_3C_2T_x$ MXene-coated cellulose yarns (including cotton, bamboo, and linen) with MXene loadings of up to 77 wt%. The capacitance along the length of the synthetic fibres employed in knitted pressure sensor textiles was 760 mF/cm, and the electrical conductivity was 440 S/cm. Furthermore, $Ti_3C_2T_x$ MXene-coated cellulose yarns (including cotton, bamboo, and linen) with MXene loadings of up to 77 wt% were produced using this process since it is easy to generate functional fibres by coating commercial fibres or yarns. They were also remarkably durable, surviving 45 washes at 80°C. Taking full advantage of MXenes' special features in fibre-based systems, researchers also used the solution spinning approach to manufacture fibres with high MXene loadings. At first, a $Ti_3C_2T_x$ MXene is mixed with graphene oxide. Moreover, the solution spinning can be used to transform spinnable forms of

GO or poly(3,4-ethylenedioxythiophene):poly(styrene sulphate) (PEDOT:PSS) into hybrid fibres. The volumetric capacitance of 586 F/cm^3 was achieved in hybrid fibres of reduced GO (rGO) and $Ti_3C_2T_x$ with MXene loadings of 95% wt%. Uzun and his co-workers [32] have reported the nematic liquid crystal (LC) phases in additive-free MXene dispersions such $Ti_3C_2T_x$ and $Mo_2Ti_2C_x$. Significant progress has been made in recent years in producing clean MXene fibres and investigating their possible applications. First, the nematic LC phase development in MXenes is discussed, and then the rheological consequences of the LC formulations in connection to solution spinning are investigated. This finding has many potential applications, including fibre-based conductors and heaters, supercapacitors, and other energy storage devices. Finally, we discuss some pressing concerns in textile operations and make suggestions for further research. Following the discussions in this chapter, the textile industry should be able to create a new generation of MXene fibres that can be utilized in a wide variety of garments and other wearables [33].

20.4 PROGRESS OF MXENES

In the process of making manufactured items, synthetic raw materials are often improperly handled, improperly produced, or accidentally discharged into the environment. The introduction of these pollutants disrupts the ecological balance due to their toxic effects on the biological system and adverse impacts on the material structure [34]. Safeguarding environmental standards necessitate the scientific management and treatment of waste discharges. There is a large variety of methods based on physical, chemical, and biological separation principles that can be used to recover and eradicate dangerous contaminants. The process efficiency has been greatly enhanced by the use of energy-saving materials and conventional methods. In order to remedy the drawbacks of traditional parts, researchers have come up with MXenes, a group of revolutionary materials still under development [35]. Graphene-based materials are extending to include other two-dimensional materials such as metal oxides, two-dimensional polymers, transition metal-based dichalcogenides, etc. [36], where M is a transition metal; X is carbon or nitrogen; T is a functional group like hydroxyl, oxygen, or fluorine; and n is an integer; the standard formula for MXene is $Mn_{+1}X_nT_x$. According to Kumar et al. [37–38], a MXene is a two-dimensional component of carbides and nitrides based on transition metals with an unequal distribution of distinct functional groups onto the metallic surface. MXene have outpaced their two-dimensional competitors in popularity since their introduction in 2011 by Young et al. [39]. They are quite conductive and can be used well for energy storage purposes due to their maximum electrical conductivity (2.105 S/m), which is comparable to that of a multi-layered graphene molecule, as well as their relatively large specific surface distribution, enriched surface mechanism due to excellent functionalization ability, easy dispersive nature in solvents like water, and most astounding electrochemical characteristics. Their 2D geometric structure allows for faster ion transport within 2D and 3D networks, their reduction/oxidation reactions create a pseudocapacitive environment for the electrically connected double layer mechanism of storing charge, and their superior mechanical features at the nanoscale are on par with, if not better than, those of larger-sized lithium batteries.

Due to their versatility, MXenes have been proven to be effective active ingredients in the development of supercapacitors and batteries [40]. Due to their superior performance characteristics, MXenes are finding increasing applications in a variety of contexts, including incorporation into conductive polymeric materials such as polypyrrole and polyaniline. Using MXene as a negative electrode to generate stronger concentration gradients in acidic electrolytic solutions and hence increase energy density has also been proposed [41]. This quality is significantly higher than that of carbon-based components, which have less capacitance and restricted potential gradient ranges of 1.0 V. For systems involving sodium/potassium ions or sulphur/lithium, however, the efficiency of carbon-based materials approaches a maximum. MXenes are selected as additives to improve these components and retain as much energy as feasible throughout the hybridization process [42]. The methods used to synthesize MXenes have evolved from those employing single-transition MXenes as research into double-transition MXenes has progressed [43]. Finding non-toxic alternatives to hydrofluoric acid-based rapid etching is the primary motivation behind the search for such technologies. Manufacturing machinery designed for uniform mass output relies on printing and stamping procedures to replace time-consuming manual steps. Therefore, research professionals are continuously attempting to create suitable topologies for storage batteries and supercapacitor materials [44], notwithstanding the challenges experienced during the material/electrode synthesis of MXenes. MXenes also have excellent chemical stability, high conductivity, and positive effects on the environment. In particular, MXenes' hydrophobicity and active functional group availability on the surface provide an adsorptive character to deal with a wide range of ionic/molecular species, suggesting their potential use in preventing environmental contamination and even in environmental sensing [45].

Research is undertaken to solve the major limitations of MXenes. Several review articles have focused on MXenes' potential usefulness in dealing with a specific category of environmental pollutants. This chapter takes a novel approach to the issue by analyzing the extensive literature on industrial processes, energy use, and environmental applications. In this chapter, we look at how different structures, synthesis methods, energy storage capabilities, and applications to different remediation strategies for the environment all work together. Current advancements concerning MXene-based materials are the topic of this in-depth review [46–47]. These materials serve a variety of purposes in the energy storage sector, including as supercapacitors and next-generation batteries.

20.5 PREPARATION OF MXENE-BASED FIBRES

Fabrics with improved performance, functionality, and sustainability can be prepared by incorporating MXene nanoparticles with their particular features into textile fibres. To successfully incorporate MXenes into the textile matrix, this novel procedure necessitates careful consideration of material production, fibre spinning, and post-processing techniques [48–49].

- **MXene Synthesis:** Producing fibres out of MXenes begins with the manufacture of the MXene nanomaterial. In order to create MXene, the A layer must be selectively etched away from MAX phases, resulting in a

2D MXene structure. The characteristics of a MXene can be modified by incorporating different transition metals (M) and carbon or nitrogen (X) combinations to suit a wide range of textile uses.

- **MXene Dispersion:** After a MXene has been synthesized, it must be dispersed equally within an appropriate solvent. For the MXene to be evenly distributed throughout the final fibre, it is essential that the dispersion process be stable.

- **Fibre Spinning:** The next stage is to use spin processes that incorporate the MXene into the fibre matrix. The kind of MXene, the fibre qualities required, and the intended purpose of the textile all dictate the technique of production, which can range from melt spinning to wet spinning to electro-spinning to solution spinning.

- **Optimization of Fibre Properties:** The mechanical, electrical, and thermal properties of the MXene-based fibre can be optimized by modifying the spinning process parameters such as MXene concentration, spinning temperature, and spinning speed.

- **Post-Processing and Drying:** After spinning, the fibres go through additional processing steps like heat treatment and chemical treatments to improve their durability and functionality. Careful drying of the fibres is essential for avoiding clumping and ensuring consistency.

- **Fibre Assembly:** After the MXene-based fibres have been processed, they can be fabricated into yarns and woven fabrics using the same methods as conventional textile production.

- **Integration with Smart Textiles:** In order to develop interactive textiles, MXene-based fibres can be fused with other smart materials, conductive elements, or electronic components. Biometric monitoring and gesture recognition are only two examples of the kinds of features that could be made possible by combining MXene fibres with sensors and microcontrollers.

- **Performance Testing:** The electrical conductivity, mechanical strength, temperature regulation, and filtration properties of MXene-based fibres are all evaluated in a battery of comprehensive performance tests. These examinations guarantee that the fibres are up to snuff for their designated uses.

- **Sustainable Practices:** Using non-toxic solvents and reducing waste are just two examples of how sustainability is prioritized throughout the cooking process. The goal is to harmonize the textile industry's efforts to lessen its environmental footprint with the manufacture of fibres based on MXenes.

Fabrics can be made smarter, more durable, and less harmful to the environment with the help of nanomaterials like MXenes, which, because of their preparation, offer a cutting-edge approach to textile manufacturing. The textile industry can pave the path for an era of textiles that meet the needs of current consumers while also addressing environmental issues by taking advantage of the special features of MXenes. Hu and co-workers [49] have reported rapidly moving cotton yarn or nylon multifilament through the MXene dispersion to produce MXene-coated yarn or fibre. To produce a MXene-based composite fibre, it is usual practice to mix a MXene with an organic solvent or polymer as a spinning dope; however, this method works only if the constituent parts are uniformly and independently dispersed. Wet spinning, electrostatic spinning, and melt spinning are the three most common techniques for spinning fibre.

20.6   MXENE TEXTILES FOR SENSING APPLICATIONS

When the spinnability or elasticity of fibres is compromised by the addition of conductive fillers, strain-sensing fabrics have problems. This is due to the fact that elastomer polymer filaments cannot be directly attributed to the conductive substance. In spite of the MXene sheets' improved conductivity, hydrophilicity, and rich terminal groups, which enable their attachment to the fibre, the composite fibre maintains its pliability. MXene-based composite fibres can detect the impact of mechanical deformation on electrical conductivity and stretchability because of their unique combination of characteristics. Wearable medical devices with MXene-based sensors can remotely monitor a patient's health, direct the patient's movements, and assist with rehabilitation [50].

The potential uses of wearable electronics in fields like medicine, electronic skin detection, the human–machine interface, smart thermotherapy, and electromagnetic interference (EMI) shielding have recently gained a lot of scientific attention. The multi-functionalization of devices can aid in the realization of high integration, low power consumption, and an increase in the number of uses for flexible electronics. In order to fabricate multifunctional devices, there have been a number of groundbreaking articles published on the topic of designing flexible substrates and various sensitive materials. Zhou et al. [50] created flexible multifunctional transparent conductive films with good EMI shielding capabilities and electro-photo-thermal performance by hot-pressing MXene and Ag NW hybrid conductive networks onto flexible polyvinyl alcohol (PVA) substrates. Zhang et al. [51] used a double-sided screen-printing technique to embellish cotton fabric with graphene, increasing its electro-thermal performance and strain-sensing capabilities. Mounting non-adhesive flexible devices on human skin surfaces with tape or bandages often results in pain and unevenness between the devices and the skin. Intricate micro- and nanostructures, such as those seen on the skin of geckos, beetles, and octopuses, can be used as adhesive coatings on electronic devices. This will ensure that flexible gadgets have just the correct amount of tackiness. Using a natural silk protein adhesive, Yao et al. [52] created a flexible skin sensor with a micropillar structure on a PDMS substrate that showed conformal and strong adherence to the surface of human skin for consistent arterial pulse measurements. Stretchable electronic devices capable of measuring electrocardiogram (ECG) signals and wrist bending movements in both wet and dry conditions were developed by Chun et al. [53] using carbon-based conductive polymer composite (CPC) sheets printed with octopus-like patterns. Incredibly strong adhesion forces have been demonstrated by this instrument on both dry and wet surfaces. Although it would be ideal to create biocompatible adhesive electronics that can incorporate several functions, planning and fabricating complex micro/nano-structural adhesive layers is time consuming and expensive.

Fabrics can be the backbone of the next generation of wearable electronics by absorbing and combining various active conductive materials using techniques like dip coating, spraying, electroless deposition, and vacuum filtration. Fabrics have many advantages due to their porous structure, pliability, and intricate network design [54]. Because of their high conductivity, superior water dispersibility, and abundant surface functional groups (-F, -O, and -OH), MXenes have been demonstrated to be suitable active materials for wearable electronics. In order to develop a conductive layer for textiles, MXenes' aforementioned qualities allow for a non-destructive solution-coating process [55]. Few works have addressed the development of skin-adhesive multifunctional electronic devices based on electronic fabrics and adhesion.

Yao and colleagues [52] created a stickable and multifunctional textile-based device using conductive MXene-decorated air-laid paper (MXene AP) encased in 50:1 polydimethylsiloxane (PDMS). This was accomplished with the use of dip coating, a straightforward and versatile coating method. Altering the dip-coating cycles of MXene-coated textiles changes their electrical conductivity. A layer of low cross-linked PDMS provides the self-adhesive capabilities of smart fabrics and protects MXenes from oxidation when exposed to moisture. The created self-adhesive smart cloth is capable of shielding against electromagnetic radiation, transforming light into heat, and detecting external pressure. The advantages of the wearable gadget were presented, including its usage as a personal warmer, an EMI shielding material, and a sensor for human motion detection, all of which bode well for the development of future multipurpose electronic devices. Pu and his co-workers have reported polyurethane fibre-based multilayer structured fibre strain sensors for wearable applications as shown in Figure 20.2a–c. Figure 20.2d shows that the AgNW-WPU layer increased the sensors' sensing range thanks to the slipping mechanism and the stabilization effect of the WPU particles as physical cross-linking sites, even under high strain [56].

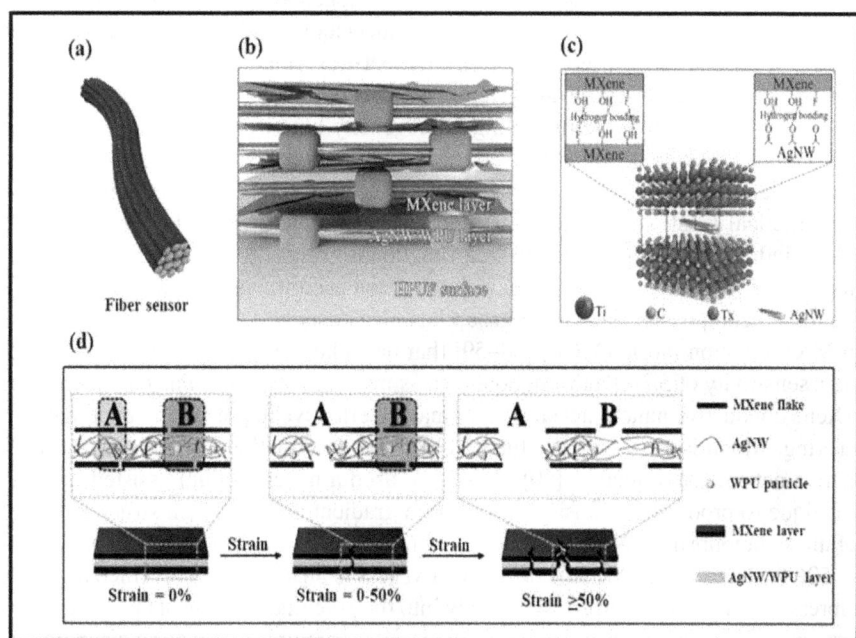

FIGURE 20.2 Schematic illustration depicts a wet-spun fibre used in a strain-sensing application with extremely high sensitivity. (a) The spinning of MXene/polyurethane fibres during their production. (b) Image of a wet-spun MXene/PU fibre that was coagulated in IPA. (c) A knitted sleeve for the straight and bent elbow. (d) Elbow sleeve strain-sensing performance under repetitive bending and straightening [56].

Reused with permission from Ref. [56]. Copyright (2019), Royal Society of Chemistry.

Wearable strain sensors are a crucial component of flexible smart electronic systems due to their potential uses in soft robots, artificial skin, human–machine interaction, and other fields. Despite their beneficial effects on society, electronic gadgets can pose risks to human health due to radiation and intense electromagnetic interference (EMI), which can compromise device accuracy. There is a pressing need for wearable electronic devices with integrated health detection and medical therapeutic capacity in physiotherapy and chemotherapy applications [57–58]. Consequently, there is an urgent requirement to develop multifunctional materials with superior strain sensing, electromagnetic interference (EMI) shielding, and Joule heating characteristics. In an effort to find a workable answer, scientists have investigated how to maximize the versatility of fabrics. Smart electronic textiles' porous fibre network structure and amazing versatility have made them useful in a wide variety of applications, including wearable strain sensors, medical heaters, thermal management, and EMI shielding materials. Fabrics' underlying network architecture is often adorned with carbonaceous conductive nanoparticles like carbon black (CB), carbon nanotubes (CNTs), and graphene to produce efficient conductive channels. Carbon nanotubes, like many other nanomaterials, tend to stack, which limits their use in smart electrical fabrics [59].

For a quick and efficient dip-coating technique to create a homogeneous and stable conductive coating on the textile skeleton, a homogenous dispersion of conductive nanomaterial is required. Additives such as surfactants and cellulose nanocrystals (CNCs) are commonly used to create a dispersion of conductive nanomaterials. Synergistic conductive material systems, created by combining several conductive materials, not only increase the filling homogeneity but also boost electrical performance without the need for additional insulated components. Synergistic conductive nanoparticles embedded in wearable conductive fabrics are currently being studied as a practical solution. Due to their high electrical conductivity, layered structure, and abundance of hydrophilic surface functional groups, 2D transition metal carbide/nitride (MXene) nanosheets have lately been used to create new types of functional nanocomposites. The same group has also described a pressure sensor based on MXene/cotton fabric (MCF) [58–59] that has a large detecting range of 160 kPa and a sensitivity of 5.3 kPa throughout a pressure range of 0–1.3 kPa. The synergistic MXene conductive nanomaterial system may be effective in preventing nanomaterial stacking, which is necessary for the construction of a continuous and uniform conductive network architecture [60]. Cao et al. used a novel vacuum-assisted filtering technique to produce composite paper with a gradient and sandwich structure out of cellulose nanofibrils (CNFs), MXene, and CNTs. The composite paper's conductivity is 2506.6 S/m, and its EMI shielding effectiveness is 38.4 dB, both of which are quite impressive. There hasn't been much study into the potential benefits of applying a uniform synergistic conductive coating to the textile structure. By creating a synergistic conductive coating from hybrid MXene/CNTs, a non-woven thermoplastic polyurethane (TPU) fabric was transformed into an electrically conductive Mxene-CNTs-TPU conductive fabric (MCT-fabric). The combined synergistic conductive coating served as the optimal active material for conductive textiles, allowing for the development of multifunctional conductive fabric with strain sensing, EMI shielding, and Joule heating characteristics. For uniformly painting the skeleton of cloth, MXene and CNTs' creation of a homogeneous dispersion is optimal. The strain sensor's sensitivity was increased by creating an ultrasensitive microcrack structure using a pre-stretching technique. The sensor's strain-detection abilities and prospective applications in motion signals and human physiology were then investigated as a function of the Mxene-CNT mass ratio. This study investigates the electromagnetic interference (EMI) shielding capabilities and shielding technique of MCT-fabric, considering its thickness, cyclic stability, long-term stability, and Joule heating properties [61].

20.7   APPLICATIONS OF MXENES IN THE TEXTILE INDUSTRY

In recent years, MXenes—a class of two-dimensional transition metal carbides and nitrides—have attracted increasing interest from a wide range of sectors, including the textile industry. MXenes, promising nanomaterials, have numerous opportunities for improving the performance, sustainability, and functionality of textiles and clothing.

Jin and co-workers [62] have reported that wearable and attachable computing systems are capable of acting as smart sensing devices with access to the mobile internet in addition to data transmission. Future HCIs may be built on the basis of these sensors. The identification of novel conductive materials is one factor influencing the advancement of textile sensors. The 2D MXene material has recently attracted a lot of interest due to its amazing abilities in a number of disciplines. The creation of conductive fabrics, the structural layout of textile sensors, and their use in wearable applications are all covered in this article. From the angles of MXene manufacture, wearability, stability, and evaluation standards, the authors also assemble and estimate the challenges and restrictions faced by MXene-based textile sensors in the realm of wearable applications. In this work, MXene materials can be included in smart textiles to hasten the creation of wearable textile sensors [62].

Sensors have made major contributions to the advancement of many sectors, including ordinary life, the intelligence industry, medical care, and other fields, and the introduction of intelligent manufacturing has sparked a wave of interest in Wang and co-workers. The observation of high electrical conductivity, structural controllability, and appealing universality with varied substrates have all brought attention to MXenes, a new family of 2D transition metal carbides/nitrides. MXenes have inspired a variety of sensors with varied uses due to their great research potential. The goal of this analysis was to speed up the process of commercializing ideal sensors by collecting the findings of a large body of research on already-existing MXene-based sensors. Case studies of MXene-based 1D, 2D, and 3D sensors are presented (as shown in Figure 20.3), along with a thorough review of the relevant literature. These various MXene syntheses can be traced back to a variety of unique starting points. The production processes, architectural differences, and potential sensing applications of MXene-based composite sensors are summarized in this work [62–63]. Due to their ability to combine the benefits of an organic–inorganic hybrid with the sustainability required for a wide variety of applications, two-dimensional (2D) MXenes have emerged as the ideal layered material. Since MXenes' surface functional groups may be finely modified by the addition of other materials, they can be utilized as a replacement for a variety of applications in the textile industry, such as the development of smart textiles and the purification of textile wastewater

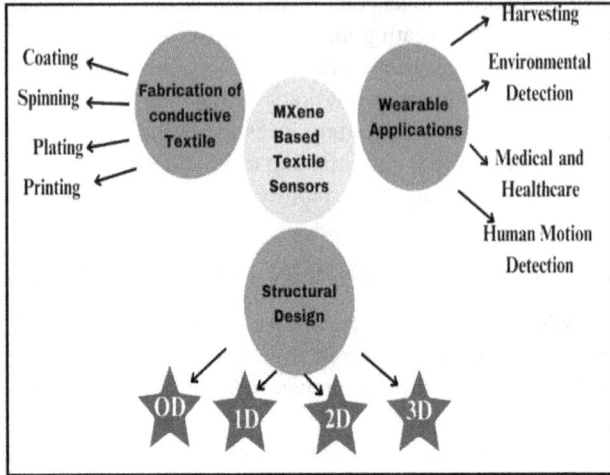

FIGURE 20.3 Different types of fabrication of conductive textiles, wearable applications, and structural design of MXene-based smart textile sensors.

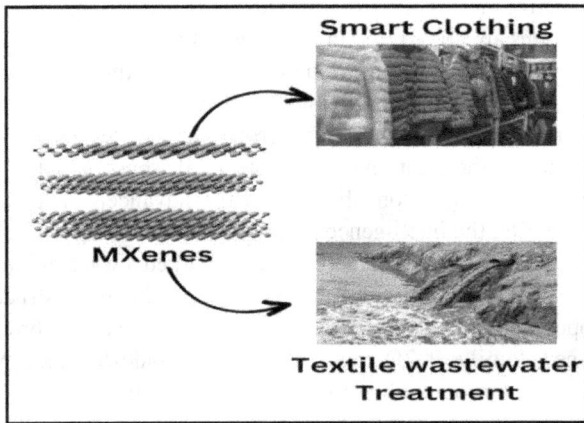

FIGURE 20.4 Improvement of MXenes in smart clothing (electronic textile) and textile wastewater treatment [64].

Copyright 2024, Elsevier.

(as shown in Figure 20.4). High-performance clothing and the elimination of textile dyes in wastewater are two areas where MXene-based textile composites shine. The role of MXenes in two subsets of the textile industry is explored in this work. The MXene electrodes have enhanced the performance of fibre, yarn, and fabric in many contexts, such as supercapacitors and pressure sensors. The increased durability of MXene-based textiles is the result of MXenes' dual qualities of adsorbent properties (treating the adsorbate) and strong binding strength to textile structures [64].

The convergence of innovations in the IoT, VR, and soft robotics has led to the emergence of wearable smart textiles as a unique platform for cutting-edge electronics. The nanofabrication of electroactive fabrics has greatly aided the development of wearable smart textile systems for applications including health monitoring, autonomous energy management, and portable sensing. The 2D transition metal carbides and nitrides known as "MXenes" have changed the field of material chemistry due to their new properties such as metallic conductivity, rich surface chemistry, programmable terminations, and improved processability. MXene-based materials have attracted the attention of the scientific community due to their new properties. This article explores the research and development of MXene-based fabrics, fibres, yarns, textiles, and composites. Before MXenes can be functionalized for use in textiles, their peculiar surface chemistry and processing procedures must be investigated. There has been a lot of research into different techniques, performance matrices, and textile functionalization strategies to increase MXenes' compatibility with textiles. MXene textiles are currently being used in a variety of applications, such as EMI shielding, flexible energy storage devices, intelligent thermotherapy, and sensors (as shown in Figure 20.5). Some of the existing challenges faced by the industry are explored, along with the possibility of future research into MXene-enabled smart textiles [65].

In order to collect reliable signals from the human body, portable medical therapy devices and monitoring systems rely significantly on flexible devices that can adhere to the skin's surface. Here, a textile-based multipurpose device that adheres to the skin is built with a simple, inexpensive, and highly scalable layer-by-layer dip-coating technique. The setup consists of sticky polydimethylsiloxane (PDMS), a highly conductive MXene-sensitive layer, and a cheap and readily available air-laid paper

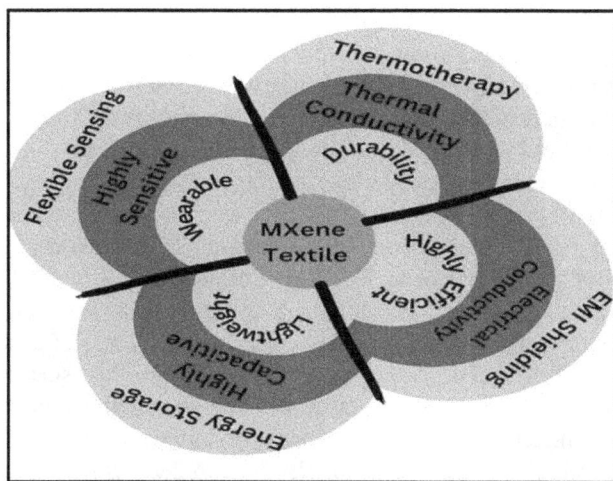

FIGURE 20.5 Variety of applications in the area including flexible sensors, energy storage systems, smart thermotherapy, and electromagnetic interference (EMI) shielding in the development of MXene textiles.

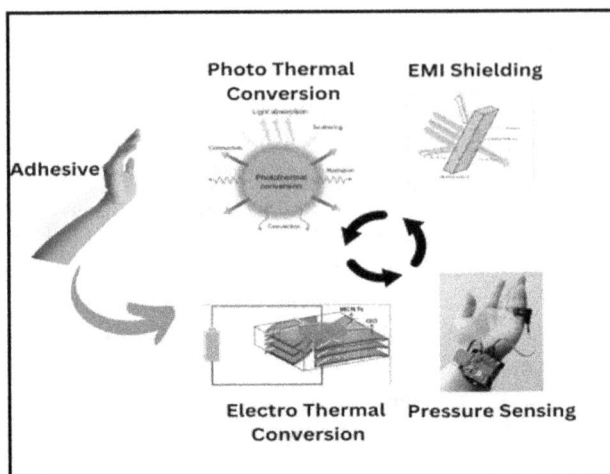

FIGURE 20.6 Preparation of self-adhesive and multifunctional smart textile devices (PDMS-MXene) showing good properties of pressure sensing, electro-thermal conversion, photo-thermal conversion, and EMI shielding.

(AP) substrate. The adhesive layer of moderately cross-linked PDMS maintains conformal contact with the skin even when the device is subjected to the bending motion of a human joint. The smart textile has excellent electro-thermal and photo-thermal conversion performance, strong cycle stability, and fine tunability. In addition to providing excellent EMI shielding, the electrical conductivity of textile electronics enables highly accurate pressure sensing for monitoring human motion (as shown in Figure 20.6). As a result, this method provides a viable paradigm for developing electronic textiles with several functions for medical applications [52].

The next generation of wearable technology would benefit greatly from electrodes manufactured from textiles that are both flexible and compliant. Making connected electron transport channels in the insulating textile substrate without changing the substrate's intrinsic porosity, flexibility, or stability is one of the toughest challenges in producing high-performance devices. The use of electrostatic self-assembly between negatively charged titanium carbide MXene flakes and positively charged polyester fabric coated with polyethyleneimine (PEI) is discussed in this article. Densified, horizontally oriented MXene flakes are deposited onto fabric fibres at a concentration of 0.8 mg cm^2. The MXene generates intergranular layers of conductive skin that may be rapidly charged/discharged at 10 V s^{-1} due to their surface chemistry, and it also provides different active places for the insertion of additional functional electrochemical materials. The conductive fabric is integrated with a conformal-coated polypyrrole (PPy) supercapacitor electrode to demonstrate the technology's viability. This supercapacitor electrode has been shown to be mechanically robust while exhibiting excellent rate performance, wide area capacitance, and cycling stability. These features demonstrate the huge potential of the existing electrodes and pave the way for the development of a new generation of textile electrodes for the smart textile industry [66]. The demand for high-performance electronic textiles that can sense strain, block EMI, and generate heat using Joules is on the rise in the realm of modern integrated smart clothing and accessories. In this work, a hybrid of a $Ti_3C_2T_x$ MXene and carbon nanotubes (CNTs) is woven into a non-woven fabric made of thermoplastic polyurethane (TPU) to produce an electronic textile with a variety of applications. A tuneable conductive MCT-fabric strain sensor boasts high sensitivity, a wide sensing range, fast response-recovery time, excellent long-term stability, and 1000 cycles of reliability due to its conductive Mxene-CNTs coating. The MCT fabric exhibits exceptional heat management capabilities, alongside its optimal synergistic conductive network and porous fibrous network structure, which contribute to its exceptional electromagnetic interference (EMI) shielding efficiency. Particularly, the MCT fabric with a thickness of 600 μm demonstrates a remarkable EMI shielding efficiency of 43 dB. This study suggests that wearable electronics and AI performance can be enhanced by enhancing heating stability, dependability, strain sensing, EMI shielding, and temperature control [67].

20.8   MXENES' EFFECTS ON THE TEXTILE INDUSTRY

Textiles are crucial to the preservation of every civilized society and one of humanity's top five requirements. Each year, almost 80 billion yards of fabric are used for things other than clothes. Amongst the many applications of electricity are devices used in industries as varied as energy storage and generation, medicine, sensing, transportation, safety, and sports. The standardization of the fabric production process throughout the years has allowed the textile revolution to give rise to clothing that is flexible in terms of function, durability, wearability, and aesthetics. Next-generation textiles, often known as smart or intelligent textiles, have emerged in response to the growing demand for miniaturization and multifunctionality [68]. The functionality of these textiles can be enhanced by using nanoparticles as nanofiber or nanocomposite substrates. Nano-finishing or nanocoating a textile substrate can confer new functional or high-performance characteristics [69]. Spray coating, dip coating, plasma polymerization, layer-by-layer coating, and sol–gel coating are some of the most effective nanocoating techniques for making smart textiles. Two-dimensional (2D) materials like graphene, MXenes, transition metal dichalcogenides, etc. have improved the usefulness of smart fabrics even more than their three-dimensional (3D) equivalents. The MXenes created by Zhang et al. [70] are a family of early transition eco-friendly metal carbides and carbonitrides that are amongst the 2D materials well suited for a number of applications, including medicine and optoelectronics. The large investments in this material for a wide variety of technologies have spurred a surge in academic interest, as seen by the exponential increase in the number of scientific and engineering research articles and the formation of a commercial sector. It was obvious that this was becoming a popular discussion point amongst academics. The MXene market is expected to grow from its current $26.4 million value in 2022 (a CAGR of 29.24%) [71] to $121.5 million by 2027. Here, we'll go through the basics of what MXenes are and how they behave. The size of the MXene flake has a profound effect on the morphology of the linked yarns and the qualities of the finished textile, and hence on the functionality of the system in which the textile is built. Larger flakes with a long distribution of 100 nm to 10 m can be generated using the MILD method, whereas smaller flakes in the range of 0.1 to 100 nm are prepared with increased sonication and density gradient centrifugation [72]. MXene flakes are used to coat the cotton fibres in textiles, allowing them to permeate the yarn and the interstice. When compared to textiles coated with little MXene flakes, the electrical conductivity of yarn is enhanced when large MXene flakes are added on its surface. The increased interfacial resistance and irregularity fractions are responsible for these findings. The electrodes' electrochemical performance can also be affected by the MXene flakes' size. The incorporation of MXene flakes into the proper solution is a crucial step in the functionalization process for improving the efficiency of smart fabrics [73]. Recent research shows that in addition to water, ethanol, dimethylformamide (DMF), and dimethyl sulfoxide (DMSO), MXene may be dissolved in a wide number of other solvents [74]. MXenes' compatibility with textiles stems from their synthesized terminal groups of similar polarity and their efficient dispersion properties. In order to demonstrate their value in textiles, it is essential that scattered MXene flakes withstand solvents. However, the textile industry is one of the most polluting sectors, making it essential to address this issue in order to achieve long-term sustainability, reduce waste, and enhance management. Membrane technology is essential for the treatment of the large volumes of oily wastewater that are common in the textile industry. Following MXene synthesis, composite membranes called Hal@MXene-PDA are made by linking halloysite nanotubes (Hal) with polydopamine (PDA) [75–76]. These membranes are capable of ultra-high oil-in-water separation. Another approach that can have a big influence on easing water constraint issues is detoxifying wastewater of harmful metals. Adsorption offers potential as a purification technology because it is cheap, simple, and effective in a controlled environment. MXene-based nanocomposites have emerged as a promising adsorbent for the removal of toxic metals from wastewater due to their hydrophilicity, large surface area, activated metallic hydroxide sites, abundant electron availability, and high adsorption capacity. The environmental pathway, characterization methodologies, benefits, advantages, and limits of MXenes when utilized in the adsorption process need constant attention when nanostructures or nanocomposites based on MXenes are developed for wastewater treatment. This research delves into the potential uses of MXenes in smart textiles and textile wastewater remediation in a variety of fields, including but not limited to athletics, showbiz, medicine, and the environment [75–76]. This research looked at the theoretical and practical applications of MXenes in the textile industry. Satisfactory summaries have been provided for each of the following topics: the analytical process flow, smart textiles, MXenes in textile materials, and MXenes in textile wastewater treatment. The first half of the paper introduces the topic at hand, smart textile clothing and wastewater treatment for the removal of textile dyes. As a second phase, the authors develop a research flowchart that outlines the seminal works that have contributed to this field of study. At last, it has been discerned that MXene nanoparticles can serve a practical purpose in the fibre, yarn, and fabric used to make smart clothing. The fourth and final argument is that MXenes have been studied extensively as an absorbent for the removal of colour from textile effluent. Academics have also debated MXenes' usefulness and worth in the textile industry [72].

20.9 CHALLENGES

Materials based on MXene present a number of unique challenges:

1. **Scalability and Cost:** MXenes are currently difficult to synthesize on a big scale, which would prevent their widespread use in the textile industry. Making MXene-based textiles economically viable requires the creation of scalable, cost-efficient production techniques.
2. **Integration into Textile Manufacturing Processes:** It might be technically challenging to incorporate MXenes into existing textile manufacturing processes while yet retaining their distinctive features. Integrating MXenes in a way that maintains the integrity of fibres and fabrics is an area that needs more study.
3. **Durability and Wash Resistance:** The durability and washability of MXene-treated textiles are essential for real-world use. Coatings made with MXenes should not degrade or wash out of the fabric, irrespective of the frequency or how harshly it is washed or exposed to the elements.
4. **Safety and Toxicity Issues:** Before MXenes can be used commercially in textiles, they must be thoroughly researched for any toxicity and safety consequences, as would be the case with any developing nanomaterial. Research is needed to determine how MXenes affects human health and ecosystems.
5. **Compatibility with Textile Dyeing and Printing Processes:** The dyeing and printing techniques commonly employed in the textile industry may be affected by the surface characteristics of MXenes. Fabrics treated with MXenes need to be compatible with each other to preserve colour saturation and freedom in design.
6. **Mechanical Flexibility:** Although MXenes' high mechanical strength has many practical applications, it can also cause fabrics to become rigid if the two benefits aren't carefully weighed. For wearable applications, it is crucial that MXene-treated fabrics maintain their pliability and comfort.
7. **Design and Application Expertise:** Many novel textile uses are possible with MXenes, but their development requires specialized knowledge and skill. To realize MXenes' full potential, partnerships between material scientists, textile technologists, and designers are essential.
8. **Consumer Acceptance:** It may be difficult to get consumers to comprehend and accept the benefits of new textile technology and materials. It will be crucial for the effective integration into the market of MXene-treated textiles to educate customers about the benefits and safety features of these products.

20.10 CONCLUSIONS

Extraordinary two-dimensional substances, MXenes have great potential to revolutionize the textile industry. MXenes' extraordinary qualities, which include high electrical conductivity, mechanical strength, and thermal stability, pave the way for a new generation of high-tech textiles. By incorporating MXene-coated fabrics into wearable electronics, flexible sensors, and energy storage devices, we can look forward to a future in which clothing is both interactive and responsive, enhancing the wearer's experience and providing new opportunities for personal expression and expression. Furthermore, MXenes' potential to improve the eco-friendliness of textiles is not to be disregarded. The increasing demand for environmentally and medically responsible textiles is fulfilled by its eco-friendly synthesis procedures and the capacity to produce self-cleaning, antibacterial fabrics. MXenes' introduction to the textile industry ushers in a new era of high-tech fabrics that seamlessly integrate with time-honoured garment construction. The combination of MXenes and textiles creates not just a more connected and integrated future but also a new standard in both style and utility. MXenes, thanks to ongoing study and development, promise to revolutionize the textile industry by allowing for the production of clothes that are not only stylish but also practical, technologically advanced, and conscientiously produced. There is much to look forward to on the road to MXenes' widespread adoption in the textile industry, and their revolutionary influence is set to reshape our understanding of and experience with textiles in the years to come.

REFERENCES

1. Jena, A., Lee, S. C., & Bhattacharjee, S. (2022). Surface-oxygen-passivation driven large anomalous Hall conductivity (AHC) in nitride MXenes: Can AHC be a tool to determine functional groups in 2D ferro (i) magnets?. *The Journal of Physical Chemistry. C, 126*(43), 18404–18410.
2. Feng, T., Li, X., Guo, P., Zhang, Y., Liu, J., & Zhang, H. (2020). MXene: Two dimensional inorganic compounds, for generation of bound state soliton pulses in nonlinear optical system. *Nanophotonics, 9*(8), 2505–2513.
3. Anasori, B., & Gogotsi, Y. (2019). Introduction to 2D transition metal carbides and nitrides (MXenes). *2D Metal Carbides and Nitrides (MXenes) Structure, Properties and Applications*, 3–12.
4. Chaudhari, N. K., Jin, H., Kim, B., San Baek, D., Joo, S. H., & Lee, K. (2017). MXene: An emerging two-dimensional material for future energy conversion and storage applications. *Journal of Materials Chemistry A, 5*(47), 24564–24579.
5. Anasori, B., Lukatskaya, M. R., & Gogotsi, Y. (2017). 2D metal carbides and nitrides (MXenes) for energy storage. *Nature Reviews Materials, 2*(2), 1–17.
6. Song, P., Liu, B., Qiu, H., Shi, X., Cao, D., & Gu, J. (2021). MXenes for polymer matrix electromagnetic interference shielding composites: A review. *Composites Communications, 24*, 100653.
7. Aziz, A., Asif, M., Ashraf, G., Iftikhar, T., Hussain, W., & Wang, S. (2022). Environmental significance of wearable sensors based on MXene and graphene. *Trends in Environmental Analytical Chemistry*, e00180.
8. Zhang, W., Miao, J., Zuo, X., Zhang, X., & Qu, L. (2022). Weaving a magnificent world: 1D fibrous electrodes and devices for stretchable and wearable electronics. *Journal of Materials Chemistry C, 10*(38), 14027–14052.
9. Rasheed, T., Kausar, F., Rizwan, K., Adeel, M., Sher, F., Alwadai, N., & Alshammari, F. H. (2022). Two dimensional MXenes as emerging paradigm for adsorptive removal of toxic metallic pollutants from wastewater. *Chemosphere, 287*, 132319.

10. Damptey, L., Jaato, B. N., Ribeiro, C. S., Varagnolo, S., Power, N. P., Selvaraj, V., . . . & Krishnamurthy, S. (2022). Surface functionalized MXenes for wastewater treatment—a comprehensive review. *Global Challenges*, *6*(6), 2100120.

11. Ibrahim, Y., Kassab, A., Eid, K., M. Abdullah, A., Ozoemena, K. I., & Elzatahry, A. (2020). Unveiling fabrication and environmental remediation of MXene-based nanoarchitectures in toxic metals removal from wastewater: Strategy and mechanism. *Nanomaterials*, *10*(5), 885.

12. Deshmukh, K., Kovářík, T., & Pasha, S. K. (2020). State of the art recent progress in two dimensional MXenes based gas sensors and biosensors: A comprehensive review. *Coordination Chemistry Reviews*, *424*, 213514.

13. Soleymaniha, M., Shahbazi, M. A., Rafieerad, A. R., Maleki, A., & Amiri, A. (2019). Promoting role of MXene nanosheets in biomedical sciences: Therapeutic and biosensing innovations. *Advanced Healthcare Materials*, *8*(1), 1801137.

14. Yang, G., Liu, F., Zhao, J., Fu, L., Gu, Y., Qu, L., . . . & Lin, Y. (2023). MXenes-based nanomaterials for biosensing and biomedicine. *Coordination Chemistry Reviews*, *479*, 215002.

15. Babar, Z. U. D., Della Ventura, B., Velotta, R., & Iannotti, V. (2022). Advances and emerging challenges in MXenes and their nanocomposites for biosensing applications. *RSC Advances*, *12*(30), 19590–19610.

16. Chen, F., Zhang, Y., & Huang, H. (2023). Layered photocatalytic nanomaterials for environmental applications. *Chinese Chemical Letters*, *34*(3), 107523.

17. Iravani, S., & Varma, R. S. (2022). MXene-based photocatalysts in degradation of organic and pharmaceutical pollutants. *Molecules*, *27*(20), 6939.

18. Tareen, A. K., Khan, K., Iqbal, M., Zhang, Y., Long, J., Mahmood, A., . . . & Zhang, H. (2022). Recent advance in two-dimensional MXenes: New horizons in flexible batteries and supercapacitors technologies. *Energy Storage Materials*, *53*, 783–826.

19. Zhang, W., Ji, X. X., & Ma, M. G. (2023). Emerging MXene/cellulose composites: Design strategies and diverse applications. *Chemical Engineering Journal*, *458*, 141402.

20. Chen, X., Yu, H., Gao, Y., Wang, L., & Wang, G. (2022). The marriage of two-dimensional materials and phase change materials for energy storage, conversion and applications. *EnergyChem*, *4*(2), 100071.

21. Adekoya, G. J., Adekoya, O. C., Sadiku, R. E., Hamam, Y., & Ray, S. S. (2022). Applications of MXene-containing polypyrrole nanocomposites in electrochemical energy storage and conversion. *ACS Omega*, *7*(44), 39498–39519.

22. Alhamada, T. F., Azmah Hanim, M. A., Jung, D. W., Saidur, R., Nuraini, A., & Hasan, W. W. (2022). MXene based nanocomposites for recent solar energy technologies. *Nanomaterials*, *12*(20), 3666.

23. Shaikh, N. S., Ubale, S. B., Mane, V. J., Shaikh, J. S., Lokhande, V. C., Praserthdam, S., . . . & Kanjanaboos, P. (2022). Novel electrodes for supercapacitor: Conducting polymers, metal oxides, chalcogenides, carbides, nitrides, MXenes, and their composites with graphene. *Journal of Alloys and Compounds*, *893*, 161998.

24. Forouzandeh, P., & Pillai, S. C. (2021). MXenes-based nanocomposites for supercapacitor applications. *Current Opinion in Chemical Engineering*, *33*, 100710.

25. Gogotsi, Y., & Anasori, B. (2019). The rise of MXenes. *ACS Nano*, *13*(8), 8491–8494.

26. Roach, A., Milhollin, R., & Horner, J. (2019). Market oppotunities for industrial hemp: Guide to understanding markets and demand for various industrial hemp plant products.

27. Wang, K., Wu, H., Meng, Y., & Wei, Z. (2014). Conducting polymer nanowire arrays for high performance supercapacitors. *Small*, *10*(1), 14–31.

28. Kumar, N., Gusain, R., & Ray, S. S. (Eds.). (2023). *Two-Dimensional Materials for Environmental Applications* (Vol. 332). Springer Nature.

29. Orangi, J., & Beidaghi, M. (2020). A review of the effects of electrode fabrication and assembly processes on the structure and electrochemical performance of 2D MXenes. *Advanced Functional Materials, 30*(47), 2005305.

30. Hu, M., Li, Z., Li, G., Hu, T., Zhang, C., & Wang, X. (2017). All-solid-state flexible fiber-based MXene supercapacitors. *Advanced Materials Technologies, 2*(10), 1700143.

31. Levitt, A., Zhang, J., Dion, G., Gogotsi, Y., & Razal, J. M. (2020). MXene-based fibers, yarns, and fabrics for wearable energy storage devices. *Advanced Functional Materials, 30*(47), 2000739.

32. Uzun, S. (2020). *Multiscale Integration of Two-Dimensional Transition Metal Carbides (MXenes) into Textile-Based Devices.* Drexel University.

33. Cao, W. T., Ma, C., Mao, D. S., Zhang, J., Ma, M. G., & Chen, F. (2019). MXene-reinforced cellulose nanofibril inks for 3D-printed smart fibres and textiles. *Advanced Functional Materials, 29*(51), 1905898.

34. Ullah, S., Shahzad, F., Qiu, B., Fang, X., Ammar, A., Luo, Z., & Zaidi, S. A. (2022). MXene-based aptasensors: Advances, challenges, and prospects. *Progress in Materials Science, 129*, 100967.

35. Pham, T. K. N., & Brown, J. J. (2020). Hydrogen sensors using 2-dimensional materials: A review. *Chemistry Select, 5*(24), 7277–7297.

36. Hu, G. (2017). *Printable 2d material optoelectronics and photonics* (Doctoral dissertation, University of Cambridge).

37. Kumar, J. A., Prakash, P., Krithiga, T., Amarnath, D. J., Premkumar, J., Rajamohan, N., . . . & Rajasimman, M. (2022). Methods of synthesis, characteristics, and environmental applications of MXene: A comprehensive review. *Chemosphere, 286*, 131607.

38. Thakur, N., Kumar, P., Sati, D. C., Neffati, R., & Sharma, P. (2022). Recent advances in two-dimensional MXenes for power and smart energy systems. *Journal of Energy Storage, 50*, 104604.

39. Young, W. H., & Young, N. K. (2010). *World War II and the Postwar Years in America: A Historical and Cultural Encyclopedia [2 volumes]: A Historical and Cultural Encyclopedia.* ABC-CLIO.

40. Mao, L., Zhao, X., Cheng, Q., Yang, G., Liao, F., Chen, L., He, P., & Chen, S. (2021). Recent advances and perspectives of two-dimensional Ti-based electrodes for electrochemical energy storage. *Sustainable Energy & Fuels, 5*(20), 5061–5113.   41. Lin, X., Liu, P., Xin, W., Teng, Y., Chen, J., Wu, Y., . . . & Wen, L. (2021). Heterogeneous MXene/PS-b-P2VP nanofluidic membranes with controllable ion transport for osmotic energy conversion. *Advanced Functional Materials, 31*(45), 2105013.

42. Yu, J., Zeng, M., Zhou, J., Chen, H., Cong, G., Liu, H., . . . & Xu, J. (2021). A one-pot synthesis of nitrogen doped porous MXene/TiO2 heterogeneous film for high-performance flexible energy storage. *Chemical Engineering Journal, 426*, 130765.

43. Shuck, C. E., Sarycheva, A., Anayee, M., Levitt, A., Zhu, Y., Uzun, S., . . . & Gogotsi, Y. (2020). Scalable synthesis of Ti3C2Tx mxene. *Advanced Engineering Materials, 22*(3), 1901241.

44. Ashishie, P. B., Louis, H., & Offiong, O. E. (2023). Tailoring Ti3C2 MXene into Ti3C2Tx, Tx= NO and alloying with M= Al, Ga, In, Tl into MTi3C2NO as electrode materials for super-capacitor devices: Perspective from first-principles density functional theory. *Journal of Physics and Chemistry of Solids*, 111468.

45. Kumar, J. A., Prakash, P., Krithiga, T., Amarnath, D. J., Premkumar, J., Rajamohan, N., . . . & Rajasimman, M. (2022). Methods of synthesis, characteristics, and environmental applications of MXene: A comprehensive review. *Chemosphere, 286*, 131607.

46. Yadlapalli, R. T., Alla, R. R., Kandipati, R., & Kotapati, A. (2022). Super capacitors for energy storage: Progress, applications and challenges. *Journal of Energy Storage, 49,* 104194.

47. Tu, T., Liang, B., Zhang, S., Li, T., Zhang, B., Xu, S., . . . & Ye, X. (2021). Controllable patterning of Porous MXene (Ti3C2) by metal-assisted electro-gelation method. *Advanced Functional Materials, 31*(31), 2101374.

48. Guo, Z., Li, Y., Lu, Z., & Liu, W. (2021, February). Applications of MXene-based composite fibers in smart textiles. In *Journal of Physics: Conference Series* (Vol. 1790, No. 1, p. 012066). IOP Publishing.

49. Hu, M., Zhang, H., Hu, T., Fan, B., Wang, X., & Li, Z. (2020). Emerging 2D MXenes for supercapacitors: Status, challenges and prospects. *Chemical Society Reviews, 49*(18), 6666–6693.

50. Zhou, B., Su, M., Yang, D., Han, G., Feng, Y., Wang, B., . . . & Shen, C. (2020). Flexible MXene/silver nanowire-based transparent conductive film with electromagnetic interference shielding and electro-photo-thermal performance. *ACS Applied Materials & Interfaces, 12*(36), 40859–40869.

51. Zhang, Y., Xia, X., Ma, K., Xia, G., Wu, M., Cheung, Y. H., . . . & Xin, J. H. (2023). Functional textiles with smart properties: Their fabrications and sustainable applications. *Advanced Functional Materials,* 2301607.

52. Yao, D., Tang, Z., Liang, Z., Zhang, L., Sun, Q. J., Fan, J., . . . & Ouyang, J. (2023). Adhesive, multifunctional, and wearable electronics based on MXene-coated textile for personal heating systems, electromagnetic interference shielding, and pressure sensing. *Journal of Colloid and Interface Science, 630,* 23–33.

53. Chen, Y., Li, J., Li, T., Zhang, L., & Meng, F. (2021). Recent advances in graphene-based films for electromagnetic interference shielding: Review and future prospects. *Carbon, 180,* 163–184.

54. Aboutalebi, S. H., Jalili, R., Esrafilzadeh, D., Salari, M., Gholamvand, Z., Aminorroaya Yamini, S., . . . & Wallace, G. G. (2014). High-performance multifunctional graphene yarns: Toward wearable all-carbon energy storage textiles. *ACS Nano, 8*(3), 2456–2466.

55. Ma, C., Ma, M. G., Si, C., Ji, X. X., & Wan, P. (2021). Flexible MXene-based composites for wearable devices. *Advanced Functional Materials, 31*(22), 2009524.

56. Pu, J. H.; Zhao, X.; Zha, X. J.; Bai, L.; Ke, K.; Bao, R. Y.; Liu, Z. Y.; Yang, M. B.; Yang, W. (2019). Multilayer structured AgNW/WPUMXene fiber strain sensors with ultra-high sensitivity and a wide operating range for wearable monitoring and healthcare. *Journal of Materials Chemistry A, 7*(26), 15913–15923.

57. Zhao, X., Wang, L. Y., Tang, C. Y., Zha, X. J., Liu, Y., Su, B. H., . . . & Yang, W. (2020). Smart Ti3C2T x MXene fabric with fast humidity response and joule heating for healthcare and medical therapy applications. *Acs Nano, 14*(7), 8793–8805.

58. Zhang, D., Yin, R., Zheng, Y., Li, Q., Liu, H., Liu, C., & Shen, C. (2022). Multifunctional MXene/CNTs based flexible electronic textile with excellent strain sensing, electromagnetic interference shielding and Joule heating performances. *Chemical Engineering Journal, 438,* 135587.

59. Liu, L., Wang, L., Liu, X., Yuan, W., Yuan, M., Xia, Q., . . . & Zhou, A. (2021). High-performance wearable strain sensor based on MXene@ cotton fabric with network structure. *Nanomaterials, 11*(4), 889.

60. Liu, L., Li, Y., Zhang, Y., Shang, X., Song, C., & Meng, F. (2023). Ni3S2 thin-layer nanosheets coupled with Co9S8 nanoparticles anchored on 3D cross-linking composite structure CNT@ MXene for high-performance asymmetric supercapacitor. *Electrochimica Acta, 439,* 141694.

61. Cao, W., Ma, C., Tan, S., Ma, M., Wan, P., & Chen, F. (2019). Ultrathin and flexible CNTs/MXene/cellulose nanofibrils composite paper for electromagnetic interference shielding. *Nano-Micro Letters, 11*, 1–17.
62. Jin, C., & Bai, Z. (2022). MXene-based textile sensors for wearable applications. *ACS Sensors, 7*(4), 929–950.
63. Wang, L., Zhang, M., Yang, B., Tan, J., Ding, X., & Li, W. (2021). Recent advances in multidimensional (1D, 2D, and 3D) composite sensors derived from MXene: Synthesis, structure, application, and perspective. *Small Methods, 5*(7), 2100409.
64. Farhana, K., Kadirgama, K., Mahamude, A. S. F., & Jose, R. (2023). Review of MXenes as a component in smart textiles and an adsorbent for textile wastewater remediation. *Chinese Chemical Letters*, 108533.
65. Ahmed, A., Hossain, M. M., Adak, B., & Mukhopadhyay, S. (2020). Recent advances in 2D MXene integrated smart-textile interfaces for multifunctional applications. *Chemistry of Materials, 32*(24), 10296–10320.
66. Li, X., Hao, J., Liu, R., He, H., Wang, Y., Liang, G., . . . & Guo, Z. (2020). Interfacing MXene flakes on fiber fabric as an ultrafast electron transport layer for high performance textile electrodes. *Energy Storage Materials, 33*, 62–70.
67. Zhang, D., Yin, R., Zheng, Y., Li, Q., Liu, H., Liu, C., & Shen, C. (2022). Multifunctional MXene/CNTs based flexible electronic textile with excellent strain sensing, electromagnetic interference shielding and Joule heating performances. *Chemical Engineering Journal, 438*, 135587.
68. Libertino, S., Plutino, M. R., & Rosace, G. (2018, July). Design and development of wearable sensing nanomaterials for smart textiles. In *AIP Conference Proceedings* (Vol. 1990, No. 1). AIP Publishing.
69. Joshi, M., & Bhattacharyya, A. (2011). Nanotechnology—A new route to high-performance functional textiles. *Textile Progress, 43*(3), 155–233.
70. Zhang, D., Shah, D., Boltasseva, A., & Gogotsi, Y. (2022). MXenes for photonics. *ACS Photonics, 9*(4), 1108–1116.
71. Parihar, A., Choudhary, N. K., Sharma, P., & Khan, R. (2022). MXene-based aptasensor for the detection of aflatoxin in food and agricultural products. *Environmental Pollution*, 120695.
72. Levitt, A., Zhang, J., Dion, G., Gogotsi, Y., & Razal, J. M. (2020). MXene-based fibers, yarns, and fabrics for wearable energy storage devices. *Advanced Functional Materials, 30*(47), 2000739.
73. Bhat, A., Anwer, S., Bhat, K. S., Mohideen, M. I. H., Liao, K., & Qurashi, A. (2021). Prospects challenges and stability of 2D MXenes for clean energy conversion and storage applications. *Npj 2D Materials and Applications, 5*(1), 61.
74. Seyedin, S., Zhang, J., Usman, K. A. S., Qin, S., Glushenkov, A. M., Yanza, E. R. S., . . . & Razal, J. M. (2019). Facile solution processing of stable MXene dispersions towards conductive composite fibers. *Global Challenges, 3*(10), 1900037.
75. Zeng, G., Wei, K., Zhang, H., Zhang, J., Lin, Q., Cheng, X., . . . & Chiao, Y. H. (2021). Ultra-high oil-water separation membrane based on two-dimensional MXene (Ti_3C_2Tx) by co-incorporation of halloysite nanotubes and polydopamine. *Applied Clay Science, 211*, 106177.
76. Lim, G. P., Soon, C. F., Al-Gheethi, A. A., Morsin, M., & Tee, K. S. (2022). Recent progress and new perspective of MXene-based membranes for water purification: A review. *Ceramics International, 48*(12), 16477–16491.

Index

A

atomic layer deposition, 38

B

biomarker, 5, 89, 99, 147, 148, 149, 154,
    156, 157, 305
black phosphorous, 114
bovine serum albumin, 73

C

capacitive deionization, 179
carcinoembryonic antigen, 89
cellulose nanocrystals, 332
CT imaging, 106, 108
CVD, 12, 14, 51

E

electrocardiogram, 330
electromagnetic interference,
    330
etching, 1–2

F

ferromagnetic, 31
field-effect transistors, 239
FRET, 93, 94

G

gas separation, 53
glassy carbon electrode, 74

H

hydrophilicity, 5, 51–53, 58, 66, 122

L

limit of detection, 76, 145–148, 151,
    263, 314

M

MAX phases, 11–22, 49, 65
metal–organic framework, 70, 115
molecular separation, 55
MXene quantum dots, 68

O

oil/water (O/W) separation, 50

P

photoacoustic, 103, 106, 108
photodynamic therapy, 5, 136, 227
photoluminescence, 123
photothermal therapy, 227, 229, 230, 231
piezoelectric, 97, 98
piezoresistive sensors, 254, 255, 256, 259, 260,
    263, 267, 268, 273, 274
plasma enhanced pulsed laser deposition, 39
polyvinylidene fluoride (PVDF), 52
prostate specific antigen, 93

R

reduced GO, 327
reverse osmosis, 177

S

Schottky junction, 309
supercapacitors, 207, 212, 215, 219

T

TENG, 312, 313

V

VOCs, 312, 313, 316

Z

ZIF-67, 194

For Product Safety Concerns and Information please contact our EU
representative GPSR@taylorandfrancis.com
Taylor & Francis Verlag GmbH, Kaufingerstraße 24, 80331 München, Germany